Crystal Growth in Science and Technology

NATO ASI Series

Advanced Science Institutes Series

*A series presenting the results of activities sponsored by the NATO Science Committee,
which aims at the dissemination of advanced scientific and technological knowledge,
with a view to strengthening links between scientific communities.*

The series is published by an international board of publishers in conjunction with the
NATO Scientific Affairs Division

A	**Life Sciences**	Plenum Publishing Corporation
B	**Physics**	New York and London
C	**Mathematical**	Kluwer Academic Publishers
	and Physical Sciences	Dordrecht, Boston, and London
D	**Behavioral and Social Sciences**	
E	**Applied Sciences**	
F	**Computer and Systems Sciences**	Springer-Verlag
G	**Ecological Sciences**	Berlin, Heidelberg, New York, London,
H	**Cell Biology**	Paris, and Tokyo

Recent Volumes in this Series

Series B: Physics

Crystal Growth in Science and Technology

Edited by

H. Arend and J. Hulliger

Swiss Federal Institute of Technology
Zurich, Switzerland

Plenum Press
New York and London
Published in cooperation with NATO Scientific Affairs Division

Proceedings of a NATO Advanced Study Institute/
Thirteenth Course of the International School of Crystallography
on Crystal Growth in Science and Technology,
held August 27–September 7, 1987,
in Erice, Sicily, Italy

Library of Congress Cataloging-in-Publication Data

Crystal growth in science and technology / edited by H. Arend and J.
  Hulliger.
        p.   cm. -- (NATO ASI Series. Series B, Physics ; v. 210)
    "September 15, 1989."
    "Published in cooperation with NATO Scientific Affairs Division."
    "Proceedings of a NATO Advanced Study Institute/Thirteenth Course
  of the International School of Crystallography on Crystal Growth in
  Science and Technology, held August 27 September 7, 1987"--CIP t.p.
  verso.
    Includes bibliographical references.
    ISBN-13: 978-1-4612-7861-0   e-ISBN-13: 978-1-4613-0549-1
    DOI: 10.1007/978-1-4613-0549-1

    1. Crystals--Growth--Congresses. 2. Crystallography--Congresses.
  I. Arend, H.   II. Hulliger, J.   III. North Atlantic Treaty
  Organization. Scientific Affairs Division. IV. Series.
  QD921.C7635 1989
  548'.5--dc20   89-28393
                                                         CIP

© 1989 Plenum Press, New York
A Division of Plenum Publishing Corporation
233 Spring Street, New York, N.Y. 10013

Softcover reprint of the hardcover 1st edition 2001

PREFACE

Science and art of crystal growth represent an interdisciplinary activity based on fundamental principles of physics, chemistry and crystallography. Crystal growth has contributed over the years essentially to a widening of knowledge in its basic disciplines and has penetrated practically into all fields of experimental natural sciences. It has acted, more over, in a steadily increasing manner as a link between science and technology as can be seen best, for example, from the achievements in modern microelectronics.

The aim of the course "Crystal Growth in Science and Technology" being to stress the interdisciplinary character of the subject, selected fundamental principles are reviewed in the following contributions and cross links between basic and applied aspects are illustrated. It is a very well-known fact that the intensive development of crystal growth has led to a progressive narrowing of interests in highly specialized directions which is in particular harmful to young research scientists. The organizers of the course did sincerely hope that the program would help to broaden up the horizon of the participants. It was equally their wish to contribute within the traditional spirit of the school of crystallography in Erice to the promotion of mutual understanding, personal friendship and future collaboration between all those who were present at the school.

For the convenience of interested research scientists and in particular also of colleagues involved in teaching of crystal growth, the editors have rearranged and completed a list of documents on a variety of crystal growth problems which is published here. The following contributions of the lecturers are all products of highly qualified specialists in their fields. The editors have, therefore gladly accepted and respected their individual presentation and focusation of the topics treated.

The Editors

CONTENTS

A THEORETICAL CRYSTAL GROWER'S VIEW OF PHASE EQUILIBRIA

R. Ghez
IBM T.J. Watson Research Center
Yorktown Heights, NY 10598
U.S.A.

ABSTRACT: This lecture first focuses on operational definitions — it is hoped without circularity — of thermodynamic systems, fields, phases, and equilibrium conditions. In particular, I will show that the latter are always the consequence of a minimum principle due to Gibbs. I will also emphasize that equilibrium does not always imply constancy of fields. Then, phase diagrams, also due to the genius of Gibbs, emerge as a graphical representation of equilibria. I will demonstate through examples (garnets and ternary III-V's) how these diagrams are used. Finally, I will indicate that the final crystal product, under certain circumstances of "slow" growth, can be predicted from equilibrium considerations alone.

I. INTRODUCTION

If crystal growth is, by its very nature, a non-equilibrium phenomenon, then how can phase diagrams — an expression of equilibria — ever be of any use? Knowledge of phase equilibria is invaluable in at least three related contexts: (i) For the prediction of final states, given an initial experimentally prepared state; (ii) To provide a "baseline" against which deviations of observable fields (temperature, stresses, compositions, motions, *etc.*) can be measured; (iii) For the intelligent construction of constitutive relations and of boundary conditions. I shall now briefly discuss some of these questions in turn.

The first context is perhaps the most useful to the experimentalist, for it is vital to know, *a priori*, which of a multitude of outcomes is possible. In other words, given an arbitrary initially prepared state — something we manufacture with chemicals, scales, crucibles, and furnaces — what is then the final product? Is it unique? The second context is of greater concern to the theoretician whose task it is to compute, if possible, the way in which driving forces dissipate. In other words, how fast do we achieve the equilibrium state? These calculations always require that one know by how much the chemical potentials, say, deviate from their equilibrium values. These deviations are generally functions of both space and time; they are thus of a dynamic nature, and they obey partial differential equations that are the expression of local conservation principles. The last context is of the greatest importance because conservation principles, alone, are empty statements. Basic chemical physics determines, nevertheless, how thermodynamic fluxes relate to field inhomogeneities at interior points of a given system and how these fields must behave at phase boundaries.

1

Of all physical sciences, Thermodynamics is perhaps most frustrating to the student because, on the one hand, it purports to describe *any* macroscopic equilibrium phenomenon [1], and, on the other hand, it seems to concern itself mainly with highly idealized physico-chemical systems. We are all more than familiar with Carnot engines, Van't Hoff reaction boxes, perfectly semi-permeable membranes, and other fictional devices. Our physical world, however, is far from ideal. In particular, it is *inhomogenous*, by which one generally means that a given system's properties vary from point to point. Indeed, we all live in the presence of external fields (*e.g.*, gravitation), and our own abode is presently hurtling through space with a motion that is not even rectilinear and uniform. How, then, can one even speak of a volume of gas at constant pressure, say? And what about the surface of a crystal, surely as gross an inhomogeneity as one can imagine?

To proceed forward requires that we hark back to Gibbs [2], for he, rather than most of his followers, clearly showed that Thermodynamics provides a general description of macroscopic equilibria [3]. It is impossible, in this space, to give anything more than a flavor of his methods, but it *is* possible to avoid fallacious or circular arguments. For example, we have all been taught that

$$dU = T\,dS - p\,dV + \sum_{i=1}^{n} \mu_i\,dN_i \qquad (1.1)$$

relates changes in internal energy U to those in entropy S, volume V, and numbers of particles N_i of n chemically distinct components. But we have rarely been told exactly to what physical systems Eq. (1.1) might apply. And then we remember endless discussions of the word "changes," reversible or not, exact or not, written variously as d, δ, or (horrors!) đ. Further, some even feel free to "integrate" Eq. (1.1) to obtain

$$U = T\,S - p\,V + \sum_{i=1}^{n} \mu_i\,N_i\,, \qquad (1.2)$$

with scant regard for constancy (and with respect to what, please?) of temperature T, pressure p, and chemical potentials μ_i. This is equivalent to saying, for example, that the integral of the differential form $dz = \cos x\,dx + \cos y\,dy$ is $z = x\cos x + y\cos y$ (up to a constant), when, in fact, it is $z = \sin x + \sin y$. A slightly more convincing argument rests on Euler's theorem for homogeneous functions of first degree [4], when applied to the "extensive" variables U, S, and N_i. But that is a tautology because one *assumes* the linearity of these variables with respect to "amount of matter" in order to prove Eq. (1.2), which is, after all, nothing but that very same statement of linearity. These remarks are not trivial, for Eqs. (1.1) and (1.2), together, yield a central result

$$S\,dT - V\,dp + \sum_{i=1}^{n} N_i\,d\mu_i = 0\,, \qquad (1.3)$$

known as the Gibbs-Duhem relation. And it is fair to say that the study of *all phase equilibria* rests on this relation. In a nutshell, it relates the changes, *for whatever reason*, of the "intensive" variables T, p, and μ_i. It is vital, therefore, that we derive this relation in as precise and general a manner as possible.

II. THERMODYNAMIC SYSTEMS, VARIABLES, AND PHASES

To free ourselves from assumptions of homogeneity requires that we take a cue from Continuum Mechanics, where by "thermodynamic system" we simply mean an object of interest that is contained in a domain $\mathcal{D}$ of physical space [5]. Now, it will come as no surprise that the mass or number of particles, within $\mathcal{D}$, of component i can be expressed as the integral

$$N_i = \int_{\mathcal{D}} C_i(\mathbf{r}) \, dV , \quad (i = 1,\ldots,n) , \tag{2.1}$$

where C_i is the corresponding mass or number "density" [6]. Less evident, perhaps, is the observation that such a representation does not require homogeneity. In other words, the densities C_i are, in general, functions of both location and time. In mathematical jargon, they are non-constant "fields" [7]. Here, for the study of equilibria, we are not concerned with temporal variations, and we only need consider how densities depend on location vector $\mathbf{r}$.

It is also no surprise that the total number of particles N, within $\mathcal{D}$, is merely the sum over all n components

$$N = \sum_{i=1}^{n} N_i \tag{2.2}$$

and that it can be represented similarly to Eq. (2.1):

$$N = \int_{\mathcal{D}} C(\mathbf{r}) \, dV . \tag{2.3}$$

It follows that the total number density is also the sum

$$C = \sum_{i=1}^{n} C_i , \tag{2.4}$$

hence, the usual definition of mole fractions

$$x_i = C_i/C , \quad (i = 1,\ldots,n) , \quad \text{with} \quad \sum_{i=1}^{n} x_i = 1 . \tag{2.5a,b}$$

The densities C_i and C are also called "concentrations," and their units depend on the units of N_i and N.

Numbers of particles and their densities are prototypes of particular kinds of physical quantities that are conserved or, at least, that can be balanced in some grand accounting scheme. Examples include mass (as above), charge, momentum, angular momentum, various forms of energy, and entropy of any macroscopic system. In these lectures we are not concerned with transport phenomena, and, therefore, the derivation of balance laws lies elsewhere. Nevertheless, such a conserved quantity — call it Q — is *additive* in the sense that its value for a collection of subsystems is the algebraic sum of its values for each subsystem.

Further, under broad assumptions about the continuity of matter, one proves that there exists a "density function" q, such that

$$Q = \int_{\mathscr{D}} q(\mathbf{r}) \, dV.$$ (2.6)

For example, internal energy and entropy have representations

$$U = \int_{\mathscr{D}} u(\mathbf{r}) \, dV \quad \text{and} \quad S = \int_{\mathscr{D}} s(\mathbf{r}) \, dV$$ (2.7a,b)

in terms of their density functions u and s. In former days, additive quantities Q were known as "extensive variables," probably to distinguish them from fields q [7,8], which were called "intensive variables." Some books make much of this difference, but, in fact, Eq. (2.6) is all there is to say. In sum, then, dealing with density functions, rather than with their integrals (technically, called "functionals"), frees us from any *a priori* assumptions regarding homogeneity.

Finally, let us turn to the notion of phase. But we must first state the obvious: The description of a given thermodynamic system depends on a *particular* choice of measurable fields, and the minimum number of these fields which provides a complete characterization is a matter of experiment. Then, by "phase" we mean a region of space, within our fundamental domain $\mathscr{D}$, throughout which *all* fields of this complete set are *continuous* [9]. A "phase boundary," on the other hand, is a surface, line, (or even, point) where, *at least one* of these fields is *discontinuous*. And that rounds out our list of definitions.

III. EQUILIBRIUM CONDITIONS

Gibbs realized that all thermodynamic equilibria rest on a single principle, namely, that the *total* energy E of a body must achieve a minimum under given constraints. These constraints always include the constancy of the total entropy and, in addition, others that depend on the system at hand. For example, constants of motion of mechanical systems (*e.g.,* total linear and angular momentum) can be included in this list of subsidiary conditions. In this form we recognize that equilibria, thermodynamic or mechanical, are fully described by a "minimum principle," and we are thus committed to variational methods [10,11]. For simplicity, we illustrate in detail a fairly simple case: *the equilibrium of an immobile, single-phase, n-component, non-reactive, fluid mixture in an external force-field.*

To carry out such calculations, we recall the flavor of the Calculus of Variations. Integrals of the type (2.6) are first parametrized by considering arbitrary displacements $\delta \mathbf{r}$ of all points of our system $\mathscr{D}$. As shown on Fig. 1, any point $\mathbf{r}$ gets mapped into a new point $\mathbf{r}_\theta = \mathbf{r} + \delta \mathbf{r}$ that depends on the real parameter θ, and the field $q(\mathbf{r})$ is then "imbedded" in the family $q_\theta(\mathbf{r}_\theta, \theta)$. Consequently, the integral

$$Q(\theta) = \int_{\mathscr{D}_\theta} q_\theta(\mathbf{r}_\theta, \theta) \, dV$$ (3.1)

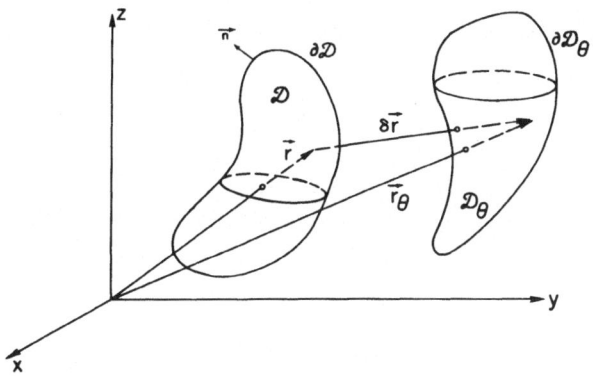

Fig. 1. Domain $\mathcal{D}$ of a given physical system, its boundary $\partial\mathcal{D}$, and the family of deformations required to establish the variational formula (3.2).

is a pure function of θ, with respect to which it can be differentiated. If $\theta = 0$ corresponds to the equilibrium state that we are seeking, then the (first) "variation" δ of any quantity is merely the expression $\delta\theta\,\partial/\partial\theta\big|_{\theta=0}$, where $\delta\theta$ is an *arbitrary* increment of θ. Carrying out this program, one finds [10,11] that the variation of the integral (2.6) is

$$\delta Q = \int_{\mathcal{D}} (\delta q + q\nabla\cdot\delta\mathbf{r})\,dV . \tag{3.2}$$

As in Hydrodynamics, δq represents the total variation $\delta q|_{\mathbf{r}} + \delta\mathbf{r}\cdot\nabla q$, which clearly separates changes due to displacements from those due to *any* other cause. These variations, however, are *virtual* (in the sense of Analytical Mechanics) because they may correspond to deformations of space and other changes that need not be realized.

We are now ready to set up our problem because a necessary condition for minimum energy is certainly $\delta E = 0$. This minimum is constrained by the subsidiary conditions

$$S = \text{const.} \quad \text{and} \quad N_i = \text{const.}, \quad (i = 1,\dots,n) . \tag{3.3a,b}$$

This first condition means that the system is adiabatically closed, and the second merely indicates that the particle numbers are conserved when the components are unreactive. As is well known, the problem of constrained minimization is best handled by introducing Lagrange multipliers — call them T' and μ'_i — to yield the variational equation

$$\delta E - T'\delta S - \sum_{i=1}^{n} \mu'_i\,\delta N_i = 0 , \tag{3.4}$$

but it is perhaps less well remembered that one *proves* [10] these multipliers to be constant fields, *i.e.*, that $\nabla T' = \nabla\mu'_i = 0$ for "isoperimetric" problems.

We still need two other pieces of information to carry out the variation (3.4): What constitutes E and what fields fully describe our system? First, the total energy includes the internal energy (2.7a) and the potential energy Φ of the system in an external force-field. The total force exerted on the components in $\mathcal{D}$ is

$$F = \sum_{i=1}^{n} \int_{\mathcal{D}} f_i(\mathbf{r}) \, dV \, , \tag{3.5a}$$

and each force density can be expressed in terms of potential gradients

$$f_i = -C_i \nabla \varphi_i \, , \quad (i = 1,\ldots,n) \, . \tag{3.5b}$$

For example, in a gravitation g- or electric E-field, we have

$$\begin{aligned} f_i &= C_i \, m_i \, \mathbf{g} \;\Rightarrow\; \varphi_i = -m_i \, \mathbf{g} \cdot \mathbf{r} + \text{const.} \, , \\ f_i &= C_i \, z_i \, \mathbf{E} \;\Rightarrow\; \varphi_i = z_i \, \phi + \text{const.} \, , \end{aligned} \tag{3.6a,b}$$

where m_i and z_i are, respectively, the mass and charge per particle (or mole) of the ith component (the C's, here, are number or mole densities) and where ϕ is the electrostatic potential. The total potential energy of our system is thus

$$\Phi = \sum_{i=1}^{n} \int_{\mathcal{D}} C_i(\mathbf{r}) \, \varphi_i(\mathbf{r}) \, dV \, . \tag{3.7}$$

There is yet one more contribution to E. Indeed, if the system is enclosed in some container, then the boundary $\partial\mathcal{D}$ of $\mathcal{D}$ produces work in any displacement. Calling $\mathbf{t}(\mathbf{r})$ the forces per unit area necessary to simulate that container, we then have

$$\delta E = \delta U + \delta \Phi + \int_{\partial\mathcal{D}} \mathbf{t} \cdot \delta\mathbf{r} \, dA \, . \tag{3.8}$$

Note that specifying reactive forces at physical boundaries frees us from requiring constant volumes. The second piece of information concerns the basic fields that form a complete set. Now it is an experimental fact that systems, such as we are describing, are fully described by $2+n$ fields u, s, and C [12]. Further, the First Principle of Thermodynamics, in local form, merely states that these fields are functionally related; for example,

$$u = u(s, C) \, . \tag{3.9}$$

This relation is not nearly as trivial as it may seem, for it implies, first, that the spatial dependence of u must occur through that of its argument (s, C), and, second, that an arbitrary variation of u is a linear form in the arbitrary variations δs and δC, written traditionally

$$\delta u = T \, \delta s + \sum_{i=1}^{n} \mu_i \, \delta C_i \, . \tag{3.10}$$

The differential coefficients T (called temperature) and μ_i (called chemical potentials) are evidently also functions of the same variables (s, C).

Inserting Eqs. (3.8), (3.7), (2.7), and (2.1) into the variational principle (3.4), and using the formula (3.2), we obtain a linear combination of terms in the independent variations δs, δC, and δr, whose coefficients must then vanish identically. Without writing out this complete expression, let us outline the results. First, the coefficient of δs yields $T(s, C) = T'$, and those of δC_i yield $\mu_i(s, C) + \varphi_i(\mathbf{r}) = \mu_i'$. Since, as mentioned above, the Lagrange multipliers are constants with respect to spatial variations, it follows that

$$\nabla T = 0 \quad \text{and} \quad \nabla(\mu_i + \varphi_i) = 0 \, , \quad (i = 1,\ldots,n) \, , \quad \text{in } \mathcal{D} \, . \tag{3.11a,b}$$

6

In other words, we have just shown that temperature and "electrochemical" potentials are necessarily constant fields at equilibrium. Next, the residual variational expression suggests the introduction of a new field

$$-p = u - T s - \sum_{i=1}^{n} \mu_i C_i,$$
(3.12)

with which, using the equilibrium conditions on T and μ, we get

$$\int_{\mathscr{D}} \left[\sum_{i=1}^{n} C_i \, \delta\varphi_i - p\nabla \cdot \delta\mathbf{r} \right] dV + \int_{\partial\mathscr{D}} \mathbf{t} \cdot \delta\mathbf{r} \, dA = 0.$$
(3.13a)

We will now see that the field p, called the thermodynamic pressure, indeed coincides with what is ordinarily meant by hydrostatic pressure in a fluid. First, note that the field p, by its very definition (3.12), is also an explicit function of the fundamental variables $(s,\mathbf{C})$. Then, we must remember that the external force-fields are "impressed" on our system (by fixed external mass or charge distributions, say) and, thus, that they are unchanged by variations other than of location. Consequently, $\delta\varphi_i = \delta\mathbf{r} \cdot \nabla\varphi_i$, and, using Green's divergence theorem, the previous expression becomes

$$\int_{\mathscr{D}} \left[\sum_{i=1}^{n} C_i \nabla\varphi_i + \nabla p \right] \cdot \delta\mathbf{r} \, dV + \int_{\partial\mathscr{D}} (\mathbf{t} - p\mathbf{n}) \cdot \delta\mathbf{r} \, dA = 0,$$
(3.13b)

where $\mathbf{n}$ is the outward-pointing normal to the closed surface $\partial\mathscr{D}$. Therefore, for arbitrary variations $\delta\mathbf{r}$, we finally get

$$\nabla p + \sum_{i=1}^{n} C_i \nabla\varphi_i = 0 \quad \text{in } \mathscr{D},$$
$$-p\mathbf{n} + \mathbf{t} = 0 \quad \text{on } \partial\mathscr{D}.$$
(3.14a,b)

These equilibrium conditions show that p has the character of a distributed force (it balances the given forces $\mathbf{t}$ on the boundary) and that it satisfies the equation for hydrostatic pressure in an external force-field.

IV. CONSEQUENCES OF THESE EQUILIBRIUM CONDITIONS

To summarize the last section, Gibbs's minimum principle for thermodynamic equilibrium implies a variational formulation that always yields the necessary conditions for equilibrium. No *a priori* assumption on homogeneity (*i.e.*, constancy of fields) is required. For the physical system at hand — an immobile, single-phase, n-component, non-reactive, fluid mixture in an external force-field — we have *proved* that the temperature field is constant, but that neither the chemical potentials nor the pressure are constant. Indeed, they are related to the spatial variations of the external force-field's potentials.

There are several important consequences and extensions of these calculations. First, we must remember that we have only dealt with the *necessary* conditions for equilibrium. In other words, *all* equilibrium states of our system must satisfy the conditions (3.11) and (3.14), but metastable states can also satisfy these same conditions. A complete classification

— clearly outside the scope of these lectures — requires the computation of second variations $\delta^2 Q$, and, in particular, the necessary conditions for *minimum* energy require that $\delta^2 E > 0$. On the other hand, the search for *sufficient* conditions for equilibrium is still a subject of investigation.

Next, a related matter is the spatial variation of energy, entropy, and mass densities. Indeed, by computing the gradients of T and of μ_i, one can show (here, we would need a discussion of stability) that ∇u, ∇s, and ∇C_i do not vanish. In other words, in an external force-field, the energy, entropy, and mass densities are not constant fields. There can be, therefore, no question of "integrating" these fields to give Eqs. (1.1-3). These results, however, remain true in *local* form. Indeed, varying Eq. (3.12) and comparing with Eq. (3.10), we get the local Gibbs-Duhem relation

$$ s\,\delta T - \delta p + \sum_{i=1}^{n} C_i\,\delta\mu_i = 0 \quad \text{in } \mathscr{D}, \tag{4.1} $$

which is always true. This relation can be compared with Eq. (1.3), which is hardly ever true [13]. Of course, it is easy to convince oneself that Eqs. (1.1-3) are correct in the absence of external force-fields, for then one shows that *all* fields are constant. In fact, Eqs. (1.1-3) then follow from Eqs. (3.10), (3.12), and (4.1) because integration of constant densities over $\mathscr{D}$ is immediate. Even in this case, however, we wish to emphasize that *no* spurious assumption on homogeneity is required: It is the consequence of a variational calculation.

Then, our local formulation of Thermodynamics also allows the introduction of densities of auxiliary energy functions through Legendre transformations. For example,

$$ g = u - T\,s + p, \tag{4.2a} $$

is the Gibbs free energy density, and, with Eq. (3.12), we have the useful result

$$ g = \sum_{i=1}^{n} \mu_i\,C_i, \tag{4.2b} $$

about which some more, later on.

And last, we must admit that the system at hand is rather simple. Chemical reactions, for example, introduce other restrictions among the various components; these are conveniently expressed in terms of "degree of advancement" [4]. If body motions are considered, then one shows that only rigid screw-motions are compatible with equilibrium. Were we also to deal with elastic bodies, then the list of variables (s,C) must be augmented by the components of the strain tensor to gain a complete description. The Gibbs-Duhem relation holds in the first two cases, but not in the last. Likewise, surface fields must be included if the thermodynamic state of the system's boundary is significant. These questions are largely beyond our scope, and an adequate treatment requires a good dose of Tensor Calculus and Differential Geometry [14-19]. In the next section, however, we will be confronted with the notion of phase coexistence, and this can hardly be divorced from an elementary consideration of boundary surfaces of thermodynamic systems. A final remark relates to the fundamental relation (3.9). Is it sufficient for the description of systems whose properties are rapidly varying in space? The theory of spinodal decomposition [20,21], for example, requires that one consider gradients ∇C as well as concentration values. In that case, Eq. (3.9) becomes the functional relation

$$u = u(s, \mathbf{C}, \nabla \mathbf{C}),\tag{4.3}$$

and the associated variational problems yield full-blown Euler equations rather than their degenerate forms.

V. GIBBS-DUHEM RELATIONS AND PHASE RULES

We have just seen that any n-component, single-phase, fluid-like system is described by the complete set of $2+n$ fields $(u, s, \mathbf{C})$, and these are related through the First Principle (3.9). Since energy and entropy densities are not directly accessible experimental quantities, it is useful to work with a different set of fields. By definition (3.10) we have

$$T = \frac{\partial u}{\partial s} \quad \text{and} \quad \mu_i = \frac{\partial u}{\partial C_i}, \quad (i = 1,\ldots,n),\tag{5.1}$$

which establishes the temperature and chemical potentials as functions of the variables $(s, \mathbf{C})$. This $1+n$ system of equations can be inverted (again, here one requires a discussion of stability) to yield the functional dependences

$$s = s(T, \mu) \quad \text{and} \quad C_i = C_i(T, \mu), \quad (i = 1,\ldots,n).\tag{5.2}$$

It follows, from Eqs. (3.9) and (3.12), that both u and p are also functions of the same variables. The Gibbs-Duhem relation (4.1) expresses that very fact: Of the $2+n$ new fields

$$T, \ p, \ \mu_1,\ldots,\mu_n,\tag{5.3}$$

only $1+n$ can be arbitrarily assigned. But this statement appears insignificant unless it is understood that Eq. (4.1) relates — in a definite way — *any arbitrary changes* in the fields (5.3). Concisely, the Gibbs-Duhem relation is a linear differential form in the variables (5.3). Its integration is the core of Thermodynamics.

Phase rules are nothing but an extension of the previous statements to systems consisting of several phases at equilibrium. We are, therefore, compelled to extend the treatment of Section III to such systems. Consider, therefore, a collection of f two-by-two contiguous phases $\mathscr{D}^{(1)}, \ldots \mathscr{D}^{(f)}$. [No equilibrium can be achieved without exchanges of matter and energy, and these occur at common boundaries — therefore, the qualifier "contiguous." The phases are also said to "coexist."] Assume, to begin, that these phases are all three-dimensional and large in extent. Consequently, we can neglect contributions from phase boundaries to the total mass, energy, entropy, *etc.* The variational equation (3.4) still holds for the system *as a whole*, and, by definition, each of these additive quantities must be summed over all parts of the system:

$$Q = \sum_{\alpha=1}^{f} Q^{(\alpha)}, \quad \text{with} \quad Q^{(\alpha)} = \int_{\mathscr{D}^{(\alpha)}} q^{(\alpha)}(\mathbf{r})\, dV.\tag{5.4}$$

Therefore, each field comes "dressed" with up to two indices. For example $C_1^{(2)}$ represents the concentration of the first component in the second phase [22]. We assume, further, that

9

a functional form $u^{(\alpha)} = u^{(\alpha)}(s^{(\alpha)}, \mathbf{C}^{(\alpha)})$, of the type (3.9), fully characterizes the First Principle in each phase (α). The development of the necessary conditions for equilibrium follows exactly as in Section III, and one shows that Eqs. (3.11), (3.14a), and (4.1) must hold in *each* phase. In addition, one finds that

$$T^{(\alpha)} = T^{(\beta)}, \quad p^{(\alpha)} = p^{(\beta)}, \quad \text{and} \quad \mu_i^{(\alpha)} + \varphi_i^{(\alpha)} = \mu_i^{(\beta)} + \varphi_i^{(\beta)}, \quad (i = 1, \ldots, n), \quad (5.5a,b,c)$$

on any surface Σ that is the common boundary of any two contiguous phases (α) and (β). In other words, the temperature, pressure, and electrochemical potentials are *continuous* at phase boundaries of large, three-dimensional phases in equilibrium [23].

The phase rule, in its usual form, is merely the enumeration of independent fields from the collection of $f(2+n)$ fields of the type (5.3):

$$T^{(\alpha)}, p^{(\alpha)}, \mu_i^{(\alpha)}, \quad (\alpha = 1, \ldots, f, \quad i = 1, \ldots, n). \tag{5.6}$$

Now, we have just seen that there are f Gibbs-Duhem relations, one for each phase, and $(f-1)(2+n)$ relations (5.5) at the $f-1$ boundaries of pairs of phases. Therefore, the number of thermodynamic variables whose values can be arbitrarily prescribed is

$$d = n + 2 - f. \tag{5.7}$$

By analogy with mechanical systems, this number d is called the thermodynamic system's *degrees of freedom*, and Eq. (5.7) is called the phase rule. This apparently trivial result is perhaps less so if we realize that, having chosen experimentally the values of *any d* variables from the list (5.6), the remaining values, as well as *all other* thermodynamic variables that depend on that list (such as the entropy densities and concentrations), are entirely determined.

The phase rule's admittedly simple derivation glosses over several important issues that can cause confusion. First, it must be remembered that it holds for phases that are in global equilibrium. It *is* often used, however, in regions that are only in local equilibrium, such as in the vicinity of plane phase boundaries. This is, in fact, an assumption on local kinetics that we will explore later. Next, the n components are assumed independent. If not, such as with chemical reactions, the number n in Eq. (5.7) must be reduced by the number of linearly independent reactions. Further, it is assumed that the list of variables (5.6) fully describes the phases' internal state. This, as was discussed at the end of the last section, may not be true, *e.g.*, for solids under non-isotropic states of stress or for electromagnetic radiation. And last, the phases are assumed to be large in extent, which means that capillarity effects are negligible. We now seek extensions of Eq. (5.7) when the phases are "small."

The general problem of surface effects is quite complex because we must learn to characterize 2-dimensional "surface phases" that form the boundaries of bulk, 3-dimensional phases. This forms the subject matter of Surface Thermodynamics [2,14,16,18,24,25] which we now briefly broach. Enumeration of surface phases is another difficulty with the general theory because they do not exist in splendid isolation from bulk phases. Since each of these (except for pathological examples like Möbius strips) must be in contact with at least two bulk phases, and since the shape of all phases is incidental, it follows that the problem of enumeration is topological in nature. Because a system of *two* bulk phases in contact through their phase boundary is our main concern, we restrict our discussion to the case

$f = 2$. Blithe application of the phase rule (5.7) would state simply that the number of degrees of freedom is equal to the number of independent components $d = n$. But is this quite correct?

Consider the two-phase system [26] depicted on Fig 2, where the second phase (*e.g.*, a vapor or liquid) entirely contains the first (*e.g.*, a crystal). We now have the decomposition $Q = Q^{(1)} + Q^{(2)} + Q^{(\Sigma)}$ of any additive quantity, where the surface integral

$$Q^{(\Sigma)} = \int_{\Sigma} q^{(\Sigma)}(\mathbf{r})\, dA \tag{5.8}$$

represents that part of Q, attributable to the phase boundary, which is necessary to satisfy overall accounting of mass, energy, entropy, *etc*. So far, so good — but the main difficulty is to find a complete set of thermodynamic fields and an adequate First Principle for the surface phase. Gibbs showed that a fluid-like surface can still be described by a fundamental relation of the type (3.9)

$$u^{(\Sigma)} = u^{(\Sigma)}(s^{(\Sigma)}, \mathbf{C}^{(\Sigma)}) \quad \text{on } \Sigma, \tag{5.9}$$

but that the surface's mean curvature K [27,28] must be considered an additional, though non-explicit, thermodynamic variable. Now, carrying out the variation (3.4), one finds exactly the same equilibrium conditions (3.11), (3.14), and (4.1) in the bulk phases $\mathscr{D}^{(1)}$ and $\mathscr{D}^{(2)}$, and the temperatures and electrochemical potentials are also continuous at the phase boundary Σ:

$$T^{(1)} = T^{(2)} = T^{(\Sigma)}, \quad \mu_i^{(1)} + \varphi_i^{(1)} = \mu_i^{(2)} + \varphi_i^{(2)} = \mu_i^{(\Sigma)} + \varphi_i^{(\Sigma)}, \quad (i = 1,\ldots,n). \tag{5.10a,c}$$

Equation (5.5b), however, must be replaced by a form of Laplace's law

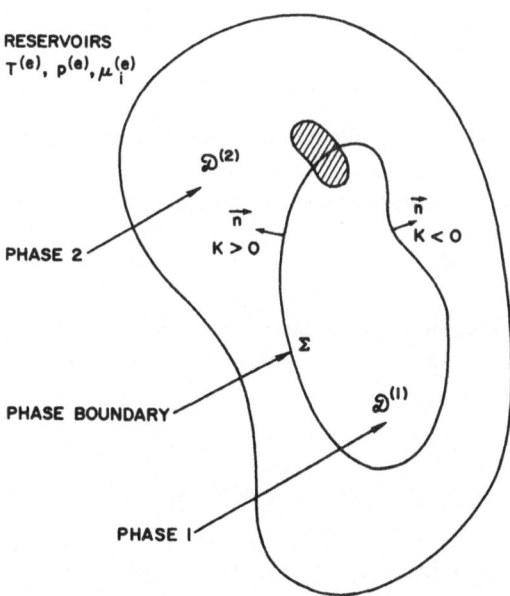

Fig. 2. Domains $\mathscr{D}^{(1)}$ and $\mathscr{D}^{(2)}$ of a physical system consisting of two phases and their phase boundary Σ.

11

$$p^{(1)} - p^{(2)} = \sigma K + \sum_{i=1}^{n} C_i^{(\Sigma)} \mathbf{n} \cdot \nabla \varphi_i^{(\Sigma)} \quad \text{at } \Sigma \,, \tag{5.10b}$$

where

$$\sigma = u^{(\Sigma)} - T^{(\Sigma)} s^{(\Sigma)} - \sum_{i=1}^{n} \mu_i^{(\Sigma)} C_i^{(\Sigma)} \quad \text{on } \Sigma \,, \tag{5.11}$$

similar to Eq. (3.12), defines the surface tension σ [29]. Equation (5.10b) specifies the discontinuity of p at the surface phase Σ, which, incidentally, *proves* that Σ is a phase boundary. The second term on its right-hand side represents the force-field on Σ due to surface mass or to surface charge.

A few words are in order regarding the mean curvature's role as a thermodynamic variable. First, in the absence of external force-fields, the pressures $p^{(1)}$, $p^{(2)}$, and the surface tension σ are all constant fields over their respective domains. This, together with Laplace's law (5.10b), implies that K must be constant over the phase boundary Σ. Therefore, the phase boundary's *shape* is characterized by *constant* mean curvature. The study of such shapes, known as the "problem of Plateau," has been the object of intense activity over the years. From the point of view of kinetics, however, one must recognize that K is a variable field over the phase boundary Σ if the adjacent phases are not in *global* equilibrium. Nonetheless, conditions (5.10) hold at boundaries within the small hatched domain on Fig. 2 if it is assumed *locally* in equilibrium, no matter what kinetic processes occur in the core of the adjacent bulk domains. Thermodynamics is thus vital for the construction of boundary conditions.

Turning to the special phase rule for small systems, there are now $3(2+n)+1$ variables because, as just seen, the mean curvature K must be appended to the list of basic observable quantities to give the new list

$$T^{(\alpha)}, \; p^{(1)}, \; p^{(2)}, \; \sigma, \; \mu_i^{(\alpha)}, \; K, \quad (\alpha = 1, 2, \Sigma, \quad i = 1.....n) \,. \tag{5.12}$$

On the other hand, there are three Gibbs-Duhem relations, one per phase, $2(1+n)$ continuity relations (5.10a,c) for T and μ, and Laplace's law (5.10b), for a total of $6+2n$ relations. Consequently, the number of independent variations in our system consisting of two bulk phases and a phase boundary is

$$d = n + 1 \,. \tag{5.13}$$

And that concludes our discussion of the phase rule's various forms. Further elaboration is available in standard books [2,4,24,25,30-33], and White's review paper on phase equilibria [34] is a model of clarity.

VI . TWO APPLICATIONS TO BINARY SYSTEMS

Phase diagrams are a graphical representation of the equilibrium conditions (5.5) or (5.10) at phase boundaries [35]. In this section we first restrict attention to a system of two large phases, each containing the same two components, and we neglect external fields and body motions. The phase rule (5.7) then indicates that there are two independent variables in the eight-member list (5.6). To see just how this is expressed analytically, then graphically, an aside on the Gibbs free energy (4.2) and its composition dependence is necessary. Indeed, chemical potentials are fine objects, but composition variables are even better.

With the definition (4.2a) and the First Principle (3.10), we obtain

$$\delta g = -s\,\delta T + \delta p + \sum_{i=1}^{n} \mu_i\,\delta C_i ,\tag{6.1}$$

which shows that the chemical potentials $\mu = \partial g/\partial C$ — like g — are functions, in each phase, of the observable fields $(T, p, \mathbf{C})$. But we have more. The result (4.2b), when differentiated with respect to concentration, yields

$$\sum_{i=1}^{n} \frac{\partial \mu_i}{\partial \mathbf{C}} C_i = \sum_{i=1}^{n} \frac{\partial \mu}{\partial C_i} C_i = 0 .\tag{6.2}$$

With Euler's theorem, this indicates merely that the chemical potentials are homogeneous functions of degree zero in the concentrations. In other words, a uniform change in C has no effect on μ, and the chemical potentials depend on composition only through ratios such as mole fractions $\mathbf{x}$. Returning now to the equilibrium conditions (5.5), and calling T and p the *values* of temperature and pressure common to both phases, we must have

$$\mu_i^{(1)}(T, p, \mathbf{x}^{(1)}) = \mu_i^{(2)}(T, p, \mathbf{x}^{(2)}) , \quad (i = 1,.....,n) .\tag{6.3}$$

These constitute a system of n equations in the $2 + 2(n-1) = 2n$ variables [36] $(T, p, \mathbf{x}^{(1)}, \mathbf{x}^{(2)})$, and thus, as predicted by the phase rule, there are exactly n free variables.

We specialize these remarks to two-component systems. Consider then a two-phase system, as was represented crudely in Fig. 2, and, for definiteness, let the "enclosed" phase (1) be a crystal, while phase (2) can be of any state of aggregation. Since composition vectors have now only one independent component, call $x^{(1)}$ and $x^{(2)}$ the composition variables in phases (1) and (2), respectively [37]. For instance, if the two components are labeled A and B, then $x^{(2)}$ can be either of the variables $x_A^{(2)}$ or $x_B^{(2)}$ in the "enclosing" phase (2). Equations (6.3) thus reduce to a system of two equations

$$f_i(T, p, x^{(1)}, x^{(2)}) = 0 , \quad (i = A,B)\tag{6.4a,b}$$

in four variables, two of which can then be chosen arbitrarily. For example, setting the pressure at some constant value, for each given value of temperature there is a unique pair of compositions $x^{(1)}$ and $x^{(2)}$ that satisfies Eqs. (6.4) identically, and this pair is in equilibrium. By varying T, one traces out the locus of equilibrium curves in the (x, T)-plane. These have names: "solidus" and "liquidus" if one is dealing with solid-liquid equilibria. Figure 3 shows such a diagram when the solid is a compound, *i.e.*, its composition is essentially constant and the solidus is a straight line at the compound composition [38]. The one-to-one corresponding equilibrium compositions at temperature T, labeled (1) and (2), are said to be connected by a "tie-line." A concrete example might be the equilibrium of solid GaAs with a liquid solution of As in the solvent Ga.

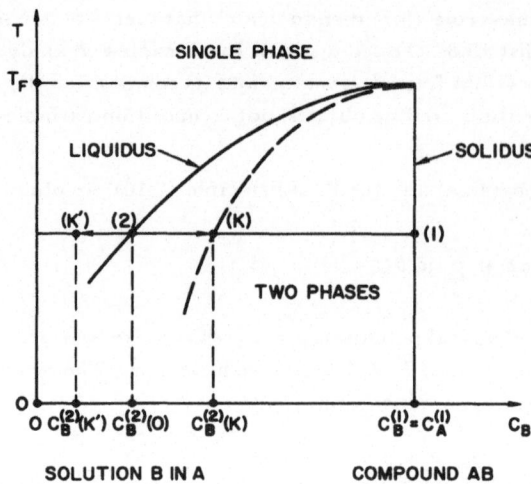

Fig. 3. Schematic portion of a binary phase diagram, showing the liquid compositions that are in equilibrium with an AB compound. Compositions are displayed in concentration units, but mole fractions are more fundamental. The dashed line corresponds to a shift in liquidus due to capillarity.

In some cases, we may even know how chemical potentials vary with temperature and composition. This is so for the III-V compound semiconductors (GaAs, InP, *etc.*), and it is then possible to calculate liquidus curves [39-41]. Suffice to say here that one assumes "regular solution" behavior of the liquid [42], which means that the chemical potentials can be written in the form

$$\mu_i^{(2)} = \mu_i^{*(2)}(p, T) + kT \ln \gamma_i^{(2)} x_i^{(2)}, \quad (i = A, B), \tag{6.5a}$$

where k is Boltzmann's constant and where the solution's non-ideality is characterized by activity coefficients γ. In the regular solution approximation, these are given by

$$kT \ln \gamma_i^{(2)} = \alpha(T) \left(1 - x_i^{(2)}\right)^2, \quad (i = A, B), \tag{6.5b}$$

in which the "interaction coefficient" α is often a linear function $a + bT$ ($a \neq 0$) of temperature [41,43-45]. Under these circumstances, one finds that the function

$$\Delta S_F (T_F - T) + kT \ln 4x(1 - x) + 2\alpha(0.5 - x)^2 = 0 \tag{6.6}$$

provides an adequate description of the liquidus curve. Here, ΔS_F and T_F are the entropy of fusion (per particle or mole) and temperature of fusion of the compound, respectively, and x stands for either of the mole fractions $(x_A^{(2)}, x_B^{(2)})$ in the liquid phase. These questions are discussed further in the excellent papers by Panish and Ilegems [45] and also by Jordan [46,47]. The sixth chapter of Casey and Panish's book [48] provides a clear tutorial and a complete list of data for binary III-V compounds. Solid-vapor equilibria and point defect structure of these compounds are examined in papers by Hurle and Mullin [49,50]. Finally, well-known books [51-53] contain extensive compilations of thermodynamic data and phase diagrams for binary and some ternary systems.

14

In sum, a large, field-free, immobile, two-phase, two-component, fluid-like system is entirely characterized by the four variables (T, p, $x^{(1)}$, $x^{(2)}$), *any two* of which can be chosen arbitrarily. In other words, it is impossible to fix, simultaneously, the values of three or more of these variables. We now ask what effects accrue when one of the phases — the crystal, say — is small. What does it mean that the number of degrees of freedom, according to Eq. (5.13), is now three? And how are phase diagrams affected by the choice of, at most, three of the variables from the 13-member list (5.12)?

To analyze such problems, it is convenient to return to Fig 2 and to imagine our two-phase system immersed in a large external phase with which it can reversibly exchange heat at temperature $T^{(e)}$, mechanical work at pressure $p^{(e)}$, and masses at chemical potentials $\mu_i^{(e)}$. This idealization requires comment. The external phase represents interactions with the world at large, *i.e.*, the experimental conditions that one can impose. This phase is also assumed very large. Therefore, changes in the enclosed phases, $\mathscr{D}^{(1)}$ and $\mathscr{D}^{(2)}$, induce negligible changes within it, and surface effects at its own external boundaries play a negligible role. Such a phase is often called a system of "reservoirs."

Consider, for example, our two-component system at constant external temperature $T^{(e)}$ and pressure $p^{(e)}$. We then seek the chemical potentials' dependence on curvature K. The situation is now very simple if we mentally travel inward, so to speak, from the external phase towards phase (1). First, constant experimentally imposed $T^{(e)}$ implies constant $T^{(2)} = T^{(e)}$ at the second phase's external boundary, because a relation of the type (5.5a) must hold there. Since $T^{(2)}$ is a constant field in phase (2), its temperature value $T^{(e)}$ "propagates," unchanged, throughout that phase. Likewise, $T^{(1)} = T^{(\Sigma)} = T^{(e)}$ by the same continuity arguments. Therefore, we have $\delta T^{(e)} = \delta T^{(1)} = \delta T^{(2)} = \delta T^{(\Sigma)} = 0$ for all conceivable variations. Then, $\delta p^{(e)} = \delta p^{(2)} = 0$ for similar reasons. Note, however, that $\delta p^{(1)} = \delta(\sigma K) \neq 0$ because of Eq. (5.10b). In other words, the phase boundary Σ is a barrier to the propagation of the pressure $p^{(e)}$ into the crystalline phase (1). The chemical potentials, on the other hand, cannot all be assigned without violating the phase rule. They are, nonetheless, constant fields over their respective domains. Moreover, by virtue of Eqs. (5.10c), their values, for each component, must be equal throughout the two-phase system and its phase boundary. If μ_A and μ_B be these values common to all three phases, and if $\delta\mu_A$ and $\delta\mu_B$ be their variations, then the above considerations show that the Gibbs-Duhem relations (4.1), in the bulk phases, reduce to

$$-\delta(\sigma K) + C_A^{(1)}\delta\mu_A + C_B^{(1)}\delta\mu_B = 0 \quad \text{in } \mathscr{D}^{(1)}, \tag{6.7a}$$

$$C_A^{(2)}\delta\mu_A + C_B^{(2)}\delta\mu_B = 0 \quad \text{in } \mathscr{D}^{(2)}, \tag{6.7b}$$

from which, for example, we can eliminate $\delta\mu_A$:

$$-\delta(\sigma K) + \left(C_B^{(1)} - C_A^{(1)}C_B^{(2)}/C_A^{(2)}\right)\delta\mu_B = 0. \tag{6.8}$$

If, now, phase (2) is a dilute solution of component B, then the second term in parenthesis must be small. If, further, the concentration of that component in phase (1) does not vary much with the curvature of that phase, then Eq. (6.8) can be integrated to give

$$\mu_B(K) - \mu_B(0) = \sigma K/C_B^{(1)}. \tag{6.9}$$

15

It must be understood that the integration is carried out from the state for which the surface is flat ($K = 0$, which means that phase (1) is effectively of infinite extent) to the state for which the curvature is finite. Moreover, that integration requires no assumption whatsoever regarding the surface tension σ's dependence on curvature or composition [54]. To go from chemical potentials to concentrations, we take advantage of our knowledge of the analytic form (6.5) in phase (2). Inserting this form into Eq. (6.9) we find the exponential behavior

$$C_B^{(2)}(K) = C_B^{(2)}(0)\, e^{\sigma K / kT C_B^{(1)}} \tag{6.10}$$

when the solution is very dilute in component B.

This formula, attributed to Ostwald, is one of many that are known generically as "Gibbs-Thomson effects." They represent shifts in equilibria due to curvature. The quantity $\sigma / kT C_B^{(1)}$ is positive and has dimensions of length; it is called the "capillary length." Therefore, formula (6.10) shows that the equilibrium concentration of the minor component above a convex surface element ($K > 0$) is larger than what would prevail over a flat element, and *vice versa* for concave surface elements. In passing, it is worth noting that Eq. (6.8) is an exact result. Its integration can be performed, without approximation, to yield extensions of Eqs. (6.9) and (6.10) that are valid for concentrated solutions, as well.

The application of Ostwald's formula (6.10) to phase diagrams is of the utmost importance. The region above the liquidus on Fig. 3 represents thermodynamic states of the liquid that are *not* in equilibrium with any finite chunk of solid. Indeed, a point such as (K') corresponds to a liquid composition $C_B^{(2)}(K')$ that is *smaller* than the liquidus composition $C_B^{(2)}(0)$ (called the liquid's "saturation" value) at the same temperature. Accordingly, the solid phase (1) that would be in equilibrium with (K') has constant *negative* mean curvature, and Differential Geometry teaches that there are no *finite* bodies of this type. Therefore, in this so-called "single-phase region" of Fig. 3, we are free to choose arbitrarily the solution's temperature and concentration — as is evident to any good cook. These liquid solutions are said to be "undersaturated" with respect to any solid. On the other hand, the region of Fig. 3 bounded by the solidus and liquidus lines represents metastable states of the liquid; it is called the "two-phase region" for the following reasons. Assume that the solid is of small extent, and therefore that its curvature is, on the whole, large and positive. It follows from Eq. (6.10) that this solid would be in equilibrium with a liquid whose concentration $C_B^{(2)}(K)$ in the minor component B is *larger* than its saturation value for a large solid. This is true at any temperature below the compound's temperature of fusion T_F. Thus, a given positive curvature K of the solid phase (1) causes a "liquidus-shift" (dashed) to the right. Reciprocally, given an arbitrary point of the two-phase region (labeled (K) on Fig. 3), then there surely exists a solid, and only one, small enough (*i.e.*, of positive and large enough curvature K) that would be in equilibrium with the liquid of composition $C_B^{(2)}(K)$. This has important consequences: A liquid mixture that is somehow prepared in the two-phase region can persist for finite periods of time because it is in equilibrium with *some* solid particles of curvature K. One says that the liquid is "supersaturated" (or "supercooled"), and the difference $C_B^{(2)}(K) - C_B^{(2)}(0)$ is just one of many measures of that supersaturation. In the study of nucleation, however, one shows that such a liquid is not stable because these small solid particles are metastable, *i.e.*, there exists a critical dimension, related to the capillary length, below which solids dissolve and above which solids grow. The supersaturation in the liquid will then slowly decay — one says that the solution "ages" — because of transfer, by diffusion, of solute to the solid precipitates.

VII. TERNARIES AND HIGHER-ORDER SYSTEMS

Our world, fortunately, is composed of thermodynamic systems other than binaries. The complexities of higher, multicomponent systems are best explored through ternaries. Indeed, ternaries are the systems of highest order whose behavior is evident from a single graph.

For two, large, coexisting, bulk phases and three components — call them A, B, and C — the phase rule (5.7) dictates that there be three degrees of freedom. As we have seen in connection with Eq. (6.3) and (6.4), phase equilibrium now reduces to the discussion of solutions of the following system of five equations in eight variables:

$$f_i(T, p, x_A^{(1)}, x_B^{(1)}, x_C^{(1)}, x_A^{(2)}, x_B^{(2)}, x_C^{(2)}) = 0, \quad (i = A, B, C), \tag{7.1a,b,c}$$

together with the normalizations [55]

$$x_A^{(1)} + x_B^{(1)} + x_C^{(1)} = 1, \quad \text{and} \quad x_A^{(2)} + x_B^{(2)} + x_C^{(2)} = 1. \tag{7.2a,b}$$

If, for example, T, p, and any single composition variable are experimentally imposed, then there is no further leeway, at equilibrium, in the choice of all other composition variables.

Another way to picture equilibria is to solve formally the three Eqs. (7.1) with respect to the composition of one of the phases — the solid's $x^{(1)}$, say:

$$x_i^{(1)} = F_i(T, p, x^{(2)}), \quad (i = A, B, C). \tag{7.3a,b,c}$$

When inserted into the first of Eqs. (7.2), they yield a functional relation between the five variables $(T, p, x^{(2)})$. This relation, together with the second of Eqs. (7.2), defines a single function

$$\Phi(T, p, x_A^{(2)}, x_B^{(2)}) = 0 \tag{7.4}$$

between temperature, pressure, and any two of the component mole fractions in the other phase — a liquid, say. If p is experimentally imposed, then Eq. (7.4) represents a surface, the "liquidus-surface," in temperature-composition space. Points of this surface represent states of the liquid that are in equilibrium with *some* solid. In fact, having chosen such a point, Eqs. (7.3) give the composition of *the* solid with which it is in equilibrium. There is thus a one-to-one correspondence between composition vectors **x** of the two phases in equilibrium, and the line that connects corresponding points is also called a tie-line.

But how can we graphically represent these surfaces without resorting to 3-dimensional perspectives? Because we are mainly concerned with compositions as a function of temperature (pressure effects are negligible for condensed systems), and because we may want to preserve the symmetry of the three components, it is expedient to introduce a "composition triangle," as follows. Imagine that mole fractions x_i (never negative!) are the components [37,56] of vectors

$$\mathbf{x} = (x_1, x_2, x_3) \tag{7.5a}$$

in the first octant of a euclidian, 3-dimensional space. Assume, moreover, that the composition vectors are "normed" according to

$$x_1 + x_2 + x_3 = 1 \, . \tag{7.5b}$$

Figure 4 shows that the tip P of an arbitrary composition vector must lie, because of the constraint (7.5b), on the equilateral "(111)-triangle" (A_1, A_2, A_3). In other words, there is a (projective) one-to-one mapping between compositions and points on this triangle. Such a representation can be used (analytically) to handle any number of components, and it helps to identify special features of the composition triangle.

First, it is clear that the vertices A_1, A_2, and A_3 correspond to the pure components A, B, and C, respectively. Next, lines parallel to an edge correspond to a constant proportion of that component which labels the edge's opposing vertex. For example, the horizontal plane (H, M_1, M_2) contains the tip of all composition vectors whose third component $x_3 \equiv x_C$ is imposed. It follows that the edges of the composition triangle represent binary mixtures. Then, lines issuing from a vertex represent fixed ratios of the two components that label the other two vertices. For example, the ratio x_1/x_2 is constant along the line (A_3, M_3) because that line is the triangle's intersection with the vertical plane (O, H, A_3, P, M_3). More generally, any straight line in the composition triangle is the intersection of that triangle with a plane through the origin, and that plane's equation is a linear combination of the mole fractions x_1, x_2, and x_3. Consequently, to find particular points (intersection of two lines) on the composition triangle, one must always solve a *linear* system of equations in the components [57]. It is often easier, however, to reason with composition vectors. Say that

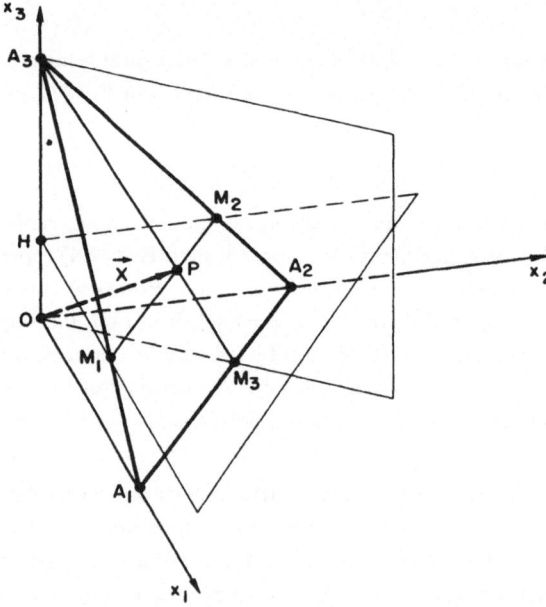

Fig. 4. Construction of the composition triangle from the 3-dimensional space of ternary compositions.

one needs the equation of a line through two known points P_1 and P_2. If $x(1)$ and $x(2)$ be the corresponding composition vectors [58], then the equation of the line in question is simply

$$x = \mu\, x(1) + (1 - \mu)\, x(2) \,, \tag{7.6}$$

where μ is a real parameter (not the chemical potential!) that varies between 0 and 1. Any point P of that line corresponds to some mixture that is not necessarily homogeneous. To interpret μ (and to derive the "lever rule") we must simply remember that the mixture N of two other mixtures, N(1) and N(2), has the overall composition

$$N = N(1) + N(2) \tag{7.7}$$

when there are no chemical reactions [59]. Comparing Eqs. (7.6) and (7.7), we find that

$$\mu = \frac{\Sigma_i\, N_i(1)}{\Sigma_i\,[N_i(1) + N_i(2)]} \tag{7.8}$$

is merely the proportion of moles of mixture (1) in the total mixture. According to Eq. (7.6), μ is also the ratio of lengths $\overline{PP_2}/\overline{P_1P_2}$, which can be read off the composition triangle. Finally, one must remember that curved lines, as well, can be traced on the composition triangle. For example, isobaric, isothermal sections of the liquidus (7.4) form families of level lines of that surface.

VIII. APPLICATION TO GARNETS AND III-V COMPOUNDS

We now end this discussion of phase equilibria with two examples of ternary systems that are of some technological interest. The first example, garnets, illustrates how several distinct compounds can be produced by compositional changes in the nutrient liquid phase.

Garnets all have the formula $Y_3Fe_5O_{12}$, or YIG for short, where any rare-earth or Gd can replace Y, and Ga or Al can replace Fe. They are grown in a high-temperature solution of molten PbO (the solvent or "flux") which dissolves the Y_2O_3 and Fe_2O_3 constituents. [One sometimes adds B_2O_3 to the flux to lower its viscosity and to reduce Pb evaporation losses.]

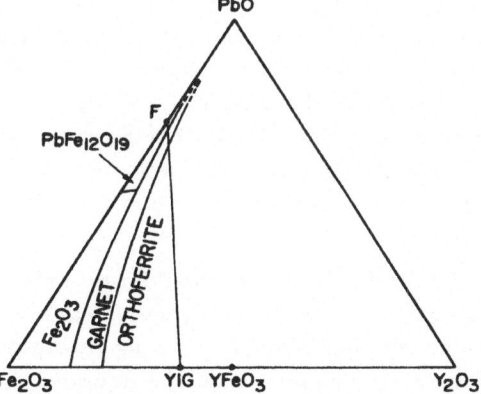

Fig. 5. Pseudo-ternary phase diagram [60] showing the primary fields which produce the labeled compounds. Liquid compositions along (a segment of) the line between points F and YIG produce yttrium iron garnet (YIG).

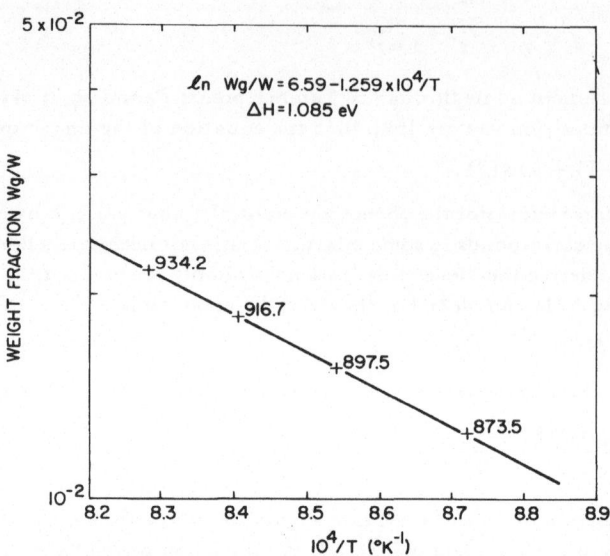

Fig. 6.   The liquidus along the pseudo-binary section F-YIG of Fig. 5, shown as an Arrhenius plot of garnet weight fraction $vs.$ T^{-1}. The labeled points correspond to experimentally determined [62] saturation temperatures (°C).

Although, strictly speaking, such systems are at least quaternaries (Y, Fe, O, Pb), it is convenient to think of them in terms of mixtures of PbO, Y_2O_3, and Fe_2O_3 *molecules.* These are the components of the phase diagram shown on Fig. 5. The properties of this system were explored by Nielsen and Dearborn [60] who showed that, depending on the initial liquid composition, one could obtain several different solids: the ternary, magnetoplumbite ($PbFe_{12}O_{19}$), the true binary, hematite (Fe_2O_3), and the "pseudo-binaries," orthoferrite ($YFeO_3$) and garnet (YIG). Indeed, the latter two compounds can be viewed as 1:1 and 5:3 mixtures, respectively, of the iron- and yttrium-bearing molecules. The regions from which, at a given temperature, one can form one or another of these compounds are called "primary-fields." For example, liquid compositions within the region of Fig. 5 labeled "garnet" will produce YIG, exclusively. Blank and Nielsen [61] explored further the range of molar ratios required for this to happen. For example, the Fe:Y ratio in the liquid must be more than seven times what is needed in solid YIG, and therefore the liquid compositions must lie to the left of the "5:3" line joining the PbO vertex to the point labeled YIG. The line F-YIG, on the other hand, is acceptable, and it forms the trace of a binary section shown on Fig. 6. This figure [62] shows a liquidus with essentially ideal solution behavior.

Precise determinations of the liquidus are essential for the understanding and modeling of growth kinetics. In these garnet systems, in particular, it was shown [62,63] that the actual concentrations C_0 at the solid-liquid interface during growth deviate from their equilibrium values $C_e(T)$. These deviations produce a supersaturated condition at the interface that is particularly noticeable (segregation and stress) in very thin films. Reference [64] gives a review of liquid-phase epitaxy, as of 1975, while references [65] and [66] review more recent developments.

20

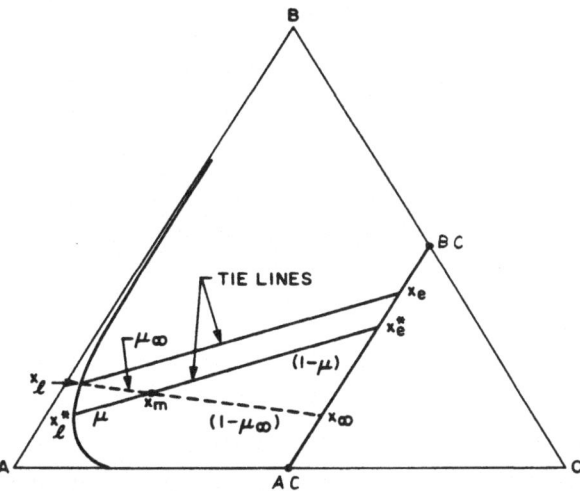

Fig. 7. A typical ternary III-III-V phase diagram showing the curved liquidus and tie-lines that connect it to solid solutions along the pseudo-binary AC-BC. The dashed line, joining initial states, is *not* a tie-line.

The second type of ternary system that we have in mind is formed from ternary mixtures of the elements (Al, Ga, In) and (P, As, Sb), in columns III and V of the periodic table. [The considerations of this section hold, as well, for II-VI systems such as (HgCd)Te.] The situation, here, is rather different from the previous example because any given ternary liquid (A, B, C) will produce a solid $A_{1-x}B_xC$ that is a member of a continuous series of solid solutions. These can be considered as mixtures, in proportion x, of the binary compounds AC and BC. As mentioned earlier, the phase diagrams of these systems have been fitted to regular solution theory [41,45-48], which means that we have *analytic* relations to compute tie-lines. Figure 7 shows such a diagram, and it shows the liquid compositions, on the curved liquidus isotherm, that correspond to solid compositions on the straight line connecting points AC to BC. Because these phase diagrams are so well understood, it is possible, in the case of III-V systems, to set up a truly ternary diffusion model of their growth and dissolution kinetics [67]. There, in the absence of better knowledge, we assumed local equilibrium at the solid-liquid interface, but the motion of that interface was computed as a full-fledged "Stefan problem." References [68] and [69] review this theory, as of 1984.

Our concern in these lectures, however, is the way in which phase diagrams can help predict how a given arbitrary liquid-solid pair will evolve after their contact [70]. Place a solid, of *arbitrary* initial composition x_∞, in contact with *any saturated* ternary solution at temperature T. Let $\mathbf{x}_\ell$ be that liquid composition, and draw, on Fig. 7, the *unique* liquidus isotherm which passes through that point of the phase diagram. We also know that there is a *unique* tie-line which connects $\mathbf{x}_\ell$ to the equilibrium solid composition x_e, and that tie-line is generally distinct from the dashed line which connects the arbitrary initial compositions. This non-equilibrium situation, if real, would be relieved by diffusion in both liquid and solid phases. Can we predict its overall direction and magnitude without having to solve a (non-linear) diffusion problem? Such a calculation is indeed quite simple. It suffices to

21

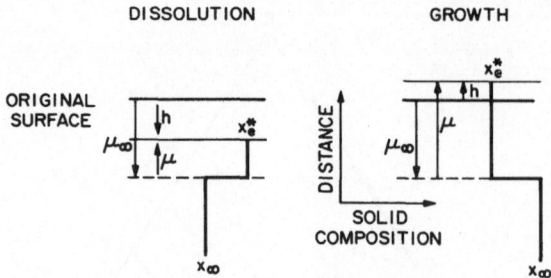

Fig. 8.   Sequence of virtual growth and dissolution steps necessary to simulate dynamic growth and
dissolution processes [70].

imagine a *virtual* process in which one mixes a small amount of the initial phases to produce
a supersaturated liquid mixture x_m. This liquid mixture relaxes, and produces (again, because a unique tie-line runs through it) a unique liquid-solid composition pair (x_l^*, x_c^*) in
equilibrium. The relative amounts of solid evolved in these virtual mixing and relaxation
processes are labeled μ_∞ and μ, and they are exactly the quantity discussed in connection
with Eq. (7.6). Now, if one finds that μ is greater than μ_∞, then the solid grows as equilibrium is approached, while dissolution is predicted when $\mu < \mu_\infty$. These two cases are explicit
on Fig. 8 which shows how the interface would travel during each virtual process. Although
the μ's are virtual, their difference $h(t)$ is the interface's *real* motion. Figures 9 and 10 (for
two typical III-III-V and III-V-V systems, respectively) demonstrate how this difference
depends on the initial liquid and solid phases. There is evidently neither growth nor dissolution when $x_\infty = x_c$ because the initial condition is at equilibrium. All other conditions,
however, are not, and one must expect mass redistribution in both phases. Finally, one can
also show [70] exactly how the difference $\mu - \mu_\infty$ is related to the real, diffusional kinetics,
and the dashed lines on these figures correspond to this calculation when the diffusivity in
the solid is small. We see that they are hardly distinct from the full lines. Consequently, for

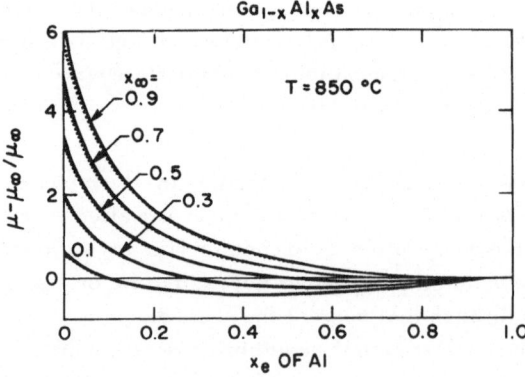

Fig. 9.   For Ga$_{1-x}$Al$_x$As, the relative amount of solid grown or dissolved as a function of equilibrium
composition in the solid. The solid curves represent thermodynamic estimates; they are labeled by the actual, initial solid composition. The dashed curves result from a calculation
of diffusion rates [70].

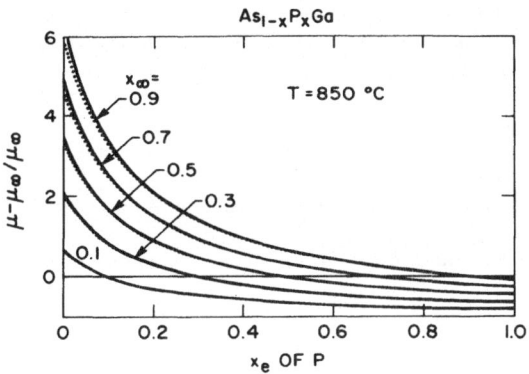

Fig. 10.   Same as Fig. 9 for the III-V-V system GaP_xAs_{1-x}.

these cases of solution growth from initially saturated solutions, we can *really* say that the system's kinetics are "close to equilibrium." All other statements to that effect, not substantiated by analytic estimates, are literature .... at best.

ACKNOWLEGMENT: It is a pleasure to thank my colleagues and friends E.A. Giess and M.B. Small (IBM Watson Research Center) and A.S. Jordan (AT&T Bell Laboratories) for their always patient and often humorous lessons during these many years of community of interest.

REFERENCES AND NOTES

1.   It is an historical accident that "Thermodynamics" does not deal with dynamics at all. Some prefer to call "Thermostatics" the study of equilibria, thus avoiding the long-winded names: "Thermodynamics of Irreversible Processes," or "Non-Equilibrium Thermodynamics" for truly time-dependent phenomena.

2.   J.W. Gibbs, "On the Equilibrium of Heterogeneous Substances," in *Scientific Papers*, Vol. 1, pp. 55-353 (Dover reprint). Often cited, rarely read........

3.   The brand of Thermodynamics that we have in mind is "naive," in the sense that mass, energy, entropy, *etc.*, are primitive, understood constructs. After all, we learn Euclidian Geometry without *really* knowing what a point or a line might be. In both cases, formalization is possible, but tedious. For more on this point of view, one may consult P.W. Bridgman's extraordinary book, *The Nature of Thermodynamics* (Harvard University Press, Cambridge, MA 1941).

4.   I. Prigogine and R. Defay, *Chemical Thermodynamics* (Longmans, Green and Co., London 1954).

5.   Such a description automatically implies continuous behavior of matter, except, perhaps, at isolated points, lines, or surfaces. Note that a thermodynamic system need not be a material system, *e.g.*, electromagnetic radiation.

6.   The units by which one measures the amount of component i are mostly arbitrary: Masses, moles, and numbers are equivalent, except for bodies in motion in any force-field for which Newton's laws require mass units.

7. Generally, a field is defined as any scalar-, vector-, or tensor-valued function of the space-time variables $(\mathbf{r}, t)$. Not all fields, however, are density functions, *i.e.*, functions such that their integral over space has a physical meaning. For example, the temperature is a (scalar) field, but its integral over space means nothing at all.

8. The alert reader will have noticed an inconsistency in notation: Whereas the energy and entropy densities in Eqs. (2.7) are denoted by lower-case letters, the mass or number densities in Eqs. (2.1) and (2.3) are, unfortunately, upper-case C's.

9. Note that the notion of continuity includes constancy: A constant field is one whose value is *independent* of location.

10. I.M. Gelfand and S.V. Fomin, *Calculus of Variations* (Prentice-Hall, Inc., Englewood Cliffs, NJ 1963).

11. C. Lanczos, *The Variational Principles of Mechanics* (Dover reprint).

12. It is convenient to think of the n densities $C_1, \ldots, C_n$ as the components of a vector $\mathbf{C}$. The same notation applies to other fields that are specific to each chemical component, *e.g.*, $\boldsymbol{\mu}$ for the set of chemical potentials, $\mathbf{N}$ for the mole numbers, and $\mathbf{x}$ for the mole fractions.

13. Likewise, Eq. (3.10) can be compared with Eq. (1.1), and Eq. (3.12) with Eq. (1.2). Only two of the three equations (1.1-3) are independent: Any one of these is a consequence of the two others (up to a constant). The same holds true for Eqs. (3.10), (3.12), and (4.1).

14. F.P. Buff, "The Theory of Capillarity," in *Handbuch der Physik*, Vol. 10, pp. 281-304 (Springer-Verlag, Berlin 1960).

15. J.C.M. Li, R.A. Oriani, and L.S. Darken, *Z. Phys. Chem. Neue Folge* 49, 271 (1966).

16. R. Ghez, *Helv. Physica Acta* 41, 287 (1968).

17. F.C. Larché and J.W. Cahn, *Acta Metall.* 21, 1051 (1973); 33, 331 (1985).

18. J.I.D. Alexander and W.C. Johnson, *J. Appl. Phys.* 58, 816 (1985).

19. W.W. Mullins and R.F. Sekerka, *J. Chem. Phys.* 82, 5192 (1985).

20. J.W. Cahn and J.E. Hilliard, *J. Chem. Phys.* 28, 258 (1958); 30, 1121 (1959); 31, 688 (1959).

21. J.S. Langer, *Ann. Phys. (NY)* 65, 53 (1971).

22. If some of these components are absent in any phase, then one must put their concentrations equal to zero. The corresponding chemical potentials generally do not vanish, however, since they represent the increase in energy with the addition of even one single particle.

23. The potentials φ_i are generally continuous throughout each phase, but they may suffer discontinuities at boundaries (*cf.* Eq. (3.6b) when there is a change in charge state at a phase boundary).

24. R. Defay and I. Prigogine, *Tension Superficielle et Adsorption* (Editions Desoer, Liège 1951).

25. V.K. Semenchenko *Surface Phenomena in Metals and Alloys* (Pergamon Press, NY 1961).

26. By which one means two *bulk* phases. The number f in Eq. (5.7) always refers to 3-dimensional phases.

27. A.J. McConnell, *Applications of Tensor Analysis* (Dover reprint).

28. The mean curvature K is defined as the sum of inverse radii of curvature of curves formed by the intersection of a surface with two orthogonal planes containing the unit normal $\mathbf{n}$. Referring to Fig. 2, that unit normal is directed from phase 1 to phase 2, and K is reckoned positive if the surface is convex towards the second phase. The mean curvature obeys an invariance property known as Bonnet's theorem: K can be computed from *any* two orthogonal sections of a surface. For example, $K = 2/R$ or $1/R$ for a sphere or cylinder of radius R. In Differential Geometry, the mean curvature, often taken with the opposite sign, is denoted by $2H$, and the symbol K, instead, usually stands for the gaussian curvature.

29. The surface tension obeys static conditions on Σ that are analogous to Eqs. (3.14). These are not needed here. Note also that Eqs. (5.9) and (5.11), together, yield a Gibbs-Duhem relation between the arbitrary variations of surface quantities $\delta T^{(\Sigma)}$, $\delta\sigma$, and $\delta\mu_i^{(\Sigma)}$. At equilibrium, one shows that $T^{(\Sigma)}$ and $\mu_i^{(\Sigma)} + \varphi_i^{(\Sigma)}$ are also constant fields over the phase boundary Σ.

30. L.S. Darken and R.W. Gurry, *Physical Chemistry of Metals* (McGraw-Hill, NY 1953).

31. E.A. Guggenheim, *Thermodynamics, an advanced treatment for chemists and physicists*, 5th rev. ed., (North-Holland, Amsterdam 1967).

32. P. Gordon, *Principles of Phase Diagrams in Materials Systems* (McGraw-Hill, NY 1968).

33. K. Denbigh, *The Principles of Chemical Equilibrium*, 3rd ed., (Cambridge University Press, Cambridge 1971).

34. W.B. White, "Phase Equilibria," in *Crystal Growth: a tutorial approach* (Edited by W. Bardsley, D.T.J. Hurle, and J.B. Mullin), pp. 17-66 (North-Holland, Amsterdam 1979).

35. Although often assumed implicitly for the construction of phase diagrams, it is *not* always required that equilibrium conditions, such as Eqs. (3.11) and (3.14), hold at *interior* points of phases.

36. Remember that there are only $n-1$ independent components of any composition vector x because of the relation (2.5b).

37. It is a happy coincidence that the word "component" is used in two related contexts: either as a chemically identifiable substance or as its proportion x_i in the composition vector x.

38. Note that a compound's composition $x^{(1)}$ can never be exactly independent of T without violating the phase rule: The conditions $p = $ const. and $x^{(1)} = $ const., if applied to Eqs. (6.4), would leave no freedom whatsoever in the choice of T and the liquid's composition $x^{(2)}$.

39. C. Wagner, *Acta Metall.* 6, 309 (1958).

40. L.J. Vieland, *Acta Metall.* 11, 137 (1963).

41. M. Ilegems and G.L. Pearson, "Derivation of the Ga-Al-As Ternary Diagram and Applications to Liquid Phase Epitaxy," in *Proc. 1968 Symp. on GaAs*, pp. 3-10 (The Institute of Physics, London 1969).

42. E.A. Guggenheim, *Mixtures* (Oxford University Press, London 1952).

43. C.D. Thurmond, *J. Phys. Chem. Solids* 26, 785 (1965).

44. J.R. Arthur, *J. Phys. Chem. Solids* 28, 2257 (1967).

45. M.B. Panish and M. Ilegems, in *Progress in Solid State Chemistry* (Edited by H. Riess and J.O. McCaldin), Vol. 7, pp. 39-83 (Pergamon Press, NY 1972).

46. A.S. Jordan, *Metall. Trans.* 2, 1959, 1965 (1971).

47. A.S. Jordan and M.E. Weiner, *J. Phys. Chem. Solids* 36, 1135 (1975).

48. H.C. Casey, Jr. and M.B. Panish, *Heterostructure Lasers*, Vol. B (Academic Press, NY 1973).

49. J.B. Mullin and D.T.J. Hurle, *J. of Luminescence* 7, 176 (1973).

50. D.T.J. Hurle, *J. Phys. Chem. Solids* 40, 613, 627, 639, 647 (1979).

51. M. Hansen, *Constitution of Binary Alloys* (McGraw-Hill, NY 1958), and supplements edited by R.P. Elliott (1965) and F.A. Shunk (1969).

52. R. Hultgren, P.D. Desai, D.T. Hawkins, M. Gleiser, and K.K. Kelley, *Selected Values of the Thermodynamic Properties of Binary Alloys* (Amer. Soc. Metals, Metals Park, OH 1973).

53. E.M. Levin, C.R. Robbins, and H.F. McMurdie, *Phase Diagrams for Ceramists*, with 1969, 1975, and 1981 supplements (Amer. Ceramic Soc., Columbus, OH 1964).

54. The surface tension's change with curvature can be reckoned from the Gibbs-Duhem relation for the surface phase Σ (*cf.* note in ref. [29]).

55. Contrary to Eqs. (6.4), here we choose to write the system derived from the conditions (6.3) in symmetric form: All composition variables are treated equivalently.

56. We use, interchangeably, literal or numerical subscript notation to label the chemical components.

57. L.A. Dahl, *J. Phys. Chem.* 52, 698 (1948).

58. The points P_1 and P_2 need not belong to different phases. To avoid confusion, the notation that follows, *e.g.*, $x(1)$, means the composition vector x at the point P_1, *regardless* of the phase to which it might belong.

59. Reactions among components may change the number (or density) of moles. In that case it is preferable to work with masses (or their density), which are conserved in chemical reactions.

60. J.W. Nielsen and E.F. Dearborn, *J. Phys. Chem. Solids* 5, 202 (1958).

61. S.L. Blank and J.W. Nielsen, *J. Cryst. Growth* 17, 302 (1972).

62. R. Ghez and E.A. Giess, *J. Cryst. Growth* 27, 221 (1974). Although irrelevant for the general argument, the liquidus on Fig. 6 refers to the garnet $EuYb_2Fe_5O_{12}$, rather than to YIG.

63. R. Ghez and E.A. Giess, *Mater. Res. Bull.* 8, 31 (1973).

64. E.A. Giess and R. Ghez, "Liquid-Phase Epitaxy," in *Epitaxial Growth* (Edited by J.W. Matthews), Part A, pp. 183-213 (Academic Press, NY 1975).

65. P. Görnert and F. Voigt, "High Temperature Solution Growth of Garnets: theoretical models and experimental results," in *Current Topics in Materials Science* (Edited by E. Kaldis), Vol. 11, pp. 1-149 (North-Holland, Amsterdam 1984).

66. E.A. Giess, "Oxide Growth from Molten Solutions," published in the *Proc. 6th Int'l. Summer School on Crystal Growth*, Edinburgh, July 5-13, 1986 (Prentice-Hall, NY 1987).

67. M.B. Small and R. Ghez, *J. Appl. Phys.* 50, 5322 (1979).

68. R. Ghez, "Theoretical Aspects of III-V Heterostructure Formation by Liquid Phase Epitaxy," in *Proc. Symp. on III-V Opto-electronics, Epitaxy, and Device Related Processes* (Edited by V.G. Keramidas and S. Mahajan), Vol. 83-13, pp. 87-97 (Electrochem. Soc., Pennington, NJ 1983).

69. M.B. Small and R. Ghez, *J. Appl. Phys.* 55, 926 (1984).

70. M.B. Small and R. Ghez, *J. Appl. Phys.* 51, 1589 (1980).

NUCLEATION

Boyan Mutaftschiev

CNRS – Laboratoire Maurice Letort
BP 104, 54600 Villers-les-Nancy
France

The name nucleation designates the ensemble of processes leading to the formation of a new, stable, phase inside an unstable mother phase. Insofar as the overall mechanism is dominated by the thermodynamic properties of the embryos of the new phase, it can be (and has been, first) considered by the methods of heterophase thermodynamics. This is also the way nucleation is taught in common courses and manuals. However, one must be aware that every phase transformation starts by the association of a very small number of molecules of the mother phase and, thus, has all the characteristics of a chemical reaction of polymerization. The purely chemical approach to nucleation is a powerful tool, which is inconvenient in only one but major point: it necessitates a large number of parameters, describing the intermediate stages of the overall reaction, associated with the formation of differently sized embryos.

The purpose of this contribution is to show how the advantages of both "phase" and "chemical" approaches can be used for the calculation of the absolute rate of nucleation. For the sake of simplicity, only cases of formation of a pure condensed (solid or liquid) phase from its (ideal) vapor will be treated. The transposition of the results to more complex systems (solution, melt), if not automatic, does not pose any fundamental problem.

PART A: THE PHASE CONCEPT

I.   THE SATURATED STATE

The equilibrium of a large condensed phase with its own vapor, at a given temperature T, is fully determined by the pressure p^0 at the saturation. Its value is obtained from the equality of the chemical potentials μ (per molecule) of the two phases. For an ideal gas:

$$\mu_g^0 = - kT \frac{d\ell nQ_g}{dN} = - kT\ell n \frac{(2\pi mkT)^{3/2}}{h^3} + kT\ell np^0 - kT\ell nkT = \tag{I.1}$$
$$= \eta + kT\ell np^0 - kT\ell nkT,$$

where Q_g is the partition function of N gas molecules in the volume V, m, k and h are the mass of the molecule, the Boltzmann and the Planck constants, respectively, and η is a function of the temperature only and is independent of the type of the phase.

For the infinite crystal of a monatomic substance, in the limits of the harmonic approximation, and at not too low temperature

$$\mu_{c,\infty} = - kT\ell n q_{c,\infty} = \eta - \phi_o - kT\ell n(\frac{kT}{2\pi m})^{3/2} \frac{1}{\nu_o^3} = \eta - \phi_o - kT\ell n\tilde{\nu}, \quad (I.2)$$

where $q_{c,\infty}$ is the partition function of an atom in the infinite crystal, ϕ_o is the work needed to extract an atom from a node of the crystal lattice and place it in the vapor, ν_o is the mean ("thermodynamic") frequency of vibration of the atoms around the lattice nodes and

$$\tilde{\nu} = (\frac{kT}{2\pi m})^{3/2} \frac{1}{\nu_o^3} \quad (I.3)$$

is the so-called "mean vibrational volume" of an atom of the crystal.

The equality,

$$\mu_g^0 = \mu_{c,\infty} \quad (I.4)$$

from equations (I.1) and (I.2) yields [1]:

$$p^0 = \frac{(2\pi m)^{3/2}}{(kT)^{1/2}} \nu_o^3 \exp(-\phi_o/kT). \quad (I.5)$$

II. THE SUPERSATURATED STATE

If a supersaturated vapor, with a chemical potential $\mu_g > \mu_g^0$ is in contact with a condensed phase, a continuous flow of matter from the first to the second results. The driving force for this transport,

$$\Delta\mu = \mu_g - \mu_g^0 = \mu_g - \mu_{c,\infty}, \quad (II.1)$$

is called "supersaturation" and the "chemical work" won during the transport of N molecules from the vapor to the condensed phase is $-N\Delta\mu$. It is clear that chemical potentials play the same role in equilibrium and mass transport of neutral matter, as electric potentials in the equilibrium and transport of electric charges.

In the absence of a condensed phase, the mother phase can remain in the metastable supersaturated state until a small portion of the new phase is formed, which serves as a "critical nucleus" for the condensation. The thermodynamic properties of this critical nucleus are treated differently from the "phase" and "chemical" points of view.

1. The Free Energy of a Small Condensed Phase

The main point of the phase approach is that the free energy of a small condensed phase is higher than that of a portion of the infinite phase having the same number of molecules. The difference between them, called "excess free energy" is commonly attributed to surface tension.

There is a two fold effect of the surface tension γ on the state of a small phase [2]. On the one hand, it is at the origin of the "surface

work" γdA spent for any infinitesimal variation dA of the surface area.
On the other hand, the curved surface of a cluster (e.g., a drop) can be
in a mechanical equilibrium with the surrounding medium, only if a pressure
difference exists on either side of it, according to the Laplace formula:

$$p'' - p' = \gamma (\frac{1}{R_1} + \frac{1}{R_2}),\qquad\qquad (II.2)$$

where R_1 and R_2 are the principal radii of the curvature. It can be easily
shown that for a closed surface of an isotropic phase, the pressure
difference of Eq. (II.2), called "capillary pressure", is equal to

$$p_\gamma = \gamma \frac{dA}{dV}\qquad\qquad (II.3)$$

and is constant throughout the phase.

Depending on the nature of the small phase, two ways of introducing
the effect of surface tension have to be considered.

a) <u>The Size Independent Chemical Potential</u>. When the small phase is
solid, it is conceivable that its chemical potential is size independent,
the effect of the capillary pressure on both potential energy and
vibrational entropy (Eq. (I.2)) being negligible, until structure is
affected. The differential work dW associated with the transport of
$di = -dN$ molecules from a gas phase containing N molecules to a cluster of
i molecules at a constant pressure is

$$dW = dG = \mu_g dN + \mu_{c,\infty} di + \gamma dA = dG_g + dG_i,\qquad (II.4)$$

where dG is the variation of the Gibbs free energy of the system and $\mu_{c,\infty}$
is the (size independent) chemical potential, per molecule, of the cluster.
The right hand side of Eq. (II.4) suggests that we can consider the Gibbs
free energy variation of the small condensed phase as consisting of a
"bulk" part, equal to the free energy of a portion of the infinite phase
having the same number of molecules, and a "surface" contribution γdA:

$$dG_i = \mu_{c,\infty} di + \gamma dA.\qquad\qquad (II.5)$$

If the upper-limit value of the surface tension is evident:

$$\lim_{i \to \infty} \gamma = \gamma_\infty,\qquad\qquad (II.6)$$

where γ_∞ is the physically measurable surface tension of the infinite
phase, no restriction applies for low values of i, but that integration of
Eq. (II.5) cannot be performed from zero to i because of the atomistic
structure of matter. In other words, for large values of i, the integrated
value of G_i from Eq. (II.5) should have the form:

$$G_i = \mu_{c,\infty} i + \gamma_\infty A + B,\qquad\qquad (II.7)$$

and the "excess free energy":

$$\Phi_i = G_i - \mu_{c,\infty} i = \gamma_\infty A + B,\qquad\qquad (II.8)$$

where the constant B can be obtained by extrapolation to i = 0. Equation
(II.8) can be considered as a definition equation of the so-called
"capillary approximation", according to which, above some size of the
small phase its excess free energy is proportional to the area of its

surface *. The constant B, which will be discussed in more detail in Section IV, takes into account, among other things, the non—continuum structure of the matter, so that G_i is not exactly an extensive variable [6].

b) The Size Dependent Chemical Potential. The next treatment, more suitable when the small phase is a liquid, leads to the same results but to a quite different physical picture. If one wants to account for the effect of the capillary pressure on the chemical potential of the liquid drop, one has to consider the variation of the Helmholz free energy of the system, dF, when $di = -dN$ molecules are transported from the vapor to the cluster [7]. If the overall process is carried out, as before, at a constant external pressure p of the gas phase, one has

$$dF = \mu_g dN - p dV_g + \mu_c di + i d\mu_c - p_c dV_c - V_c dp_c + \gamma dA, \qquad (II.9)$$

where the chemical potential μ_c is now pressure dependent, p_c being the pressure inside the condensed phase, V_g and V_c, the respective volumes of the gas and of the small phase. Since the reversible infinitesimal work dW is still related to the Gibbs free energy,

$$dW = dG = dF + pdV = dF + p(dV_g + dV_c),$$

one has:

$$dG = [\gamma dA - (p_c - p)dV_c] + [id\mu_c - V_c dp_c] + \mu_g dN + \mu_c di. \qquad (II.10)$$

The terms inside the brackets of Eq. (II.10) are both nil: the first by virtue of Eq. (II.3) and the second by definition, since the change takes place at constant temperature. Then

$$d\mu_c = v_c dp, \qquad (II.11)$$

where $v_c = V_c/i$ is the volume of a molecule in the condensed phase. It follows from Eq. (II.10), which is in this approximation the equivalent of Eq. (II.4), that

$$dG = \mu_g dN_g + \mu_c di = (\mu_c - \mu_g)di. \qquad (II.12)$$

Therefore, when the pressure dependence of the chemical potential is considered, the excess free energy of the cluster is already included in this chemical potential. The equivalence of both treatments is seen by integration of Eq. (II.11):

$$\mu_c = \mu_{c,\infty} + \int_{p^0}^{p_c} v_c dp = \mu_{c,\infty} + v_c (p_c - p^0).$$

If one neglects the difference between p and the pressure p^0 of the saturated vapor, one has:

$$\mu_c = \mu_{c,\infty} + v_c (p_c - p) = \mu_{c,\infty} + v_c p_\gamma. \qquad (II.13)$$

Substituting μ_c from Eq. (II.13) into Eq. (II.12) and taking into account Eq. (II.3) (note that $dV_c = v_c di$), one obtains Eq. (II.4) as promised.

---

* Different attempts have been made [3-5] to deduce the size dependence of the surface tension for very small clusters. Since such a (formal) dependence is clearly related to intermolecular forces and structure, the obtained results lack universality. For this reason, we believe that the capillarity approximation has a sense insofar as macroscopic surface tension is used. When excess free energy deviates from that of Eq. (II.8), exact microscopic calculations give better results than artificial manipulations of surface tension.

2. Equilibrium

The condition for equilibrium of the small phase with the super-saturated vapor can be deduced by either Eqs. (II.4) or (II.12). In both cases $(dG)_{p,T} = 0$, which yields:

$$\mu_g - \mu_{c,\infty} = \mu_g - \mu_g^0 = \Delta\mu = \gamma \frac{dA}{di} , \tag{II.14}$$

in the case of size independent chemical potential of the condensed phase, and

$$\mu_g = \mu_c , \tag{II.15}$$

in the case of size dependent chemical potential μ_c. These two equations, which are nothing else but different forms of the famous Thomson-Gibbs formula, merit equal attention. The first, because it leads directly to an explicit expression, e.g., for the equilibrium pressure of a liquid drop $(A = 4\pi r^2$, $i = (4\pi/3v_c)r^3$, where r is the droplet radius) in an ideal vapor:

$$\Delta\mu = kT\ln \frac{p}{p^0} = \frac{2\gamma v_c}{r} . \tag{II.16}$$

The second, because it allows calculation of the equilibrium pressure of any i-sized cluster by the methods of the statistical thermodynamics [8]. If the partition function of the cluster Q_i is known, then Eq. (II.15) transforms to

$$\mu_g = - kT \frac{d\ln Q_i}{di} , \tag{II.17}$$

or, if one considers that, according to Eqs. (I.4) and (I.2),

$$\Delta\mu = \mu_g + kT\ln q_{c,\infty} ,$$

one obtains the Thomson-Gibbs formula in the alternative form:

$$\Delta\mu = - kT(\frac{d\ln Q_i}{di} - \ln q_{c,\infty}) . \tag{II.18}$$

3. The Nucleation Work

The integration of Eqs. (II.4) or (II.12) gives the reversible work of formation of an i-sized cluster from the supersaturated vapor. From Eq. (II.4) one has:

$$\Delta G(i) = \int dG_i + \int dG_g = \int dG_i - \mu_g i \tag{II.19}$$

$(dN = -di)$. We have already performed the integration of the first term of the right hand side of Eq. (II.19). According to Eq. (II.7), we have for not too small clusters:

$$\Delta G(i) = \gamma_\infty A - (\mu_g - \mu_{c,\infty})i + B = \gamma_\infty A - \Delta\mu i + B. \tag{II.20}$$

Two comments:

(i) Equation (II.20) contains the integration constant B which cannot be evaluated from classical arguments or, better, according to the classical thermodynamics of continua, its value should be zero. Therefore, at this point of our considerations, we will consider the "classical" part of the free energy variation of Eq. (II.20) only:

$$\Delta G_{cl}(i) = \gamma_\infty A - \Delta\mu i, \tag{II.21}$$

and shall keep in mind

$$\Delta G(i) = \Delta G_{cl}(i) + B, \tag{II.22}$$

when we will be able to calculate B by the means of statistical thermo-dynamics (Section IV);

(ii) Since the surface area of a homothetically growing cluster is proportional to $i^{2/3}$, it is clear that the value of $\Delta G_{cl}(i)$ might pass through a maximum for a cluster size i = i* determined by the condition:

$$\left(\frac{d\Delta G_{cl}(i)}{di}\right)_{i=i*} = \gamma_\infty \left(\frac{dA}{di}\right)_{i=i*} - \Delta\mu = 0 \tag{II.23}$$

(note that the exact free energy variation $\Delta G(i)$ from Eq. (II.20) has a maximum for the same cluster size!). The identity of Eqs. (II.23) and (II.14) shows that the equilibrium cluster of i* molecules, considered in the previous Section, is the one that necessitates a maximum work of formation, $\Delta G_{cl}*$. This means also that the equilibrium is a labile one, and that $\Delta G_{cl}*$ must be overcome if a cluster of any dimension i > i* has to be formed. For those reasons, clusters having the exact size i* are called "critical nuclei" and the associated free energy variation $\Delta G_{cl}*$ is the activation energy of the nucleation process. The exact size dependence of the work of formation $\Delta G_{cl}(i)$ can be obtained by substituting $\Delta\mu$ from Eq. (II.23) into Eq. (II.21):

$$\Delta G_{cl}(i) = \gamma_\infty \left[A - i\left(\frac{dA}{di}\right)_{i=i*}\right]. \tag{II.24}$$

If, as mentioned, one writes down

$$A = \Omega i^{2/3} \tag{II.25}$$

(Ω is a geometric constant, which also contains the "area" a of one molecule on the surface of the cluster), one has:

$$\Delta G_{cl}(i) = \frac{1}{3}\gamma_\infty A*\left[3\left(\frac{i}{i*}\right)^{2/3} - 2\frac{i}{i*}\right], \tag{II.26}$$

where $A* = \Omega i^{*2/3}$. The factor before the brackets of this equation represents the maximum value of $\Delta G_{cl}(i)$ when i = i*:

$$\Delta G_{cl}* = \frac{1}{3}\gamma_\infty A*. \tag{II.27}$$

It is important to know the dependence of $\Delta G_{cl}*$ on the supersaturation. From Eqs. (II.23) and (II.25) one has:

$$\Delta\mu = \gamma_\infty\left(\frac{dA}{di}\right)_{i=i*} = \frac{2}{3}\gamma_\infty\Omega i^{*-1/3} = \frac{2}{3}\gamma_\infty\frac{\Omega^{3/2}}{A*^{1/2}}. \tag{II.28}$$

Then:

$$\Delta G_{cl}* = \frac{1}{3}\gamma_\infty A* = \left(\frac{\gamma_\infty\Omega}{3}\right)^3\left(\frac{2}{\Delta\mu}\right)^2. \tag{II.29}$$

Hence, at increasing supersaturation $\Delta\mu$, both the size of the critical nucleus and the activation energy for nucleation $\Delta G_{cl}*$ decrease, as can be seen in Figure 1.

To resume, the increase in size of a cluster is first associated with an increase of the free energy, due to the development of a free surface, and then with a decrease of the free energy, due to the gain of chemical

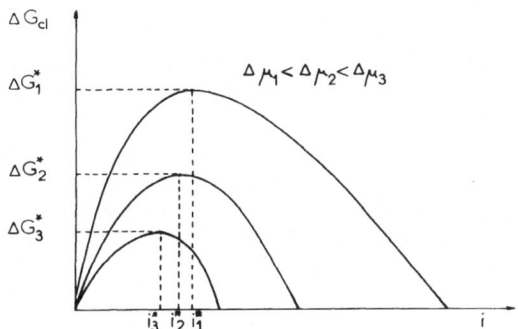

Fig. 1. Variation of the Gibbs free energy $\Delta G_{cl}(i)$ associated with the formation of i-sized clusters, at different supersaturation $\Delta\mu$. Note that when $\Delta\mu$ increases, both size $i*$ of the critical nucleus and nucleation work $\Delta G*$ decrease.

work $-\Delta\mu i$. A cluster which is in (labile) equilibrium with the super-saturated mother phase has a maximum work of formation and serves as critical nucleus for further condensation. The size of the critical nucleus and the height of the energetic barrier, that is the variation of the free energy associated with its formation, decrease when super-saturation increases.

Another way of calculating the nucleation work is suggested by the integration of Eq. (II.12):

$$\Delta G_{cl}(i) = \int_{0}^{i} \mu_c di - \mu_g i, \qquad (II.30)$$

where now μ_c is size dependent. As pointed out in the discussion of Eqs. (II.15) to (II.18), the relationship between μ_c and i is given by the Thomson-Gibbs formula. On Figure 2, where this relation is plotted, the constant chemical potential μ_g of the supersaturated vapor is represented by the dashed line. The value of $\Delta G_{cl}(i)$ from Eq. (II.3) is equal to the area of the surface enclosed between the μ_c-curve and the μ_g = const-line. One can notice that until the μ_c-curve crosses the μ_g-line, the area is positive and increases by increasing i. For $i = i*$, $\mu_c = \mu_g$. The further increase of i adds a negative contribution to $\Delta G_{cl}(i)$, because μ_c is now smaller than the chemical potential μ_g of the vapor.

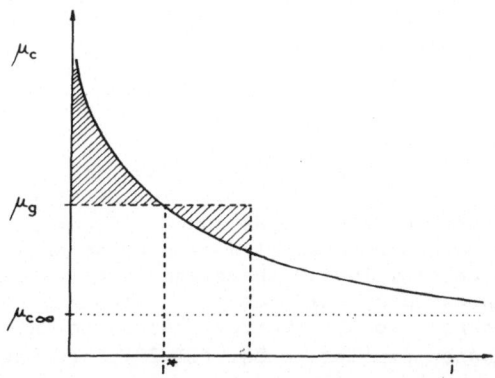

Fig. 2. Graphical representation of Eq. (II.30) (see text).

III. THE EQUILIBRIUM SHAPE

Although the problem of the equilibrium shape does not enter into the scope of these lectures, we will briefly mention its impact on the activation energy for nucleation.

It is understandable that insofar as the value of the maximum nucleation work ΔG_{c1}^* is just a fraction of the surface free energy (Eq. (II.27)), the shape of the cluster must be of prime importance. Since the overall nucleation kinetics depends exponentially on the value of ΔG^*, as in the case of any activated process, only critical nuclei having the smallest total surface free energy for a given number of molecules i (or for a given constant volume V_c) can be taken in consideration. Hence, the equilibrium shape condition,

$$(\gamma A)_{V_c} = \text{minimum},\qquad\qquad\text{(III.1)}$$

for an isotropic phase, or:

$$(\Sigma_j \gamma_j A_j)_{V_c} = \text{minimum},\qquad\qquad\text{(III.2)}$$

for an anisotropic phase (a polyhedron, the j-th face of which has an area A_j and a surface tension γ_j), must be introduced in some way in the calculation of the nucleation work, in order to find the "easiest path" for overpassing the energetic barriers. In the case of isotropic phase, e.q., a liquid drop, the solution is trivial: the shape with the minimum area for a given volume is the sphere. When the cluster is crystalline, Eq. (III.2), known as the Curie-Wulff condition [9,10] for the equilibrium shape, leads to a quite simple result. If, starting from an imaginary point in space, one constructs vectors normal to the different faces of the crystal, the lengths h_j of which are proportional to the surface tensions γ_j of those faces, the condition of Eq. (III.2) yields

$$\frac{\gamma_1}{h_1} = \frac{\gamma_2}{h_2} = \ldots = \frac{\gamma_j}{h_j} = \text{const.}\qquad\qquad\text{(III.3)}$$

Now, while all possible imaginable faces may participate to this construction, the smallest closed polyhedron possesses only a limited number of faces (at least at temperatures considerably lower than the melting point) and its shape is the equilibrium one.

The above result, known as "Wulff theorem" is completed by thermo-dynamic considerations [11,12] showing that the constant of Eq. (III.3), related to the moduli of the vectors h_j or, otherwise, to the volume of the crystal, has the value $\Delta\mu/2v_c$, where, as before, $\Delta\mu$ is the thermodynamic supersaturation and v_c is the volume of a molecule in the condensed phase. Equation (III.3) transforms then to:

$$\Delta\mu = \frac{2\gamma_1 v_c}{h_1} = \ldots = \frac{2\gamma_j v_c}{h_j}.\qquad\qquad\text{(III.4)}$$

Comparison of Eq. (III.4) with Eq. (II.16) shows the link with the previous considerations on the thermodynamic equilibrium of clusters and their work of formation in a supersaturated mother phase. Equation (III.4) says that the Thomson-Gibbs formula can be applied also to an anisotropic cluster, provided that it has an equilibrium shape and that its characteristic linear dimension, expressed by the vector h_j is conjugated to the surface tension of the face normal to this vector. As to the nucleation work, it is given by an expression similar to Eq. (II.27), accounting however for the anisotropy of the surface energy:

$$\Delta G_{c1}^{*} = \frac{1}{3} \Sigma \gamma_j A_j^{*}. \qquad (III.5)$$

PART B: THE CHEMICAL CONCEPT

IV. THE EQUILIBRIUM DISTRIBUTION OF CLUSTERS

We saw that the phase treatment of nucleation has some difficulties in conciliating the results of classical thermodynamics, which deals typically with continua, with the atomic structure of very small clusters. Furthermore, as shown in Part C of this course, the mathematical methods developed for the kinetic treatment of chemical chain reactions are more powerful in calculating the absolute rate of nucleation than the early kinetic treatment of Volmer and Weber [13], based on the probabilities for heterophase fluctuations.

The starting point of the "chemical" treatment of nucleation [14] is that gas molecules, as well as clusters of any size, are considered as molecules of different chemical species. Therefore:

- clusters of a given size i, where i = 1,2,3,..., are all undistinguishable from one another,
- clusters of all sizes are always present in the gas mixture.

It is clear that the hetrophase equilibrium between a cluster and its vapor, discussed in the preceding sections, is now replaced by the homogeneous equilibrium of a gas mixture of molecules at different stages of polymerization. Chemical thermodynamics gives the condition for equilibrium when molecules of the chemical species A_i are formed from i molecules of the chemical species A_1, according to the reaction

$$iA_1 \rightleftarrows A_i. \qquad (IV.1)$$

This condition is the well-known law of mass action:

$$\frac{n(i)}{n_1^i} = \frac{q_i}{q_1^i}, \qquad (IV.2)$$

where n(i) and n_1 are the concentrations (n = N/V) of the i-mer and of the monomer respectively, while q_i and q_1 are the so-called "internal partition functions" of the species (i.e., the partition function of one isolated molecule of the species moving freely in a volume V = 1). If one takes into account that the chemical potential per molecules of any given species i (i = 1,2,3,...) is:

$$\mu_i = - kT\ln \frac{q_i}{n(i)}, \qquad (IV.3)$$

Eq. (IV.2) transforms to

$$\mu_i = i\mu_1, \qquad (IV.4)$$

which is the general condition for chemical equilibrium, substantially different from the phase equilibrium conditions of Eqs. (I.4) and (II.15). Since the chemical treatment does not recognize the individuality of the clusters, it is understandable that the chemical potential μ_i of Eq. (IV.3) accounts for the appearing or the disappearing from the system of an entire cluster of a given size, and thus depends on the concentration of the species. This is not the case in the heterophase treatment, where a cluster in equilibrium with the vapor exchanges molecules and varies continuously in size, without losing its individuality. For these reasons, following the phase concept, the chemical potential per molecule of the cluster is equal to that of the gas molecules and the equilibrium is independent of the cluster's concentration.

If one wants to take advantage of the progress done in the phase treatment, as regards the size dependence of the excess free energy of a cluster and the determination of its equilibrium shape, we have to associate formally the internal partition function q_i of a molecule of the i-th species with the free energy F_i,

$$F_i = - kT\ln q_i,\tag{IV.5}$$

of a small i-molecular phase moving alone in a volume of unity. The law of mass action, Eq. (IV.2), can then be written down in one of the alternative forms:

$$n(i) = q_i/(q_1/n_1)^i = \exp[-(F_i - i\mu_1)/kT];\tag{IV.6a}$$

$$n(i) = (q_i/q_{c,\infty}^i)\exp(i\Delta\mu/kT) = \exp[-(\Phi_i - i\Delta\mu)/kT];\tag{IV.6b}$$

$$n(i) = \exp[-\Delta G(i)/kT],\tag{IV.6c}$$

where the approximation $F_i \simeq G_i$, valid for uncompressible condensed clusters, has been made, and the parameters F_i, $q_{c,\infty}$ and Φ_i have been introduced by virtue of Eqs. (IV.5), (II.1), (I.2) and (II.7). The small phase is assumed to be a solid (localized state).

The size distribution of the clusters in the gas mixture is qualitatively well understandable from Eq. (IV.6b). The partition function q_i of the cluster is larger than that of i-molecules inside the infinite crystal, $q_{c,\infty}^i$, and the ratio $q_i/q_{c,\infty}^i$ is monotonically decreasing with increasing i. When $\Delta\mu$ is nil or negative (saturation or undersaturation), the size distribution function $n(i)$ is decreasing also continuously (cf. Figure 3, curve A). When the mother phase is supersaturated ($\Delta\mu > 0$), $n(i)$ passes through a minimum (Figure 3, curve B), defined by the condition:

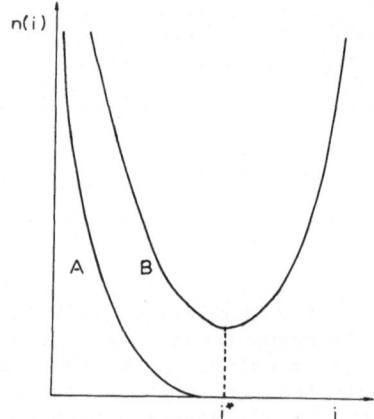

Fig. 3. Equilibrium size distribution function n(i) of clusters at the saturation, curve A, and at a supersaturation, curve B.

$$\left(\frac{d\ln n(i)}{di} \right)_{i=i*} = \left(\frac{d\ln q_i}{di} - \ln q_{c,\infty} + \frac{\Delta\mu}{kT} \right)_{i=i*} = 0 \tag{IV.7}$$

The existence of a minimum in the size distribution function is not satisfactory for the mind. It is clear that for i > i* the distribution is unrealistic (it forsees an infinite number of infinite sized clusters!). This is due to the fact that for sizes i > i*, irreversible processes take place throughout the system. Therefore, at supersaturation the entire equilibrium distribution must be considered formally as nothing but a reference state of the system. One may try, however, to imagine the conditions in which at least a part of n(i) is physically realizable. This can happen at not too high supersaturations, when the size i* is large and n* << 1, i.e., when the time for the establishment of equilibrium distribution for all clusters having sizes 1 < i < i* is shorter than the average time of formation of the first i*-sized cluster. These conditions are similar to those of the phase treatment which ignores the existence of subcritical nuclei and considers that the state of metastable equilibrium of the non-associated vapor is only maintained during the time preceding the formation of the first critical nucleus. Consequently, one should not wonder that the condition for minimum of the size distribution function, Eq. (IV.7), is identical in form with the Thomson-Gibbs formula, Eq. (II.18), which defines the critical nucleus, whereas the notion of critical nucleus is not needed at all in the chemical treatment.

In the further discussion of the size distribution law we shall use the form of Eq. (IV.6b). The excess free energy Φ_i of the i-molecular cluster entering this equation is equal to:

$$\Phi_i = G_i - \mu_{c,\infty} i \simeq F_i - \mu_{c,\infty} i, \tag{IV.8}$$

and only for sufficiently large clusters, is given by Eq. (II.8), according to the capillarity approximation,

$$\Phi_i = \gamma_\infty A + B.$$

If the i-molecular cluster is crystalline, its internal partition function can be written as [14,15]

$$q_i = \frac{(2\pi mkT)^{3i/2}}{h^{3i}} q_{tr} q_{rot} q_{vib} \exp[-U_i/kT], \tag{IV.9}$$

where U_i is the potential energy of the cluster, all atoms at rest, and q_{tr}, q_{rot}, q_{vib} are respectively the configurational integrals (the partition functions in the phase space of the positional coordinates only) corresponding to three degrees of freedom of translation, to three degrees of freedom of rotation and to 3i-6 degrees of freedom of vibration.

The partition function of an atom in the infinite crystal being equal to

$$q_{c,\infty} = \frac{(2\pi mkT)^{3/2}}{h^3} q_{0,vib} \exp(-u_0/kT) = \frac{(2\pi mkT)^{3/2}}{h^3} \tilde{v} \exp(-u_0/kT) \tag{IV.10}$$

(where $q_{0,vib}$ is the configurational integral of vibration along three directions, equal to the mean vibrational volume $\tilde{v}$, cf. Eq. (I.3), and u_0 is the potential energy of the atom at rest in the lattice node), the ratio of the partition functions in Eq. (IV.6b) becomes:

$$\frac{q_i}{q_{c,\infty}^i} = \frac{q_{tr} q_{rot} q_{vib}}{q_{0,vib}^i} \exp[-(U_i - iu_0)/kT]. \tag{IV.11}$$

Using a suitable reference system for the computation of q_{tr}, q_{rot} and q_{vib}, one can show that:

- the configurational integrals q_{tr} and q_{rot} are independent from the cluster size. The first is equal simply to the volume V of the vessel (which has been taken here equal to unity), and the second is equal to [16]

$$q_{rot} = \frac{4\pi r_o^2 2\pi s_o}{\sigma},$$ (IV.12)

where r_o and s_o are of the order of magnitude of the interatomic distance in the solid and σ is a symmetry number;

- the configurational integral corresponding to the degrees of freedom of vibration is equal to

$$q_{vib} = (\frac{kT}{2\pi m})^{(3i-6)/2} \prod_1^{3i-6} \frac{1}{\nu_j},$$ (IV.13)

where ν_j is the frequency of one of the 3i-6 vibrational modes of the cluster. From Eqs. (IV.11), (IV.10) and (IV.13) one obtains:

$$\frac{q_i}{q_{c,\infty}^i} = \frac{q_{tr}q_{rot}}{q_{o,vib}^2} \prod_1^{3i-6} \frac{\nu_o}{\nu_j} \cdot \exp[-(U_i - iu_o)/kT]$$

$$= Q_{rep}\exp[-(U_i - iu_o - kT \sum_{j=1}^{3i-6} \ln \frac{\nu_o}{\nu_j})/kT],$$ (IV.14a)

or

$$\frac{q_i}{q_{c,\infty}^i} = \exp[-(U_i - iu_o - kT \sum_{j=1}^{3i-6} \ln \frac{\nu_o}{\nu_j} - kT\ln Q_{rep})/kT],$$ (IV.14b)

where Q_{rep} is the so-called "replacement partition function" [17,18] which accounts for the fact that in respect to the same amount of molecules in the infinite crystal, in the i-molecular cluster six degrees of freedom of vibration are replaced by three degrees of freedom of translation and three degrees of freedom of rotation.

The comparison of Eq. (IV.14b) with Eqs. (IV.6b) and (II.8) shows that:

(i) The first three terms of the excess free energy Φ_i, that is the exponent of Eq. (IV.14b),

$$\Phi_i = U_i - iu_o - kT \sum_{j=1}^{3i-6} \ln \frac{\nu_o}{\nu_j} - kT\ln Q_{rep},$$ (IV.15)

must be identified with the surface free energy of the cluster and the fourth term is size independent. The linear dependence of Φ_i on $i^{2/3}$ for not too small clusters would then justify the capillarity approximation, Eq. (II.8), fix its lower limit of validity and permit the determination, by extrapolation at i = 0, of the constant

$$B = -kT\ln Q_{rep}.$$ (IV.16)

(ii) The replacement partition function, equal for our model of monatomic crystal and high temperature harmonic approximation to [19]

$$Q_{rep} = \frac{V4\pi r_o^2 2\pi s_o}{\sigma \tilde{v}^2}$$ (IV.17)

38

(cf. Eqs. (IV.14a), (I.3) and (IV.12)) is dominated by the ratio $V/\tilde{v}$ [20], which is of the order of magnitude of 10^{25} (for $V = 1$ cm^3), while the contribution of the other terms is of the order of unity.

(iii) According to Eq. (II.22), the size distribution law, Eq. (IV.6c) can be written down as

$$n(i) = \exp[-\Delta G(i)/kT] = Q_{rep}\exp[-\Delta G_{cl}(i)/kT]. \tag{IV.18}$$

V.   THE MERITS OF THE CAPILLARITY APPROXIMATION

1.   Crystalline Clusters

The difference of the first two terms of Eq. (IV.15) is proportional to the number of broken bonds on the surface of the clusters and, hence, roughly to its area. Therefore, it should be identified with the potential part of the total surface free energy of the cluster. A first nearest neighbor model can be used for its calculation. For tetrahedral clusters with fcc structure, it is equal to [16]

$$U_i - iu_o = 3\phi L(L + 1),$$

where ϕ is the binding energy of two nearest neighbors and L is the number of atoms in the edge of the tetrahedron. In the outlines of the same model, the potential part of the surface tension for the (111)-face of the infinite crystal is

$$\gamma_{u,\infty} = 3\phi/2a,$$

where a is the area of an atomic site on the (111)-face. Taking into account that the total number of such sites is equal to:

$$n_s = 2L(L + 1),$$

one obtains that even for the smallest clusters, the excess potential energy is exactly equal to the potential part of the total surface energy calculated with the help of (the constant) $\gamma_{u,\infty}$:

$$U_i - iu_o = \gamma_{u,\infty}n_s a = \gamma_{u,\infty}A.$$

The third term of Eq. (IV.15) is clearly the entropy part of the surface free energy. Calculation has been performed on models of two- and three-dimensional clusters in the harmonic approximation [16]. Figure 4 shows the dependence of the excess entropy S^e/k on $i^{2/3}$ (curve A) and on the true number n_s of surface atoms (curve B) for a tetrahedral cluster with fcc structure. Two remarks could be made:

- an almost perfect linear dependence is observed for the S^e/k vs. $i^{2/3}$-plot, and not for the S^e/k vs. n_s plot. This is due probably to the fact that edges and corners play for those cluster sizes a non negligible role which is accentuated by the non-equilibrium shape of the clusters (the equilibrium shape being a cubo-octahedron). The use of the average number of surface atoms $i^{2/3}$, instead of the true number of atoms n_s on surface + edges + corners, seems to attenuate this effect;

- the linear dependence is already achieved for cluster sizes $i = 35$ atoms (5 atoms on the edges of the tetrahedron!).

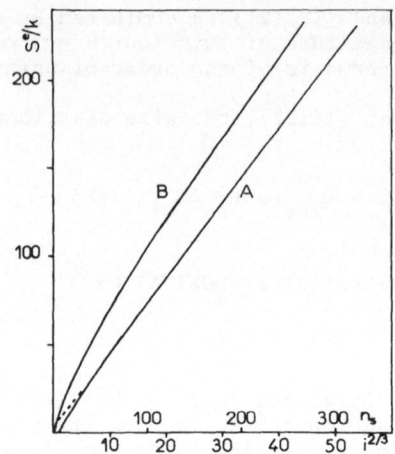

Fig. 4. Excess vibrational entropy of i-molecular tetrahedral clusters
drawn from the calculations of Hoover, Hindmarsh and Holian [16]
after transformation from fixed center of mass to 6 fixed atomic
coordinates: curve A, as a function of $i^{2/3}$; curve B, as a
function of the true number of surface (edge, summit) atoms n_s.

Figure 5 shows in double logarithmic form the same dependence for
cluster sizes between 4 and 20000 atoms. The slope of the straight line,
0.655, is not very different from the expected value of 2/3.

2. Liquid Drops

In the absence of an exact treatment of the partition function of a
liquid drop, its free energy can be determined by computer simulation. A
very detailed study on this subject has been performed some 15 years ago
[21]. It started from the state of ideal gas and proceeded to the
calculation of the free energy by numerical integration when volume and
temperature were decreased, until a liquid cluster with the desired size
remained alone (no gas phase!) in a volume slightly larger than its own
volume. The computed free energy is that corresponding to all internal
degrees of freedom (mostly of translation) compatible with the existence

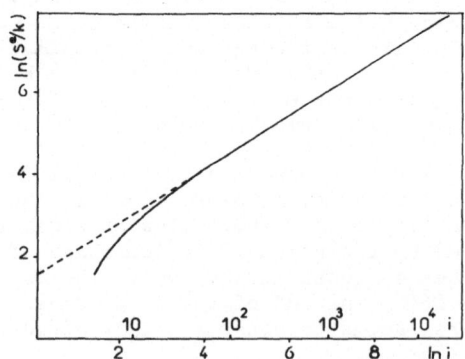

Fig. 5. Excess vibrational entropy issued from the same calculations as
Figure 4 [16] but extended to large cluster sizes, in double
logarithmic plot. The slope of the linear part is 0.655 ≃ 2/3.

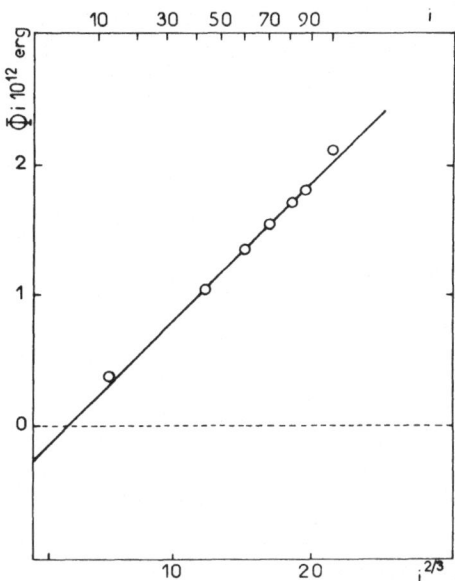

Fig. 6. Excess free energy of liquid clusters drawn from Monte Carlo
simulation after Lee, Barker and Abraham [21]. From the slope
of the line, with the assumption of (statistically) spherical
shape, one calculates γ_∞ = 16.28 dyn/cm. Theoretical value for
the same model γ_∞ = 16.18 dyn/cm.

of the cluster, and to three external degrees of freedom associated with
the precession of its center of mass inside the volume of confinement V_{cf}.
The excess free energy resulting from this study is plotted on Figure 6 as
a function of $i^{2/3}$. As in the case of crystalline clusters, the validity
of the macroscopic capillarity concept is manifested down to sizes of 10
to 20 atoms. Furthermore, the value of the surface tension, calculated
from the slope of the line, Figure 6, with the assumption of a spherical
cluster shape, is in very good agreement with the value of the macroscopic
surface tension, calculated for the same model substance [21].

The constant B, appearing in Eq. (II.8) as the ordinate at the origin,
can also be determined from Figure 6. The replacement partition function,
Q_{rep}, calculated from Eq. (IV.16), will be that of a liquid cluster
confined in the volume V_{cf}. The replacement partition function Q_{rep} of a
cluster moving in a volume of unity is obtained by division of Q_{rep} by
V_{cf}. Its value is then very nearly equal to that of crystalline clusters
of the same substance ($\simeq 10^{29}$).

As a conclusion of this thermodynamic part of this paper, we should
stress the surprising precision with which the capillarity approximation
permits the calculation of the size dependence of the excess free energy
in both considered cases of solid clusters and liquid drops of monatomic
substances. As to the constant B, which from a strictly classical thermodynamic point of view must be equal to zero, it can be exactly calculated
in the case of simple models of solid clusters, and can be determined by
extrapolation in the case of liquid clusters. Since this constant enters
in the pre-exponential factor of the size distribution function, Eq. (IV.18),
one can state that classical thermodynamics is quite successful in
calculating the exponential term of the Eq. (IV.18). This explains its
ability to interpret qualitatively nucleation experiments on the basis of

macroscopic parameters, such as surface tension. The pre-exponential factor of the same equation which can be, we already saw it, as large as 10^{29}, must be calculated from microscopic models or determined by Monte Carlo or Molecular Dynamics simulation techniques.

PART C: HOMOGENEOUS NUCLEATION KINETICS

VI. STEADY STATE

Steady state nucleation kinetics should be considered as a mathematical model rather than as a common physical phenomenon. In nature, the formation of the very first critical nucleus disturbs already the mother phase in its immediate neighborhood. Depending on the rate of matter transport towards the growing overcritical cluster, assured by diffusion, convection or stirring, the perturbation can be confined, in some cases, in a small volume around the cluster. In the rest of the mother phase, nucleation can continue for a while, with the same constant rate, until the volumes of perturbation, the so-called "diffusion halos", start to overlap. The nucleation rate goes then down from the steady state value to zero.

If providing matter for the growth of overcritical clusters signals the end of the steady state nucleation period, its beginning does not coincide with the establishment of the supersaturation either. The change in the equilibrium cluster size distribution when passing from saturation to supersaturation needs time, and also this distribution is modified when the first critical nuclei appear, to result in a "steady state cluster size distribution", $z(i)$. If this "time lag" in nucleation, which will be discussed in the next section, is longer than the time necessary for the diffusion halos to overlap, steady state nucleation cannot be reached at all.

The mental procedure of realizing a steady state while nucleation is running, is to introduce continuously in the system the same amount of monomer (gas) molecules, as that consumed by the polymerization reaction of condensation [22-24]. Simultaneously, we have to withdraw from the system all clusters larger than some maximum size i_{max}. A constant flow, from smaller to larger clusters, which must be identified with the steady state nucleation rate, is then established, according to the kinetic equation:

$$J_o = \alpha(i-1)z_1 z(i-1) - \beta(i)z(i), \tag{VI.1}$$

where $\alpha(i-1)$ is the rate constant of the bimolecular reaction of association:

$$A_{i-1} + A_1 \underset{\beta(i)}{\overset{\alpha(i-1)}{\rightleftarrows}} A_i, \tag{VI.2}$$

while $\beta(i)$ is the rate constant of the inverse, monomolecular reaction of dissociation.

The consideration of bi-molecular reactions between polymer and monomer molecules only, (Eq. (VI.2)), is justified for two reasons:

- polymolecular reactions are highly improbable
- bi-molecular encounters between two polymer molecules are rare, because the concentrations of all other species but the monomers are extremely low.

If, in some way, one can force the net flux of matter, J_o of Eq. (VI.1), to be zero, as regards the supply by monomers but also in their reactions with clusters of larger sizes, the system will tend to the state of equilibrium discussed in section IV of this course [14]. Equation (VI.1) can then be written as:

$$0 = \alpha(i-1)n_1 n(i-1)0 - \beta(i)n(i),$$ (VI.3)

where the rate constants remain the same and the equilibrium concentrations $n(i)$, $i = 1,2,\ldots,(i-1),i,\ldots$ are identical with those given by Eqs. (IV.6). Equilibrium and steady state concentrations are in principle different. However, since the total concentration of polymers is considerably lower than the concentration of the monomers alone, the latter is unaffected by the difference between $n(i)$ and $z(i)$ for $i \geqslant 2$. Thus one can set:

$$z_1 = n_1.$$ (VI.4)

From Eqs. (VI.1) and (VI.3) one obtains:

$$J_o = \alpha(i-1)n_1 n(i-1)\left[\frac{z(i-1)}{n(i-1)} - \frac{z(i)}{n(i)}\right]$$ (VI.5a)

$$\simeq -\alpha(i)n_1 n(i)\frac{\partial}{\partial i}\left[\frac{z(i)}{n(i)}\right] = -\alpha(i)n_1\frac{\partial z(i)}{\partial i} - \frac{\alpha(i)}{kT}n_1 z(i)\frac{\partial \Delta G(i)}{di}.$$ (VI.5b)

In the right-hand side of the latter equation $n(i)$ has been substituted from Eq. (IV.6C). Integration of Eq. (VI.5b) results in the value of the ratio z/n:

$$\frac{z(i)}{n(i)} = \frac{J_o}{n_1}\int_i^{i_{max}}\frac{di}{\alpha(i)n(i)} = \frac{J_o}{n_1}\int_i^{i_{max}}\frac{\exp[\Delta G(i)/kT]}{\alpha(i)}\,di.$$ (VI.6)

Here the upper limit, i_{max}, has been chosen large enough to satisfy the boundary condition $z(i) = 0$ for $i = i_{max}$.

We already saw in section II.3 that the variation of the Gibbs free energy associated with the formation of an i-molecular cluster passes through a maximum for $i = i^*$ (Eq. (II.26)). Its development in series in the vicinity of i yields:

$$\Delta G(i) = \Delta G^* - \lambda\frac{(i-i^*)^2}{2},$$ (VI.7)

where $\lambda = -(d^2 G/di^2)_{i=i^*}$. On the other hand, because of the sharp maximum of the function to be integrated in Eq. (VI.6) for $i = i^*$, the size dependent rate constant $\alpha(i)$ can be replaced by its value α^* corresponding to i^*, and taken out of the integral. One has:

$$\frac{z(i)}{n(i)} = J_o\frac{\exp(\Delta G^*/kT)}{\alpha^* n_1}\int_{i=i^*}^{i_{max}-i^*}\exp[-\frac{\lambda}{2kT}(i-i^*)^2]d(i-i^*).$$ (VI.8)

The term $\sqrt{(kT/\lambda)}$ is the half width (standard deviation) of the Gaussian function in Eq. (VI.8). For $i < i^* - \sqrt{(kT/\lambda)}$ and, since $i_{max} \gg i^*$, the integration limits can be extended from $-\infty$ to $+\infty$. The integration of Eq. (VI.8) yields finally:

$$\frac{z(i)}{n(i)} = J_o\frac{\exp[\Delta G^*/kT]}{\alpha^* n_1}(\frac{2\pi kT}{\lambda})^{1/2} = \text{const.}$$ (VI.9)

43

This result is valid also for i = 1, a size for which the value of the ratio z/n is already known to be unity (Eq. (VI.4)). Replacing by this value the constant of Eq. (VI.9), one obtains directly the expression for the steady state nucleation rate:

$$J_o = \alpha^* n_1 (\frac{\lambda}{2\pi kT})^{1/2} \exp(-\Delta G^*/kT). \tag{VI.10}$$

This expression has a very simple physical meaning. If one replaces in it the exponential term by the concentration n(i) for i = i*, from Eq. (IV.6c), one has:

$$J_o = \alpha^* n_1 n(i^*) (\frac{\lambda}{2\pi kT})^{1/2}. \tag{VI.11}$$

The steady state nucleation rate is thus equal to the rate of the irreversible bi-molecular reaction between simple (monomer) molecules and i*-sized molecules, whose concentration in the gas mixture is the smallest of all the chemical species. The A_{i^*} species acts, therefore, as an "activated complex" [25] in the chemical reaction of polymerization. The last multiplier in Eq. (VI.11), which is a measure of the width of the function of 1/n(i) in the vicinity of i*, takes into account that chemical species of sizes i ≃ i* are also participating to the nucleation kinetics.

According to Eq. (II.26), $\lambda = -d^2\Delta G/di^2 = 2\Delta G_{c1}^*/3i^{*2}$. Thus, it follows from Eq. (VI.10):

$$J_o = \frac{\alpha^* n_1}{i^*} (\frac{\Delta G_{c1}^*}{3\pi kT})^{1/2} \exp(-\Delta G^*/kT). \tag{VI.12}$$

Finally, if one looks for the link with classical thermodynamics, one has to remember Eqs. (II.22) and (IV.16) and write down:

$$\begin{aligned}
J_o &= \frac{\alpha^* n_1}{i^*} (\frac{\Delta G_{c1}^*}{3\pi kT})^{1/2} \exp[- (\Delta G_{c1}^* + B)/kT] \\
&= \frac{\alpha^* n_1}{i^*} (\frac{\Delta G_{c1}^*}{3\pi kT})^{1/2} Q_{rep} \exp(-\Delta G_{c1}^*/kT).
\end{aligned} \tag{VI.13}$$

This up-to-date version of the absolute rate of steady state nucleation, subsequent to the extended analysis of all parameters playing a role in the thermodynamics of embrios, done in the preceding sections, necessitates some final remarks.

(i) If one expresses the classical nucleation work ΔG_{c1}^* of Eq. (VI.13) through the supersaturation of the mother phase, according to Eq. (II.29), one has in logarithmic form:

$$\ell nJ_o = C_1 - \frac{C_2}{(\Delta\mu)^2} , \tag{VI.14}$$

where

$$C_1 = \ell n \frac{\alpha^* n_1}{i^*} (\frac{\Delta G_{c1}^*}{3\pi kT})^{1/2} Q_{rep}$$

is practically constant, and

$$C_2 = \frac{4}{27} \frac{(\gamma_\infty \Omega)^3}{kT} .$$

In numerous experimental works, the linear dependence of ℓnJ on $1/\Delta\mu^2$ has been checked [26-27]. In general, the slope of the line can be fairly well fitted with the calculated constant C_2 which makes use of such macroscopic notions as surface tension γ_∞ and equilibrium shape (through

Ω), even if the calculated critical nucleus size turns out to be a few tens of molecules only. This fact should not astonish if one remembers the conclusion of section IV on the applicability of the capillarity approximation down to molecular sizes.

(ii)   No unanimity has been reached as to the experimental fit of the constant C_1 of Eq. (VI.14), especially because, until recently, the replacement partition function Q_{rep} in Eq. (VI.13) was not generally recognized and, since Becker-Doering [24] and Frenkel [14], has been substituted by the concentration of monomers n_1. For this reason, and depending on the system under consideration, the estimated pre-exponential factor of Eq. (VI.13) sometimes differed from the experimentally measured one by up to ten orders of magnitude or even more.

We believe that only detailed model calculations on small crystalline clusters, or computer simulations on small liquid clusters can give access to the value of Q_{rep} (or of the constant B in Eq. (IV.16)). Since it happens to be size independent, the further use of capillarity approximation for larger clusters (or lower supersaturations) is fully justified, as far as the same substance is concerned.

(iii)   To avoid confusion, we must notice that in the kinetic part of this paper we used the rate constant $\alpha(i)$, as introduced in chemical kinetics. In the ideal gas approximation, it is proportional to the collision cross section and to the reduced mass of the couple i-mer – monomer.

Most nucleation treatments [28] use the growth frequency, $f(i)$, of an i-molecular cluster. In the original Becker-Doering paper [24], $f(i)$ is proportional to the "area" of the cluster and to the impinging frequency of the gas molecules. The correspondence of these two parameters is:

$$f(i) = \alpha(i)n_1.$$

VII. THE TIME LAG IN NUCLEATION

This is a rather complicated mathematical problem which will be only sketched here.

We saw in the preceding section that when steady state was established, the ratio of the steady state to the equilibrium size distribution functions was a constant equal to unity in the whole size range between i = 1 and $i_1 = i* - \sqrt{(kT/\lambda)}$. From the manner integration of Eq. (VI.8) has been performed, it is clear that the constant of Eq. (VI.9) is equal to one half for i = i*, and that it is equal to zero for $i_2 > i + \sqrt{(kT/\lambda)}$. The ratio z/n vs. i has, therefore, the idealized shape shown on Figure 7, curve A, while its real shape is that of curve B of the same Figure [27].

Now, if one calls $c(i,t)$ the time dependent concentration of the i-sized nuclei in the gas mixture, where t = 0 is the moment in which supersaturation is established, it is obvious that $c(i,t)$ will rapidly reach the steady state concentration $z(i)$ for low values of i and will stay below the $z(i)$ value when i approaches i*. Thus, the ratio $c(i,t)/n(i)$, also plotted on Figure 7, curves C,D,E, after increasing time periods, is quasi-immediately equal to $z(i)/n(i) = 1$ outside a "critical region" which, as before, extends roughly from i_1 to i_2, while inside this region it tends asymptotically to $z(i)/n(i)$ after some time.

Mathematically, now the non-constant flow $J(i)$ of matter through the clusters of different size classes is given by an equation similar to (VI.5b):

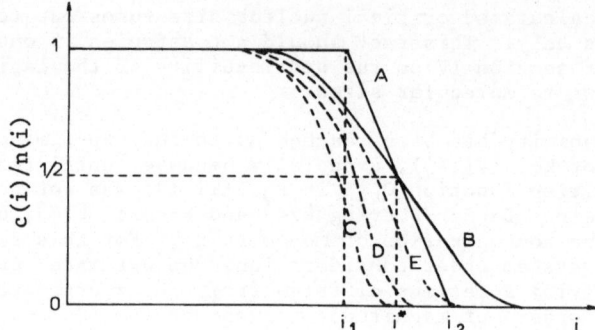

Fig. 7. Curves A and B are respectively the idealized and the exact ratio between steady state ($z(i)$) and equilibrium ($n(i)$) cluster concentrations as a function of the size. i^* is the size of the critical cluster while i_1 and i_2 are the boundaries of the critical region. Curves C, D and E give the ratio between the non-steady state concentration $c(i,t)$ and $n(i)$ after increasing time intervals.

$$J(i) = -\alpha(i)n_1 n(i) \frac{\partial}{\partial i}\left[\frac{c(i,t)}{n(i)}\right], \qquad (VII.1)$$

together with the continuity condition:

$$\frac{\partial c(i,t)}{\partial t} = J(i) - J(i+1) \simeq -\frac{\partial J(i)}{\partial i}, \qquad (VII.2)$$

which yields:

$$\frac{1}{n_1 n(i)}\frac{\partial c(i,t)}{\partial t} = \left[\frac{\partial \alpha(i)}{\partial i} - \frac{\alpha(i)}{kT}\frac{\partial \Delta G(i)}{\partial i}\right]\frac{\partial}{\partial i}\left[\frac{c(i,t)}{n(i)}\right]$$
$$+ \alpha(i)\frac{\partial^2}{\partial i^2}\left[\frac{c(i,t)}{n(i)}\right]. \qquad (VII.3)$$

Here, as in Eq. (VI.5b), the value of $n(i)$ in the right-hand side has been substituted from Eq. (IV.6c).

The already discussed approximation done at this point [29] assumes that only the time dependence of $c(i,t)/n(i)$ in the size range from i_1 to i_2 is relevant for the nucleation rate. Therefore both terms in the first brackets of the right-hand side of Eq. (VII.3) are nil; the first, because $\alpha(i) = \alpha^* = \text{const.}$, the second, because $\Delta G(i)$ has a maximum for $i = i^*$. Equation (VII.3) reduces then to:

$$\frac{1}{n(i)}\frac{\partial c(i,t)}{dt} = \alpha(i)n_1 \frac{\partial^2}{\partial i^2}\left[\frac{c(i,t)}{n(i)}\right], \qquad (VII.4)$$

which admits the approximate solution:

$$\frac{c(i,t)}{n(i)} = 1 - \frac{i - i_1}{\Delta i^*} - \frac{2}{\pi}\sum_{s=1}^{\infty}\frac{1}{s}\sin(s\pi\frac{i - i_1}{\Delta i^*})\exp\left[-\frac{s^2\pi\alpha^* n_1 t}{(\Delta i^*)^2}\right], \qquad (VII.5)$$

with

$$\Delta i^* = i_2 - i_1 = 2\sqrt{(kT/\lambda)}.$$

When steady state is reached, $t \to \infty$, Eq. (VII.5) yields:

$$\frac{c(i,\infty)}{n(i)} = \frac{1}{2} - \frac{i - i^*}{\Delta i^*}, \qquad (VII.6)$$

46

valid for the critical region only ($i_1 < i < i_2$), in agreement with the curve A of Figure 7. The substitution of $c(i,\infty)/n(i)$ from Eq. (VII.6) into Eq. (VII.1), gives for the steady state nucleation rate the expression:

$$J'_o = \alpha^* n_1 n(i^*) \frac{1}{2} \left(\frac{\lambda}{kT} \right)^{1/2}, \tag{VII.7}$$

different by a factor $\sqrt{2\pi}/2 = 1.25$ from Eq. (VI.11), because of the applied approximations [30].

By substituting $c(i,t)/n(i)$ from Eq. (VII.5) into Eq. (VII.1), putting $i = i^*$ and taking into account Eq. (VII.7), one obtains the final result:

$$J = J'_o [1 + 2 \sum_{s=1}^{\infty} (-1)^s \exp(-s^2 t/\tau)], \tag{VII.8}$$

where

$$\tau = \frac{8kT}{\pi \lambda \alpha^* n_1} = \frac{12}{\pi} \frac{kT i^{*2}}{\Delta G^* \alpha^*} \tag{VII.9}$$

is the characteristic time lag, i.e., the time necessary to the system to reorganize itself from the state of stable equilibrium to the steady state of nucleation (Figure 8). It can be shown, from Eqs. (II.28) and (II.29), that the ratio $i^{*2}/\Delta G^*$ is proportional to $1/(\Delta\mu)^4$. The time lag is, thus, shorter the higher the supersaturation, which is easily understandable. Very important, however, is the rate of transport of matter toward the growing clusters, measured by the product $\alpha^* n_1$. So, in the cases of nucleation of droplets or crystals from the vapor phase, the

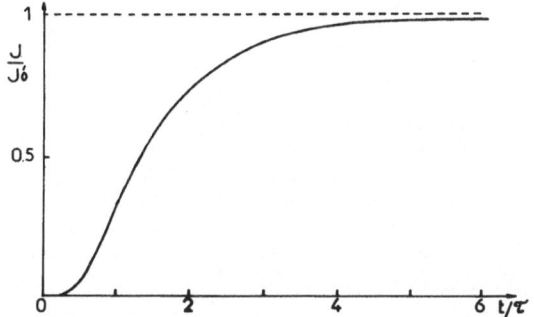

Fig. 8. Scaled steady state nucleation rate J/J'_o from Eq. (VII.8) as a function of reduced time t/τ.

establishment of steady state is practically instantaneous, while nucleation of crystals in silicate glasses, during devitrification, has such a great time lag that the decrease of nucleation rate, due to advanced growth, starts before steady state is reached.

ACKNOWLEDGEMENT

The author is grateful to Mrs D. Strub for precious help in preparing the manuscript.

47

REFERENCES

1. O. Stern, Physikal. Z., 14:629 (1913).
2. J. W. Gibbs, "The Scientific Papers", Dover, NY, 55-349 (1961).
3. R. C. Tolman, J. Chem. Phys., 17:333 (1949).
4. J. G. Kirkwood and F. P. Buff, J. Chem. Phys., 17:338 (1949).
5. S. Ono and S. Kondo, in: "Handbuch der Physik", Springer Verlag, Berlin, vol. 10 (1960).
6. T. L. Hill, "Thermodynamics of Small Systems", Benjamin, NY, part 1, 27-58 (1963).
7. S. Takagi, J. Appl. Phys., 24:1453 (1953).
8. A. Bonissent and B. Mutaftschiev, J. Chem. Phys., 58:3727 (1973).
9. P. Curie, Bull. Soc. Miner. de France, 8:145 (1885).
10. G. Wulff, Z. Kristallogr., 34:449 (1901).
11. M. Volmer, "Kinetik der Phasenbildung", Steinkopf, Dresden, 87-91(1939).
12. R. Kaischew, Bull. Bulg. Acad. Sci. (Phys. Ser.), 2:191 (1951).
13. M. Volmer and A. Weber, Z. Physik. Chem., 119:277 (1926).
14. J. I. Frenkel, "Kinetic Theory of Liquids", Dover, NY, 366-400 (1955).
15. J. Lothe and G. M. Pound, J. Chem. Phys., 36:2080 (1962).
16. W. C. Hoover, A. C. Hindmarsh and B. L. Holian, J. Chem. Phys., 57: 1980 (1972).
17. F. F. Abraham, "Homogeneous Nucleation Theory", Academic Press, NY, 139-149 (1974).
18. K. Nishioka and G. M. Pound, in: "Advances in Colloid and Interface Science", A. C. Zettlemoyer, ed., Elsevier, Amsterdam, 205 (1977).
19. B. Mutaftschiev, in: "Interfacial Aspects of Phase Transformations", B. Mutaftschiev, ed., Reidel, Rotterdam, 63 (1982).
20. H. Reiss, J. Stat. Phys., 2:83 (1970).
21. J. K. Lee, J. A. Barker and F. F. Abraham, J. Chem. Phys., 58:3166 (1973).
22. L. Farkas, Z. Phys. Chem., 125:236 (1927).
23. R. Kaischew and I. N. Stranski, Z. Phys. Chem., 26B:317 (1934).
24. R. Becker and W. Doering, Ann. Phys., 24:719 (1935).
25. S. Glasstone, K. J. Laidler and H. Eyring, "The Theory of Rate Processes", McGraw Hill, NY, 153-201 (1941).
26. J. P. Hirth and G. M. Pound, "Condensation and Evaporation", Pergamon, Oxford, 29-40 (1963).
27. S. Toschev, in: "Crystal Growth: an Introduction", P. Hartman, ed., North-Holland, Amsterdam, 1-49 (1973).
28. J. L. Katz, in: "Interfacial Aspects of Phase Transformations", B. Mutaftschiev, ed., Reidel, Rotterdam, 261-286 (1982).
29. B. Y. Lyubov and A. L. Roitburd, in: "Problems Metallovedeniya i Fiziki Metallov", Metallurgisdat, Moscow, 91 (1958).
30. D. Kashchiev, Surf. Sci., 14:209 (1969).

BASIC MECHANISMS OF CRYSTAL GROWTH

D. Aquilano

Dipartimento di Scienze della Terra
Università di Torino
Via S. Massimo 24, 10123 Torino, Italy

INTRODUCTION

The early stage of crystallization is the formation of 3D nuclei in the mother phase (vapor, melt, solution). A second step is represented by the growth of every face of the supercritical 3D nuclei.

The scheme drawn in Figure 1 shows the factors affecting the advancement rate of a given {hkl} form, when the crystallization temperature (T_C) and the thermodynamic supersaturation ($\Delta\mu = KT \ln(\sigma_v + 1)$) are given.

These factors belong roughly to two categories:

Factors depending on the crystal structure: a) surface structure of the {hkl} form, and its corresponding character, according to the Hartman and Perdok theory [1]; b) perfection (or not) of the surface. If the surface is penetrated by one (or more) screw dislocation lines, the growth is favored at very low supersaturation.

Factors depending on the mother phase: a) the structure of the mother phase (vapor, solution, melt) which affects the equilibrium properties of the crystal interface. For a given surface structure of a {hkl} form, the values of both surface specific free energy (γ_{hkl}) and edge specific free energies (ρ_{hkl}) depend on the interaction between the surface and the surrounding medium; b) transport properties of the growth medium strongly affect the growth rate of a face. When considering vapor or solution growth some parameters must be taken into account: density (ρ), viscosity (η) and the relative velocity (υ) of the flowing matter with respect to the growing face; the size (1) of the face and the thickness of the boundary layer (δ) through which the growth units diffuse (D_v) to the interface.

Especially in growth from the vapor phase the diffusion of the growth units onto the growing surface plays an important role, the diffusion coefficient (D_s) depending both on the surface structure and on the nature of the growth units.

All these factors determine the so-called growth mechanisms of a given crystal face: they are summarized in Figure 1.

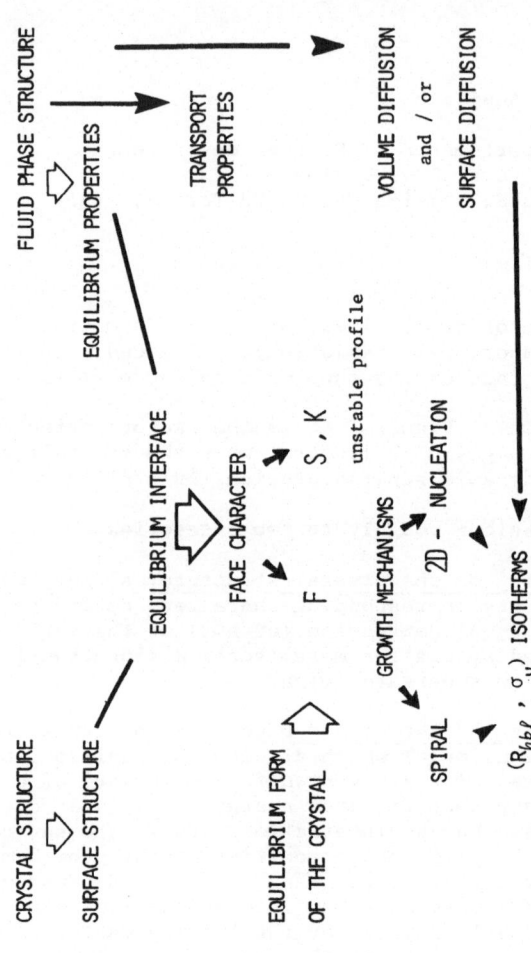

Fig. 1. Main factors affecting the growth mechanisms

The relative growth rates (R_{hkl}) determine, in a steady state, the growth habit of a crystal.

Experimentally we can measure the growth isotherms (R_{hkl}, σ_v) at different crystallization temperatures (T_C) and hence we can try to establish the mechanism whose theoretical law agrees in the best way with the experimental data.

2. STRUCTURE OF A CRYSTAL FACE

Figure 2a shows an idealized picture of a perfect simple cubic crystal.

Each small cube represents a growth unit; the interaction (ϕ) energy is confined to the nearest neighbors.

Three kinds of faces may be distinguished:

— {100} faces: on the uppermost layer two uninterrupted and periodic chains of ϕ bonds are easily seen, along two perpendicular directions [100] and [010]. The {100} slices (having thickness d_{100}) are compact and and very stable. According to Hartman and Perdok [1] we call these faces flat faces (F), or equilibrium faces (see section 2.1.4).

— {110} faces: growth units form only one kind of uninterrupted chain, along the [010] direction, within a slice of thickness d_{110}. We call these less compact faces, stepped face (S), as it results from their profile.

— {111} faces: no bond exists between two successive growth units, within a slice d_{111}. These are kinked faces (K).

This distinction is a rough one, but it is useful to understand the equilibrium and growth behavior of a face, when its surface structure is considered.

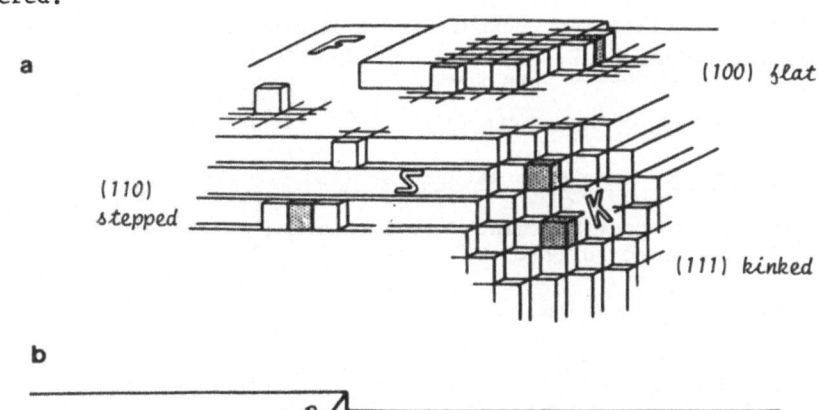

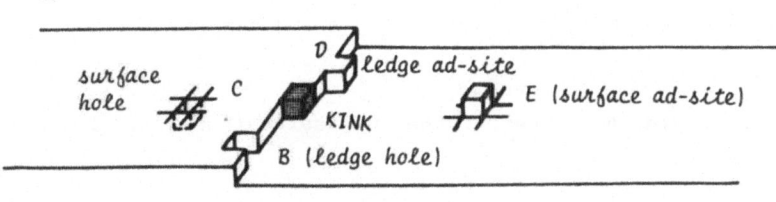

Figure 2

2.1 Crystal-Vapor Equilibrium: a Simple Atomistic Treatment [2]

Let us consider the elementary condensation-evaporation processes on a F surface, as represented in Figure 2b; each cube has a vibration frequency ν and a mass m.

2.1.1 Equilibrium between a vapor (v) and a crystal of infinite size

At the equilibrium the chemical potential (μ) of an atom (cube) is the same in the vapor and in the crystal.

$$\mu_v = - \phi_v - kT \ln(p_e/kT) + kT \ln(2\pi mkT)^{3/2}h^{-3}$$

$$\mu_{c\infty} = - \phi_{c\infty} + 3kT \ln(kT/h\nu). \tag{1}$$

Hence the equilibrium pressure p_e is:

$$p_e = (2\pi mkT)^{3/2}\nu^3 (kT)^{-2}\exp\{-(\phi_v - \phi_{c\infty})/kT\} \tag{2}$$

where the difference $(\phi_v - \phi_{c\infty}) = \Delta H_{evap}$ (per atom).

If we assume that the potential energy of an atom in the vapor is zero $\rightarrow \Delta H_{evap} = - \phi_{c\infty}$, then from the footnote*:

$$\langle \Delta H \rangle_{evap}^{\infty} = 3\phi = - \phi_{c\infty}. \tag{3}$$

2.1.2 Equilibrium between a vapor (v) and a crystal of finite size (C)

Expressions (1) have to be considered with $p_e^{(n)}$ instead of p_e, ϕ_c instead of $\phi_{c\infty}$, and μ_c instead of $\mu_{c\infty}$.

The ratio between the two equilibrium pressures

$$\ln \frac{p_e^{(n)}}{p_e} = \frac{\phi_c - \phi_{c\infty}}{kT} \tag{4}$$

shows that the equilibrium pressure for a finite crystal is higher than the saturation pressure p_e, being $\phi_c > \phi_{c\infty}$, as it may be obtained (from the footnote) when considering that

$$\langle \Delta H \rangle_{ev} = - \phi_c = 3\phi\{1 - (1/n)\}. \tag{5}$$

2.1.3 The thermodynamic supersaturation $(\Delta\mu)$

Expression (4) may be related to a chemical potential difference $(\mu_\infty - \mu_{c\infty})$ as defined in (1). We call

$$\Delta\mu = \mu_{c\infty} - \mu_c = \phi_c - \phi_{c\infty} = kT \ln \frac{p_e^{(n)}}{p_e}$$

and, putting $\beta = p_e^{(n)}/p_e$, one obtains

$$\Delta\mu = kT \ln\beta \tag{6}$$

where $\Delta\mu \gtrless 0$ according to whether the supersaturation ratio is $\gtrless 1$.

---

* For a cube of finite size (n^3 atoms) the evaporation work is $3\phi(n^3 - n^2)$; the corresponding value per atom is $\langle \Delta H \rangle_{ev} = 3\phi\{1 - (1/n)\}$. For a cube of infinite size ($n \rightarrow \infty$) it turns out that $\langle \Delta H \rangle_{ev} = 3\phi$.

2.1.4   Surface sites and equilibrium surfaces

From Figure 2b one can obtain the potential energy of each type of surface sites:

$$E \rightarrow - \phi \qquad A \rightarrow - 3\phi \qquad B \rightarrow - 4\phi$$
$$D \rightarrow - 2\phi \qquad\qquad\qquad C \rightarrow - 5\phi$$

Site A (dotted position in Figure 2b) is the only one in equilibrium with a saturated vapor, both for finite and infinite crystals, having its evaporation energy equal to $<\Delta H>_{evap}$. This is a kink-site.

Hence the probability for a kink to be occupied is 1/2, being the kink equilibrium site.

Other sites are not equilibrium sites; their probability to be occupied is < 1/2 (E, D) or > 1/2 (B, C). A face like the one drawn in Figure 2a has a surface structure which is able to maintain its profile, even if the coverage degree of ad-atoms (E) increases. In the following we will show that these equilibrium faces grow layer by layer until a barrier is overcome when the face may grow as a K-face (uncorrelated-diffuse growth) [3].

2.1.5   Non-equilibrium surfaces

From Figure 2a one can see that every site on {111} faces has an occupancy probability equal to 1/2. The surface profile will depend on the homogeneity of the flow of the growth units. Furthermore the profile will be unstable, in respect to every small fluctuation of the chemical potential of the mother phase.

{110} faces show a growth behavior similar to that of K-faces, for no correlation exists between the growth of two adjacent chains [010].

2.2   Crystal-Vapor Equilibrium: a Phenomenological Treatment [4]

In Figure 3 a polyhedral convex crystal is represented; its volume V is limited by faces having areas A_i and surface free energy γ_i.

At thermodynamical equilibrium the total surface free energy reaches a minimum value for any given crystal volume:

$$\delta \Sigma \gamma_i A_i = 0 \qquad \delta V = 0 \tag{7}$$

Expression (7) may be written in terms of the distances (h_i) between a central point P of the polyhedron and the corresponding faces (A_i). A

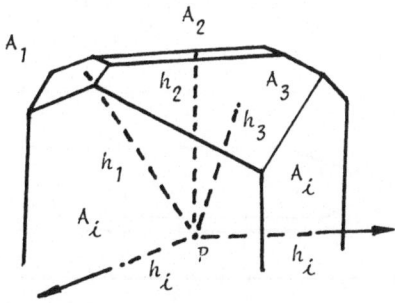

Figure 3

simple calculation is presented in [5] and the Gibbs-Wulff theorem is obtained:

$$\frac{\gamma_1}{h_1} = \frac{\gamma_2}{h_2} = \ldots = \frac{\gamma}{h_i} = const., \text{ that means:}$$

$$\gamma_1 : \gamma_2 : \ldots : \gamma_i = h_i : h_2 : \ldots : h_i$$

(8)

Two important consequences arise from (7) and (8): the equilibrium shape of a crystal is a convex polyhedron; and this polyhedron is limited only by faces having the lowest γ values.

When considering the crystal model drawn in Figure 2a, a simple calculation allows to obtain the specific surface free energy, within the Born-Stern approximation: $\gamma = (W/2A_i)$, where W is the separation work of the crystal along the area A_i.

Labelling with a the area of a cube face, $\gamma_{100} = \phi/2a^2$; $\gamma_{110} = 2\phi/2\sqrt{2}a^2$; $\gamma_{111} = 3\phi/2\sqrt{3}a^2$ and hence, according to (8):

$$\gamma_{111} : \gamma_{110} : \gamma_{100} = \sqrt{3} : \sqrt{2} : 1$$

which shows, from a simple drawing, that both forms {110} and {111} do not belong to the equilibrium shape of this crystal-model. On the contrary {100}, which is a flat face, according to 2.1.4, is an equilibrium surface and belongs to the equilibrium shape.

The Gibbs-Wulff theorem applies also for two-dimensional crystals formed on the top of 3D-crystals. The equilibrium condition is then: $\delta\Sigma\rho_i l_i = 0$ and $\delta S = 0$ (9) where ρ_i is the specific free energy of the ith edge, l_i its corresponding length and S is the area of the 2D-crystal represented in Figure 4.

It can be shown that (9) is satisfied by the analogous relation:

$$\rho_1 : \rho_2 : \ldots \rho_i \ldots = h'_1 : h'_2 : \ldots : h'_i \ldots$$

3. GROWTH KINETICS OF K-FACES

As we mentioned in 2.1.4 a kinked face, as the {111} in Figure 2a, looks like an infinite population of kink-sites. In a real case $\alpha \sim 1$ represents the fraction of kinks on {111}.

Let us call $\dot{n}\downarrow$ and $\dot{n}\uparrow$ the number (per unit time) of atoms entering a kink and going out of it, respectively; the net balance will be: $\dot{N} = \dot{n}\downarrow - \dot{n}\uparrow = \dot{n}\downarrow (1 - (\dot{n}\uparrow/\dot{n}\downarrow))$. The normal growth rate of {111} results to be: $R_{111} = \alpha \cdot d_{111} \cdot \dot{N}$ where d_{111} is the equidistance of the reticular planes 111.

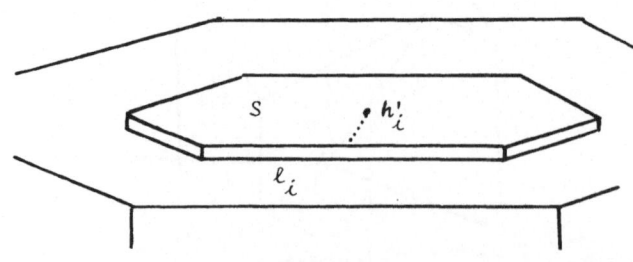

Figure 4

At equilibrium $\dot{n}\downarrow = \dot{n}\uparrow$. The equilibrium constant of the process is $k_e = \dot{n}\downarrow/\dot{n}\uparrow = \exp(-\Delta\mu/kT)$ and hence

$$R_{111} = \alpha d_{111}\, \dot{n}\uparrow \{1 - \exp(-\Delta\mu/kT)\}$$

Remembering that expression (6) in 2.1.3 has to be written, for vapor growth:

$$\Delta\mu = kT \ln(p/p_e)$$

where p is the pressure in the growth medium ($p = p_e + \Delta p$).

If $\Delta p \ll p$ (practical case) $\rightarrow \ln(p/p_e \sim \Delta p/p)$ and hence $1 - \exp(-\Delta\mu/kT) \sim (\Delta p/p)$; when considering that $\dot{n}\downarrow = a^2 \cdot p \cdot (2\pi mkT)^{1/2}$ one obtains:

$$R_{111} = \{\alpha a^2 d_{111} p_e (2\pi mkT)^{-1/2}\}\sigma_v \tag{9}$$

which shows that R_{111} depends linearly on the relative supersaturation σ_v.

It may be easily seen that also for the stepped (S) faces the growth rate depends linearly on the supersaturation.

4. GROWTH KINETICS OF FLAT (F) EQUILIBRIUM FACES

Let us consider an F face (Figure 2b): species E, D are absorbed on the surface and ledge respectively (C and B are holes).

θ is the coverage degree (number of occupied sites over the total number of available sites: $0 \leqslant \theta \leqslant 1$, so we label

$\theta_s^+ \quad \theta_1^+$ (coverage degrees for absorbed sites on the surface and ledge respectively)

$\theta_s^- \quad \theta_1^-$ (for holes)

E, D, C, B are all in thermodynamical equilibrium among them and with the vapor at the pressure p: the corresponding adsorption isotherms, according to the Mutaftschiev model [3] are:

$$\frac{\theta_s^+}{\{1 - (\theta_s^+ + \theta_s^-)\}} = \frac{p}{p_e} \exp\{-2\phi(1 - 2\theta_s^+)/kT\}$$

$$\frac{\theta_s^-}{\{1 - (\theta_s^+ + \theta_s^-)\}} = \frac{p_e}{p} \exp\{-2\phi(1 - 2\theta_s^-)/kT\}$$

$$\frac{\theta_1^+}{\{1 - (\theta_1^+ + \theta_1^-)\}} = \frac{p}{p_e} \exp\{-\phi(1 - 2\theta_1^+)/kT\}$$

$$\frac{\theta_1^-}{\{1 - (\theta_1^+ + \theta_1^-)\}} = \frac{p_e}{p} \exp\{-\phi(1 - 2\theta_1^-)/kT\} \tag{10}$$

At equilibrium $p = p_e$ and $\theta_s^+ = \theta_s^- = \theta_s$; $\theta_1^+ = \theta_1^- = \theta_1$. Hence expressions (10) reduce to:

$$\theta_s/(1 - 2\theta_s) = \exp\{-2\phi(1 - 2\theta_s)/kT\}$$

$$\theta_1/(1 - 2\theta_1) = \exp\{-\phi(1 - 2\theta_1)/kT\} \tag{11}$$

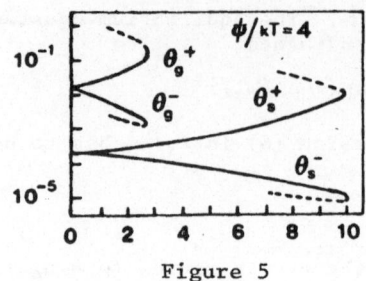

Figure 5

Figure 5 represents the adsorption isotherms (10) for a typical value of $\phi/kT = 4$.

θ^+ is a continuously increasing function of $(\Delta\mu/kT)$ until a critical supersaturation value is reached, over which metastable states appear (dotted lines) and the face can grow without any supersaturation increase (diffuse growth, roughening transition).

However it must be mentioned that such critical supersaturation values correspond to $(p/p_e \sim 10^5$ for the ledges and to $(p/p_e) \sim 10^{10}$ for the surface, both values being meaningless for normal growth conditions. We must conclude that F faces cannot grow through this mechanism and, having considered that they can grow even at very low supersaturation $(p/p_e \sim 1.001)$, we must find some "less expensive" mechanism, i.e., a process acting at low supersaturations.

4.1 Growth Mechanism of Perfect F Faces: 2D-Nucleation

Just at the beginning of the 20's Volmer [4] proposed a simple theory for the growth of a perfect face.

Let us consider a very low coverage degree for holes ($\theta_s \to 0$) and a very dilute adsorption layer ($\theta_s^+ \ll 1$) on a flat perfect face. Remembering that $p/p_e = \exp(\Delta\mu/kT)$ the first of expression (10) becomes

$$\theta_s = \exp\{-(2\phi - \Delta\mu)/kT\} \tag{12}$$

An equilibrium exists on the surface between n^2 ad-atoms and 2D-"polymers" built-up by n^2 atoms (square-shaped nuclei, for the sake of simplicity). The relation is:

n^2 (ad-atoms) $\overset{\rightarrow}{\leftarrow}$ "polymer" of n^2 atoms.

To create this "polymer" the energy required is

$$\Delta E = (n^2 - n)2\phi \tag{13}$$

and the corresponding equilibrium constant of this reaction is

$$(\theta_{n^2}/\theta_s^{n^2}) = \exp(\Delta E/kT) = \exp\{(n^2 - n)2\phi/kT\}. \tag{14}$$

By putting (12) in (14) one obtains the coverage degree of polymers (θ_{n^2}) as a function of their size (n^2) and of the supersaturation $\Delta\mu$:

$$\theta_{n^2} = \exp\{-(2n\phi - n^2\Delta\mu)/kT\}. \tag{15}$$

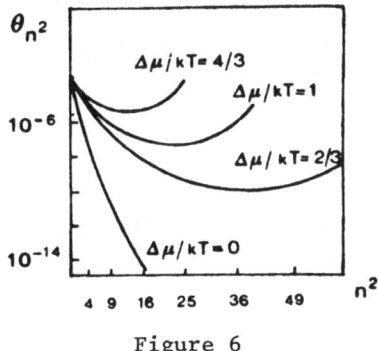

Figure 6

Expression (15) is represented in Figure 6, for $\phi/kT = 4$ where θ shows a minimum when

$$n^* = \phi/\Delta\mu. \tag{16}$$

n^* depends on the free energy variation due to the formation of 2D-nuclei.

As a matter of fact, for a square nucleus having n atoms on its side:

$$\Delta G_n^{2D} = 2n\phi - n^2\Delta\mu \tag{17}$$

which shows a maximum for $n^* = \phi/\Delta\mu$, the same value at which θ_{n^2} has a minimum. This means that, for n^*, the system reaches an unstable equilibrium between critical nuclei (n^* size) and the atoms adsorbed on the surface. The nuclei, once the n^* size is reached, can grow spontaneously when a single atom is added to the critical size (super-critical nuclei). When

$$n = n^0 \rightarrow \Delta G_{n^0}^{2D} = \phi^2/\Delta\mu = n^0\phi \tag{18}$$

which represents the activation barrier for two-dimensional nucleation.

From (18) and (15) we may obtain the number N^* of critical nuclei for unit area, in a steady state, remembering that n_s is the number of available sites (per unit area):

$$N^0 = n_s\theta_{n^2}^0 = n_s\exp\{-\phi^2/(kT\Delta\mu)\}. \tag{19}$$

To calculate the normal growth rate of a perfect face, through the 2D-nucleation mechanism, the nucleation frequency $J(n^*)$ is required ($J(n^*)$ being the number of critical nuclei formed per unit time).

The assumption we make is the simplest one: a 2D-nucleus, once the supercritical size is reached, extends and covers all the available area (A) before a second nucleus is formed on the spreading layer (mono-nucleation model). In this case the maximum growth rate of the face will be:

$$R_{2D}^{hk1} = d_{hk1}AJ(n^0)N^0. \tag{20}$$

For a better understanding of the $J(n^*)$ formulation, the reader may refer to the contribution given by B. Mutaftschiev (in this book). We limit ourselves to use the $J(n^*)$ expression obtained by Stranski-Kaischew [6] and Becker-Döring [7].

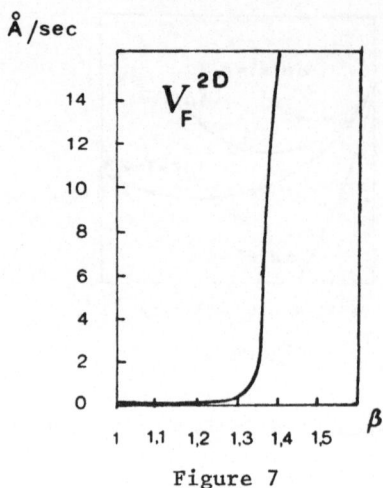

Figure 7

According to these authors we can write:

$$R_{2D}^{hkl} = d_{hkl} A n_s \nu_{sl} (\Delta\mu/kT) \exp(-3\phi/kT) \exp(\Delta G_{n_0}^{2D}/kT) \quad (21)$$

where ν_{sl} indicates the transition frequency of an atom from the surface to the ledge of the spreading nucleus.

Expression (21) is represented in Figure 7, where $\phi/kT = 4$, $A = 1$ cm^2, $d_{hkl} = 10^{-8}$ cm, $n_s = 10^{14}$ cm^{-2} and $\nu_{sl} = 10^{10}$ sec^{-1}.

It must be outlined that only for $\beta = (p/p_e)$ values exceeding 1.35 R_{2D} is not null.

This means that 2D-nucleation mechanism is less expensive than "diffuse growth" for a perfect F face and that, on the other hand, a critical supersaturation has to be overcome in order to observe an appreciable growth rate.

As we mentioned at the end of section 4 growth can occur at very low supersaturations; in the next section we will show how only the hypothesis that the crystal dislocations can "interact" with crystal growth, allows to explain why a face can advance, under conditions close to the equilibrium, and that, consequently, new mechanisms with a very low activation energy exist.

4.2 Growth Mechanisms of Defective F Faces: Spiral Growth

A real crystal is not perfect: it contains a large number of dislocations (edge, screw and mixed) [8]. For our purposes it is sufficient that one screw dislocation crosses a flat face and that its Burgers vector has a non-null component perpendicular to the face: in this case an exposed ledge, of length 1, anchored in P will be observed on the face. Around P the lattice planes are strongly deformed (PE) but, far from it, they become undeformed again (EM = l_e) (Figure 8a).

Burton, Cabrera and Frank [9] calculated the activation energy ΔG_{l_e} required for growth of the exposed ledge (EM), as a function of the ratio $(l_e/2r^*)$, where r^* is the radius of the critical 2D-nucleus in equilibrium with the supersaturated vapor:

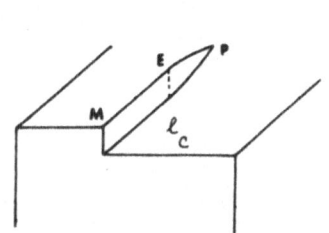 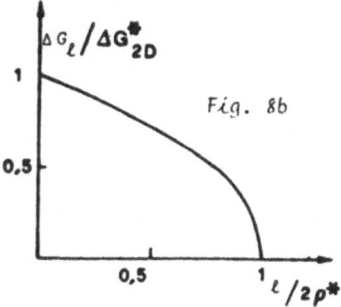

Figure 8

$$r^* = (\rho a^2/\Delta\mu^*) \quad \text{(Gibbs-Thomson expression)} \tag{22}$$

ρ being the specific ledge energy (erg/cm) and a^2 the area of a growth unit along the ledge.

Figure 8b shows the ratio $\Delta G(l_c)/\Delta G^*_{2D}$ as a function of $l_c/2r^*$; it is easily seen that no activation energy is required when the length of the exposed ledge equals that of a 2D-nucleus in equilibrium at $\Delta\mu^*$, according to (22).

Two situations can occur on the face, according to whether $l_c \overset{>}{<} 2r^*$. a) – $l_c < 2r^*$: the exposed ledge <u>cannot advance</u> at $\Delta\mu = \Delta\mu^*$, but only when $\Delta\mu > \Delta\mu^*$. b) – $l_c > 2r^*$: no activation energy is needed and the step can advance with its own velocity (depending on $T, \Delta\mu, \ldots$) even if $\Delta\mu$ slightly exceeds the equilibrium value.

In the case b) the step advances, so generating new steps (all belonging to the equilibrium form of the 2D-nucleus on the surface), according to the sequence drawn in Figure 9; the pattern is a growth spiral which winds up covering the surface with a new layer (of thickness b_n) after each rotation of 2π, as it may be seen in Figure 10.

In a steady state the spiral mechanism is self-perpetuating; the face can grow layer by layer; the equidistance y_0 between two successive steps depends on the shape of the critical nucleus and on its size (hence on $\Delta\mu$):

$$y_0 = A \frac{\rho a^2}{\Delta\mu} \tag{23}$$

where A is a shape factor.

Far from the spiral center each step advances with velocity V_T (velocity of the step train).

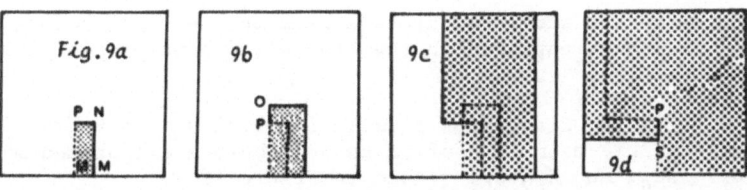

Figure 9

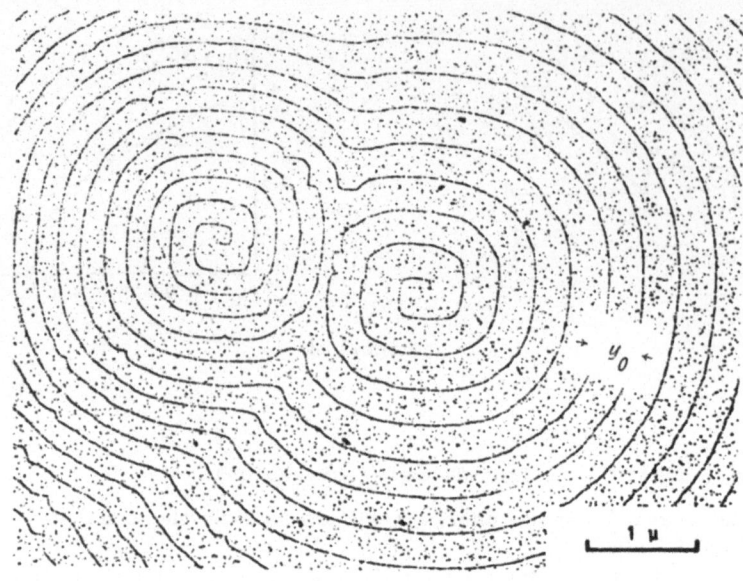

Figure 10

The growth rate of a flat face (hkl) result in:

$$R_{hkl} = \frac{v_T d_{hkl}}{y_o} .$$ (24)

The question arises about the calculation of V_T.

How does V_T depend on the surface diffusion of ad-atoms? What is the role of diffusion of growth units through the volume around the growing face?

Shortly speaking, are we able to predict the shape of the growth isotherms ($R_{hkl}, \Delta\mu$) for a given (hkl) flat face on the ground of a given mechanism?

The theory of Burton, Cabrera and Frank (BCF) (1951) represents up to date the most comprehensive answer to these questions, even if during the last 30 years many improvements have been reached on these topics.

Let us consider the elementary processes described in Figure 11. An atom coming from a supersaturated vapor impinges on to the surface after a diffusion path in the vapor (diffusion coefficient D_v). The step existing on the surface creates a surface diffusion field around it, through the concentration gradient of adsorbed atoms (diffusion coefficient D_S). Ad-atoms statistically move towards the step, describing a mean free path x_s, before desorbing in the bulk phase*.

If the diffusion atom encounters a kink on the step, before its desorption, it may be captured in the step which advances for a length a.

---

* $<x_s^2>$ = $D_s \tau_s$ is the Einstein's formula in which $D_s = a^2 \nu \exp(-\Delta G_{sd}/kT)$, $\tau_s = \nu \exp(E_{sv}/kT)$ mean life of an ad-atom before evaporated again into the vapor. ΔG_{sd} and E_{sv} are quoted in Figure 12.

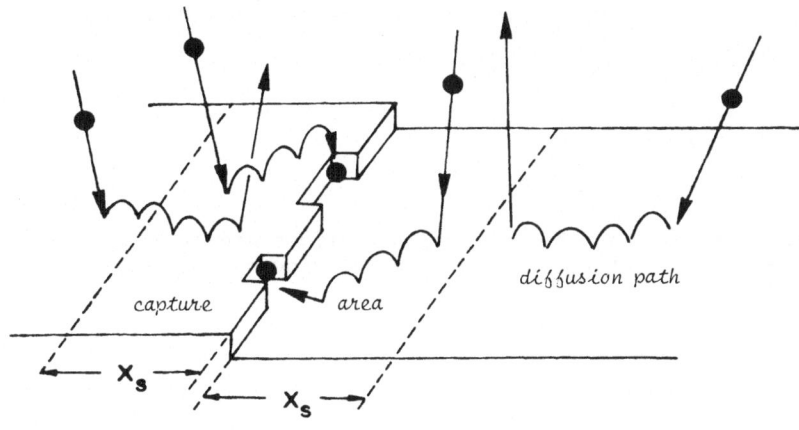

Figure 11

The two diffusion paths (in the bulk and onto the surface) are consecutive and the kinetics are determined by the slowest process.

In the vapor phase $D_V \sim 10^{-5}$ cm^2 sec^{-1} and $D_S \sim 10^{-10}$ cm^2 sec^{-1}. BCF considered that a reasonable assumption is to neglect the path in the bulk phase and then to limit the problem to the surface diffusion, as the rate determining process.

4.2.1 Surface diffusion to the steps: BCF theory

I. <u>A single step on the surface.</u> With reference to Figure 12, we define

$$\sigma_v = \beta_v - 1 = (p/p_e - 1) \quad \text{bulk supersaturation}$$

$$\sigma_S = \beta_S - 1 = (n_s/n_{se} - 1) \quad \text{surface supersaturation}$$

where n_s, n_{se} are the actual and equilibrium concentrations of the adsorbed atoms respectively.

The current on the surface, of diffusing atoms, will be:

$$j_s = -D_s \, \text{grad} \, n_s = n_{se} D_s \, \text{grad} \, \psi \tag{25}$$

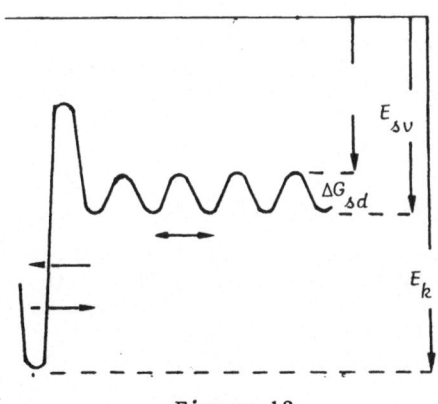

Figure 12

where $\psi = \sigma_V - \sigma_S$, and the matter flowing from the bulk to the surface is

$$\Phi_v = (\sigma_v - \sigma_s)n_{se}/\tau_s = n_{se}\psi/\tau_s \tag{26}$$

under the assumption that the step movement can be neglected in the diffusion problem, BCF assessed the continuity equation on the surface:

$$\text{div } j_s = \Phi_v \rightarrow <x_s^2>\nabla^2\psi = \psi. \tag{27}$$

Remembering that the boundary conditions are σ_S (step) = 0, $\sigma_S = \sigma_V$ very far from the step, and that the solution must be symmetrical with respect to the step, one obtains:

$$v_{step} = 2x_s \nu exp(-E_k/kT)\sigma_v \tag{28}$$

which shows that the advance of the step is due to the atoms condensing from the vapor on a strip of width x_s at both sides of the step ($2x_s$ = = capture area).

II. A step train on the surface. A growing spiral may be represented by a sequence of parallel and equidistant (y_o) steps (Figure 10).

In this case the solution of the continuity equation (27) must take into account the periodicity of the boundary conditions:

$$\psi = \sigma_v \frac{\cosh(y/x_s)}{\cosh(y_o/2x_s)} \tag{29}$$

and the train velocity V_T will be:

$$v_T = 2x_s \nu exp(-E_k/kT)\sigma_v \tanh(y_o/2x_s) \tag{30}$$

which reduces to (28) when $y_o \rightarrow \infty$ (a single step on the surface).

Remembering that, for low σ_V values, ($\sigma_V < 1$) $\Delta\mu = kT \ln\beta = kT \ln(\sigma_V + 1) \sim kT\sigma_V$ and that $y_o = A\rho a^2/\Delta\mu$, putting $\sigma_1 = A\rho a^2/kTx_s$ expression (24) when substituting expression (30) for V_T, can be written down:

$$R_{hk1} = C \frac{\sigma_v^2}{\sigma_1} \text{Tanh} \frac{\sigma_1}{\sigma_v}. \tag{31}$$

Two interesting limiting cases must be considered to get an impressive view of (31):

i)   very low supersaturations: $\sigma \ll \sigma_1 \rightarrow y_o \gg 2x_s$

(parabolic law) $R_{hk1} \sim (C/\sigma_1)\sigma_v^2$ \tag{32}

ii)   high supersaturations: $\sigma \gg \sigma_1 \rightarrow y_o \ll 2x_s$

(linear law) $R_{hk1} \sim C\sigma_v$ \tag{33}

Parabolic and linear laws intersect at $\sigma = \sigma_1$.

In the i) case, when the steps are very far from one another, no superposition occurs between the diffusion fields (capture areas) around the steps, and then each step advances, independently of its neighbors (case of a single step $\rightarrow$ 4.2.1 I).

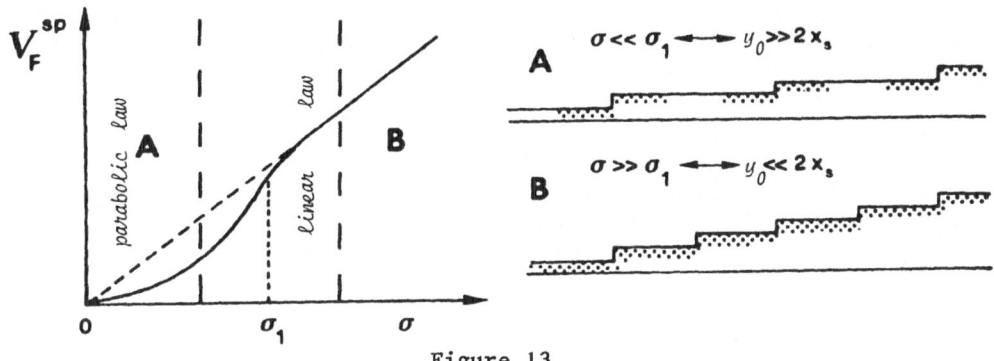

Figure 13

In the ii) case, when the steps approach, due to the supersaturation increase, the capture areas overlap and then any growth unit arriving on the surface has a chance of entering one of several steps.

It must be pointed out that the parabolic region of the (R, σ_V) curve will be shifted to lower values of σ_V and the curve linearized if σ_1 becomes small. This is possible if ρ is small (foreign atom adsorption on the ledges of steps) or large x_s. We may conclude that the value of σ_1 is a measure of the interaction of diffusion fields, so determining the shape of (R, σ_V) isotherms (Figure 13).

4.2.2 The volume diffusion as the rate determining step: the GGC model (Gilmer, Ghez and Cabrera) [10]

When considering the growth from solution, we must remember that from a qualitative point of view there is no essential difference with respect to the growth from vapor: nevertheless a quantitative theory of the rate of growth from solution is much more difficult.

Three paths of solute growth units are to be taken into account:

a)   direct diffusion of units from the bulk to the kinks (through a diffusion layer δ);
b)   diffusion over the surface;
c)   diffusion along the step-ledge (before the growth unit enters a kink).

It is difficult to determine the relative importance of these three currents.

BCF gave a simplified solution of this problem: they supposed that the contributions from the surface and ledge diffusions can be neglected [9] and the only remaining path is the one of a growth unit in the bulk of the mother phase.

Later on Gilmer, Ghez and Cabrera (GGC) having neglected the ledge diffusion path, treated the coupled volume and surface diffusion equation. A simplified calculation was proposed by Bennema and Gilmer [11] and we limit ourselves to give the final result and its interpretation. The growth rate of the face is

$$R = \frac{\sigma_v \Lambda N_o \Omega}{\tau_{desolv}} \left\{ \frac{\delta \Lambda}{\tau_{desolv} D_v} + \frac{y_o}{2x_s} \coth \frac{y_o}{2x_s} + \frac{\tau_k}{\tau_{de-ads}} \frac{y_o}{2a} \right\}^{-1} \tag{34}$$

63

where Λ is the mean free path of a growth unit in the solution, N_o the concentration of growth units (cm^{-3}), τ_k relaxation time of a growth unit for entering the step from the surface, Ω the volume of a growth unit in the crystal, and

$$\tau_{desolv} = \nu^{-1}\exp(\Delta G_{desolv}/kT); \quad \tau_{de-ads} = \nu^{-1}\exp(\Delta G_{de-ads}/kT$$

The largest of the three terms (the largest impedance for growth) in (34) determines the rate limiting process.

a) If the first term dominates, it means that the volume diffusion process is the slowest and limits the growth rate. In this case (34) reduces to

$$R = \frac{D_v}{\delta} N_o \frac{\sigma_v}{(\sigma_v + \sigma^*)} \tag{35}$$

where the constant $\sigma^* = \sigma_1 \Lambda \tau_{desolv} (\delta\tau_{vol.diff.})^{-1}$: i) At low σ values ($\sigma_v < \sigma^*$) we obtain a parabolic law:

$$R \sim \frac{D_v}{\delta\sigma^*} N_o \Omega\sigma_v^2$$

ii) At higher σ values ($\sigma \gg \sigma^*$) a linear law is

$$R = D_v\delta^{-1}N_o\Omega\sigma_v$$

b) If the second term in (34) is the largest, then the surface diffusion and/or entry into the adsorbed layer is rate limiting and the equation reduces to the classical BCF relation (31) (surface diffusion model).

c) If the third term is the largest, then the direct incorporation of growth units in the kinks is the rate determining process and (34) reduces to

$$R = C'\sigma_v^2 \quad \text{(second parabolic law)} \tag{36}$$

where $C' = \dfrac{N_o\Lambda}{\tau_k'} \dfrac{kT}{\rho} \Omega$, τ_k' being the relaxation time for entering the step from the solution bulk.

4.2.3 The volume diffusion as the rate determining step: Chernov's model

Chernov [12] (1961) proposed a theory of crystal growth from solution based on a calculation of the flow to a system of parallel steps, assuming no surface or ledge diffusion (Figure 14).

The calculation gives a similar result for the growth rate to the model mentioned above. For low σ values a quadratic law is predicted and a linear one at high σ values. But, the linear law does not pass through the origin, intersecting the R axis in the negative part (Figure 15).

Within this model the slope depends on face orientation only in the non-linear region. In the linear region the straight lines $R(\delta)$, for different faces, and identical δ, have identical slopes.

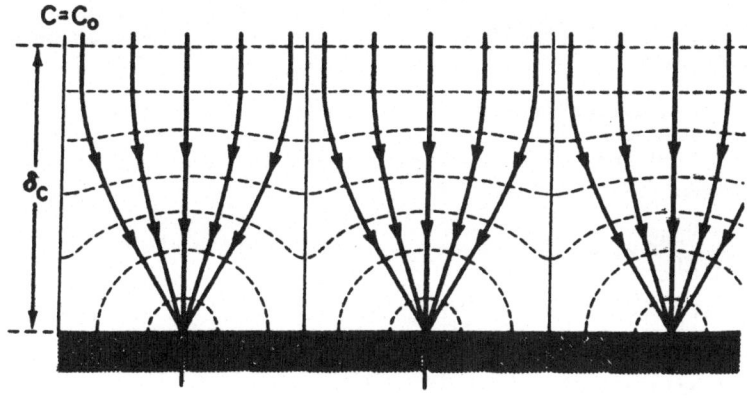

Figure 14

This should not occur if particles adsorbed on crystal faces play an essential role in the growth process (surface diffusion).

Therefore experimental information concerning the anisotropy of dislocation-induced growth rates at different supersaturations could be important in studying the effect of self-adsorption and surface diffusion on the kinetics of growth from solution.

5. HOW CAN YOU DETERMINE THE GROWTH MECHANISM(S)
   FROM GROWTH ISOTHERMS (R, σ_V)?

When, during a growth experiment, it is not possible to check the microtopography of the growing surface, it is sometimes difficult to assess the growth mechanism. Nevertheless a good way to do it is to measure (R, σ_V) curves at different temperatures of crystallization.

Let us suppose that, at each T_c, you get a certain $R = b\sigma_V^n$, where b is the kinetic coefficient depending on T_c. The exponent n is easily

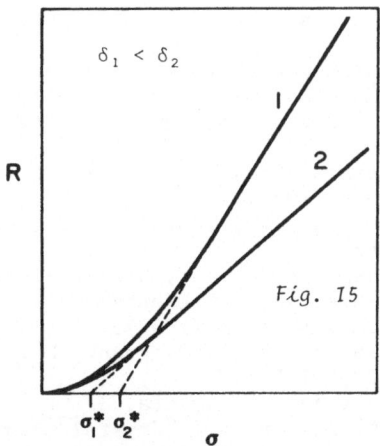

Figure 15

obtained from $(\ln R, \ln \sigma_v)$. If $n = 2$ the function $(R^{1/2}, \sigma_v)$ allows to determine the kinetic coefficient b and its standard deviation with a good precision.

You must consider $\ln b$ as a function of $(1/T_c)$ and hence you can control the self-consistency of the growth law with respect to the temperature, and determine the experimental value of the crystallization enthalpy (E_{exp}), remembering that

$$- R \frac{\partial \ln b_{exp}}{\partial (1/T_c)} = E_{exp}. \tag{37}$$

From all possible $R = b\sigma_v^n$ coming from the different theoretical mechanisms (BCF surface diffusion, BCF volume diffusion, GGC coupled surface and volume diffusion, Chernov isotherms ...), you may obtain the theoretical value:

$$- R \frac{\partial \ln b}{\partial (1/T)} = E_{th} \tag{38}$$

Finally you compare E_{exp} with E_{th} and choose the mechanism whose E_{th} gives the best fit with E_{exp}.

A last consideration: as it may be seen from different kinetic laws we mentioned in Sections 3 and 4, the kinetic coefficient b is a function of equilibrium and kinetic parameters which depend on the growth mechanism and whose dependence on T may be expressed as $\sim \exp(\Delta G_i/RT)$; ΔG_i may refer either to a variation of the Gibbs free energy between equilibrium states or to an activation energy. In the simplest cases the activation energy for crystallization (E_{th}) depends both on T and ΔH_i values corresponding to the processes describing a particular growth mechanism.

The possibility to get the calculation of E_{th} depends on the experimental system and on the availability of a "good" model of adsorbed layer. A good review on the application of crystal growth theories to crystallization from solution is given by J. Garside [13] in "Current Topics in Materials Science" vol. 2, E. Kaldis, ed., North-Holland (1976).

REFERENCES

1. P. Hartman and W. G. Perdok, Acta Cryst., 8:49 (1955).
2. I. N. Stranski and R. Kaishew, 8, Physik. Chem., B26:100,114,312 (1934).
3. B. Mutaftschiev, in: "Adsorption et Croissance Cristalline", Colloques Internationaux du CNRS, n. 152:231 (1965).
4. M. Volmer, Kinetik der Phasenbildung, Th. Steinkopf Verlag, Dresden, Leipzig (1934).
5. S. Toschev, in: "Crystal Growth. An Introduction", P. Hartman, ed., North-Holland, 328 (1973).
6. I. N. Stranski, Z. Kristall., 105:91 (1943); R. Kaischew, Ann. Univ. Sofia, 10:2 (1943).
7. R. Becker und Doring, Ann. Physik, 24:719 (1935).
8. J. Weertman and J. Weertman, "Elementary Dislocation Theory", Mcmillan Series in Materials Science (1964).
9. W. K. Burton, N. Cabrera and F. C. Frank, Phil. Trans. Roy. Soc., (London), A243:299 (1951).
10. G. H. Gilmer, R. Ghez and N. Cabrera, J. Crystal Growth, 8:79 (1971).
11. P. Bennema and G. H. Gilmer, in: "Crystal Growth. An Introduction", P. Hartman, ed., North-Holland, 263 (1973).
12. A. A. Chernov, Soviet Physics Uspekhi, vol. 4, n. 1:116 (1961).

13. J. Garside, in: "Current Topics in Materials Science", vol. 2, p 484,
       E. Kaldis, ed., North-Holland (1977).

BIBLIOGRAPHY

To get a better and deeper understanding on growth mechanisms, some
fundamental books may be suggested to the reader:

P. Hartman, "Crystal Growth. An Introduction", North-Holland (1973).
M. Ohara and R. C. Reid, "Modeling Crystal Growth from Solution", Prentice-
     Hall Int. (1973).
B. Pamplin, "Crystal Growth", Pergamon Press (1975).
D. Elwell and H. J. Scheel, "Crystal Growth from High-Temperature
     Solutions", Acad. Press (1975).
R. Ueda and J. B. Mullin, "Crystal Growth and Characterization", ISSCG2,
     North-Holland (1974).
B. Mutaftschiev, "Interfacial Aspects of Phase Transformations", NATO
     Advanced Study Institutes Series, D. Reidel Publ. Co. (1982).

FUNDAMENTALS OF MELT GROWTH

J. J. Favier and D. Camel

CEA/IRDI/DMECN/DMG/SEM
Laboratoire d'Etude de la Solidification
Centre d'Etudes Nucleaires
85X, 38041 Grenoble Cedex, France

PART I - THE IMPORTANCE OF SOLIDIFICATION

Solidification is and will remain for many years to come the most common method of material preparation.

From time immemorial, man has continually searched for ways of improving the performance of the materials he needs. Real innovation occurred as soon as he had mastered the rudiments of preparing low-melting point alloys (Copper Age), followed by the development of foundry techniques (introduction of the first low-hearth furnaces) and high melting point alloy preparation techniques (Iron Age). The science of metallurgy was born.

Today, metallurgical processes still represent the principal method of preparation of materials. Although metallurgy has been practised for centuries, it is only over the past hundred years or so that it has been transformed into a truly industrial activity. Indeed, it was only at the beginning of the 20th century that metallurgy was really to become a science.

Most of the techniques used in recent decades, and their subsequent development, were based on know-how, with continual refinements being provided by experimental workers. Today, while most of the phenomena governing the control and monitoring of material structures are fully understood, solidification remains a subject where many questions are still unanswered. These are of a more general nature than the simple production of alloys or crystals and encompass the wider field of modern solid-state physics. How can the size of dissipative structures caused by the instability of a solidification front be foreseen? No answer has yet been found to this question which is of similar nature to that of predicting the shape of combustion flame fronts, or the formation of patterns in the physical chemistry of reactive media.

However, recent advances in the theory of solidification have led to the design of industrial crystallogenetic facilities for producing high-quality crystals such as silicon without structural defects or chemical heterogeneities. Thanks to this considerable progress, it is now possible to consider constantly, new applications in electronics. Strict control of solidification processes is leading to a new industrial revolution and it is clear that the next stage will involve the control of growth of more complex crystals such as GaAs with a significantly higher performance.

The present book is aimed precisely at closing the gap separating fundamental crystal growth from the control of preparation techniques. The purpose of this presentation on the fundamental aspects of melt growth is to show that the solidification process is governed by mechanisms at different scales (atomic scale at the interface, cf. Part II, macroscopic scale in the vicinity of the interface, diffusion fields and the liquid sample or casting scale, convection effects, cf. Part III), and also by the coupling of these mechanisms, as encountered in morphological interface stability problems, cf. Part IV. All these mechanism studies have experienced rapid development in recent times although more stress appears to have been placed on demonstrating their relative importance than in giving an exhaustive explanation.

PART II – MICROSCOPIC ASPECTS OF CRYSTAL GROWTH

There are currently two types of theoretical approaches to the atomic behavior of an evolving solid–liquid interface: a) structural approaches, which endeavor to describe precisely the structure of the interface in order to predict its mode of displacement, and b) kinetic approaches based on more general assumptions respecting the interface structure but attempting to establish a growth model compatible with the conditions imposed on the system. At the same time a relationship between the interface advance rate and the applied thermodynamic driving force is deduced.

II.1 THE STRUCTURAL APPROACH

The main aim of these theories is to study the structure that the interface is likely to adopt under given solidification conditions. The underlying idea is quite simple, and was developed initially by Jackson [1] for the case of a pure substance (Figure 1a) and an equilibrium condition. The interface, assumed to be a single layer, is characterized by a degree of overlap θ, which is a proportion of solid "adatoms" on the interface layer. Jackson calculates the free energy of the system formed by a part of the liquid, the interface and a part of the solid as a function of θ when the two bulk phases are in thermodynamic equilibrium. Mutaftschiev [2] performs the calculations for the case where a given driving force $\Delta\mu^{\ell s}$ is applied to the interface, that is to say when there is a difference in chemical potential between the two phases (Figure 1b). Depending on whether or not there are stable equilibrium states (for Jackson) or metastable equilibrium states (for Mutaftschiev) with low values of θ, the interface may be considered as smooth or rough. The existence of metastable states depends on the value of an energy parameter referred to as α by Jackson and F by Mutaftschiev

$$\alpha = F = \frac{\ell n}{2} \frac{\Delta\phi}{kT} \geqslant 2 \qquad (I.1)$$

where ℓ is the number of direct neighboring atoms on the interface plane compared with the total number of direct neighboring atoms in the crystal; n is the number of direct neighboring atoms of an atom within the crystal; and k is Boltzmann's constant. $\Delta\phi = \phi_C + \phi_\ell - 2\phi_{C\ell}$ is the difference in interatomic bonding energy, respectively: ϕ_C in the crystal, ϕ_L in the liquid and $\phi_{C\ell}$ between a "liquid" atom and a "solid" atom.

70

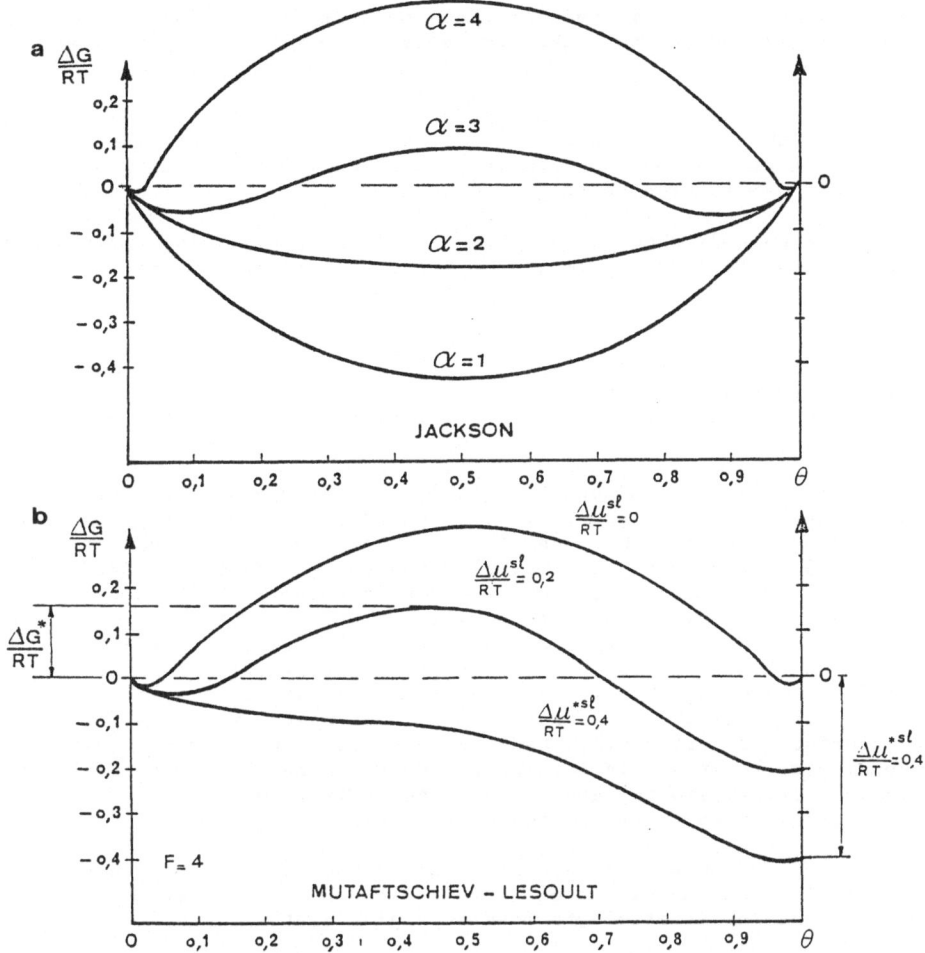

Fig. 1. Free energy of the liquid/interface/solid system as a function of
the degree of overlap θ.
1.a. Jackson's model [1].
1.b. Mutaftschiev's model [2].

The interface morphology may be obtained by means of the previous
expression provided that $\Delta\phi$ can be evaluated.

Jackson assumes that the liquid wets the crystal perfectly (hence
$\phi_L = \phi_{CL}$) and therefore proposes:

$$\Delta\phi = \phi_C - \phi_L = \frac{Lf}{N_{Av}} \qquad (I.2)$$

based on the latent heat of fusion. Mutaftschiev has shown that Lf is
often a very poor approximation of $\Delta\phi$. Eustathopoulos and Desre [3] have
proposed evaluating $\Delta\phi$ from the liquid-crystal interfacial tension,
demonstrating the importance of the "wetting" term. Lesoult et al. [4]
have follwoed Mutaftschiev's approach in the case of binary solid
solutions. Their formulation is identical. It is difficult to evaluate
$\Delta\phi$ in the case of solid solutions as this involves pair interchange
energies, which are known only approximately. It was possible to make an

evaluation in case of diluted solutions by accepting certain approximations that do not take into account possible adsorption at the interface.

Jackson's theory is of limited interest. it is based on very rough hypotheses regarding the energy characteristics of the interface and provides information only on the interface equilibrium structure, and is hence of no use in predicting growth.

Mutaftschiev's theory [3] gives one major result. The interface can always be transformed from smooth to rough by applying sufficient driving force, even if the faceting factor F is greater than 2.

Temkin [5] gives a more precise description of the interface by assuming that it consists of several layers, and he demonstrates that there is a two-way transition. There is thus a critical value of $\Delta\mu^{\ell s}$ that will cause the structure of the interface to change from one type to the other.

For Mutaftschiev [3] and Lesoult [4], the foregoing results may be used to predict crystal growth modes, but only with extreme precaution. These authors consider that when F is greater than 2, the peak of the curve $\Delta G(\theta)$ is the energy barrier that has to be overcome in order to fill an atomic layer completely, i.e., to cause the interface to move an interatomic distance. When this barrier exists, growth is difficult. In contrast, when F is less than 2, there is no barrier, and when F is greater than 2 and the applied driving force is greater than the critical value $\Delta\mu*$, the energy barrier disappears and growth is easier. It is hardly possible to affirm more than this. At most, it may be felt that if the interface is very rough, then a general movement of the interface perpendicular to itself is possible, and growth may be "normal"; if the interface is very smooth, growth can occur only laterally. No conclusion may be reached with regard to the other cases.

The major drawback of these structural theories is that they provide no information on the extent of non-equilibrium required at any instant to cause any type of interface to move at a given rate.

II.2 KINETIC APPROACH

This drawback is overcome by the kinetic growth theories. So far, however, discussion has concentrated on "physical" models, i.e., models that are dependent on the atom stacking model. Developed for vapor-phase growth, they may be partially applied to molten-bath growth. Here again, two extreme cases call for specific models.

Wilson [6] and Frenkel [7] assume that all "liquid" atoms hitting the interface become "solid". This is tantamount to considering all interface sites as equivalent and thus having the crystal grow at any point on its surface. The energy arrangement proposed by Wilson and Frenkel is as follows (Figure 2): solid and liquid atoms at the interface have different energy levels. To change from liquid to solid, atoms must pass through an activated state. By virtue of thermal agitation alone, liquid atoms are able to overcome the activation barrier, which is very weak. This model produces a linear relation between the overall rate of interface movement (R) and the deviation from the interface equilibrium temeprature ΔT_k, at least provided that ΔT_k is not too great;

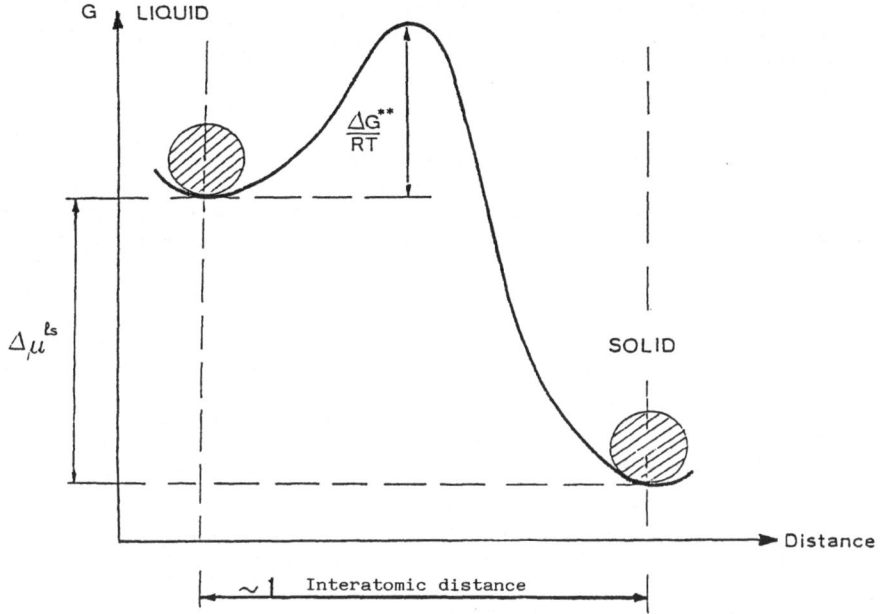

Fig. 2. Energetic scheme of the solid/liquid interface in the Wilson and Frenkel model [6,7].

$$R = \mu_1 \Delta T_k \qquad (I.3)$$

where μ_1 is the kinetic constant.

Volmer [8] and Becker and Döring [9] have adopted an opposite view. They have developed a model which assumes that the liquid-solid interface consists of a completely filled atomic layer with no point defects. For any progression to occur, a two-dimensional nucleus must form on the interface. This will then give rise to an entire new atomic layer. The same problem of nucleation arises for each layer. In this case, the interface rate of advance depends chiefly on the stage of nucleation, the driving force required for the layer to grow from the nucleus being negligible in comparison with that needed to form the nucleus itself. When the nucleation rate is evaluated on the basis of the homogeneous nucleation theory, a relation is used involving an exponential variation of the deviation from equilibrium. The kinetic relationship is written in the form:

$$R = \mu_2\, e^{-(\mu_3/\Delta T_k)} \qquad (I.4)$$

in which μ_2 and μ_3 are two constants.

From a physical point of view, it is obvious that this type of growth is not easy if there is a slight deviation from equilibrium. Because of this, the least defect on the interface may encourage the system to change, with less driving force being needed. This is the case, for example, with growth that occurs from dislocations or contiguous twin crystals. These defects may act as repeated and non-saturable sources for the incident atoms and thus bypass the nucleation phase. Other types of relation, for example parabolic ones, are thus obtained.

While these physical models give rise to kinetic relations, they do not define the experimental conditions in which they would be valid for a given material and in particular for an applied driving force.

Cahn [10] has attempted to answer this question by drawing up a phenomenological theory of crystal growth. Cahn [11] characterizes any moving interface by means of its degree of roughness and by a kinetic parameter, and states that it may not move from a block perpendicular to itself if the applied driving force is sufficiently weak. He thus clearly identifies the principle of interface resistance. For every substance, there is a critical driving force below which crystallization occurs by means of a lateral growth mechanism and beyond which growth is normal.

This is to be compared with the results obtained by Mutaftschiev [3] and Temkin [5], who also proved the existence of a critical driving force which caused the interface structure to alter from smooth to rough. The question is, whether the same critical driving force is involved in both cases? So far, this question cannot be answered. All that can be affirmed is that the existence of a critical driving force is, to date, the only link between the two atomic growth theories.

If research into this aspect of growth is progressing only very slowly, it is partly because there are enormous difficulties involved with experimental observations of the interface on such a small scale, as is the case with kinetic law measurements, particularly in metals. The theories cannot therefore be backed up with experimental evidence.

II.3 THEORY OF PLANAR FRONT SOLIDIFICATION IN ABSENCE OF CONVECTION

In this section solidification in the absence of gravity is considered.

When liquids freeze, a wide variety of structures can be formed. Theoretical models are available for some of the steady-state or nearly steady-state structures. A solution must be obtained to the partial differential equations describing heat and solute flow, which also satisfy certain conditions at the solid-liquid interface. Since the limitations of the models rest on the approximations involved, the equations and conditions will be briefly considered.

II.3.1 Interface Conditions

When a liquid freezes, the interface temperature T_I is related to the interface liquid composition C_I, the surface curvature and the departure of the interface from local equilibrium. This may be expressed as an undercooling equation [12]:

$$T_I = T_o - \Delta T_s - \Delta T_c - \Delta T_k \tag{II.1}$$

where T_o is a convenient reference temperature, ΔT_s, ΔT_c and ΔT_k are the undercooling values resulting respectively from solute, curvature and kinetics.

If T_o is the freezing temperature for an alloy of composition C_o:

$$\Delta T_s = m(C_o - C_I) \tag{II.2}$$

where m is the liquidus slope. The local liquid composition C_I departs from the bulk composition because solute is rejected at the solid-liquid interface. The curvature undercooling ΔT_c arises because the shape of the

74

interface defines a particular surface energy term. It is given for isotropic energies by:

$$\Delta T_c = A \ (1/r_1 + 1/r_2) \tag{II.3}$$

where A is the Gibbs Thomson coefficient, r_1 and r_2 are the principal radii of curvature. The kinetic undercooling term ΔT_k arises because of non-equilibrium effects ($\Delta T_k \to 0$ when the velocity $R \to 0$). For non-faceting materials, this term is considered to be small and reasonably isotropic:

$$\Delta T_k = BR \tag{II.4}$$

where B is a constant ($B = 1/\mu_1$ see Eq. I.3). For a faceting material, the normal velocity of a facet depends on the rate of production of steps on the face and thus depends in a much more complex fashion on the local undercooling kinetics.

Most of the discussion which follows is concerned with non-faceted growth, thus ΔT_k is generally neglected except for very rapid growth rates. In non-faceted growth ΔT_k and ΔT_c should be reasonably isotropic. However, certain crystallographic features can be explained only by including a slight anisotropy in these terms.

Both heat and solute are rejected at the solid-liquid interface leading to the final two interface equations. For solute:

$$R(C_s - C_T) = D_{Lc} \ \frac{\delta C}{\delta n} - D_{Sc} \ \frac{\delta C}{\delta n} \tag{II.5}$$

where D_{Lc} and D_{Sc} are the liquid and solid diffusion coefficients, n is the normal to the interface. C_s is the solid composition and, for low velocities, it is expected to be given by C_S where k is the equilibrium distribution coefficient. During fast growth, such as in splat quenching, C_s may be very different from this value (i.e., $k = f(R)$).

For heat flow, a similar equation applies:

$$RL = K_s \ \frac{\delta T}{\delta n} - K_L \ \frac{\delta T}{\delta n} \tag{II.6}$$

where L is the latent heat per unit volume, K_s and K_L are the conductivities.

II.3.2 Heat and Solute Transport

In the simplest situation, where there is no fluid motion, heat and solute flow occurs by conduction and diffusion [13]. The equations:

$$D_{iT} \nabla^2 T = \frac{\delta T}{\delta t} \tag{II.7}$$

and

$$D_{iC} \nabla^2 C = \frac{\delta T}{\delta t} \tag{II.8}$$

must be solved in both the solid and liquid phases (t is time, D_{iT} and D_{iC} the thermal and solute diffusivities in the relevant phases). A solution consists in solving (II.7) and (II.8), satisfying equations (II.1), (II.5) and (II.6). Provided the shape of the interface is allowed to evolve freely, actual growth structures should be produced.

Because of the difficulty in carrying out time-dependent analyses, many solidification processes are assumed to be steady-state. Clearly, planar front, cellular and lamellar eutectic growth are nearly steady-state. Other structures, such as dendrites, may be considered to be approximately steady-state, at least in the vicinity of the tip. For steady-state growth, equations (II.7) and (II.8) may be transformed to coordinate moving with the interface. The equations then become [14]:

$$D_{iT}\nabla^2 T + R\,\frac{\delta T}{\delta x} = 0 \tag{II.9}$$

and

$$D_{iC}\nabla^2 C + R\,\frac{\delta C}{\delta x} = 0 \tag{II.10}$$

where x is the steady-state growth direction. A steady-state solution thus consists in solving (II.9) and (II.10) for an interface shape which satisfies equations (II.1), (II.5) and (II.6).

When fluid motion occurs, the complex effect of convection should be included in the transport equations. This is extremely difficult, so that much of the work in the past has ignored convection. The effect of fluid motion will be considered elsewhere.

II.3.3 Eutectic Growth

II.3.3.1 Regular structures. Regular lamellar or rod-like structures are formed in eutectics where both phases grow in a non-faceted fashion. Lamellar or rod-like solidification processes are other examples of almost steady-state growth. During eutectic growth, the solute rejected from one phase is transported across the interface to the other.

Again, the steady-state equations (II.9) and (II.10) have to be solved for an interface shape which also satisfies (II.1), (II.5) and (II.6). Steady-state solutions have been obtained (e.g., [12,15]). Undercooling ($\Delta T = T_0 - T_I$, where T_0 is in this case the eutectic temperature) is plotted against lamellar spacing in Figure 3. At small spacings, the undercooling is large because of the curvature undercooling and at wide spacings it is large because of the larger diffusion distances affecting ΔT_s in equation (II.1). Again, an additional condition is necessary to define the growth conditions. Early work suggested an

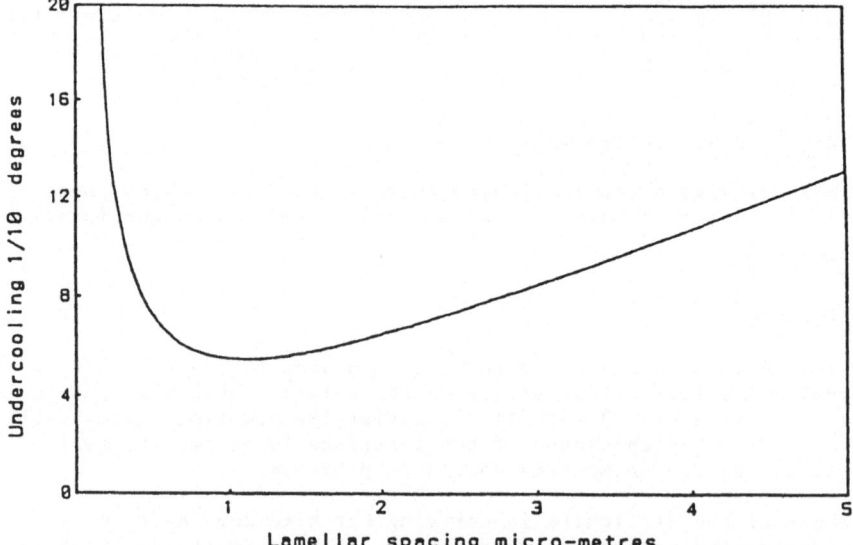

Fig. 3. A plot of interface undercooling as a function of lamellar spacing.

optimizing condition, in this case minimum undercooling [15]. In other work, it was suggested [12] and later shown [16] that there is a maximum spacing above which an individual lamella becomes unstable. This leaves a small range of possible spacings just above the spacing corresponding to the minimum undercooling. By considering the spacing change mechanism [17], it was concluded that lamellae grow just below the minimum undercooling value. Thus, in the case of the eutectic, consideration of stability in fact leads to the optimizing conditions. Experimental work [18,19,20] indicates that the theoretical predictions are obeyed rather well in practice and that the spacing very nearly corresponds to the minimum undercooling spacing.

In regular eutectic growth, the solute layers are similar to those for array dendrites. There is a layer which decays in approximately a lamellar spacing away from the interface, and an exponential layer which decays with D_L/R [12]. This spacing is never much greater than a few microns, with the result that, usually, $D_L/R \gg \lambda$. If the exponential layer is disturbed by convection, the average solid composition may be changed and the fluid flow might be expected to have a minor effect on the lamellar spacing. This aspect is to be considered in the next section. It is perhaps worth pointing out that by a suitable choice of alloy composition and growth direction, it should be possible to stabilize the exponential layer on earth. The eutectic temperature varies little with alloy composition so that eutectics do not suffer from a steeple-like instability. If the exponential layer is stabilized, the lamellar spacings are probably small enough so that errors induced by neglecting fluid flow are probably small, even at low growth rates.

II.3.3.2 <u>Irregular eutectics</u>. Irregular structures are usually formed when at least one phase is faceted. Although the structures are not steady-state, it is suggested that a plot of undercooling against spacing should be similar to that shown in Figure 3. It is found in practice that the undercooling and spacings are much larger than the minimum undercooling spacing. This is explained [21] by suggesting that lamellar plates grow penetrating each other, thus making the structure coarser.

The final spacing is then a measure of the ability of the structure to branch or form new plates.

In irregular eutectics, the spacings are generally much wider than in regular eutectics. As opposed to regular eutectics, the spacing is also affected by temperature gradients (e.g., [21]). Any convection in the exponential layer will thus have a marked effect on both spacing and segregation.

II.3.4 <u>Ripening</u>

When finely-divided solids (e.g., secondary dendrite arms) are held in a liquid at a constant temperature, it is found that the average particle size d increases according to:

$$d^3 - d_o^3 = \alpha t \qquad\qquad (II.11)$$

where d_o is the average spacing at time $t = 0$ and α is a constant. Coarsening occurs because it reduces the total surface energy. Concentration gradients exist between the bulk liquid and the liquid in contact with the solid when the particle has other than the average spacing. This is apparent from equation (II.1) where, if the temperature is constant, $T_L = T_o$ and since the velocity is low, $\Delta T_K \to 0$, then $\Delta T_S = \Delta T_C$. This gives:

$$m(C_o - C_L) = 2A/r \qquad \qquad \text{(II.12)}$$

The problem has been treated assuming that only particles growing at the maximum growth rate exist [22] and secondly that an almost steady-state distribution of sizes exists [23]. Both analyses give expressions of the correct form, the latter being more accurate.

During dendritic growth coarsening occurs with similar but more complex kinetics than given in equation (II.11).

PART III - DIFFUSION LAYER, INFLUENCE OF CONVECTION

III.1 INTRODUCTION

Owing to the low chemical diffusivity values D_s involved (i.e., low with respect to the thermal diffusivity values of metals), solute transport in liquids associated with solidification has two common features:

- firstly, it occurs over time scales that are usually shorter than those of the liquid bath and may vary considerably.
- and secondly, it is easily affected by convection in a number of different ways.

A number of these convective effects have already been identified. Some may cause serious defects in the products (macro- and micro-segregation in single-crystals [24-33] or ingots [34-35]), while others make it impossible to check the growth theories (modification of growth conditions at the tip of the dendrite or of inter-cellular or inter-dendritic spacing [29,36]). However, it is still uncertain how these experimental parameters should be adjusted in order to determine what effects they have.

To make any progress at all, three additional theoretical tools should be used in conjunction with the results of the experimental studies. These are:

- an order-of-magnitude analysis (or scaling law analysis) in order to obtain an approximate idea of the effects of the various parameters over a large range of corresponding values;
- simplified models, for which analytical solutions may possibly be obtained, and which enable the phenomena to be represented by means of a limited number of parameters;
- finally, in a necessarily limited number of cases, numerical simulations based on complete physical representation.

This section describes some typical experimental data relating to the effect of convection on crystal growth on a flat interface, and on eutectic and columnar solidification. The list is not exhaustive. The application of scaling law analysis to these phenomena will be illustrated, and the quantitative models simply referred to. The basic concepts used in the examples are outlined first of all.

III.2 BASIC CONCEPTS INVOLVED IN SCALING LAW ANALYSIS

III.2.1 Boundary Layer Phenomena

The boundary layer concept may be defined by an order-of-magnitude analysis. This involves not only the relationships between the dimensions

78

Table 1. Order of Magnitude of the Various Terms in the Balance Equations

Type	Diffusive Transport	Convective Transport	Production
Mass	---	$\dfrac{u^*}{x^*}$, $\dfrac{w^*}{z^*}$	---
Solute	$\dfrac{1}{x_c^{*2}}$, $\dfrac{1}{z_c^{*2}}$	$Sc\,\dfrac{u(x_c^*)}{x_c^*}$, $Sc\,\dfrac{w(x_c^*)}{z_c^*}$	---
Momentum	(Viscous forces) $\dfrac{u^*}{x_v^{*2}}$, $\dfrac{u^*}{z_v^{*2}}$ $\dfrac{w^*}{x_v^{*2}}$, $\dfrac{w^*}{z_v^{*2}}$	(Inertia forces) $\dfrac{u^{*2}}{x_v^*}$, $\dfrac{u^*w^*}{z_v^*}$ $\dfrac{u^*w^*}{x_v^*}$, $\dfrac{w^{*2}}{z_v^*}$	(External forces: gravity) Gr

of the values governing a phenomenon (which is the object of dimensional analysis) but also the form of the equations used to solve the problem. The method involves defining a minimum number of reference values with which it is possible to evaluate the mathematical order of magnitude of each term in the equations. In particular, it may be necessary to define specific length scales for lines normal and tangential to a given boundary (respectively X and Z), as well as for the corresponding components of fluid velocity (respectively U and W) [37,38,39]. Table 1 shows the various coefficients associated in this case with the different terms of the balance equations relating to total mass, the mass of a solute (with a concentration c in the liquid) and the momentum. The following are used in this Table:

- lower case letters indicate the dimensionless variables defined by:

$$x = X/H \quad z = Z/H \quad u = UH/\nu \quad v = WH/\nu;$$

- asterisks indicate the scales of the values relating to solute or momentum transport factors (indices C and V respectively);
- Gr and Sc are the standard dimensionless numbers defined by:

$$Gr = \frac{(\Delta\rho/\rho)gH^3}{\nu^2} \qquad Sc = \frac{\nu}{D}$$

- H is the geometric dimension of the liquid bath;
- $\Delta\rho$ is a horizontal density deviation;
- g is the acceleration due to gravity;
- ν is the kinematic viscosity.

When the convection speed is sufficiently high, convective transport prevails over diffusive transport throughout the medium (that is to say if it is assumed that $x^* \simeq z^* \simeq 1$ in the above Table). The diffusion term may therefore be ignored, which reduces the order of the equations. However, general compatibility with the boundary conditions can no longer be guaranteed. There is thus a boundary layer x^* near the interface, where diffusion is not negligible. The extent of x^* is given by:

diffusive transport $\simeq$ convective transport. (III.1)

It can be seen that the amount of convective transport of solute and the extent of the solute boundary layer depend on Sc.

In cases where the interface is flat and a non-slip condition is applied,

$$z^* \simeq 1$$

and in order to conserve mass (in conformity with the continuity equation):

$$u^* = w^* \cdot x^*. \qquad (III.2)$$

The momentum is thus expressed by re-writing (III.1) as follows:

$$w^* \cdot x_v^* = u^* \cdot x_v^* = 1 \qquad (III.3)$$

and, for the solute:

$$Sc \; w(x_c^*) \cdot x_c^{*2} = Sc \; u^*(x_c^*) \cdot x_c^* = 1. \qquad (III.4)$$

When $x \ll x_v^*$, i.e., within the viscous boundary layer, the development of the velocity components as a function of the distance x from the interface gives the following:

$$w(x) = \frac{w^*}{x_v^*} \cdot x$$

$$u(x) = w(x) \cdot x = \frac{w^*}{x_v^*} \cdot x^2. \qquad (III.5)$$

The breakdown of balance equations into normal and tangential components of the various vector quantities may be extended to the case of an interface of random boundary [40,41]. This reveals additional terms involving the local radius of curvature r of the interface (cf. the simple case of Fourier's equation broken down into cylindrical or spherical components). If r is sufficiently small, then:

$$x^* = z^* = r \text{ and } u^* = w^*. \qquad (III.6)$$

III.2.2 The Diffusion Layer during Solidification

In the absence of convection, the barycentric speed in the reference frame associated with the solidification interface is simply:

$$U = -V \text{ or } u = -Re \qquad (III.7)$$

where V is the solidification rate.

If the solidification front is planar, the following equations are obtained by applying equation (III.1):

$$X_c^* = D/V \text{ or } x_c^* = Pe^{-1}. \qquad (III.8)$$

This boundary layer characterizes the extent of the average rejection area in front of a macroscopically flat front, regardless of whether single-phase growth, off-eutectic coupled growth, cellular solidification or dendritic solidification is involved.

In the last two cases, the solidification front is not planar locally. A local rejection area must therefore be distinguished, the extent of which will be of the order of magnitude of the local radius of

Table 2. Scales of X_C^* and Δc for Various Cases of Solidification

	X_C^*	Δc
Planar front:		
Eutectic front:	$\dfrac{D}{V}$	$\dfrac{1 - k_O}{k_O} C_L$
– local rejection:	λ	$\dfrac{V\lambda}{D}(1 - k)C_L$
– average rejection:	$\dfrac{D}{V}$	$C_L - C_E$
Dendritic front:		
– local rejection:	R	$\dfrac{VR}{D}(1 - k)C_L$
– average rejection:	$\dfrac{D}{V}$	$\dfrac{DG}{m}$
– inter-dendritic:	$\dfrac{T_L - T_E}{G}$	$C_L - C_E$

curvature, i.e., respectively the inter-lamellar spacing λ and the radius R of the dendrite tip.

The scale of concentration deviations Δc in the liquid is determined by means of the formula expressing the solute balance at the solidification interface. Locally, this balance is always expressed as follows:

$$D \frac{\partial c}{\partial X} = V(1 - k_o)\, c(0). \tag{III.9}$$

With off-eutectic coupled growth and dendritic solidification, the following respective boundary conditions are considered at macroscopic scale:

$$c(0) = c_E \tag{III.10}$$

$$\left(\frac{\partial c}{\partial x} \right) = \frac{G}{m} \tag{III.11}$$

where G is the temperature gradient and m the liquidus slope.

Finally, in the space between the dendrite arms, the concentration at all points is imposed by the equilibrium condition at the solid-liquid interface.

The scales of X_C^* and Δc are summarized in the form of dimensional quantities in Table 2 above, for the various cases of solidification.

III.3 SEGREGATION OF AN IMPURITY OR DOPANT DURING CRYSTAL GROWTH

The interaction of convection-induced bulk movement with the diffusion layer in front of a planar solidification front causes the various components to be transported by convection. This results in segregation in the solid formed. Typical examples of this mechanism are Bridgman-type solidification (illustrated on Figure 4a) and the Czochralski method of growing doped crystals.

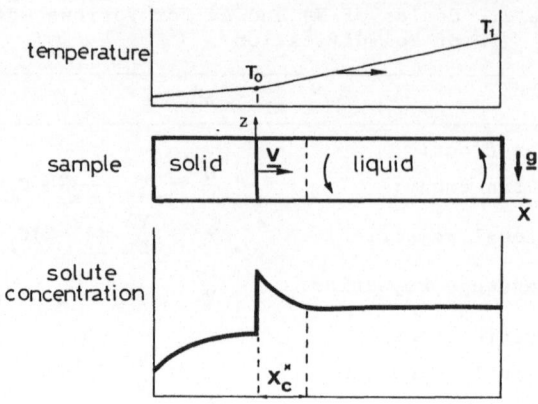

Fig. 4. Diagram of Bridgman-type solidification of an impure liquid.

III.3.1 Underline{General Behavior of Scaling Laws for Solute Boundary Layer (or K_{eff})}

The criterion for defining the boundary layer of a solute may easily be established if liquid movement is not coupled with the concentration field. In the area where boundary layer conditions prevail ($x^*_c \ll 1$, $x^*_v \ll 1$), the criterion is written as follows, taking into account (III.3), (III.4) and (III.5):

$$\begin{cases} Sc\ u(x^*_c) \cdot x^*_c = 1 \\ u(x^*_c) = \max(Re,\ \dfrac{1}{x^{*3}_v} \cdot x^{*2}_c) \end{cases} \qquad (III.12)$$

These are solved to give:

$$\text{or} \quad \begin{aligned} x^*_c &= \min(Pe^{-1},\ x^*_v Sc^{-1/3}) \\ \Delta^* &= Pe\ x^*_c = \min(1,\ Pe\ x^*_v Sc^{-1/3}) \end{aligned} \qquad (III.13)$$

$\Delta^* \simeq 1$ is characteristic of purely diffusive transport in the liquid and $\Delta^* \to 0$ of convective transport.

There is no difficulty in extending this analysis to cases where the viscous and solute boundary layers are not necessarily both at a smaller scale than the smallest dimension of the liquid bath. This point is considered in [7].

Boundary condition (III.9) gives the new concentration scale when convection is present. Given that:

$$\left(\frac{\partial c}{\partial x} \right)_o \sim \frac{C(0) - C(\infty)}{x^*_c} \qquad (III.14)$$

then the following can be deduced immediately:

$$k_{eff} = \frac{k_o}{1 - (1 - k_o)\Delta^*} \qquad (III.15)$$

with, in particular:

$$\text{and} \quad \begin{aligned} k_{eff} &\to 1 \text{ when } \Delta^* \to 1 \\ k_{eff} &\to k_o \text{ when } \Delta^* \to 0. \end{aligned}$$

III.3.2 Segretation Models

From the previous relations it is possible to evaluate the composition of the crystal formed by solidification of a liquid of given composition when convection occurs under assumed steady-state conditions. However, this composition is not constant with time (i.e., longitudinal segregation takes place within the crystal), for two reasons:

- firstly, solidification begins in a liquid of uniform composition, and there is thus a transition stage, corresponding to accumulation of solute in the diffusion layer.
- Secondly, the liquid bath has finite dimensions, which means that removal of solute from the bath by means of convective transport results in progressive enrichment of the remaining liquid (in accordance with a characteristic Scheil-type behavior).

These two phenomena combined produce a general S-shaped longitudinal segregation curve (see Figure 5). Parameters may be defined for these curves by using the same type of model with a stagnant layer [25], which gives a single analytic solution as a function of the two parameters: extent Δ of the stagnant layer and concentration value at the boundary of this layer.

III.3.3 Czochralski Method

With the Czochralski method, the tangential convective speed is imposed by the crystal or crucible rotation speed:

$$w^* = \frac{\omega H^2}{\nu} = Re_\omega. \tag{III.16}$$

The classic result for forced convection may be deduced from (III.3) and (III.16):

$$x^*_v = Re_\omega^{-1/2}, \tag{III.17}$$

(III.13) therefore becomes:

$$x^*_c = \min(Pe^{-1}, Re^{-1/2}Sc^{-1/3}). \tag{III.18}$$

The scaling law for x^* obtained for convection conditions is thus identical to that provided by the classic model with a stagnant layer of the Burton, Prim and Slichter type [24]. However, it should be noted that the solute boundary layers defined by order-of-magnitude analysis (Δ^*) and by the stagnant-layer model (Δ) are equivalent only when $\Delta \to 0$, and are in fact connected by the following relation (cf. [18]):

$$\Delta^* = 1 - e^{-\Delta}. \tag{III.19}$$

Relation (III.18) is often fairly well confirmed. However, an analysis of the possible contribution made by thermal convection (cf. para. 3.4 and [32]) would seem to indicate that, in many experimental cases, it may be far from negligible.

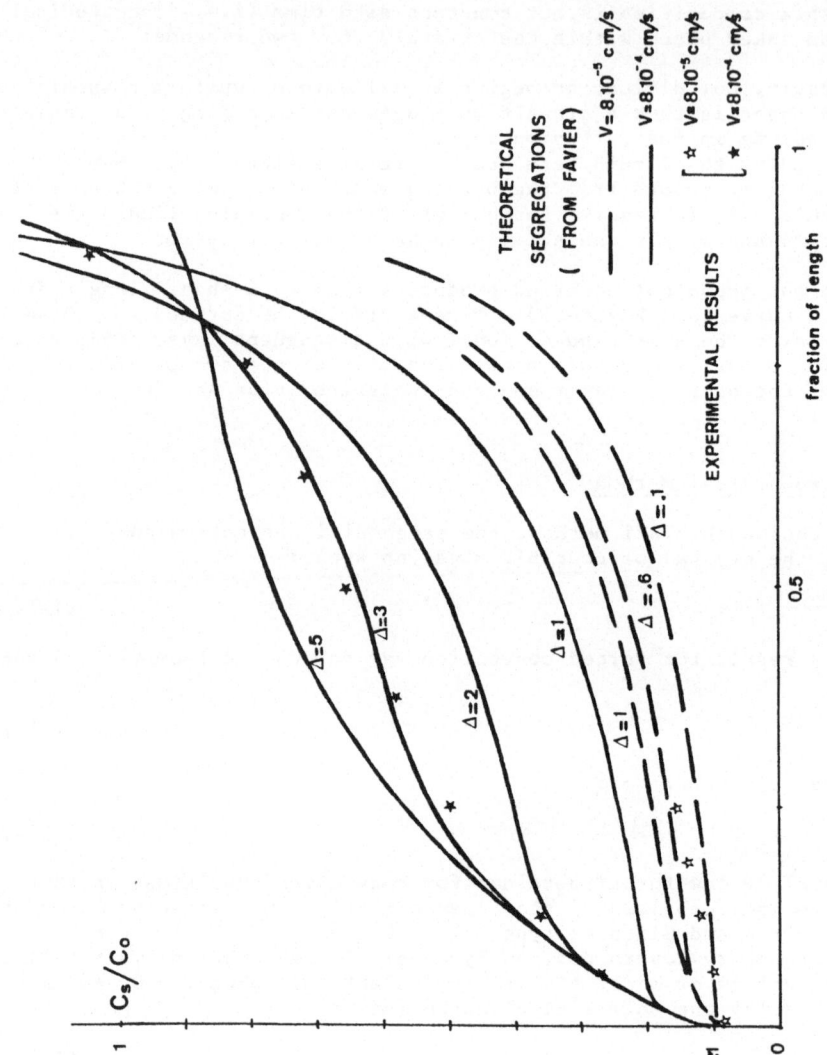

Fig. 5. Longitudinal segregation profiles for given values of Δ (model from [25] and experimental results from [33]).

III.3.4 Bridgman Solidification

When Bridgman solidification occurs, the initial cause of convective movement is the density differences that arise as a result of temperature gradients imposed within the liquid. The convective speed scale is given by balancing the Archimedes and inertia forces, in the case of boundary-layer conditions, i.e.:

$$w*^2 = Gr \tag{III.20}$$

hence, using (III.3), the classic relation:

$$x_v^* = Gr^{-1/4} \tag{III.21}$$

and, according to (III.13):

$$\Delta^* = \min(1, Pe\ Gr^{-1/4}Sc^{-1/3}). \tag{III.22}$$

Experimental results from a number of sources may be correlated by relation (III.22) (cf. Table 3 and [30]). It would appear that, when horizontal Bridgman solidification occurs under normal gravity conditions, natural convection usually results in virtually perfect mixing of the liquid, except for the case of solidification of very thin liquid samples (i.e., of the order of 1 mm). In addition, the law to the power -1/4 shows that a very great drop in gravity is needed in order to increase Δ^* significantly (e.g., $g < 10^{-4}\ g_o$ in order to grow a Ga-doped Ge crystal in pure diffusion conditions in a space laboratory).

With vertical Bridgman solidification, the imposed temperature gradient is in theory a stabilizing factor (cold at the bottom). However, because of technological limiting factors, it is impossible to eliminate horizontal temperature gradients completely. The resulting residual convection is sufficient to cause convective transport of the solute. In this arrangement, purely diffusive transport can be caused only by associating a stabilizing density gradient with the concentration gradient in the diffusion layer itself.

Table 3. Values of the Segretation Parameter Δ for Various Experiments of Bridgman Growth of Doped Crystals, and Comparison with the Values Calculated Below (III.22)

Ref.	System	H (cm)	G (K cm^{-1})	V (cm.s^{-1})	g/g_o	Δ^*_{exp}	Δ^*_{calc}
[4]	Sn(Ag)	0.22	46	10^{-3}	1	0.6	0.3
	Sn(Ag)	0.05	46	10^{-3}	1	1	1
[5]	Ge(Ga)	1	50	8.10^{-4}	1	0.142	0.05
	Ge(Ga)	1	50	8.10^{-4}	$\sim 10^{-5}$	1	1
[10]	Ge(Ga)	0.4	50	8.10^{-5}	1	0.1	0.01
	Ge(Ga)	0.4	50	8.10^{-4}	1	1	0.1
[6]	Al(Cu)	0.6	13.5	$5.7.10^{-4}$	1	0.6	0.2
	Al(Cu)	0.6	13.5	$5.7.10^{-4}$	$\sim 10^{-5}$	1	1

III.4 CONVECTIVE EFFECTS AT A NON-PLANAR SOLIDIFICATION FRONT

III.4.1 Eutectic Solidification

When externally-induced convection is imposed, the velocity field near a eutectic front is virtually the same as that ahead of a planar front. In contrast, the tangential diffusion length is of the order of the lamellar spacing λ: $z^* \simeq \lambda$.

The condition given by (III.1) may then be written:

$$\frac{V^*(\lambda)}{\lambda} \simeq \frac{D}{\lambda^2} . \tag{III.23}$$

Convection thus contributes to transporting the component rejected in front of one lamella to the neighboring lamella. It is thus clear that convection results in an increase in the inter-lamellar spacing, as shown in [42] and [43]. However, in view of the low values of λ, there has to be fairly intense agitation for the effect to be significant.

III.4.2 Columnar Dendritic Growth

In columnar dendritic growth, convection in the bulk liquid has an effect on segregation, even if there is intense agitation, given the slight differences in concentration in the liquid ahead of the front. On the other hand, any convective movement in the dendritic area has an effect as soon as the speed of movement is comparable to the growth rate. A movement of this kind is generally set up by density gradients associated with concentration gradients existing in the area [34]. Given the fact that the inter-dendritic channels are so small, a movement of this kind is governed by Darcy's law of viscous flow in a porous medium.

Sophisticated models [35] may be used to calculate the radial segregation associated with inter-dendritic solute convection. However, these models do not take into account possible exchanges between the inter-dendritic liquid and bulk liquid. These exchanges, which depend on orientation with respect to gravity, have a pronounced effect on longitudinal segregation, resulting in the deformation of the dendritic front (Figure 6). This confirms that the liquid ahead of the front is not of uniform composition [44].

PART IV - MORPHOLOGICAL STABILITY OF THE SOLIDIFICATION FRONT

So far, the solidification front has been considered simply as a particular macroscopic surface of the solid/liquid system, coinciding with the liquidus isotherm. However, this surface may adopt a different morphology, to a scale of 10 or 100 μm, depending on the forces acting on it during the solidification process. Local temperature fluctuations, mechanical shocks and convection currents within the liquid are perturbations that may cause momentary folding of the interface (Figure 7). Depending on the nature of the boundary conditions imposed on the system, and the growth rate in particular, morphological modifications of the interface may be attenuated or amplified naturally. The structure of the solid is then affected by the morphology of the front. This is how the well-known cellular or dendritic structures are formed.

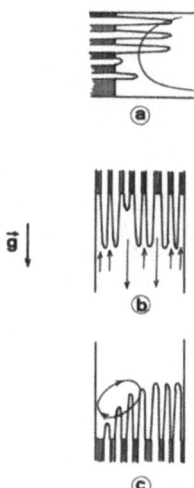

Fig. 6.   Sketch of different mushy zone configurations illustrating non
          uniform composition of the liquid ahead of the front.
          a: Horizontal growth with a solute heavier than the liquid bulk.
          b: Vertical growth downwards (solutally unstable configuration).
          c: Vertical growth upwards (solutally stable configuration).

Two questions are raised by this change in interface morphology as a
result of the conditions imposed during solidification, namely:

   - what is the stability threshold of the planar interface?
   - once this threshold has been crossed, what is the size of the
     dissipative structure formed?

IV.1   DRIVING FORCES INVOLVED

The description of steady-state local growth conditions given above
indicated that temperature and solute concentration gradients occur in
the vicinity of the solidification front. The solid-liquid interface was
characterized during the nucleation phase by its equilibrium energy

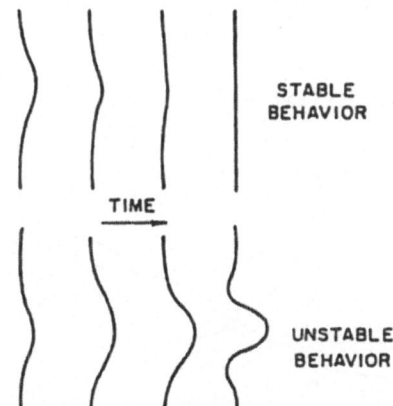

Fig. 7.   Schematic diagram of the response of an interface to a
          morphological perturbation.

properties, i.e., the interface tension. During growth, other determining features are the interface kinetics and anisotropy which reflect the difficulty experienced by the interface to move in a given direction.

The stabilizing role of the interface tension as regards small morphological perturbations is intuitive, as this will oppose the formation of a greater specific area of interface. The respective effects of the thermal fields and solute fields were originally identified by Chalmers et al. [45] who developed the notion of constitutional undercooling. Although these fields are sensitive to convection effects and instabilities, morphological stability theories for a long time considered only purely diffusive flow of heat and matter.

IV.2 MORPHOLOGICAL STABILITY IN THE ABSENCE OF CONVECTIVE INSTABILITIES IN THE LIQUID

IV.2.1 The Principle of Constitutional Undercooling

In 1953, Rutter and Chalmers [45] observed that cells disappeared when the ratio between temperature gradient and growth rate increased. In order to interpret the effect of G/v on the stability of a smooth front, their reasoning, based on thermodynamic principles (Figure 8) was as follows: as it advances, the interface rejects impurities in such a way that the local solid-liquid equilibrium temperature at any point in the liquid may rise above the real temperature at the same point. This is what occurs in the shaded area in Figure 2. Rutter and Chalmers [45] postulate that, in this specific case, a protuberance of the solidification front will tend to grow spontaneously.

In the same year, Tiller, Jackson, Rutter and Chalmers [14] quantified this idea. In this case of solute transfer by means of pure diffusion in the liquid, the familiar instability condition is obtained:

$$G < m \, G_c \qquad \qquad (IV.1)$$

or
$$\frac{G}{v} < - \frac{m_L \, C_o \, (1 - k_o)}{D_L \, k_o} \qquad \qquad (IV.2)$$

or again
$$\frac{G}{V} < \frac{\Delta T_s}{D} \qquad \qquad (IV.3)$$

in which ΔT_s is the solidification interval.

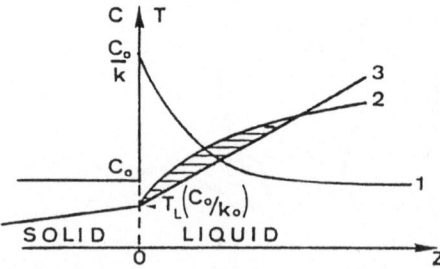

Fig. 8. Chemical undercooling of the front according to [45].
    1. Liquid concentration profile.
    2. Corresponding equilibrium temperature.
    3. Real temperature.

The stabilizing role of the thermal field and destabilizing role of the chemical field are clearly apparent.

The previous theory, based on the principle of equilibrium, neglects all dynamic phenomena such as volume heat transfer, release of latent heat at the interface, etc. In addition, it is obviously of no use in providing a description of changes in the disturbed interface with time.

Mullins and Sekerka [46] and Voronkov [47] attempted successively to fill in these gaps.

IV.2.2 Linear Analysis of Perturbations

IV.2.2.1 Analysis Principle. Assuming a solidifying system with well-defined boundary conditions. Sekerka investigated whether exact analytical solutions to the problem are stable when there are slight morphological perturbations in the interface.

Considering an elementary perturbation in the interface morphology, which is isotropic in the interface plane (x,y) and advancing through the liquid (in the direction Oz) by a distance $\delta(t)$ as a function of time, then:

$$z = \delta(t)\sin\omega x$$

where ω is the spatial frequency corresponding to a wavelength $\lambda = 2\pi/\omega$.

The principle involved in such an analysis is to evaluate perturbations throughout the steady-state fields (in this case temperature and concentrations) and then to examine the conditions in which these perturbations attenuate or increase. Only the first-order terms are kept in the equations, in order to carry out a linear analysis.

The perturbed fields T_S, T_L, C_L, C_S obey the various equations and certain boundary conditions. Mullins and Sekerka assume that these perturbations cancel out at any instant when far from the interface:

$$T_S(x,y,\infty,t) = T_L(x,y,\infty,t) = C_S(x,y,\infty,t) = C_L(x,y,\infty,t) = 0. \qquad (IV.5)$$

It is convenient to represent the temporal and spatial interdependence of these perturbed quantities in the form $\exp\{\sigma t + i\omega(x + y)\}$ (Fourier's analysis). σ is generally a complex quantity $\sigma = \sigma_r + i\sigma_i$. If σ_r is

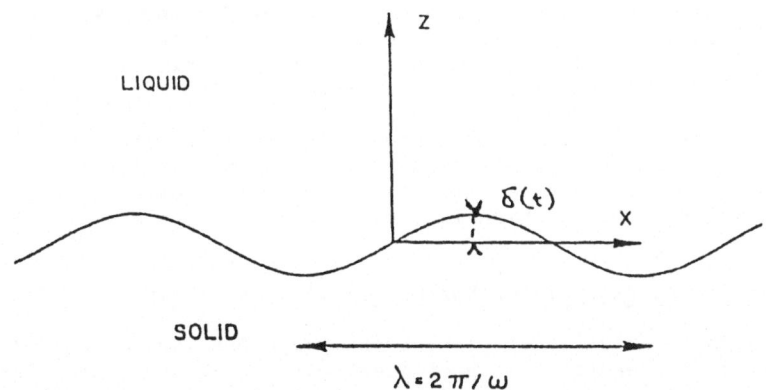

Diagram of sinusoidal perturbation of the solidification front in the (x,z) plane.

positive for all values of ω, perturbed field solutions increase towards
infinity and the situation is unstable. If σ_r is negative for all values
of ω, the system is stable. It is therefore necessary to calculate the
relation between σ and ω, referred to as the dispersion relation, using
the relevant equations and boundary conditions in order to determine the
stability limit of the system and then deduce the critical wavelength at
the threshold $\lambda* = 2\pi/\omega*$. This is the stability exchange principle and is
written as follows:

$$\sigma_r(\omega) < 0; \; Max(\sigma_r(\omega)) = 0 \; and \; \sigma_i(\omega) = 0 \quad V\omega.$$

IV.2.2.2 Results of the Mullins and Sekerka model [46]. The
dispersion relation obtained by these authors is as follows:

$$\sigma(\omega) = \left[\frac{K_S G_S + K_L G_L}{K_S + K_L} + \frac{D}{V} S(\omega) \right] \cdot \frac{1}{\Delta T_s} \qquad (IV.6)$$

in which K_S and K_L are the solid and liquid heat conductivities, G_S and G_L
the temperature gradients of the interface and $S(\omega)$ a function between 0
and 1, dependent on k_o, the partition coefficient, and on:

$$A = \frac{k^2}{1-k} \frac{T_M}{mCo} \frac{V}{D} \frac{\gamma_{SL}}{L}$$

referred to as the capillary parameter (L_α: latent heat of fusion) as it
contains the interfacial tension γ_{SL}. The marginality or threshold
stability is calculated for the Fourier wavelength $\omega*^{-1}$ for which
$\sigma(\omega*) = 0$ and $\sigma(\omega) < 0$ for every other value of ω.

Sekerka's morphological instability criterion then assumes the
following form, which is very similar to that of Chalmers:

$$\frac{K_S G_S + K_L G_L}{K_S + K_L} < m \, G_c \, S(A, \, k_o). \qquad (IV.7)$$

The mean temperature gradient at the interface, weighted by the liquid
and solid thermal conductivity values, replaces the temperature gradient
on the liquid side. The chemical gradient G_c is multiplied by the
stability function $S(A,k_o)$, shown on Figure 9 as a function of the
parameter A for an Al-Cu alloy ($k_o = 0.14$). The interfacial tension of
liquid alloys is generally small compared to the latent heat released at
the solidification rates that occur in classic metallurgical and foundry
processes. In this case, S is close to 1 and the criteria formulated by
Chalmers and Sekerka combine. This explains why Chalmers' simple
criterion is generally sufficient to predict instabilities in the
solidification front. It will be shown later that this simplification is
no longer valid for high solidification rates.

Finally, it should be recalled that, in contrast with Chalmers'
theory, which by its very nature cannot predict the wavelength
corresponding to interface instability, Sekerka's theory predicts a
critical wavelength $\omega^{-1}*$, which is likely to develop in a stable manner at
the interface.

IV.2.3 Non-Linear Analysis of Morphological Instability

Only linear analysis can predict interface behavior at the transition
between stable and unstable morphologies, as it takes into consideration
only small perturbations in the system. In practice, it is difficult or
even impossible to remain in this marginal state. The threshold is always
exceeded, making it necessary to consider changes in the system after the

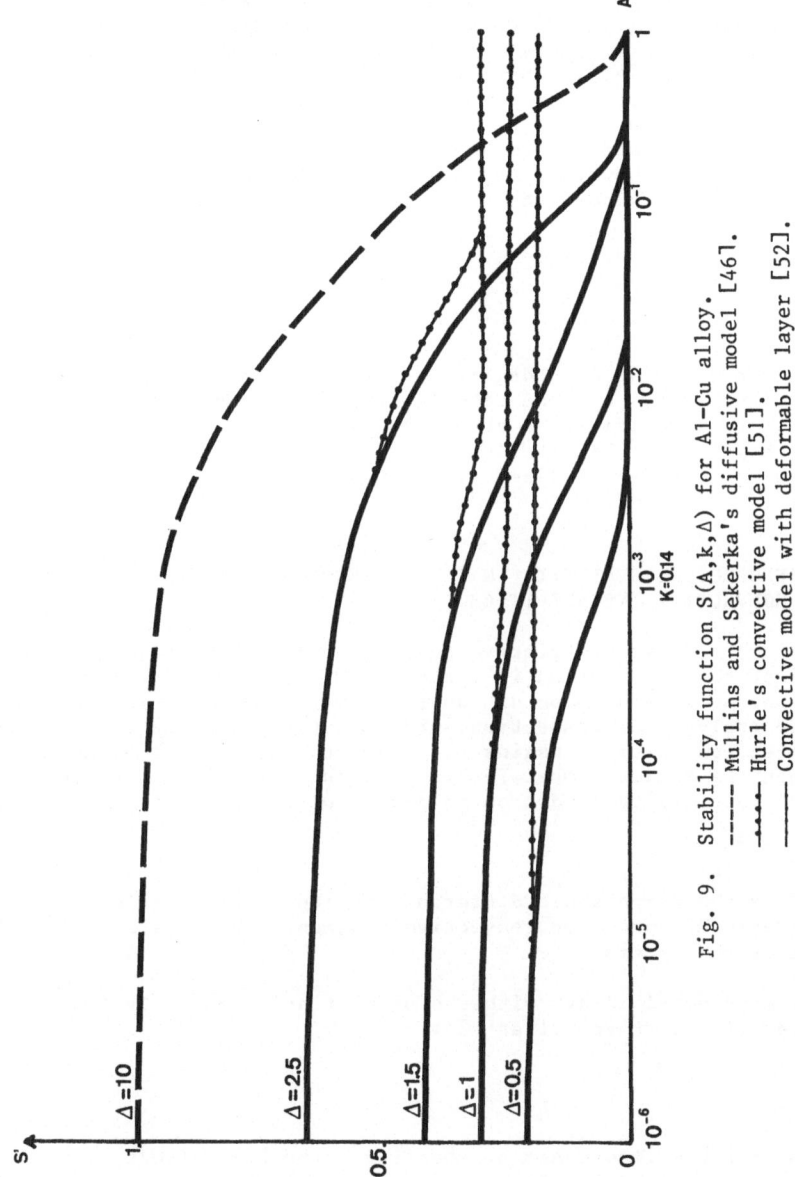

Fig. 9. Stability function $S(A,k,\Delta)$ for Al-Cu alloy.

---- Mullins and Sekerka's diffusive model [46].
⋯⋯ Hurle's convective model [51].
—— Convective model with deformable layer [52].

split has occurred. This is possible by using non-linear analysis techniques.

The higher-order terms are kept in the perturbed state equations. this involves writing the amplitude equation for the perturbations $\delta(t)$ that will show time-dependent changes in the perturbation once the stability threshold has been crossed. The morphology of the front may be predicted as a function of deviation from the threshold. The calculations involved are extremely complex and require numerical analysis. Wollkind et al. have, however, proposed an analytical solution that is valid under certain conditions [48]. These authors have identified the sequence of morphologies observed experimentally by several other authors, for increasing deviations from the threshold, namely nodes (point depressions), elongated cells and hexagonal cells. Their theory only gives this result, however, provided that the interfacial tension γ_{SL} is assumed to vary in accordance with concentration levels along the interface:

$$(\frac{\delta \gamma_{SL}}{\delta C_L} \neq 0)$$

which is surprising.

Other non-linear methods, tested in hydrodynamic problems, have also been used recently. For the moment, it will be assumed that these new approaches to solidification phenomena, which have been proposed in the past three years, may in the long term provide an exact description of the morphology of the destabilized front, which linear analysis is unable to do.

IV.3 MORPHOLOGICAL STABILITY IN THE PRESENCE OF CONVECTIVE INSTABILITIES

By far the most restrictive feature of the theories described above is the fact that convection is ignored. Without even considering forced convection, which always occurs during casting operations, the order-of-magnitude analysis of convection, given in the preceding chapter, shows that convection movement begins to affect the solute field ahead of the front when the Peclet number Pe* is less than or equal to $(GrSc)^{1/4}$. Chalmers' instability condition in equation (IV.3) may then be written:

$$Pe_c > \frac{\Delta T}{\Delta T_s} \qquad \qquad (IV.3')$$

where ΔT is the temperature difference in the liquid gradient over the characteritic distance for convective movement, which is normally the diameter of the crucible.

For most metal alloys with the concentration normally found in metallurgy (i.e., above 0.1 at %):

$$
\begin{array}{ccc}
Pe^* & >> & Pe_c \\
10 \text{ to } 100 & & 10^{-2} \text{ to } 1
\end{array}
\qquad (IV.8)
$$

This confirms the importance of coupling convective effects with morphological instabilities.

The convection modes that are likely to develop ahead of the front are complex and highly dependent on specific solidification conditions. These include the forced convection in the bath, due to mechanical agitation, or natural convection due to thermal or solute effects connected with solute rejection and subsequent interaction. The literature deals with various cases corresponding to increasingly complex situations.

IV.3.1 Undercooling in the Presence of Bulk Convection

Hurle [49] was the first to extend the criterion of undercooling to solidification with a solute boundary layer Δ of the same type as that introduced by Burton, Prim and Slichter [24]. The concentration profile at the interface is modified so that G_c, the solute diffusion gradients at the front, becomes $k_{eff}G_c$, where k_{eff} is the effective distribution coefficient with a value between k_0 and 1. The instability criterion from equation (IV.3) then becomes:

$$\frac{G}{V} < \frac{k_{eff}\Delta T_s}{D} \, . \tag{IV.9}$$

It can be seen that bath agitation increases planar interface stability when k_0 is less than 1 and reduces it when k_0 is greater than 1.

IV.3.2 Linear Analysis of Perturbations with a Steady-State Convective Velocity Field

Linear analysis with perturbations of the Mullins and Sekerka type was extended to the case of convection with a boundary layer first by Hurle [50] and then by Hurle and Coriell [51]. These authors applied the analysis technique to the non-perturbed situation illustrated in Figure 10a. This is a realistic case taking into account a general movement of the liquid (convective-diffusive regime as described previously). The

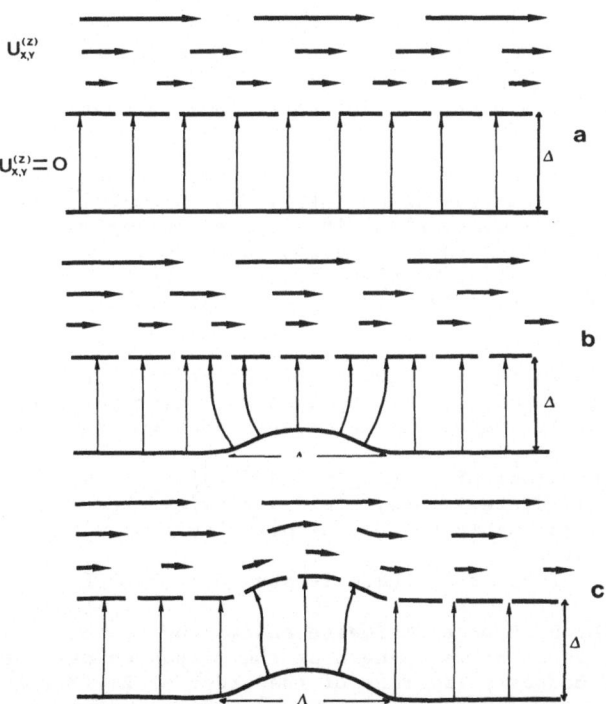

Fig. 10. Various boundary layer assumptions.
a. Burton, Prim and Slichter's basic state [24].
b. Hurle's rigid boundary layer [49,50].
c. Deformable boundary layer [52].

velocity field is assumed to be zero within the Δ layer, and concentration uniform outside this layer. When a morphological perturbation occurs at the interface, the solute field within the layer is perturbed. In contrast, these authors assume that the perturbation is cancelled out at the Δ boundary: $e(X,Y,\Delta,t) = 0$ which simplifies the calculation considerably (Figure 10b).

While this assumption is acceptable for low-amplitude perturbations compared to the extent of the boundary layer, it produces unacceptable interference reflections when the perturbation wavelength approaches the order-of-magnitude of the boundary layer.

Favier and Rouzaud [52] removed this limitation by assuming that, in diffusion conditions, the boundary layer deforms in accordance with the morphological perturbation of the interface (Figure 10c). This is a realistic assumption if one considers the respective response times of chemical diffusion field deformation and viscous velocity. Their ratio gives the Schmidt number $Sc = \nu/D$, which is always very large as compared to 1 for metal alloys ($10 < Sc < 100$). The boundary condition in Δ for the perturbed solute is as follows:

$$C(X,Y,Z,t) = -\delta(X,Y,t)\left(\frac{\partial C}{\partial z}\right)_\Delta \qquad (IV.10)$$

where δ is the perturbation amplitude.

A new stability function $S'(A,k_o,\Delta)$ in the stability criterion is defined from this calculation. This function depends on convection intensity through Δ.

$$\frac{K_S G_S + K_L G_L}{K_S + K_L} < m\, G_c\, S'(A,k_o,\Delta) \qquad (IV.11)$$

$S'(\Delta)$ tends towards S when Δ is infinite, that is when convection becomes negligible.

Figure 9 shows the change in S' as a function of A for an Al-Cu alloy ($k_o = 0.14$) and for various values of Δ. The prevailing influence of convection on morphological stability is clearly apparent. $\Delta = 10$ corresponds to Sekerka's pure diffusion function S. By assuming a deformable boundary layer, absolute stability may be obtained with high values of the capillary parameter A. With Hurle's assumption of a rigid boundary layer, this is not possible.

Figure 11 shows the wave number of the critical perturbation in the marginal state as a function of the boundary layer parameter Δ, for various values of A. Three characteristic patterns can be clearly seen:

- with high values of Δ, chemical diffusion predominates,
- with intermediate values, fluid viscosity predominates; Hurle and Coriell were unable to obtain this situation with the assumption of a rigid layer,
- with low values of Δ, there is absolute stability ($\lambda \to \infty$).

The transition between diffusive and viscous conditions occurs when the relation between the wavelength of the morphological instability and the size of the boundary layer is of the order of 2: ($2\pi(\Delta/\lambda) \sim 12$), which gives a posteriori confirmation of the assumption of a deformable boundary layer.

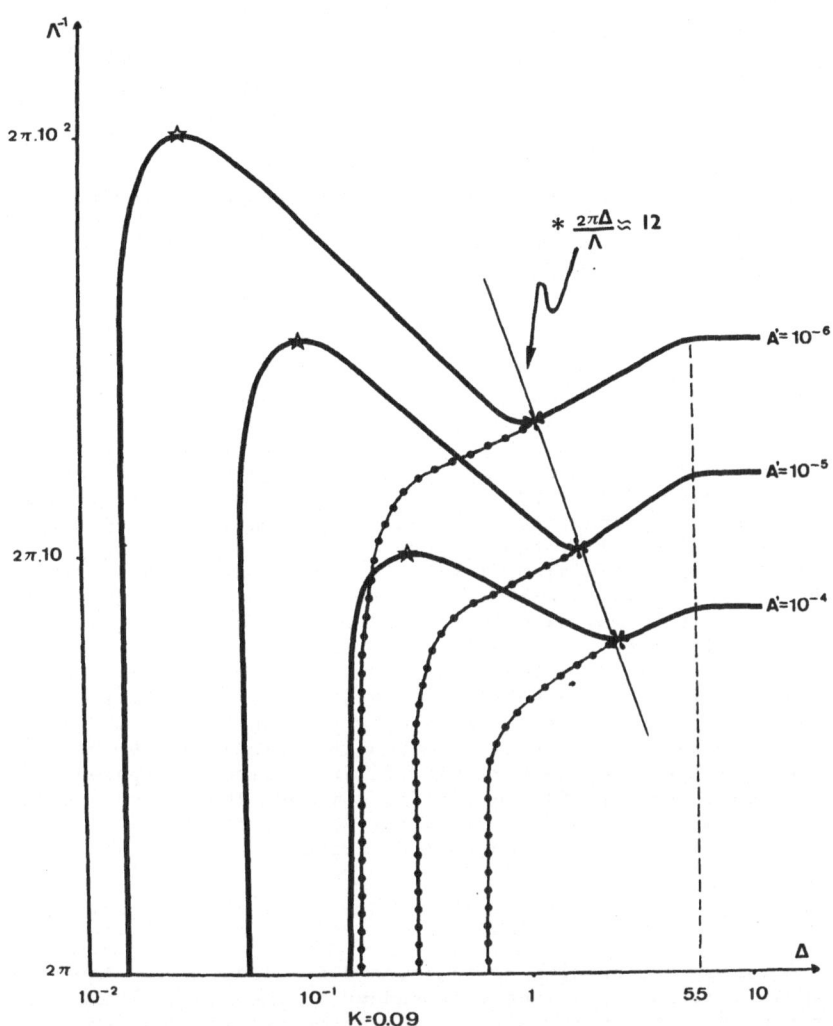

Fig. 11. Spatial frequency of the first stable perturbation along the
front as a function of convection (Δ decreases when the level of
convection increases).

IV.3.3 Connection between Morphological and Solute Instabilities

The simultaneous existence of concentration and temperature gradients
in the immediate vicinity of the interface may give rise to hydrodynamic
instability. Indeed, as the local density of a bulk component is a
function of both its temperature and concentration $\rho = \rho(C,T)$, this
component of the fluid will tend to be moved by local fluctuations in the
gravity field (Figure 12) until it reaches an area with different
temperature and concentration. As heat is diffused more rapidly than
chemical components (characteristic diffusion times being in the ratio of
the Lewis number Le = K/D >> 1), the component will be in local thermal
equilibrium but not in chemical equilibrium and will therefore tend to
return to its original position, creating general movement within the

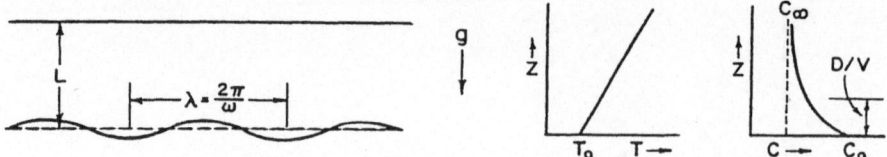

Fig. 12. Principle of double diffusion ahead of the front.

liquid. This is the so-called double diffusion phenomenon. As this
hydrodynamic instability develops in the immediate vicinity of the
interface, it is obvious that it will affect the interface morphology.

Boettinger, Coriell and Sekerka [53] were the first to define the
conditions in which an instability of this type would occur in the
exponential concentration profile perpendicular to a solidification front.
Their resolution of the problem is purely numerical. Hurle [54] examined
the same problem from an analytical standpoint, ignoring the thermal
aspect of convection (Ra_T = 0), and thus simplifying the calculation.
Both cases treat coupling with the usual morphological instability.
Caroli et al. [55] consider that, apart from conditions resulting in
morphological λ_M and solutal λ_C instability wavelengths of the same order
of magnitude (generally obtained with a gentle temperature gradient
$\simeq 1$ Kcm^{-1}), coupling remains only slight, which means that the phenomena
may be treated separately. Finally, they introduce the effect of a change
in volume during solidification (advection) which, in their view, has a
greater influence on the instability threshold [56]. The main result of
these theories, taken from the article by Boettinger et al. [53], is
shown on Figure 13. The stability domain is represented in the
concentration-velocity plane for solidification with an imposed gradient.
Two curves bound the interface stability domain. The descending straight
line is the usual morphological stability limit, which is well represented
by the generalized chemical undercooling criterion:

$$(\frac{K_S G_S + K_L G_L}{K_S + K_L} \text{ instead of } G_L$$

as, at low rates, the stabilized function S is close to 1. The rising
curve, whose position is directly dependent on the force of gravity, is
the thermosolutal instability limit. With imposed rate and gradient, the
critical wavelength is a function of concentration only.

Two families of wavelengths occur at the instability threshold
(Figure 14), namely:

- short wavelengths (large wave number ω), originating in chemical
  undercooling instability,
- long wavelengths (small wave number ω), of hydrodynamic origin.

Near the cross-over point (point A in Figure 13), oscillatory modes may
occur, corresponding to the combination of the two instabilities.

While these models have the advantage of illustrating a new type of
interface instability - that of deformation of a solidification front,
accompanied by localized hydrodynamic instability just ahead of it - the
boundary conditions they adopt are often unrealistic when it comes to
applying them to standard metallurgical experiments. It has in fact been
shown that even when great care is taken during directional solidification
[57], it is extremely difficult to eliminate all cooperative movement of
the liquid on the ground. For this reason, Hennenberg et al. [58]

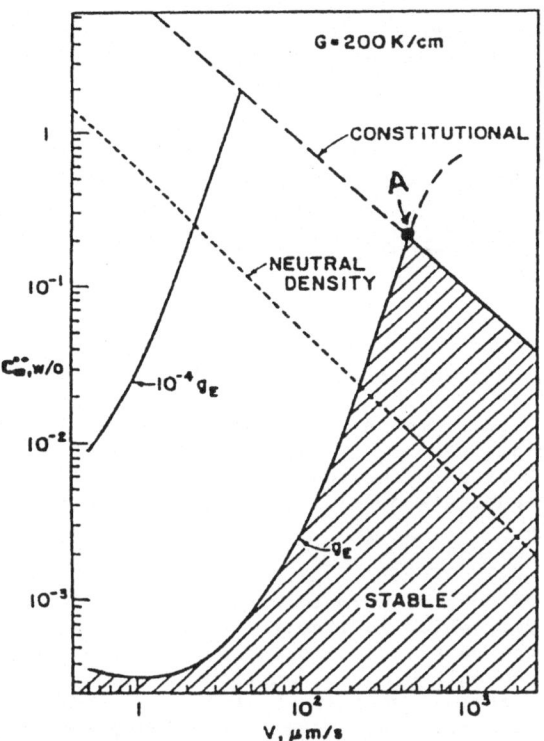

Fig. 13.   Morphological instability in a Pb-Sn alloy with thermosolutal
instabilities.

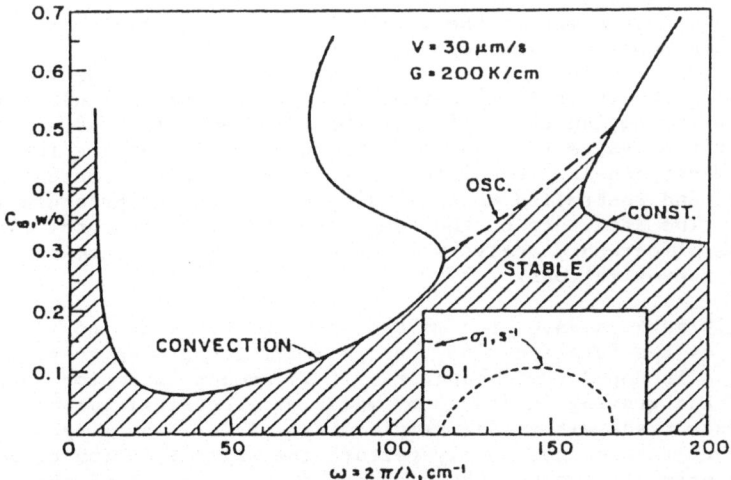

Fig. 14.   Occurrence of first stable perturbation with thermosolutal
instabilities (σ_i may no longer be zero).

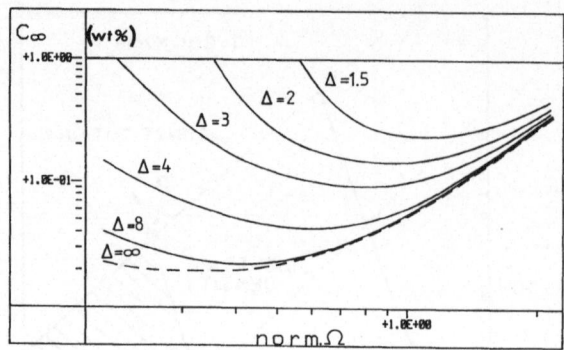

Fig. 15.   Critical concentration in the bulk C_∞ as a function of
          normalized wave number Ω for different convection level in the
          liquid bulk. Hydrodynamic branch (as on Figure 14).
          --- Coriell's model [53] and Hennenberg's model for $\Delta = \infty$ (pure
          diffusion in the bulk) [59].
          ——Hennenberg's model for different levels of convection in the
          bulk.

recently introduced double thermosolutal diffusion into the deformable
boundary-layer model, which is itself necessary on account of convection
movements within the bulk bath. They calculated [59] the critical
concentration in the bulk C_∞ as a function of the normalized wave number
for different convective levels in the bulk characterized by the non-
dimensional boundary layer thickness Δ in which thermosolutal
instabilities occur. Figure 15 clearly indicates the stabilizing
influence of bulk convection on the hydrodynamic branch of the stability
diagram, as compared with the case ($\Delta = \infty$) treated by Coriell et al. when
the bulk is governed only by diffusion [53]. Bulk convection also
decreases the wave length of hydrodynamic perturbation on the interface.

IV.4 ANALYSIS OF MAIN EXPERIMENTAL RESULTS

     Experimental studies have attempted to characterize the instability
threshold and wavelength of the marginal state. Available literature [60-
64] describes numerous experiments, only a few of which meet the
conditions required for precise determination. This is due to the fact
that crossing the threshold is a transitory event, which could only be
monitored by a continuous display of the interface or in-situ measurement
of a significant value (for example the undercooling at the interface).
The experiments consist in solidiying an alloy of given concentration in a
directional and controlled manner (using an imposed temperature gradient
and rate). The structure of the sample is observed after the event by
means of metallographic analysis. The transition is determined by
successive approximations.

     Certain authors have used melt quenching techniques to fix the
morphology of the interface [65], or pull-off techniques to reveal the
front [66]. In doped semiconductors, it is common to employ instantaneous
marking of the growing interface by means of electrical pulses to reveal
the interface trace within the sample mass. This is an elegant technique
provided that it does not unduly perturb the solidification process [67].
Several experimental workers have observed the movement of the interface
and changes in its morphology in transparent materials (organic materials)
under the microscope [68]. Finally, a technique involving the

thermoelectric effect has been developed for continuous measurement of undercooling at the interface in certain alloys. This precisely indicates the break in chemical undercooling at the instability threshold [69]. Most of the studies show close agreement (to within a few percent) between experimental and theoretical values of the stability threshold. It is not possible with the conditions used (i.e., low or moderate rates) to differentiate the chemical undercooling criterion from Sekerka's theory (with the value of the stability function remaining close to 1). Experiments with gentle temperature gradients should indicate whether Sekerka's generalized temperature gradient is more suitable than Chalmers' use of the temperature gradient in the liquid alone. Experiments performed to date are inconclusive in this respect [70] and it may be assumed that the simple criterion of chemical undercooling provides a good indication of the planar front's stability threshold. One of the reasons for the difficulties encountered in all these experiments is certainly the effect of convection, which may be considerable, as we have already seen. Answers might be provided by experiments under strictly controlled, purely diffusive or convective conditions, such as could be achieved with microgravity [71].

There is less agreement in the case of semi-metals (e.g., Bi). Hecht and Kerr [72] have found that the interface remained stable whereas it should have been instable according to the theory. The particular properties of solid-liquid interfaces in semi-metals (anisotropic interface tension, slow growth kinetics) are perhaps the reason for this discrepancy (cf. section IV.5.2). In contrast, there is distinctly less agreement between the wavelengths of the instabilities observed at the threshold and Sekerka's theory. The most significant results have been obtained by Morris and Winegard [73], who have shown the sequence of morphologies during transition at the cross-over point, by means of pulling-off and autoradiography of Pb-Sb alloys. Other experiments [74] have been carried out using the same technique with Sn(Pb) alloy (Figure 16). Though initially smooth (16a), the interface develops a network of microdepressions (pockmarks or nodes - 16b); grooves then form between the depressions, to form elongated cells (16c,16d). Finally, a regular and generally hexagonal network is formed (16e). The characteristic dimension of these cells is almost an order of magitude lower ($\simeq$ 50 µm instead of 300-400 µm) than that predicted by Sekerka's diffusive theory. Convection, which still occurs in these experiments, may tend to decrease the size of the cell network and this could partially explain the discrepancy (Figure 11). Finally, experiments reproduce the full decrease in cell size in accordance with the drop in growth rate and temperature gradient values, as predicted by theory [64] (Figure 17).

IV.5 APPLICATION OF MORPHOLOGICAL STABILITY
      THEORIES TO RAPID SOLIDIFICATION

Recent preparation and structural modification techniques involving fast solidification have led to renewed interest in the problems of morphological stability. This section deals simply with the consequences that a high solidification rate will have on the stability of the interface.

The solidification rate is involved in morphological destabilization in three ways. High rates will tend:

- to increase chemical undercooling (G_c is proportional to V) and hence to destabilize the interface; this is the only effect that is considered in Chalmers' theory;

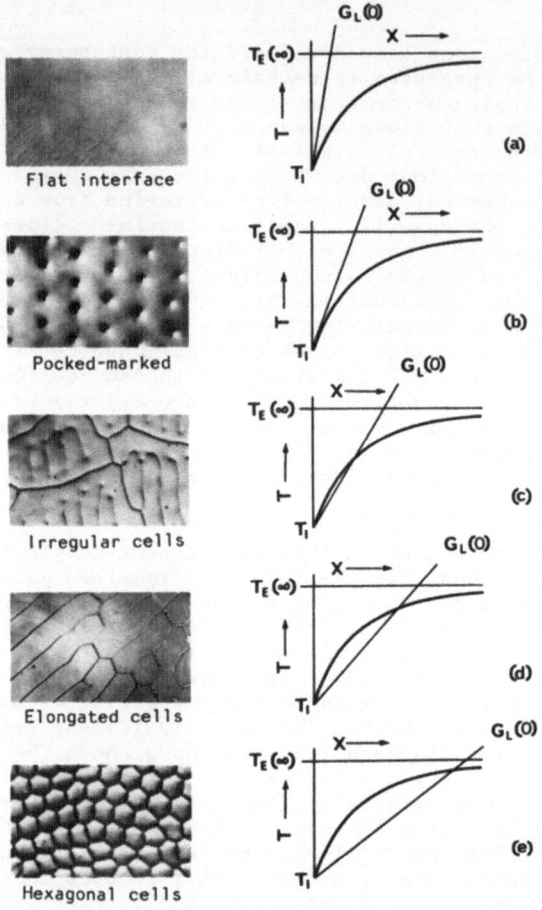

Fig. 16.   Sequence of interface structures near the split in Sn(Pb) alloy, according to [74].   Interface obtained by pulling off the liquid.

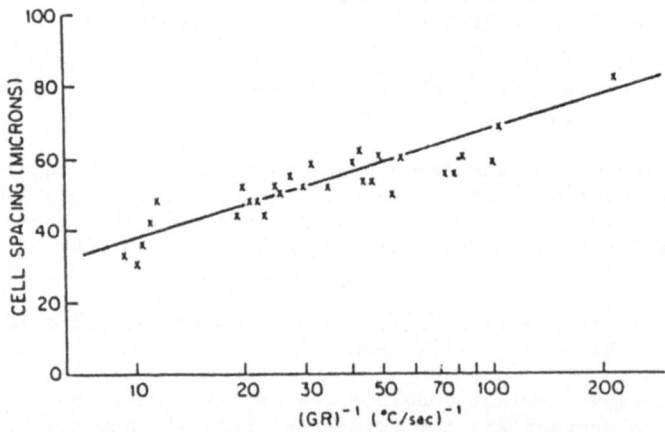

Fig. 17.   Variation in cell size as a function of the gradient rate product (GR) in Sn(Pb), according to [62].

- to increase interfacial tension γ_{SL} contribution; Sekerka's capillary term A is proportional to the product $\gamma_{SL} \cdot V$;
- to reveal deviation effects from thermodynamic equilibrium.

IV.5.1 Effect of Interfacial Tension

Interface deformation is prevented by a high capillary term as the stability function S tends towards 0. Above a critical rate V*, the interface remains planar. This is the <u>absolute stability</u> phenomenon. An expression for V* may be derived from Sekerka's theory:

$$V^* = \frac{mD(1 - k_o)C}{k_o^2 T_M SL} \qquad\qquad (IV.13)$$

Figure 18 shows the competition between increasing chemical undercooling and capillary effect in an Al-Ag alloy. A destabilized structure should no longer occur with rates of more than a few cm/s in a 0.1 wt % alloy.

This effect of restoring the planar front at high solidification rates has been tested experimentally on several alloys [75,76] by surface treatment of massive samples heated locally by electron beam heating and cooled passively. Freezing rates of the order of 1 to 100 cm/s have been obtained. Figure 19 summarizes the results obtained for Ag-Cu alloy with 1 and 5wt % of Cu, in the form of a plot of growth rate versus composition. The transition between destabilized and planar front concurs quite closely with the absolute stability theory, given the uncertainties that exist with regard to thermal conditions.

The results obtained with Al-Ag and Al-Mn alloys are less encouraging [75]. The critical restabilization rates V* calculated are very low (10^{-2} m/s for Al-Ag, 5×10^{-4} for Al-Mn) and were considerably exceeded during the actual experiments, without the destabilized structures disappearing. Other effects need to be examined.

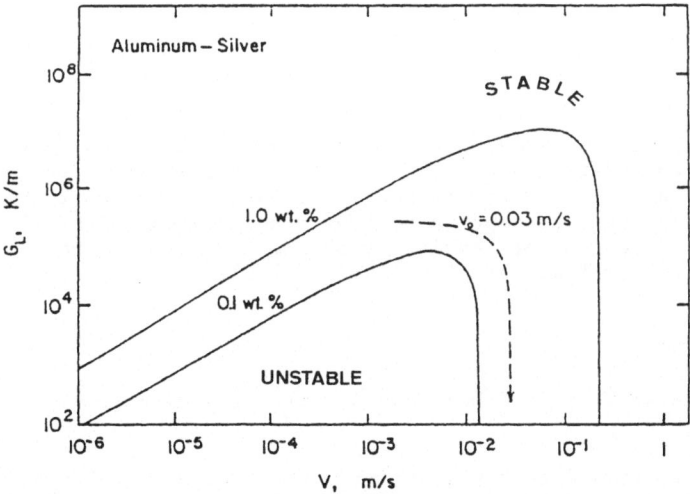

Fig. 18. Stability in the G-V plane near absolute stability.

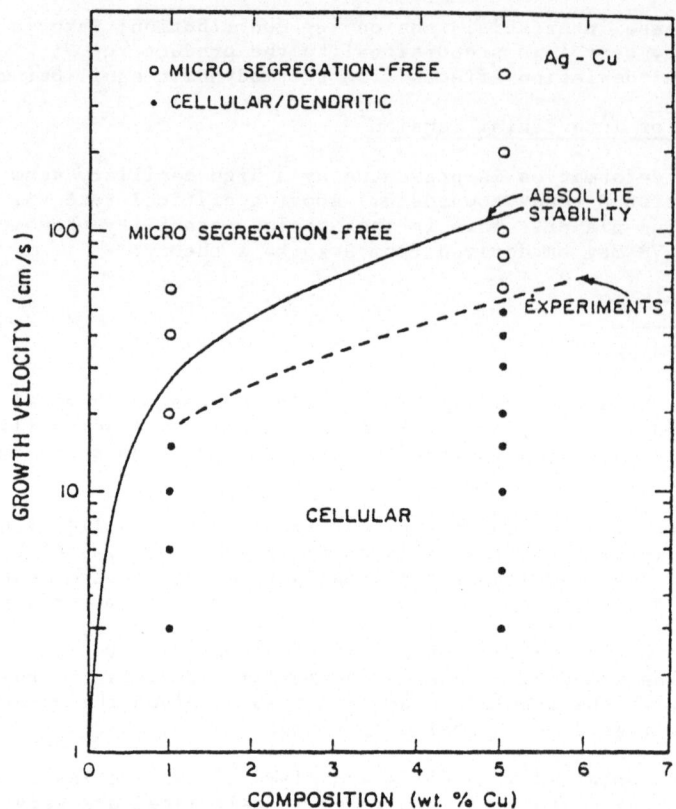

Fig. 19. Transition from absolute stability in Ag-Cu alloys with 1 and 5
          wt % Cu (according to [76]).

IV.5.2 Kinetic Effects

Sekerka's theory assumes that solidification occurs at thermodynamic
equilibrium. The distribution coefficient k_0, for example, is deduced
from the phase diagram. With the rates involved in fast solidification
(1 m/s to 10 m/s) this assumption no longer holds good. Coriell et al.
[77] have introduced an "off-equilibrium" coefficient into the stability
theory, deduced from Baker and Cahn's model [78]:

$$k = \frac{k_0 + \beta_0 V}{1 + \beta_0 V}$$

(IV.14)

β_0 being a constant. This tends towards 1 at very high growth rates (line
T_0).

Introducing $k(V)$ into the stability theory gives rise to a different
dispersion relation, in which stability exchange no longer necessarily
applies. Oscillatory modes may then occur at the origin of the band
structures observed with concentrations of the order of 10 wt % of copper
in Ag-Cu alloy [76].

102

IV.6 CONCLUSIONS

Destabilization of the solidification front is a vital aspect of metal structure control. It involves several driving forces whose respective roles are sometimes difficult to identify. From a practical point of view, the problem is made even more complex as a result of combined instabilities of hydrodynamic origin, as convection always occurs in all preparation processes. Detailed analysis of liquid-phase transport, as described in Part 3, is therefore absolutely essential and should be pursued further.

It is surprising to see that Chalmers' simple criterion is more often than not adequate for predicting the stability threshold of a solidification front. For a long time to come, it will continue to be the metallurgist's basic tool in this respect. In view of the erroneous theoretical predictions of the wavelengths of front perturbations further detailed experimental and theoretical investigations in this field are needed.

REFERENCES

1. K. A. Jackson, Liquid metals and solidification, Am. Soc. Met., Novelty, Ohio, 174 (1958).
2. B. Mutaftschiev, Adsorption et croissance cristalline, Colloque CNRS, Paris, 231 (1965).
3. N. Eustathopoulos and P. Desré, Met. Chem., 1:45 (1975).
4. G. Lesoult, Thesis, Nancy (1976).
5. D. E. Temkin, "Crystallization Processes", English translation published by Consultants Bureau, New York (1968).
6. H. A. Wilson, Phil. Mag., 50:238 (1900).
7. J. Frenkel, Physik Z. Sovjet Union, 1:498 (1932).
8. M. Volmer, "Kinetic der Phasenbildung", Dresden and Leipzig, Steinkopff (1939).
9. R. Becker and W. Döring, Ann. Physic., 24:719 (1935).
10. J. W. Cahn, Acta Met., 8:554 (1960).
11. J. W. Cahn, W. B. Hillig and G. W. Sears, Acta Met., 12:1421 (1964).
12. K. A. Jackson and J. D. Hunt, Trans. Met. Soc. AIME, 236:246 (1966).
13. H. S. Carslaw and J. C. Jaeger, "Conduction of Heat in Solids", Oxford University Press (1959).
14. W. A. Tiller, K. A. Jackson, J. W. Rutter and B. Chalmers, Acta Met., 1:428 (1953).
15. W. A. Tiller, Liquid metals and solidification, Am. Soc. Met., Cleveland, Ohio (1958).
16. S. Strässler and W. R. Schneider, Phys. Cond. Matter, 17:153 (1975).
17. D. D. Double, Mat. Sci. and Eng., 11:325 (1973).
18. M. Tassa and J. D. Hunt, J. Crystal Growth, 32:38 (1976).
19. J. N. Clark and R. Elliott, J. Crystal Growth, 33:167 (1976).
20. J. N. Clark and R. Elliott, Met. Trans. AIME, 7A:1197 (1976).
21. D. J. Fischer and W. Kurz, Acta Met., 28:777 (1980).
22. G. W. Greenwood, Acta Met., 4:243 (1956).
23. J. M. Lifshiftz and V. V. Slyozow, J. Phys. Chem. Solids, 19:35 (1961).
24. J. A. Burton, R. C. Prim and W. P. Slichter, J. Phys. Chem. Solids, 21:1987 (1963).
25. J. J. Favier, Acta Met., 29:197,205 (1981).
26. V. I. Polezhaev, in: "Crystals: Growth, Properties and Applications", 10:87-150, Springer-Verlag, Berlin (1984).
27. F. Weinberg, Trans. Met. Soc. AIME, 227:231 (1963).
28. A. F. Witt et al., J. Electrochem. Soc., 125:1832 (1978).
29. J. J. Favier et al., Acta Astronaut., 9:255 (1982).

30. D. Camel and J. J. Favier, J. Crystal Growth, 67:42,57 (1984); J. Phys., 47:1001 (1986).
31. L. D. Wilson, J. Crystal Growth, 44:247 (1978).
32. A. D. W. Jones, "Progress in Crystal Growth and Characterization", 9:139 (1984).
33. A. Rouzaud, D. Camel and J. J. Favier, J. Crystal Growth, 73:149 (1985).
34. R. Mehrabian, M. Keane and M. C. Flemings, Met. Trans., 1:1209 (1970).
35. S. D. Ridder, S. Kou and R. Mehrabian, Met. Trans., 128:435 (1981).
36. J. T. Mason, J. D. Verhoeven and R. Trivedi, Met. Trans., 15A:1665 (1984).
37. H. Slichting, "Boundary Layer Theory", 7th Ed., McGraw Hill, New York (1979).
38. S. Ostrach, Ann. Rev. Fluid Mech., 14:313 (1982).
39. L. G. Napolitano, Acta Astronautica, 9(4) (1982).
40. L. G. Napolitano, "L'Aerotecnica Missili e Spacio", 183 (Dec. 1977).
41. L. G. Napolitano, Acta Astronaut., 6(9):1093 (1979).
42. J. M. Quenisset and R. Naslain, J. Crystal Growth, 54:465 (1981).
43. V. Baskaran and W. R. Wilcox, J. Crystal Growth, 67:343 (1984).
44. M. D. Dupouy, D. Camel and J. J. Favier, Accepted in Acta Met. (1988).
45. B. Chalmers and J. N. Rutter, Can. J. Phys., 31:15 (1953).
46. W. W. Mullins and R. F. Sekerka, J. Appl. Phys., 34:323 (1963).
47. V. Voronkov, Sov. Phys. Solid State, 5:2378 (1965).
48. R. Sriranganathan, D. J. Wollkind and D. B. Oulton, J. Crystal Growth, 62:265 (1983).
49. D. T. J. Hurle, J. Solid State Electron, 3:37 (1961).
50. D. T. J. Hurle, J. Crystal Growth, 5:162 (1969).
51. S. R. Coriel, D. T. J. Hurle and R. F. Sekerka, J. Crystal Growth, 32:1 (1976).
52. J. J. Favier and A. Rouzaud, J. Crystal Growth, 64:367 (1983).
53. S. R. Coriell, M. R. Cordes, W. J. Boettinger and R. F. Sekerka, J. Crystal Growth, 49:13 (1980).
54. D. T. J. Hurle, E. Jakeman and A. A. Wheeler, J. Crystal Growth, 58:163 (1982).
55. B. Caroli, C. Caroli, C. Misbah and B. Roulet, J. de Physique (1984).
56. B. Caroli, C. Caroli, C. Misbah and B. Roulet, J. de Physique (1985).
57. A. Rouzaud, Thesis, Grenoble (1984).
58. M. Hennenberg, J. J. Favier, A. Rouzaud and D. Camel, J. de Physique, 148:173 (1987).
59. M. Hennenberg, A. Rouzaud, D. Camel and J. J. Favier, J. Crystal Growth, 85:49 (1987).
60. R. A. Jackson and J. D. Hunt, Acta Met., 13:12 (1965).
61. G. S. Cole and W. C. Winegard, J. Inst. Met., 92:322 (1963).
62. T. S. Plasket and W. C. Winegard, Can. J. Phys., 37:1555 (1959).
63. W. Bardley, J. M. Callan, H. A. Chedzey and D. T. J. Hurle, Solid State Electron., 3:142 (1961).
64. J. D. Coulthard and R. Elliott, The Solidification of Metals, Iron and Steel Inst., Publ. No. 110, 61 (1968).
65. R. M. Sharp and A. Hellawell, J. Crystal Growth, 6:334 (1970).
66. L. R. Morris and W. C. Winegard, J. Crystal Growth, 5:361 (1969).
67. D. E. Homes and H. C. Gatos, J. Appl. Phys., 52(4):2971 (1981).
68. L. R. Morris and W. C. Winegard, J. Crystal Growth, 6:61 (1969).
69. J. J. Favier, Doctoral Thesis, University of Grenoble (1977).
70. K. G. Davis and P. Fryzuk, J. Crystal Growth, 8:57 (1971).
71. J. J. Favier, Le Projet Spatial MEPHISTO, in-house report CEA-DMG (1985).
72. M. V. Hecht and H. W. Kerr, J. Crystal Growth, 7:136 (1970).
73. L. R. Morris and W. C. Winegard, J. Crystal Growth, 5:361 (1969).
74. Y. Malméjac, Doctoral Thesis, University of Grenoble (1972).

75. R. J. Schaefer, S. R. Coriell and F. S. Biancaniello, in: "Rapidly Solidified Amorphous and Crystaine Alloys", B. H. Kear, B. C. Giessen and M. Cohen, eds., Vol. 8, North Holland (1982).
76. W. J. Boettinger, D. Shechtman, R. J. Schaefer and F. S. Biancaniello, Met. Trans., A15:55 (1984).
77. S. R. Coriell and R. F. Sekerka, J. Crystal Growth, 61:499 (1983).
78. J. C. Baker, Ph.D. Thesis, Chapitre V, MIT (1970).

FUNDAMENTALS OF CRYSTAL GROWTH FROM VAPORS

Franz Rosenberger

Center for Microgravity and Materials Research
University of Alabama in Huntsville
Huntsville, AL 35899, USA

1. INTRODUCTION

Crystallization from vapors has gained great importance in the preparation of epitaxial and polycrystalline layers of semiconductors, insulators and metals, as well as for the growth of some semiconductor bulk crystals. The vapor growth of bulk materials is typically conducted in closed or semi-closed systems. If the vapor pressure of a material at a desired growth temperature exceeds 10^{-2} torr or so, it may be efficiently grown by Physical Vapor Transport, PVT (sublimation-condensation), Figure 1a. If the vapor pressure is too low for practical PVT rates, one may utilize a reversible reaction between the solid material and a gaseous "transport agent" that results only in volatile products, for Chemical Vapor Transport, CVT, Figure 1b. The growth of solid layers, on the other hand, is typically conducted in open flow systems, either by Physical Vapor Deposition, PVD, Figure 1c, where the transport from the subliming source to the growth region is enhanced by a flowing carrier gas, or by Chemical Vapor Deposition, CVD, Figure 1d, where a reactive gas mixture is blown over a hot substrate. Occasionally, open flow systems have also been used for the growth of bulk crystals. There exists a large number of compilations of experimental techniques and theoretical models for these and other vapor growth processes in reviews [1-9] and proceedings of international conferences [10-15].

The widespread application of vapor growth is due to various advantages over, say, crystallization from melts. Since the growth temperatures are typically considerably lower, many materials that decompose before melting or possess high temperature solid-solid phase transitions can be prepared with less difficulties from the vapor. At lower temperatures high purity conditions are also easier to attain due to: (a) reduced diffusion of impurities from containers, (b) reduced reaction rates with container walls, (c) enhanced segregation of impurities at the growing interface [16], and (d) effective separation of low-vapor-pressure particulates (carbon!) during transport. Furthermore, lower temperature gradients and, hence, lower stress gradients and dislocation densities can be more readily obtained. Lower growth temperatures result also in lower concentrations of thermally created (Schottky, Frenkel) defects. In addition, vapor-solid interfaces are often strongly faceted, i.e., the growth kinetics is highly anisotropic; see also Section 3. This can lead to considerably higher morphological

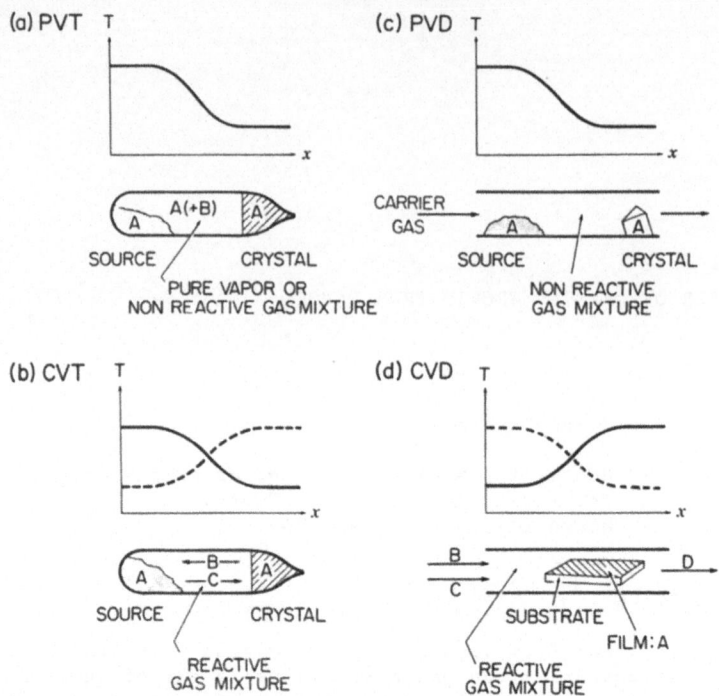

Fig. 1. Widely used methods for vapor growth and corresponding
temperature profiles. (a) Physical vapor transport: sublimation
and recrystallization of the source material A. (b) Chemical
vapor transport: reaction of solid A with vapor B to volatile
product(s) C and back-reaction to solid A in zone of different
temperature; solid and dashed T(x) for endothermic and exothermic
transport reaction, respectively. (c) Physical vapor deposition:
transfer of subliming A by (saturated) carrier gas to colder
growth area. (d) Chemical vapor deposition: reaction of gases
(vapors), e.g., B and C, blown into heated reaction zone to form
solid A and gaseous products, e.g., D; solid and dashed T(x) for
endothermic and exothermic deposition reaction, respectively.

(growth shape) stability than in the melt growth of the same material.
Due to the low mass density of vapors, the nutrient flux to a growing
interface can be rapidly changed. This is particularly important for thin
film growth and the preparation of abrupt compositional changes
(semiconductor junctions).

    Low growth rates and uncontrolled, parasitic nucleation are often
referred to as major drawbacks in vapor crystallization. Low growth rates
may be a consequence of interfacial kinetics limitations. These can arise
from the lower growth temperature and/or an intrinsically slower mechanism
of molecule addition to the lattice (Section 3). Reactive transport or
deposition processes may further complicate and possibly retard growth
kinetics. Low growth rates may also result simply from the relatively
large latent heat flux that must be dissipated from growing solid-vapor
interfaces. Note in this context that heats of condensation are typically
one order of magnitude larger than corresponding heats of solidification.
But in many situations, low growth rates as well as uncontrolled
nucleation reflect a lack of macroscopic transport control, rather than
intrinsic limitations. Hence, a quantitative understanding of the

transport conditions is particularly desirable for optimal utilization of a vapor growth technique.

An even stronger need for insight into the heat and mass transport conditions of (vapor) growth processes comes from a more applied aspect. Most technologically interesting properties of solids are extrinsic, i.e., are obtained through "dopants". Device applications often require a very uniform distribution of these electronically or optically active impurities or deviations from stoichiometry in the host lattice. For dopant introduction during growth one must take into account that most dopants are incorporated into the lattice with a different concentration than the one offered in the nutrient to the interface. They are either partly rejected or preferentially incorporated. Not too far from equilibrium, this interfacial redistribution or segregation is governed by the thermodynamics of the host-impurity system [16]. The local dopant concentration, however, that is offered to the interface, is governed by the fluid dynamics that prevails in the nutrient. Non-uniform as well as time-dependent interfacial dopant supply lead to a non-uniform distribution in the solid. For applications often particularly undesirable are the dopant concentration striations that result from oscillatory heat and mass transfer in the nutrient.

In the following sections we will discuss fundamental aspects of the transport in the nutrient and the molecular interface kinetics of vapor growth. Thereby we will emphasize the characteristic differences to growth from condensed nutrients, such as melts and solutions.

2. MACROSCOPIC TRANSPORT

For the description of the macroscopic transport behavior, i.e., of the "fluid dynamics" of vapor crystal growth, it is important to realize that vapors possess quite different properties than condensed fluids. In particular, the mass density of vapors under typical crystal growth conditions is at least three orders of magnitudes lower than that of solutions and melts. Consequently, as illustrated in Table 1, the diffusivities for momentum, concentration and heat are orders of magnitudes higher than in condensed phases. Also, since they are essentially based on the same molecular mechanism, these diffusivities are all of same magnitude. Furthermore, the thermal expansion coefficients of vapors, depending on the process temperature, are one to two orders of magnitude larger than in condensed phases. As a consequence of these particular properties of vapors, quantitative transport descriptions of vapor crystal growth processes are often considerably more difficult than, for instance, in melt growth. In both, closed ampoule (PVT, CVT) as well as open flow (PVD, CVD) systems, free or forced convection is rarely strong enough to restrict

Table 1. Characteristic Diffusivities for Concentration, D_{AB}, Momentum (Kinematic Viscosity), ν, and Heat (Thermal Diffusivity), κ, in Various Crystal Growth Nutrients [16,17].

Nutrient	D_{AB} [cm^2/s]	ν [cm^2/s]	κ [cm^2/s]
Silicon melt	10^{-4}	3×10^{-3}	5×10^{-2}
Aqueous solution	10^{-5}	2×10^{-2}	10^{-3}
Vapor mixture (no H$_2$)	0.2	0.2	0.3

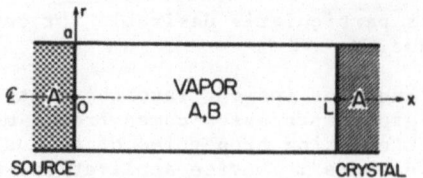

Fig. 2. Model for PVT in cylindrical ampoule; cross-sectional view.

interfacial gradients of concentration and temperature to regions that
are narrow compared to the overall dimensions of the vapor space about
the growing crystal. Hence, mathematically convenient boundary layer
descriptions, that have been advantageously used in melt and solution
growth models, can, at best, give only semi-quantitative insight in
vapor growth. Also, the pronounced thermal expansion of vapors couples
heat and mass transfer so strongly, that transport descriptions based on
constant density ("incompressible flow") lead to large errors. In the
following we will illustrate these features with two examples.

Figure 2 defines a model for the physical vapor transport of material
A through another vapor component B (inert gas or impurity, assumed to be
rejected by the crystal). Interfacial equilibrium vapor pressures are
assumed for A corresponding to the interfacial temperatures, with $T_O > T_L$.

Figure 3 presents steady-state solutions of the diffusion equation
for the (reduced) concentration distribution of A across the vapor space
[18]. One sees that the concentration descends rather uniformly from the
source to the crystal except for very high interfacial concentration
ratios (N_{Pe}'s) - that are rarely reached in actual PVT systems - where a
concentration boundary layer develops. Various experimental and numerical

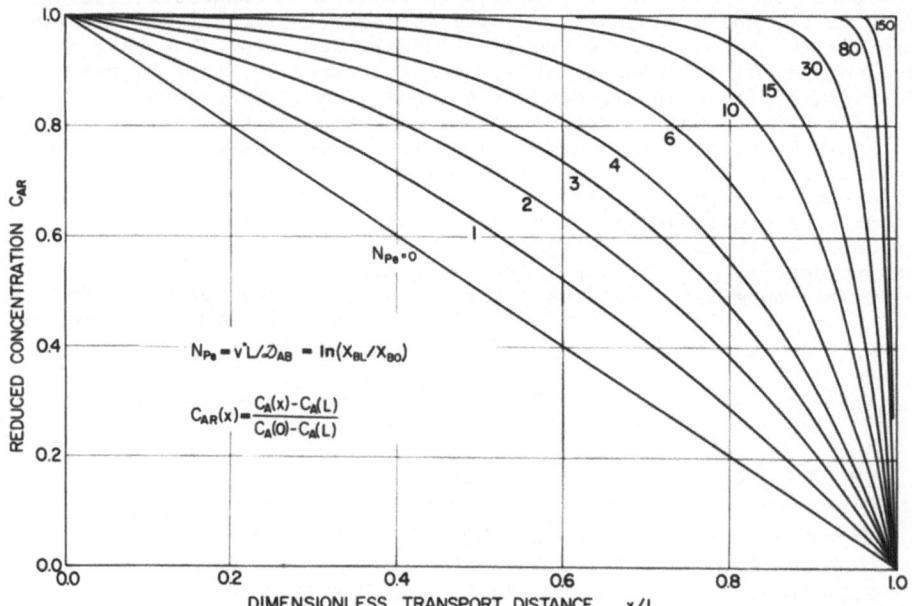

Fig. 3. Reduced concentration of vapor component A versus dimensionless
         transport distance x/L (see Figure 2) for various ratios of the
         mole fractions of component B at the crystal and source,
         respectively. For details see [18].

110

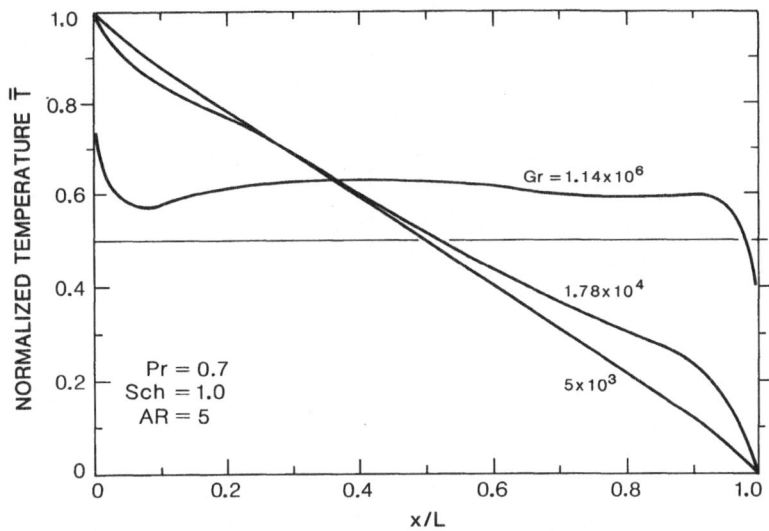

Fig. 4. Temperature profiles on the axis of the PVT model of Figure 2 for various $T = T_0 - T_L$ (i.e., convection velocities and Grashof numbers). For details see [20].

investigations of this problem including thermal and solutial convection [19-23] have shown that convective flows deform these concentration profiles significantly, yet, under practical crystal growth conditions never to an extent that would correspond to the "completely mixed" bulk fluid case often found in melt growth. This is further illustrated by the temperature profiles of Figure 4 that were calculated for horizontal orientation of the PVT system of Figure 2. Again, only for buoyancy-driven flow velocities (Grashof numbers, Gr) that are typically far above those obtained in PVT and CVT, can the convective flow drastically deform the straight conduction profile that results from the (relatively rapid) thermal diffusion between the hot and cold ampoule ends.

The above considerations show that in typical closed ampoule growth, convective transport exceeds diffusion only over parts of the transport path and, consequently, transport rates are often only about twice as high as mere diffusion rates. Such essentially diffusion-limited transport results in relatively low growth rates. The large width of the concentration profiles of Figure 3, however, - even if somewhat "compacted" by convection - can be readily utilized for transport (and purity) enhancement: if the interior of the ampoule, or rather the vicinity of the growing crystal, is "opened" to vacuum through small holes (Figure 5), drastic increases of transport and growth rates are experienced. This is because the concentration of segregating vapor components that accumulated and form a diffusion barrier in front of the growing crystal, is reduced by the effusion flux. Of course, the holes have to be sufficiently small to keep the effusion losses for the crystal material to an acceptable level. For an experimental and theoretical investigation, as well as application references for this rapid, "diffusionless" vapor growth technique, see [25]. In this context one must realize that even with theoretically pure and congruently vaporizing material, closed ampoule growth will always be diffusion-limited and, hence, slow, because of the outgassing (CO, H_2O) that is experienced even with semiconductor grade quartz ampoules, in particular during the sealing procedure [26,27]. For distinctly incongruently vaporizing materials, of

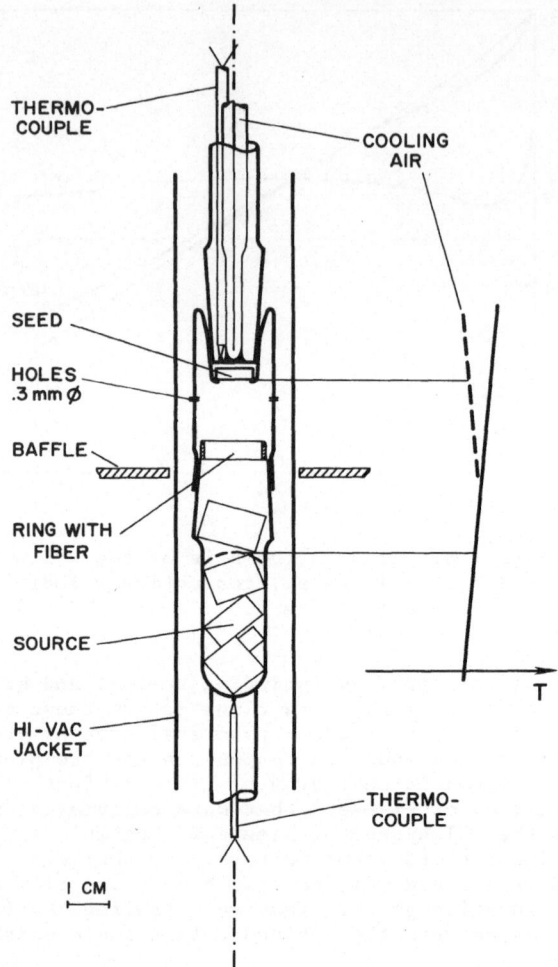

THERMO-
COUPLE

COOLING
AIR

SEED

HOLES
.3 mm ⌀

BAFFLE

RING WITH
FIBER

SOURCE

HI-VAC
JACKET

THERMO-
COUPLE

T

I CM

Fig. 5.   Experimental set-up for diffusionless physical vapor transport
(schematic). For details see [24].

course, the diffusionless technique cannot be applied because eventually
the crystal components with the higher fugacity will be sufficiently
depleted from the vapor phases that phase separation in the growing solid
sets in.

In vapor deposition processes (PVD, CVD, Figure 1) one might expect
that due to the forced nature of the flow, much better mixing of the vapor
phase and more rapid transport to the interface can be obtained. This
expectation is reflected in numerous boundary layer treatments of the
transport in CVD; for a recent review of CVD fluid dynamics see [28].
However, in order to obtain reasonable yield, i.e., economical conversion
of, say, silicon containing vapor species to solid silicon layers, the
flow velocities have to be chosen small enough that diffusion can
establish significant concentration gradients throughout the whole reactor
space, rather than only in a thin layer above the substrate. Hence, it is
not surprising that in-situ concentration distribution studies and
temperature profile measurements [29], optical investigations [30] as well

as most recent modelling [31-33] have shown that there are no boundary
layers formed under typical CVD flow conditions.

This, and another essential difference to melt growth, namely the
importance of Soret diffusion [34,35] in CVD, are well illustrated in
Figure 6. Soret or thermal diffusion tends to enrich heavier components
in the colder parts of a temperature gradient that is imposed on a gas or
vapor mixture. The combination of large temperature gradients and large
differences in molecular weight prevalent in CVD lead to strong
contributions of thermal diffusion to mass transfer. Figure 6 shows
numerical results for SiH_4 concentration profiles in a CVD reactor,
obtained without and with Soret diffusion. Besides the absence of a
boundary layer in the concentration, one sees that thermal diffusion
counteracts Fickian diffusion to the substrate to an extent that, in the
vicinity of parts of the cold top wall, the SiH_4 inlet concentration is
exceeded by more than 40%. The importance of this mechanism for layer
thickness uniformity has been recognized only very recently [33].
Specifically it was found that, depending on the distance from the leading
edge of the substrate (susceptor), thermal diffusion causes growth rate
decreases and increases of up to 40% and 140%, respectively; see Figure 7.

3. INTERFACE STRUCTURE AND ATOMIC ROUGHNESS

The kinetic details of a specific crystal growth process are largely
determined by the atomic roughness of the interface between the crystal
and its nutrient. Until recently it was assumed that vapor-solid
interfaces are generally atomically smooth and, hence, growth occurs by
(intrinsically slow) layer-spreading mechanisms, i.e., occurs
macroscopically as faceted. In addition to the widespread observation of
facets in vapor growth, this contention of atomic smoothness was based on
the results of earlier models.

Burton, Cabrera and Frank (BCF) [36], expanding work by Kossel,
Stranski and Bethe, devised a model to estimate the atomic roughness of
surfaces. In a single layer, two-level model, adatoms were allowed to
occupy lattice sites above a perfect plane of a Kossel crystal. Employing
the exact Onsager solution for the spin-1/2 Ising model, BCF postulated
surface roughening (which they termed surface melting) to occur for a
square surface lattice at the temperature T_r defined by

$$(kT_r)/\phi = 0.567, \qquad (1)$$

where ϕ is the bond energy. Based on (1), BCF concluded that thermal
roughening for solid-vapor interfaces occurs typically above the melting
point of the model and, hence, will rarely be observed in reality.

Some account of interaction of a fluid phase with the solid phase,
and its consequences for surface roughness was given by Jackson [37]. He
calculated the change in Gibbs free energy ΔG of an initially planar
interface upon addition of extra molecules to form a single layer from its
coexisting fluid. Assuming random distribution of the added molecules
(Bragg-Williams approximation) Jackson derived the widely used criterion
for interface roughness based on the relation

$$\alpha = qf/kT, \qquad (2)$$

where q is the internal energy change per atom (often approximated by the
"latent heat", i.e., the enthalpy) for the phase transition considered,
and the anisotropy factor f is the fraction of nearest neighbors that lie
in a bulk plane parallel to the interface layer. For $\alpha < 2$, Jackson's

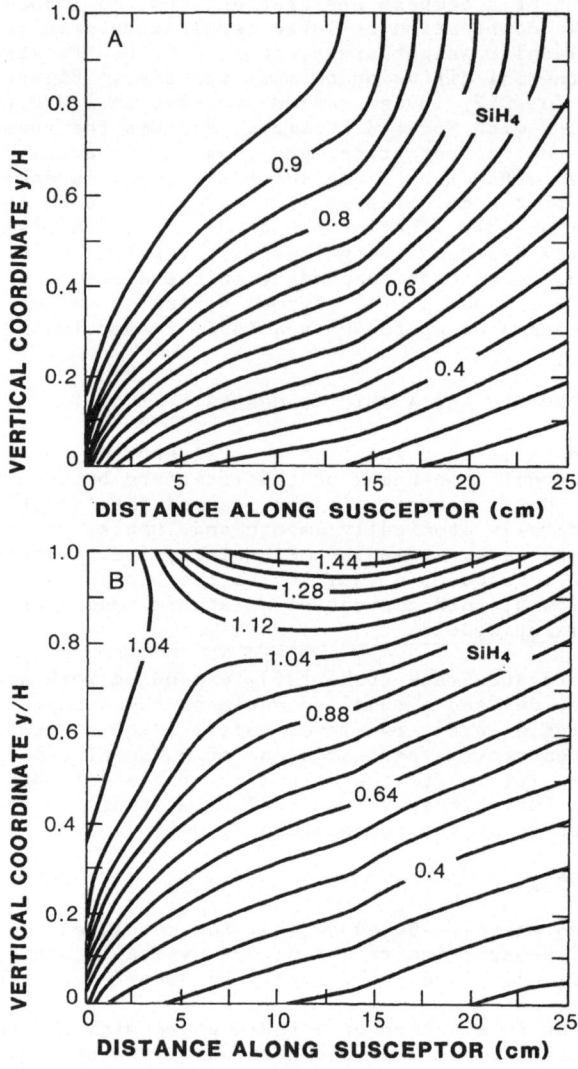

Fig. 6. Numerical results for isoconcentration lines of silane in a CVD
reactor with substrate temperature 1323°K, (upper) cold wall
temperature 300°K and reactor height 2 cm. Concentration values
normalized to inlet concentration. (a) without, and (b) with
thermal diffusion. From [33].

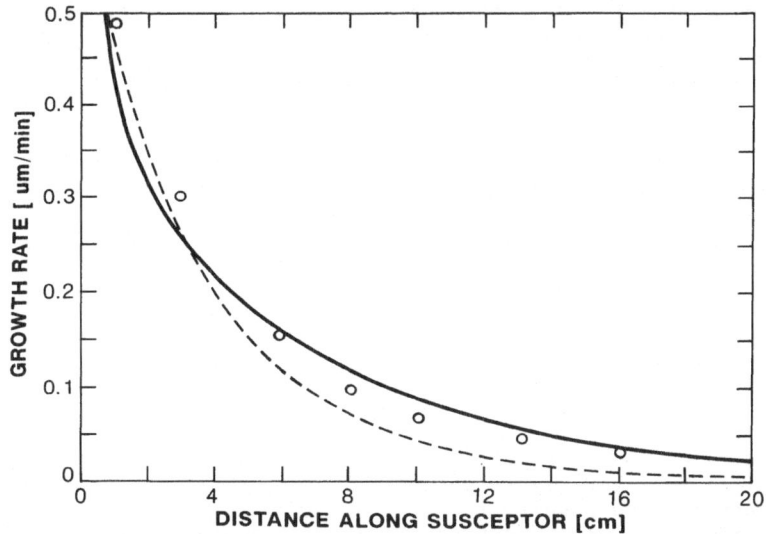

Fig. 7. Comparison of numerical results for CVD growth of GaAs vs.
distance from leading edge of susceptor. Solid and dashed curves
with and without thermal diffusion, respectively. Points are
experimental data. From [33].

model predicts high interface roughness, for systems with $\alpha > 2$ atomically
smooth interfaces are to be expected. Again, this model when applied to
solid-vapor equilibria, predicts exclusively atomically smooth interfaces.

At this point it should be emphasized that all these treatments have
been based on the classic assumption made by Kossel, that both the lattice
parameters and the bond energy on the interface are the same as in the
bulk of the solid. Intuitively one would expect that this "constant bond"
approach is better met by liquid-solid than by vapor-solid interfaces. In
liquids the number and spacing of nearest neighbors are quite comparable
to the coexisting solid. However, in equilibrium with vapor (of typically
three orders of magnitude lower mass density than the solid) the molecules
in the solid surface encounter a pronounced asymmetry in the force field
across the interface, that tends to lead to surface relaxation and
reconstruction, and the bond strength that is derived from bulk properties
will not be representative of the surface conditions.

With this in mind, it is not surprising to find that the model
predictions, for instance based on Jackson's alpha-factor, hold very well
for most melt-solid equilibria. However, recent experimental results
suggest that the constant bond approach fails to predict realistic
surface roughness for numerous vapor-solid systems [38]. The limited
value of such an approach can be demonstrated without resorting to any
details of a specific model, simply based on readily accessible,
macroscopic (averaging) energy considerations: complete vaporization of a
crystal will occur on supply of the cohesive energy (essentially the heat
of sublimation) of the solid to the system. This allows one to ignore
specific surface effects. Yet, in the constant bond model, atoms in a
kink site possess an internal energy equal to the cohesive energy. Note
that for bulk vaporization as well as vaporization from a kink one has to
break per atom $\nu/2$ bonds, where ν is the number of nearest neighbors. On
the other hand, the creation of bulk vacancies, i.e., the removal of an
atom from the bulk to a surface site of "average irregularity", which,

again, is the kink site [39], requires also the net breaking of $\nu/2$ bonds. Hence, if constant bond considerations were realistic, the cohesive energy E_{coh} and the bulk vacancy creation energy E_{bv} of a given system should be equal. However, a detailed survey of metallic, ionic, covalent and van der Waals systems shows that typically $E_{coh} > E_{bv}$; or in other words, in most systems bonds on the surface are considerably stronger than in the bulk [38]. Though quantum mechanical in nature, one can give various atomistic arguments for this "bond strengthening" phenomenon on surfaces; for details see [38].

Utilizing this insight, Ming, Chen and Rosenberger have introduced a site-dependent bond strength ("variable bond") approach into the treatment of atomic surface roughness [38]. The site dependence of the internal energy of an atom, ψ, with n nearest neighbors of a specific site, was cast into the simple form

$$\psi(n) = - (a + bn)n, \qquad (3)$$

where the material constants a and b can be evaluated from heats of sublimation and bulk vacancy formation energies. Note that in the earlier, constant bond models, b = 0, i.e., the internal energy of an atom is simply n times the constant bond energy, ϕ. Besides this new ansatz, the treatment follows essentially the earlier statistical models. Examples for results of the variable bond treatment are given in Figure 8 in the form of ΔG of an initially planar interface upon addition of adatoms. One sees that for both Cu and Pb, for which thermal roughening had been experimentally observed for the (110) planes, only the variable bond model gives realistic results, i.e., shows a lower G for a partly occupied (atomically rough) surface.

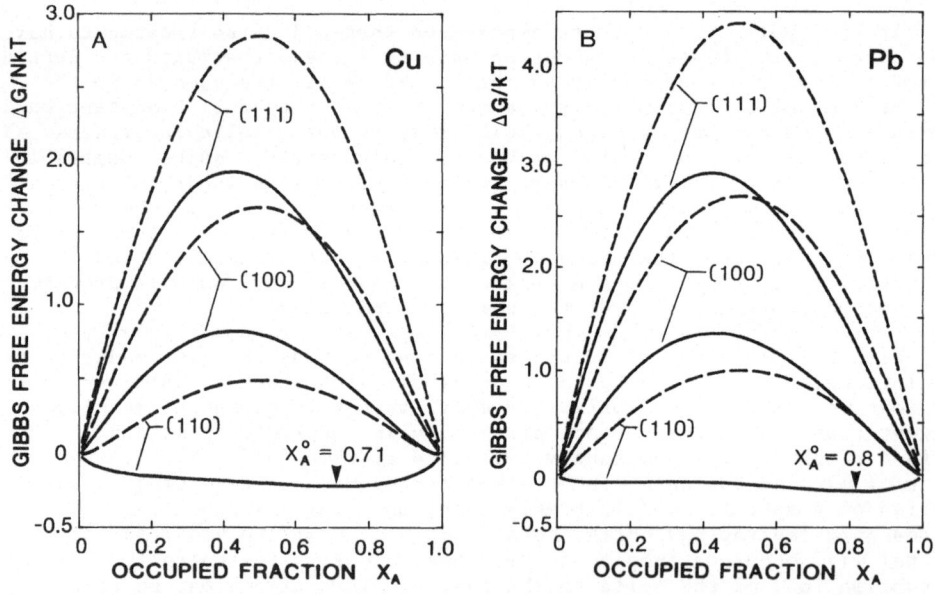

Fig. 8. Normalized Gibbs free energy vs. fraction of surface sites occupied by solid atoms at temperatures somewhat below the bulk melting point, calculated from site-dependent bond model for various surface orientations of (a) copper and (b) lead. From [38].

116

Summarizing one can say that both experimental and theoretical work give strong evidence that surface roughening or melting occur in a large number of vapor-crystal systems. This contention may appear rather recent. One should, however, recall that already in 1942 Stranski [40] had pointed out that for crystal faces that are wetted by their melt, it should be thermodynamically favorable for a thin liquid film to exist on the surface below the bulk melting temperature. Since wetting is observed for various faces of many materials, surface melting (roughening) can be expected to be a widespread phenomenon. Bulk melting, according to Stranski, consists then in the "dissolution" of the lattice in the growing regions of surface melt. This was already envisioned by Tammann [41] in clear distinction from Lindemann's contention [42] that bulk melting occurs as a "catastrophic collapse" of the whole lattice once the lattice vibration amplitudes reach a critical magnitude.

ACKNOWLEDGEMENTS

The author owes much to the stimulating interaction with his co-workers. In particular, J. R. Abernathey, P. Bontoux, J.-S. Chen and K.-C. Chiu. D. W. Greenwell, B. L. Markham, N.-B. Ming, T. A. Nyce, J. Ouazzani and G. H. Westphal have made essential contributions to the material presented here. Research support by the National Science Foundation under Grant DMR-8408398, the Microgravity Science and Applications Division of the National Aeronautics and Space Administration under Grants NSG 1534 and NAG1-733, and the State of Alabama are gratefully acknowledged. Additional thanks are due to T. A. Nyce for proof-reading the manuscript.

REFERENCES

1. M. M. Faktor and I. Garrett, "Growth of Crystals from the Vapour", Chapmann and Hall, London (1974).
2. E. Kaldis, in: "Crystal Growth", C. H. L. Goodman, ed., p. 49, Plenum, London (1974).
3. D. K. Wickenden and R. W. Brandner, in: "Crystal Growth", B. R. Pamplin, ed., p. 397, Pergamon, Oxford (1975).
4. M. E. Jones and D. W. Shaw, in: "Treatise in Solid State Chemistry", N. B. Hannay, ed., Vol. 5, p. 283, Plenum, New York (1975).
5. E. Schönherr, in: "Crystals: Growth, Properties and Applications", Vol. 2, p. 51, Springer, Berlin-Heidelberg-New York.
6. J. Mercier, J. Crystal Growth, 56:235 (1982).
7. R. Cadoret, in: "Interfacial Aspects of Phase Transitions", B. Mutaftschiev, ed., p. 453, Reidel, Dordrecht (1982).
8. A. A. Chernov, Modern Crystallography III, "Crystal Growth", Chapter 8, Springer, Berlin (1984).
9. D. W. Hess, K. F. Jensen and T. J. Anderson, Rev. Chem. Engrg., 3:97 (1985).
10. Proceedings of the First International Conference on Vapor Growth and Epitaxy, J. Crystal Growth, 9 (1971).
11. Proceedings of the Second International Conference on Vapor Growth and Epitaxy, J. Crystal Growth, 17 (1972).
12. Proceedings of the Third International Conference on Vapor Growth and Epitaxy, J. Crystal Growth, 31 (1975).
13. Proceedings of the Fourth International Conference on Vapor Growth and Epitaxy, J. Crystal Growth, 45 (1978).
14. Proceedings of the Fifth International Conference on Vapor Growth and Epitaxy, J. Crystal Growth, 56 (1981).
15. Proceedings of the Sixth International Conference on Vapor Growth and Epitaxy, J. Crystal Growth, 70 (1984).

16. F. Rosenberger, "Fundamentals of Crystal Growth", Section 6.2, Springer, Berlin-Heidelberg-New York (1979).

17. F. Rosenberger and G. Muller, J. Crystal Growth, 65:91 (1983).

18. D. W. Greenwell, B. L. Markham and F. Rosenberger, J. Crystal Growth, 51:413 (1981).

19. B. L. Markham, D. W. Greenwell and F. Rosenberger, J. Crystal Growth, 51:426 (1981).

20. B. L. Markham and F. Rosenberger, J. Crystal Growth, 67:241 (1984).

21. J. C. Launay, J. Miroglio and B. Roux, J. Crystal Growth, 51:61 (1981).

22. J. C. Launay and B. Roux, J. Crystal Growth, 58:354 (1982).

23. J. Ouazzani, P. Bontoux, F. Elie, R. Peyret and B. Roux, Proceedings of the Sixth European Symposium on Materials Sciences Under Microgravity Conditions, Bordeaux 1986, ESA SP-256 (1987).

24. F. Rosenberger and G. H. Westphal, J. Crystal Growth, 43:148 (1978).

25. J. R. Abernathey, D. W. Greenwell and F. Rosenberger, J. Crystal Growth, 47:145 (1979).

26. G. J. Russell and J. Woods, J. Crystal Growth, 46:323 (1979).

27. G. Schmidt and R. Gruehn, J. Crystal Growth, 57:585 (1982).

28. D. W. Hess, K. F. Jensen and T. J. Anderson, Rev. Chem. Engrg., 3:97 (1986).

29. S. Berkman, V. S. Ban and N. Goldsmith, in: "Heteroepitaxial Semiconductors for Electronic Devices", G. W. Cullen and C. C. Wang, eds., Chapter 7, Springer, New York-Heidelberg-Berlin (1978).

30. L. G. Giling, J. Electrochem. Soc., 129:634 (1982).

31. H. K. Moffat and K. F. Jensen, J. Crystal Growth, 77:144 (1986).

32. H. K. Moffat and K. F. Jensen, J. Electrochem. Soc., 135:459 (1988).

33. J. Ouazzani, K.-C. Chiu and F. Rosenberger, in: "Proceedings of the Tenth International Conference on Chemical Vapor Deposition", G. W. Cullen, ed., p.165, Electrochem. Society, Pennington, NJ (1987).

34. K. E. Grew and T. L. Ibbs, "Thermal Diffusion in Gases", Cambridge University Press (1952).

35. J. O. Hirschfelder, C. F. Curtiss and R. B. Bird, "Molecular Theory of Gases and Liquids", Wiley, NewYork (1954).

36. W. K. Burton, N. Cabrera and F. C. Frank, Philos. Trans. Roy. Soc. London, A243:299 (1951).

37. K. A. Jackson, in: "Liquid Metals and Solidification ", M. Maddin, ed., ASM, Cleveland (1958).

38. J.-S. Chen, N.-B. Ming and F. Rosenberger, J. Chem. Phys., 84:2365 (1986).

39. C. P. Flynn, "Point Defects and Diffusion", p. 7, Oxford University Press, London (1972).

40. I. N. Stranski, Naturwissenschaften, 30:425 (1942).

41. G. Tammann, Z. Phys. Chem., 68:205 (1910).

42. F. A. Lindemann, Phys. Z., 11:609 (1910).

FUNDAMENTALS OF SOLUTION GROWTH

William R. Wilcox

Clarkson University
Potsdam
New York 13676, USA

GROWTH KINETICS AND MORPHOLOGY

Growth Processes

In order for growth to occur the solute species must move from the bulk solution to the crystal surface, be adsorbed on the crystal surface, move to a step, and finally move to a kink on the step. If ionic species are involved then dehydration and chemical reaction may also be required. The latent heat of crystallization must also move back into the solution, since in solution growth the crystal is generally surrounded by the solution and there is no heat sink within the crystal. The net effect is that the surface of the crystal must be at a slightly different temperature from the bulk of the solution. In the following discussion we ignore the thermal effect, which is generally small albeit not zero.

The movement of solute from the bulk solution to the crystal surface occurs by a combination of molecular diffusion and convection (solution movement) under the general title of mass transfer. The processes that occur on the surface we may lump together under the general title of surface integration. If the growth steps on the crystal surface are sufficiently close together, we may consider that the mass transfer is perpendicular to the crystal surface and steady, i.e., one-dimensional steady state. This is shown schematically in Figure 1. Under these conditions the rate of mass transfer and surface integration must be equal:

$$\text{Growth rate} = \text{surface integration rate} = \text{mass transfer rate} \quad (1)$$

$$V_c = g(C_i - C_e) = D(dC/dx)_i/C_c(1 - F_i) \quad (2)$$

where V_c is the growth rate normal to the surface in m/s, g is a function depending on the growth mechanism, C_i is the interfacial solute concentration in mol/m^3, C_e is the equilibrium solute concentration corresponding to solubility at the growth temperature, D is the solute diffusion coefficient in the solution, x is the distance into the solution from the surface of the crystal, C_c is the solute concentration in the crystal (equal to density divided by molecular weight when no solvent is

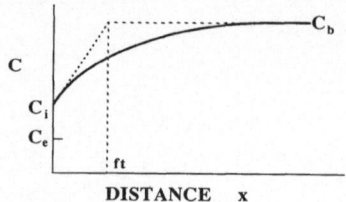

DISTANCE x

Fig. 1. Solute concentration vs. distance into solution. Here ft is
fictitious stagnant film thickness.

incorporated in the crystal), and F_i is the interfacial volume fraction of
solute in the solution (equal to C_i times partial molar volume) when no
solvent is incorporated in the crystal.

Mass Transfer

It has been customary in the crystal growth literature on mass
transfer to speak of an unstirred layer, diffusion layer, boundary layer,
or stagnant film. This stems from the concept put forth by Nernst in the
last century to explain the growth rate of crystals from solution. As
shown in Figure 2a it is imagined that there is no convective mixing in a
thin layer of solution adjacent to the crystal, while mixing is complete
beyond the stagnant film thickness ft. If this were true, and the
diffusion coefficient were constant in the solution, the solute
concentration would increase linearly with distance away from the crystal,
as indicated by the dashed line in Figure 1.

Fluid mechanicians have known for decades that the stagnant film
concept is totally untrue. It simply does not exist, no matter what it is
called. It is true that the fluid velocity must vanish as one approaches
a solid surface, as shown in Figure 2b. This is called the no-slip
condition. Thus there is a fluid velocity <u>parallel</u> to the crystal
surface, due to whatever stirring device one is using in the growth
system. The actual concentration profile, shown by the solid line in
Figure 1, is linear only near the crystal surface. It bends smoothly to
approach the bulk concentration. The stagnant film thickness is defined
as the thickness of stagnant film that would give the correct
concentration gradient at the crystal surface if there really were a
stagnant film. The film thickness can be determined only if the
interfacial concentration gradient is found from experiment or from
solution of the mass transfer and fluid mechanics differential equations.
Furthermore, one danger of the concept is that one may conclude that the
mass transfer rate is proportional to the diffusion coefficient; in
reality the mass transfer goes as some fractional power of the diffusion
coefficient. That is, the stagnant film thickness is a function not only
of the hydrodynamics, but also of the diffusion coefficient! The crystal
grower really ought to outgrow this 19th century concept.

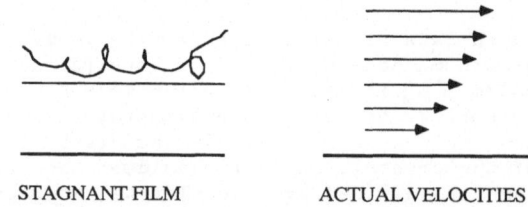

STAGNANT FILM   ACTUAL VELOCITIES

Fig. 2. Comparison of stagnant film model with actual velocity profile.

120

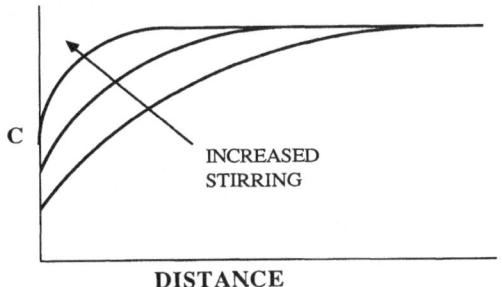

C

INCREASED
STIRRING

DISTANCE

Fig. 3.   Influence of stirring on concentration gradient.

Increasing the stirring rate causes several changes, as shown in
Figure 3. The region in the solution over which the concentration changes
decreases. The interfacial concentration gradient increases. And the
interfacial concentration increases. Thus both the surface integration
rate and the mass transfer rate increase, in accordance with Equation (2).
Another way of looking at this is to consider the effect of changing the
interfacial concentration on the surface integration rate and the mass
transfer rate separately, as shown in Figure 4. As C_i is increased the
surface integration rate increases because the driving force, $(C_i - C_e)$,
increases. On the other hand, if the stirring is kept constant the mass
transfer rate decreases because its driving force, $(C_b - C_i)$, decreases.
The observed growth rate is that at which the surface integration rate and
the mass transfer rate are equal, i.e., when the two curves in Figure 4
intersect. Figure 5 shows the effect of increasing the stirring. The
curve for surface integration remains unchanged while the mass transfer
curve swings upward, hinged at C_b. This causes the intersection of the
two curves to move, increasing both C_i and V_c.

You may have noticed that the mass transfer portion of Equation (2)
does not exactly correspond to Fick's first law; an additional term,
$(1 - F_i)$, appears in the denominator. This term, which increases the mass
transfer rate, arises from the fact that the solubility is finite. One
can argue that such a term <u>must</u> be present by considering the limiting
case of melt growth, wherein no solvent is present, i.e., $F_i = 1$. In that
case a finite growth rate occurs even in the absence of a concentration
gradient. Equation (2) shows that dC/dx approaches zero as F_i approaches
1 only if the $(1 - F_i)$ term is present in the denominator.

Another way of understanding the presence of the $(1 - F_i)$ term in
Equation (2) is to regard the crystal surface as being stationary.
Because growth is taking place, there is a flow of crystal away from the

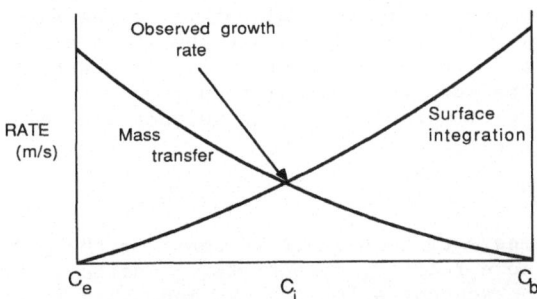

Fig. 4.   Surface integration and mass transfer kinetics.

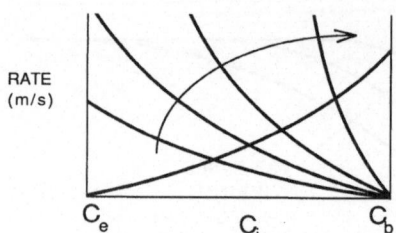

RATE
(m/s)

C_e C_i C_b

Fig. 5. Influence of stirring. Arrow shows increased stirring.

crystal surface. Conservation of mass requires that there be an equal
molar flow into the crystal surface from the solution. This flow has been
called the Stefan flow, interfacial flow, and crystallization flow. It
carries solute with it, unless the solute concentration is negligibly
small. The $(1 - F_i)$ term corrects the diffusion equation for its
presence. It is sad to note that papers continue to be published in the
crystal growth literature that do not take it into account. Periodically
someone discovers the "anomaly" that applying Fick's first law (without
the interfacial flow correction) yields lower growth rates than observed.

Surface Integration

In the previous paragraphs we concentrated on mass transfer and
implicitly assumed that the surface integration kinetics are fixed.
Actually the interface kinetics depend on the perfection of the crystal.
In experiments the growth rate of a face is often observed to fluctuate,
and sometimes even to stop for some time. The causes are uncertain, but
speculation has involved impurity poisoning and dislocation interactions.

It has also been assumed that the interface kinetics at a point is
controlled by the local interfacial supersaturation at that point.
Actually the local supersaturation controls only the movement rate of the
steps, except at the points where steps originate. It should be
understood that faceted growth is a consequence of stepwise growth.
Furthermore, for a face to remain relatively flat it is necessary for step
propagation to be rapid compared to step generation. Steps are generated
either at the points of highest supersaturation by two-dimensional
nucleation or at the sites of emergence of screw dislocations. If a screw
dislocation is at a region of relatively high supersaturation it will
dominate step generation. Otherwise a competition will exist between step
generation by screw dislocations and by two-dimensional nucleation. This
competition plays an important role in morphological stability during
solution growth, as discussed later.

Thus the equations and methods described above really only apply at
the point where steps are generated. Elsewhere on the face the growth
rate must be the same as at the point where the steps are generated or the
face will change shape. The mass transfer rate is constant over the
entire face, and so the surface concentration gradient is inversely
proportional to the volume fraction of the solvent according to Equation
(2).

Multicomponent Growth

Everything becomes more complex if we consider the growth of a
crystal from two or more reacting components. Consider, for example, the
growth of NaCl from an aqueous solution. We know that in the solution we
have hydrated ions of Na^+ and Cl^-. An ion moves not only in response to

122

its own concentration gradient but also to any electric field in the
solution and the concentration gradients of other species. An electric
field will arise not only from the fact that one ionic species will
diffuse faster than another, but also because of preferential adsorption
of ions on the crystal surface and orientation of solvent molecule dipoles
near the crystal surface. If we have a stoichiometric solution we may
treat it as a binary and use the equations developed above. However, if
the solution deviates from stoichiometry or contains ions that are not
incorporated in the crystal then the mass transfer is multicomponent. The
diffusion of a component depends not only on its own concentration
gradient but also on the gradients of the other components in the
solution. For example, it is possible for one component to move from a
region of low concentration to a region of high concentration, contrary to
the behavior of binaries and our intuition. Multicomponent diffusion
coefficients tend to be much more concentration dependent and are
difficult to measure. The mathematics of mass transfer becomes very
difficult, even assuming the multicomponent diffusion coefficients are
known.

Surface integration also becomes more complicated when the crystal
grows from reacting species. The different species are adsorbed
differently, diffuse at different rates, and yet must be incorporated in
the stoichiometric ratio. Consequently, the surface integration kinetics
becomes a function of all of the species concentrations at the crystal
surface. With rapid stirring and relatively slow surface integration
kinetics, the surface concentrations become nearly equal to the bulk
concentrations, allowing this functional dependence to be determined
experimentally.

Widely-spaced Steps

In the beginning we assumed that the mass transfer is one-dimensional
and steady because the growth steps are sufficiently close together. How
close must they be and what are the consequences if they are not?

Figure 6 is a diagram of a train of steps moving to the right at
velocity V_s. During the time t it takes for a step to move by a fixed
point diffusion in the solution is felt over a distance $(Dt)^{1/2} =$
$(Dd/V_s)^{1/2}$, where d is the distance between steps. The macroscopic growth
rate $\bar{V} = V_s h/d = D(dC/dx)_i/C_C(1 - F_i)$. Thus the ratio of diffusion
distance in the solution to step spacing d is $(Dt)^{1/2}/d = (Dh/V)^{1/2}/d =$
$(hC_C(1 - F_i)/(dC/dx))^{1/2}/d$. If this quantity is much less than one then
the mass transfer is one-dimensional and steady. If not then the
diffusion field distorts as each step passes by. For mono-atomic steps,
the ratio of diffusion distance to step spacing is generally large.
However, macro steps often form that move slowly (and often become
unstable, as described later). In spite of this a steady state treatment
of combined diffusion and surface integration can be made by moving the
coordinate system along with the step train. Treatments of coupled steady
state volume diffusion and surface integration have appeared in the
literature, but only for very low solubility (F_i implicitly assumed
negligible).

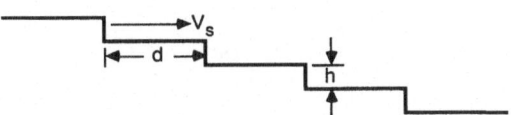

Fig. 6.   Step train. V_s is velocity of step, d is distance between steps,
          and h is step height.

Morphology

The morphology of a crystal is its shape. The habit is its usual shape under given growth conditions. It is often true that the faces exhibited by a crystal are those with the greatest number of bonds within the plane of the surface. Because these planes have the lowest surface energy it has frequently been assumed that the crystal grows so as to minimize its surface energy, i.e., that crystal growth occurs at thermodynamic equilibrium. This is not true. Crystal growth is a kinetic phenomenon and the faces exhibited by a crystal are those that grow the slowest.

The lowest energy faces do usually grow the slowest, because the edge energy of steps is higher; there are more broken bonds on such steps than for those on planes with fewer bonds within the plane. With a higher edge energy, step formation both by two-dimensional nucleation and by screw dislocations is greatly reduced. However, there are many exceptions to the above morphological rule. Crystal habit often depends on super-saturation, pH, solvent, and impurity content. A variety of morphologies are often observed even in a batch of crystals grown together.

The difference between kinetic morphology and equilibrium morphology can be demonstrated experimentally. A small crystal confined in a small drop of solution will slowly change to its equilibrium habit.

You may convince yourself that the slowest growing faces are the largest faces by examining Figure 7. Two facets are shown at three time intervals. The distance between successive positions of a face equals the product of growth rate and time; the larger the growth rate, the larger the distance between successive positions. In this figure face A is growing much slower than face B. Notice that face B is becoming smaller, and will eventually disappear. There is a range of ratios of growth rates for the two faces that will allow both faces to survive, although with different sizes. The calculation of this range is not difficult and is left as an exercise for the student.

MORPHOLOGICAL STABILITY

Inclusions

The most serious problem in crystal growth from solution is the occlusion of solution to form inclusions inside the crystal. The problem' becomes more severe as supersaturation increases, crystal size increases, stirring decreases, concentration gradients along the crystal surface increase, step height increases, and temperature fluctuations increase.

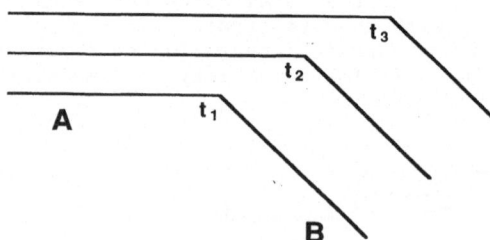

Fig. 7. Facet positions vs. time t. Face B is growing over twice as fast as A and is decreasing in size.

Inclusion formation is strongly influenced by solvent, growth temperature, impurities, etc.

Inclusions occur in many forms. They may be large, equidimensional and isolated. They may occur in groups elongated along the crystal surface. Many tiny inclusions may lie in a flat or curved sheet. (This is called a veil.) They may be only at the centers of the faces or may be at the corners of edges. They may consist only of solution, or may contain both solution and gas in a ratio that varies widely throughout a given crystal.

Inclusions are undesirable because they are a source of impurity. Sometimes, as in pharmaceuticals crystallized from organic solvents, the solvent itself may be undesirable. Because impurities are usually rejected by a growing crystal, the inclusions may contain virtually all of a given impurity within a crystal. Inclusions scatter and absorb light in optical crystals. Great numbers of dislocations emerge from inclusions, especially the larger ones, because of misalignment as the crystal closes over a cavity during growth.

Constitutional Supercooling

It is instructive to examine solution growth from the viewpoint of the classical constitutional supercooling model, as shown in Figure 8. Because the crystal is a sink for solute, the solute concentration increases with distance into the solution, as we have discussed previously. Corresponding to this concentration increase the equilibrium temperature increases (for a positive latent heat of crystallization), as shown in 8B. Liberation of the latent heat at the crystal surface causes the temperature to increase slightly there. Corresponding to this, the solubility is slightly increased as shown in 8A. Thus the supersaturation and the supercooling increase as we move out into the solution from the crystal surface. That is, we always have constitutional supercooling in solution growth (unless we locate an effective heat sink inside the crystal).

Reasoning in analogy to melt growth one would always expect morphological instability in solution growth. It should be impossible to grow a crystal without inclusions and even dendrites forming. And yet large relatively perfect crystals are grown from solution routinely. Why? How?

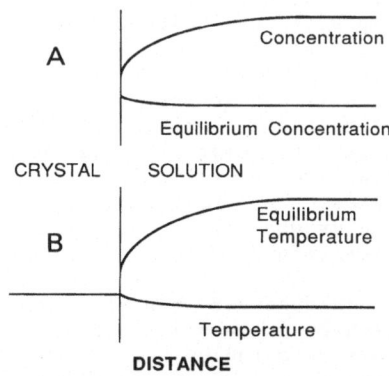

Fig. 8. Constitutional supercooling in solution growth.

Solution growth is stabilized by being stepwise. We know that it is stepwise because we observe facets. In fact in those few cases in which facets are not observed, the growth readily becomes dendritic (ammonium chloride in water is an excellent example). Now we must ask ourselves why stepwise growth sometimes becomes unstable and traps solution. The answer to this is not so clear.

Some Mechanisms by which Inclusions may Form

In one viewpoint, crystals become unstable when steps become so close together that the surface appears atomically rough and the steps are scarcely discernible. Adatoms need not diffuse far to find a kink on a step. The growth rate is controlled by mass transfer and the constitutional supercooling concept is applicable.

In another view, a step train becomes unstable when it slows down. We make an analogy with traffic flow in and out of a large city. Indeed the mathematics of step train movement and automobile movement is similar. In the morning, when traffic is moving into the city, the average velocity diminishes as one approaches the city. This is unstable and causes large velocity fluctuations; a car may be stopped one moment and moving fast the next. In the evening, when traffic is moving out of the city, the average velocity increases as one moves away from the city. This is stable and the traffic moves smoothly.

In crystal growth it is common for the dislocations to be located near the center of a face. Generation of new steps at the dislocations competes with generation of new steps by two-dimensional nucleation at the face corners where the supersaturation is higher. The rate of two-dimensional nucleation is much more supersaturation dependent than the rate of forming new steps at a screw dislocation. Because of this, screw dislocation generation predominates at small supersaturations and two-dimensional nucleation predominates at large supersaturations.

In the absence of convection, the interfacial supersaturation is higher at the corner of the crystal face than at the center. When the bulk supersaturation is low, steps are formed at the screw dislocations near the centers of the faces. As they propagate outward, they encounter an increasing interfacial supersaturation and so accelerate. This is stable and no inclusions form. On the other hand, when the bulk supersaturation is high, steps are formed by two-dimensional nucleation at the corners. As the steps propagate inward they encounter a decreasing interfacial supersaturation and decelerate. This is unstable and inclusions form. This explains why inclusions are more common with higher supersaturations and larger crystals.

In the presence of gentle convection, the supersaturation diminishes as solution flows along the crystal surface. Steps will tend to form where the flow first contacts the crystal. They will propagate in the same direction as the flow, and so will decelerate and become unstable. Vigorous convection can cause the interfacial concentration to become relatively uniform and the growth rate to be controlled by surface integration, which is stabilizing.

Two phenomena have been observed when steps decelerate. In some cases they form waves which break much like those at the beach. Under the waves are formed either discrete inclusions or tubular inclusions, elongated in the direction of step movement. In other cases macrosteps are formed, and these can occlude solution in three ways.

A macrostep may stop moving forward and grow taller as new atomic steps arrive on top of it. Finally the interfacial supersaturation at the top edge becomes sufficiently high that a layer grows out over the top of the cavity. When it reaches the other side, the cavity is sealed and a large inclusion has been formed.

The base of a macrostep may stop growing while the top continues, forming an overhang. If the overhang is thin, gravity or convective stresses may cause it to bend down and contact the underlying crystal, thereby forming an inclusion. Or the overhang may continue until it meets crystal growing from the other direction.

In some cases the macrostep itself becomes unstable and develops waves or cells. The troughs of the waves trap solution either as a series of small inclusions or as tubes.

Macrosteps and inclusions also result from dissolution. Since dissolution is largely mass transfer controlled, the edges of an initially faceted crystal become rounded. When growth is reinitiated, macrosteps and inclusions form at these rounded edges. If dislocation etch pits form during dissolution, these can give rise to small inclusions during regrowth. It is common practice in solution growth to begin with the solution above the equilibrium temperature, so as to dissolve any crystalline powder floating in the solution or on the surface of the seed crystal, and to remove damage from the seed crystal. Consequently it is common for a layer of inclusions to form at the boundary between the seed crystal and the grown material, thus neatly outlining the seed.

It appears that very small temperature fluctuations can lead to formation of inclusions. At first it might be supposed that this is a manifestation of the known tendency of inclusion formation to increase as supersaturation is increased. However, the temperature fluctuations observed to cause inclusions are much less than a steady supercooling required to produce inclusion formation. Thus this appears to be a transient phenomenon. Let us propose an explanation.

At low supersaturations the steady state spacing between growth steps is larger than it is at high supersaturations. If we are growing at a low supersaturation which is suddenly increased, the step spacing will correspond to the previous lower supersaturation rather than the new higher supersaturation. The steps will accelerate and may very well bunch and form a wave or macrostep. On the other hand, it may be during a sudden decline in supersaturation that instabilities develop. The steps will be close together already and will decelerate. A theoretical treatment of such transient behavior is needed.

Veils appear to form by cracking of a crystal, either during growth or later. When a crack contains solution, it heals to leave many tiny inclusions. In this way the surface energy of the crack is reduced. Cracks may arise from stress due to temperature or composition variations throughout the crystal. They may also arise from stress generated by large inclusions. For example, the vapor pressure of water in aqueous inclusions becomes sufficient to crack crystals when they are heated.

Characteristics of Inclusions

There are a number of interesting features about inclusions formed as above. One is that dislocations are frequently formed during closure of a cavity, due to mismatch between the two sections of the crystal growing together. This is particularly common with large inclusions. These dislocations can have a significant effect on the growth kinetics, morphology, and morphological stability.

Another interesting aspect of inclusions is that they frequently contain gas bubbles. These were first reported in mineral specimens and attributed to growth from higher temperatures and pressures. If the thermal contraction of the solution is greater than the crystal, bubbles would be expected upon cooling to room temperature. Mineralogists supposed that the temperature at which the bubbles redissolve is the original growth temperature. However, room temperature experiments have shown that gas bubbles are incorporated in inclusions even in the absence of a temperature change after being trapped. Furthermore, the ratio of gas to liquid varies widely from one inclusion to another in a given crystal; indeed it may range from zero to one, from pure gaseous inclusions to purely liquid inclusions.

Gases are generally much more soluble in liquids than in solids. Since the solutions used for crystal growth normally contain dissolved gases (e.g., air), it might be expected that gas bubbles would form. However, if one calculates the concentration of dissolved gas expected, it hardly seems large enough to nucleate bubbles. Thus the presence of gas bubbles remains a little surprising. There is no doubt that they form, and they probably play an important role in formation of some inclusions. For example, since surface tension depends strongly on concentration, it might be expected that vigorous Marangoni convection would occur around a bubble. Unfortunately no observations have been reported of gas bubble formation and behavior during solution growth, so we remain quite ignorant.

Movement of Inclusions

Solvent inclusions can be moved through and out of crystals by the application of a temperature gradient. The crystalline material dissolves at the hot end of an inclusion and crystallizes out at the cold end (assuming solubility increases with temperature). This causes the inclusion to move toward higher temperatures. In the process a variety of interesting phenomena occur.

The end of the inclusion that is dissolving develops facets while the crystallizing end is rounded, which is just the opposite of behavior observed in growth. Similarly if a crystal is heated isothermally, the inclusion develops facets. That is, an inclusion behaves much like a negative crystal.

If a gas bubble is present, vigorous Marangoni convection can occur about it. The bubble and some or all of the solution will migrate through the crystal away from the high temperature regions, which is is just the opposite direction of that for 100% solvent inclusions. This occurs because solvent evaporates from the hot end of the bubble and condenses out on the cold side, where it dissolves fresh crystalline material.

When a solvent inclusion moves to an exterior surface of the crystal, some of the solvent is replaced by a gas bubble, which proceeds to cause the remaining solvent to move back into the crystal.

IMPURITY EFFECTS AND INCORPORATION

Impurities have been observed to have a large influence on growth kinetics, morphology, and morphological stability. These effects are attributed primarily to adsorption of impurity on the growing crystal surface. This may either lower the growth rate or increase it. If the impurity adsorbs strongly on the steps and lowers their energy, then it becomes easier to generate new steps by two-dimensional nucleation or a screw dislocation. On the other hand, impurity species can interfere with addition of solute species to growth steps. The net effect of impurity adsorption depends on which of these two phenomena predominates.

Another consequence of impurity adsorption is impurity incorporation in the crystal. At slow growth rates adsorbed species have time to desorb before they are buried by growth steps. Equilibrium is approached between the bulk crystal and the bulk solution. At very high growth rates adsorbed species have no time to desorb and the impurity concentration in the bulk crystal approaches that adsorbed on the crystal surface. Since adsorption, step velocity and height, etc. vary from face to face, impurity incorporation also varies from face to face. This can lead to an hourglass effect.

Impurity striations are observed in many crystals grown from solutions. The impurity concentration varies primarily normal to the crystal faces, producing the appearance of bands demarking the crystal shape at the time they were formed. Since impurity incorporation depends strongly on growth rate, striations are taken as evidence of a fluctuating growth rate. While poor temperature control is the primary cause of a fluctuating growth rate, interfacial processes are sometimes responsible. For example a dislocation may move from one face to another, causing the growth rate to drop on one face and increase on the other. One can distinguish between striations caused by phenomena in the bulk solution and interfacial processes. If a striation covers all crystal faces, it must be caused by something that happened throughout the solution and affected all faces at the same time. On the other hand if a striation is only on one face, then whatever caused it must have occurred on that face and on no other.

TECHNOLOGY OF CRYSTAL GROWTH FROM SOLUTIONS

There is an unlimited number of combinations of ways of growing crystals from solution. Simply cooling a solution will lead to uncontrolled nucleation and poor quality crystals. Consequently over the years crystal growers have developed many techniques of seeding, controlling supersaturation, and stirring. A few of these are discussed below.

Seeding

In research laboratories a seed may be selected from spontaneously nucleated crystals. In industrial practice a seed is cut from a preceding crystal. Orientation is selected to give maximum growth rate or desired morphology. A hole may be drilled into the seed for insertion of a pin to restrain it, or it may be held by clamps or cement. This cutting and drilling produces a great deal of damage and dust, which may be removed by dissolving part of the seed prior to insertion in the solution. In addition some dissolution is carried out in the growth solution prior to lowering the temperature to cause growth.

The transition from dissolution to growth normally produces inclusions about much of the seed crystal. A recent experiment in Spacelab 3 on triglycine sulfate showed that a very gradual transition from dissolution to growth may avoid these inclusions.

During growth it is desirable to avoid spurious nucleation of unwanted new crystals, which may settle on growing seeded crystals and ruin them or may merely absorb solute and lower the yield. Some solution-crystal systems are extremely subject to spurious nucleation while problems are seldom encountered in others. Typically one speaks of a metastable supersaturation limit, below which one can grow crystals without nucleation and above which nucleation occurs. In practice the metastable limit is influenced by a variety of factors. For example, some foreign particles may act as nucleation sites, so that filtering of the solution prior to growth is wise. Vigorous agitation produces nuclei in some systems. Impurities may inhibit or enhance nucleation.

Production of Supersaturated Solutions

Below are some methods available for producing a supersaturated solution. In all cases, supersaturation should be carefully controlled and varied only slowly in order to avoid formation of inclusions. For example a temperature control on the order of $0.001^\circ C$ is generally desired.

1. Temperature programming.
2. Circulation of solution between crystal at one temperature to solid source of solute at a temperature producing a larger solubility. Circulation may be by natural convection (buoyancy) or by means of a pump. (Hydrothermal growth uses natural convection between source below and growing crystals above, with a baffle plate in between to control circulation.)
3. Evaporation of solvent and condensation with flow through solid source.
4. Removal of solvent by evaporation or ultrafiltration through a membrane.
5. Electrodialysis.
6. Changing the solubility of the solvent by slowly mixing in solvent with lower solubility for crystal. This may be done by diffusion through a gel or in a microgravity environment.
7. Changing the pressure. At sufficiently high pressure the solubility depends on pressure. Unlike temperature changes, a pressure change propagates virtually instantly throughout a liquid. (This technique was suggested by P. J. Shlichta at the Jet Propulsion Laboratory.)
8. Chemical reaction to form insoluble crystal. Mixing of reagents must be slow or excessive nucleation occurs. Seeding is generally not employed. To avoid convection either a gel or a microgravity environment is used.

Stirring

Vigorous stirring generally allows a higher growth rate while maintaining crystalline perfection. However, it must be uniform so as not to produce stagnant regions or recirculating flows, or inclusions will be produced. And it must not be so vigorous as to produce spurious nucleation. Stirring techniques are:

1. Buoyancy-driven free convection. This occurs in the absence of forced convection and results in inclusions unless the growth rate is very low.
2. Rotation of rack containing crystals. Alternate rotation in one direction and then the other avoids inclusions on downstream facets due to recirculating eddies.
3. Rotation of crystal.

4. Accelerated crucible rotation technique. Container is rotated for a short time, stopped for a time, rotated again, etc. (The hydrodynamicists call this spin-up/spin-down. It has been successfully used for flux growth, and probably produces a fluctuating growth rate or even dissolution alternating with growth.)
5. Rotating impellers of various designs and locations, e.g., a propeller below the crystal. Baffles along the walls increase mass transfer without rapid rotation of the solution.
6. Circulating pump, either internal or external.

CONCLUSION

While one can state many general principles and write textbooks full of equations, successful solution growth of any new crystal will require careful experimentation. Even with crystals grown commercially for many years, room remains for creative improvements to improve crystal quality and growth rates.

Need remains for improvements in both the art and the science of solution growth.

REFERENCES

Below are listed only a few of the many recent publications on solution growth. The serious practitioner should make use of Chemical Abstracts and Science Citations Index to locate papers of particular applicability.

1. H. K. Henisch, "Crystal Growth in Gels", Pennsylvania University Press, University Park (1970).
2. W. R. Wilcox, A generalized treatment of mass transfer in crystal growth, Volume 2 of "Preparation of Properties of Solid State Materials", Dekker, New York (1976).
3. I. Sunagawa and P. Bennema, Morphology of growth spirals: theoretical and experimental, Volume 7 of "Preparation of Properties of Solid State Materials", Dekker, New York (1976).
4. M. Ohara and R. C. Reid, "Modelling Crystal Growth Rates from Solution ", Prentice-Hall, Englewood Cliffs, NJ (1973).
5. I. Tarjan and M. Matrai, eds., Laboratory Manual on Crystal Growth, Akademiai Kiado, Budapest, Hungary (1972).
6. A. Tsuchiyama, M. Kitamura and I. Sunagawa, Distribution of elements in growth of $(Ba,Pb)(NO_3)_2$ crystals from the aqueous solutions, J. Crystal Growth, 55:510 (1981).
7. A. Yokotani et al., Fast growth of KDP single crystals by electrodialysis method, J. Crystal Growth, 67:627 (1984).
8. J. F. Cooper and M. F. Singleton, Rapid growth of potassium dihydrogen phosphate crystals, Proc. Int. Conf. Lasers 1984, 567 (1984).
9. D. Shiomi, T. Kuroda and T. Ogawa, Thermal analysis of a growing crystal in an aqueous solution, J. Crystal Growth, 50:397 (1980).
10. M. Kitamura et al., Growth and dissolution of $NaClO_3$ crystal in aqueous solution, Mineral. J., 11:119 (1982)
11. G. M. Loiacono, The industrial growth and characterization of KD_2PO_4 and CsD_2AsO_4, Acta electronica, 18:241 (1975).
12. R. Boistelle, Crystal growth from non aqueous solutions, in: "Interfacial Aspects of Phase Transformations", B. Mutaftschiev, ed., Reidel (1982).
13. A. E. Nielsen, Electrolyte crystal growth mechanisms, J. Crystal Growth, 67:289 (1984).

14. R. Boistelle, The concepts of crystal growth from solution, in: "Advances in Nephrology", J. P. Greenfeld et al., eds., Vol. 5, p. 173, Year Book Medical Publishers, Chicago (1986).

15. W. Kolasinski, On the formation of liquid inclusions in some solution-grown crystals: (I) Sodium-cadmium formate, Mat. Res. Bull., 19:867 (1984).

16. W. J. P. van Enckevort and L. A. M. J. Jetten, Surface morphology of the (010) faces of potassium hydrogen phthalate crystals, J. Crystal Growth, 60:275 (1982).

17. R. Rodriguez and M. Aguilo, Unstable growth of ADP crystals, J. Crystal Growth, 47:518 (1979).

18. W. R. Wilcox, Morphological stability of a cube growing from solution without convection, J. Crystal Growth, 38:73 (1977).

19. P. Slaminko and A. S. Myerson, The effect of crystal size on occlusion formation during crystallization from solution, AIChE J., 27:1029 (1981).

20. L. A. M. J. Jetten, V. van der Hoek and W. J. P. van Enckevort, In situ observations of the growth behavior of the (010) face of potassium hydrogen phthalate, J. Crystal Growth, 62:603 (1983).

21. H. Narayanan, G. R. Youngquist and J. Estrin, Non-uniform solution growth of potassium aluminium sulfate, J. Colloid Interface Sci., 85:319 (1982).

22. P.-S. Chen, W. R. Wilcox and P. J. Shlichta, Free convection about a rectangular prismatic crystal growing from a solution, Int. J. Heat Mass Transfer, 22:1669 (1979).

23. F. H. Mischgofsky, Observations on step bunching for the layer perovskite $(C_3H_7NH_3)_2MnCl_4$ using a compound holographic interference and conventional microscope, J. Crystal Growth, 43:549 (1978).

24. I. Sunagawa and P. Bennema, Modes of vibrations in step trains: rhythmical bunching, J. Crystal Growth, 46:451 (1979).

25. W. R. Wilcox, Influence of convection on the growth of crystals from solution, J. Crystal Growth, 65:133 (1983).

FUNDAMENTALS OF FLUX GROWTH

Dennis Elwell*

Elwell Associates
2245 Nob Hill Drive
Carlsbad, CA 92008, USA

1. INTRODUCTION

"Flux growth" is the name given to crystal growth from high temperature solutions, where the "flux" is typically a molten salt or oxide used as a solvent. The use of this word "flux" is derived from its more general usage for a substance which reduces the melting temperature, or dissolves oxides as in soldering. The principles discussed here can normally be applied to metallic solvents also. Another term which overlaps with flux growth is growth from non-stoichiometric melts, where a substantial excess of one component of a melt can be considered to be the solvent for the crystallizing phase. This article will concern the growth of bulk crystals from oxide or molten salt solutions where the solute concentration is normally from 1% to 30% (molar or weight fraction) and the temperature of crystal growth is from 300-1800°C.

The use of the solvent leads to both the main advantage and the main disadvantage of this method. The advantage is its flexibility - it can be stated with reasonable confidence that a high temperature solvent can be found for any material so that flux growth in its broadest sense is the most widely applicable of all techniques of crystal growth. The limitations in terms of material applicability are therefore only those of the user's imagination, provided of course that the phase required does exist in the appropriate temperature range.

The main disadvantage of this method is that the solvent provides a large pool of impurity which can be incorporated into the crystals grown, either substitutionally or as solvent inclusions. Therefore, although crystals having a high degree of perfection can be grown, it is difficult to attain purity levels comparable with those possible from a pure melt. The size of crystals which can be grown is also limited compared with melt growth, and flux growth is normally applied only to those materials which cannot be grown from a pure melt.

Crystals grown from fluxed melts are used both in research and by industry. One of the most important applications of flux growth is in the synthesis of new materials. Many examples can be found in the literature

---

* Present address: Hughes Aircraft Company, 500 Superior Avenue, Newport Beach, CA 92658, USA.

of new materials which were first formed as small crystals in a flux
growth experiment, either by accident or during systematic studies of
novel systems. This is especially true of rare-earth or transition metal
compounds, particularly complex oxides. In addition, materials known from
studies of reacted powders have often been crystallized for the first time
from fluxed melts. Materials which have been crystallized for commercial
use include yttrium iron garnet $Y_3Fe_5O_{12}$ (known as YIG) and its
derivatives, also emerald, ruby and alexandrite for use in the gem
industry.

There are many reviews and articles which may be consulted for
further details on flux growth. The literature up to 1975 is summarized
comprehensively by Elwell and Scheel (1975), whose book remains the only
specialized text on this subject in English. Russian-speaking students
are referred to the book by Timofeeva (1978). Recent reviews include
those by Tolksdorf (1977), Elwell (1980), Scheel (1982), Wanklyn (1983),
Gornert and Sinn (1985) and Tolksdorf (1985).

2. CHOICE OF A FLUX

The multi-component liquids which are typically used to grow crystals
by the flux method are extremely complex and it would be a difficult task
to characterize all the species which are present close to the
crystallization temperature. There have been, however, a small number of
studies aimed at clarifying the nature of the solutions so we now have the
beginnings of an understanding of the nature of some fluxed melts which
have been found in practice to yield large, high quality crystals.

In general, the solutions which are used in flux growth show a
tendency towards compound formation rather than towards immiscibility.
For example, HfO_2 crystals can be grown from PbO solutions only below
$1200°C$ because $PbHfO_3$ crystallizes at temperatures above this. Although
this tendency to compound formation is common, the substitutional
incorporation of solvent ions must be minimized, so it is important to
choose a flux which differs in some major respect from the material
crystallized. Differences of valence and/or of cation and anion radii
should be substantial. As an example, PbF_2 differs from Al_2O_3 in respect
of cation and anion valence and of cation radii. The incorporation of Pb^{2+}
or F^- ions in the alumina crystals should therefore be very small unless
other impurities are present in the solution which could compensate for
these crystal-chemical differences.

Table 1 lists some fluxes which have been found to be capable of
yielding large and/or high quality crystals. In general, the choice of a
flux has been made empirically but based on experience with related
materials. Fluxes which have at least one ion in common with the crystal
are preferred where possible since they reduce the number of
substitutional impurity ions.

Table 1. Examples of Flux Systems

Type of Flux	Example	Examples of Crystals
Lead compounds	PbO/PbF_2	YIG, Al_2O_3, $MgAl_2O_4$
Borates	BaO/B_2O_3	$Y_3Al_5O_{12}$, $RFeO_3$
Vanadates, molybdates	Li_2O/V_2O_5, $K_2Mo_2O_7$	$ZrSiO_4$, $DyKMo_2O_8$
Halides	NaCl, KF	$CaWO_4$, $BaTiO_3$

Both the physical and chemical properties of the flux are important, and the earliest trends in selecting a flux tended to emphasize physical properties. Low melting point, low viscosity and low volatility are particularly important for most applications. In addition, apart from the obvious requirement of a high solubility for the material to be crystallized, the flux should not react with some suitable crucible material and should be easy to remove from the crystals after the end of the growth process.

In recent years there has been a stronger emphasis to try to understand the behavior of fluxes and the nature of their solutions, especially at the atomic level. An early review of this topic was given by Elwell (1975) and more recent work is well described in the reviews by Wanklyn (1983) and Gornert and Sinn (1985). Cryoscopy and heat of solution measurements have been the main experimental methods used, and van Erk (1979) has developed a model of the species present in Y_2O_3 + Fe_2O_3 solutions in PbO/B_2O_3 fluxes which is of particular interest.

Wanklyn (1977) has studied the dependence of the size and habit of complex rare earth oxide crystals on the composition of the melt. She noted that, in a system such as $PbO/V_2O_5/R_2O_3$, larger and more equi-dimensional crystals could be obtained when the excess of the acidic component (e.g., V_2O_5) needed to produce the required phase was reduced to a minimum. It was also found that replacing part of the basic oxide by the corresponding fluoride would increase the solubility and favor the growth of fewer, larger crystals.

It is often found beneficial in choosing a flux composition to add a few percent of an "additive" which influences the number of crystals grown without substantially changing the solubility or the phase diagram. The most popular of these additives is B_2O_3. Elwell and Coe (1978) found using the simplified "hard sphere" liquid model that B_2O_3 addition to PbO/PbF_2 fluxes tends to cause additional complexing. How much complex formation is optimum is an interesting and unresolved question. The situation we require is that crystallization should occur easily, at low supersaturation, on existing crystals while the nucleation of new crystals should require much higher supersaturation. These conditions will favor the growth of fewer and therefore larger crystals.

At present it is not possible to make reliable predictions of optimum fluxed melt compositions using the above condition as a goal. Wanklyn (1987) has argued that a second competing cluster species in the solution is helpful and this agrees with the model of "additive" behavior outlined by Elwell and Scheel (1975).

Raman spectroscopy or laser Doppler velocimetry which might improve our understanding of fluxed melts and interface processes are difficult to apply to high temperature solutions. My own dream is a tool like scanning tunneling microscopy which could explore surface structure on an atomic scale while the crystal is growing.

3. MECHANISMS OF FLUX GROWTH

The theory of flux growth is essentially that of solution growth which is reviewed in this book by Wilcox. Here we focus on a few essentials and describe some important experimental techniques for understanding growth mechanisms.

In designing an experiment, it is useful to know whether solute transport or interface kinetics are rate-determining. The rate-limiting step can be best explored using a thermobalance to measure the crystal growth rate, with an arrangement to stir the solution, for example by rotating the crystal (see Elwell et al., 1975). If solute diffusion to the crystal is rate-limiting, the growth rate will be proportional to the supersaturation but will change as the solution is stirred. If interface kinetics are rate-limiting, the growth rate will vary with the square of the supersaturation as predicted by the Burton-Cabrera-Frank theory. Solute diffusion is often rate-limiting in unstirred solutions and interface kinetics become rate-limiting in rapidly stirred solutions, but these are broad generalizations. Heat transport is relatively unimportant under typical flux growth conditions.

How fast can a crystal be grown? There are two important aspects to be considered. First, if we consider a plane surface growing in a solution, the supersaturation gradient condition (Scheel and Elwell, 1973) leads to an expression for the maximum stable growth rate:

$$v_{MAX} = \frac{D\Delta H n_e}{\rho R T^2} \frac{dT}{dz} \tag{1}$$

where D is the solute diffusion coefficient, ΔH the heat of solution, n_e the equilibrium solute concentration at temperature T, ρ the density of the crystal and dT/dz the temperature gradient normal to the surface. In practice, most crystals grow in a negative temperature gradient because the heat of solution is removed through the solution and not through the crystal. Crystals can therefore only grow with planar facets because of an inherent stabilization mechanism associated with growth on habit faces.

Of even greater importance in choosing crystal growth conditions is the supersaturation inhomogeneity, the variation in supersaturation between the center and edge of a crystal face. This inhomogeneity is the inevitable consequence of the faceted crystal shape; the corners and edges of a crystal "see" a greater volume of solution than the face center, and so will have a higher supersaturation. A faceted crystal can therefore maintain its shape only if this inhomogeneity is compensated by more efficient growth kinetics at the face centers (Chernov, 1972). Studies of crystal defects by X-ray topography and similar techniques (see, for example, Roberts and Elwell, 1981) show that each face is dominated by one or more active growth centers - points where dislocations with some screw character intersect the surface and act as sources for layers which spread across the surface towards the edges. Each face of a crystal has its individual character, and quantitative generalizations are difficult. The general remark can be made, however, that stirring the solution helps promote stable growth of large crystals by reducing the solute gradient across the crystal faces.

Wanklyn and co-workers (Wanklyn, 1983; Wanklyn et al., 1983) have used thermogravimetry to measure supercooling in fluxed melts by comparing the equilibrium solubility curve with that obtained under dynamic conditions which simulate growth by slow cooling. The aim is to find a melt composition capable of yielding just one crystal, as explained in the previous section. Another powerful but simple technique which can yield vital information is the induced striation method (ISM). Striations are induced in crystals by negative temperature pulses of a few degrees, which cause rapid growth over a short time. A higher than average entrapment of impurity occurs during this rapid growth, forming the bands which can be revealed by polishing and etching the crystals after growth. Since the pulse sequence is known, the crystal growth rate and changes in crystal shape can be followed even when the crystals were grown in a totally enclosed crucible. This method was applied to flux growth particularly effectively by Gornert and co-workers (Gornert and Wende, 1976) and the

review by Gornert and Sinn (1985) lists 15 studies using ISM. In addition
to yielding data on growth rate and growth history, ISM can be used to
measure the supersaturation which is normally unknown, for example during
slow cooling. In this case the direction of the temperature pulses is
reversed and heating pulses of varying amplitude are applied. Small
positive pulses simply slow down the growth rate. However, if the local
temperature exceeds the equilibrium value in the heating transient, there
will be redissolution of the crystal. Dissolution occurs preferentially
at the corners, which will become rounded. Subsequent regrowth on
continuation of cooling fills in these corners, but the rounding can be
seen on etching the crystal. The minimum temperature pulse to cause "edge
rounding" is the supercooling, which can be readily correlated with
supersaturation if the solubility curve is known. Of course the
temperatures should ideally be measured at the site of the crystal, which
may not be easy experimentally.

4. EXPERIMENTAL TECHNIQUES

A. Generating the Supersaturation

Although other means of generating a supersaturation are known, most
flux-grown crystals are produced by slow cooling, solvent evaporation or
temperature gradient transport. Expressions for the growth rate in terms
of the system parameters are listed in Table 2. The crucible volume V
varies from 5-10 ml for small quantities of very expensive materials to
around 10 liters in commercial growth. The linear growth rate is of the
order of mm/day and the area A of a crystal face varies as the crystals
become larger.

In slow cooling, the variable which is under the control of the
experimenter is the cooling rate, which is typically around 1 deg/hr. In
growth by solvent evaporation, the flux evaporation rate dV/dt is
controlled by the temperature and by the size of the hole through which
evaporation occurs. The advantage of this alternative is that growth
occurs at constant temperature, which can be important if the material
exhibits a phase transition at some temperature which would be in the
normal range for slow cooling. Some flux evaporation tends to occur even
in growth by slow cooling, unless a sealed crucible is used. In growth by
gradient transport, a temperature difference ΔT is maintained between a
hotter region where polycrystalline nutrient material is held and a cooler
region where a seed or cooled "finger" is located and the crystal is

Table 2. Crystal Growth Rate v in Terms of System Parameters

Slow cooling	$v = \dfrac{V}{\rho A} \left(\dfrac{dn_e}{dT} \right) \left(\dfrac{dT}{dt} \right)$
Flux evaporation	$v = \dfrac{n_e}{\rho A} \left(\dfrac{dV}{dt} \right)$
Gradient transport	$v \left(\dfrac{\rho \delta_c}{D} + \dfrac{\rho \delta_N A}{DA_N} \right) + \left(\dfrac{v}{B} \right)^{1/q} = \dfrac{n_e \Delta H \Delta T}{RT}$

V = crucible volume; ρ = crystal density; A = crystal surface
area; n_e = solubility; T = temperature; t = time; δ_c = crystal
boundary layer thickness; δ_N = nutrient boundary layer thickness; D = solute diffusion coefficient; B,q = kinetic
coefficients; ΔH = heat of solution.

grown. Also in this case the crystal can be maintained at constant temperature and the method is most appropriate for growth of solid solution crystals. This method normally uses fluxes of low volatility such as BaO/B_2O_3, and their high viscosity is a severe problem since solute transport is often inadequate to achieve even slow growth rates. A variant of this method involves the use of thin solvent zones in place of bulk solution; solute transport occurs across a narrow zone which is moved either using a strip heater or by moving the sample through the gradient*.

Electrolysis of molten salt solutions can be used to grow crystals of conducting materials. Quite large crystals can sometimes be grown by this method, for example of tungsten bronzes or LaB_6 (see the review by Elwell, 1977).

B.   Some Practical Considerations

In all cases except electrocrystallization, the driving force for crystal growth involves temperature and it is of great importance for crystal quality to use close temperature control. Gornert and Sinn report that temperature variations above $0.1\,^{\circ}C$ will result in visible striations in the crystals, so this should be the minimum target for the temperature controller. Fortunately the latest generation of microprocessor furnace controllers allows this kind of stability to be achieved provided that precautions are taken to optimize the control parameters.

Furnace design and the attainment of appropriate temperature distributions for restricting growth to the central region of the crucible base are discussed in the review by Tolksdorf (1977). Figure 1 shows the construction of a furnace used for some years by Tolksdorf and co-workers for the small-scale production of garnet crystals. The furnace must be of tight construction to avoid drafts, and locating it in an enclosure as shown in the Figure is desirable.

Several thermocouples are used for temperature sensing so that averaging may be used. Pt-6% Rh versus Pt-30% Rh thermocouples are preferred since these have the lowest sensitivity to changes in the cold junction temperature. It should be noted also that the furnace enclosure is swept by an airflow which is stabilized to $\pm$ $0.5\,^{\circ}C$. Other features of Figure 1 will be described later.

The crucibles used for flux growth are often made of platinum, which has the disadvantage of high cost but is relatively easy to fabricate, by shaping, welding etc. It is desirable for any substantial flux growth operation to train at least one member of the group to become skilful in platinum fabrication.

When seed crystals are not available, a platinum needle can be used to localize crystal growth as described by Elwell and Morris (1979). The condition of the platinum can be vital for the nucleation of relatively few crystals, as was demonstrated by the study by Wanklyn et al. (1984). This investigation showed that a platinum surface exposed to a fluxed melt can develop preferred nucleation sites by local dissolution, so the history of the internal surface of a platinum crucible is also of importance when nucleation is not restricted to a "cold finger".

Seeding is highly desirable even in systems where it is difficult to restrict subsequent growth to a single crystal. The problems of introducing a seed into a saturated melt are discussed by Tolksdorf

---

* Temperature gradient zone melting (TGZM) is a further possibility.
  (Remark added by the editors.)

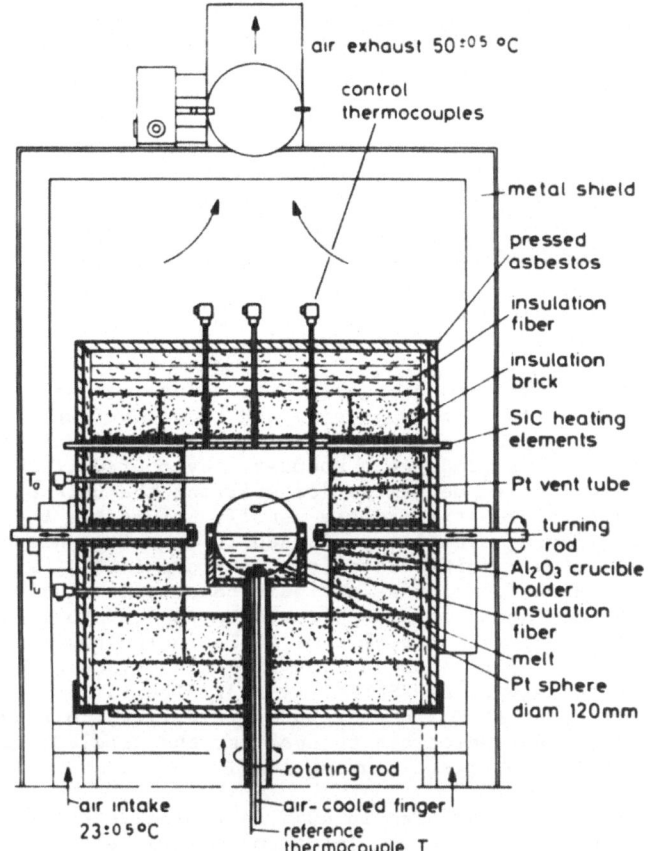

air exhaust 50±0.5 °C

control thermocouples

metal shield

pressed asbestos

insulation fiber

insulation brick

SiC heating elements

Pt vent tube

turning rod

Al₂O₃ crucible holder

insulation fiber

melt

Pt sphere diam 120mm

T_o

T_u

rotating rod

air intake 23±0.5°C

air-cooled finger

reference thermocouple T

Fig. 1.   Schematic design of a furnace used for flux growth (Tolksdorf, 1977).

(1977). In order to minimize these problems, Tolksdorf developed a spherical crucible as can be seen in Figure 1. The seed is initially mounted above the melt surface and is moved to a position at the bottom of the fluxed melt by a 120° rotation of the crucible about a horizontal axis. In the position where the crystal is grown, local cooling is provided by an air-cooled finger.

Elwell and Scheel (1975) drew attention to the importance of optimizing the cooling program during flux growth by slow cooling. Cooling and reheating programs are described which can be used when seeding is not available; the idea here is to redissolve small crystals which nucleate during the initial cooling through the solidification temperature, so as to select one or only few crystals for further growth. If a crystal is to be grown to large dimensions, the linear growth rate should ideally be just below the maximum stable value at all times. In the beginning where the crystal area is small (see the first equation in Table 2), the cooling rate must be low. Subsequently the cooling rate can be increased, but care must be taken to avoid exceeding the growth rate at which the supersaturation inhomogeneity leads to growth instability. Non-linear cooling programs, which were difficult to implement in 1975

but can be realized very simply with the latest controllers, can be
designed around the principle of avoiding the supersaturation
inhomogeneity.

Elwell and Scheel proposed an equation of the form

$$v_{MAX} = \left(\frac{0.2Du\sigma^2 n_e^2}{Sc^{1/3}\rho_c^2 L} \right) \tag{2}$$

with σ the relative supersaturation, u the solution flow rate and Sc the
Schmidt number. L is the length of crystal grown at any time, and Eq. (2)
can be used to calculate the cooling rate program which will minimize the
time needed to grow inclusion-free crystals.

Stirring the solution to homogenize the solute concentration, to
reduce the supersaturation inhomogeneity and to allow faster growth, is of
great value. In top-seeded growth, stirring is achieved by rotating the
crystal. In theory this produces a boundary layer of uniform thickness
but it is not an effective mechanism for stirring the bulk of viscous
melts. Immersed platinum stirrers are usually unsuitable because platinum
is so soft, and alloying elements to harden it are often dissolved by the
flux. The accelerated crucible rotation technique (ACRT), in which
stirring is achieved by continuous periodic changes in crucible rotation
rate, is particularly valuable since it can be applied to sealed
crucibles. For a fairly recent review of ACRT, including a short
description of the principles and a list of large crystals which have been
grown using this technique, see Scheel (1982). The largest crystals grown
from fluxed melts using ACRT are over 200 g in weight. Tolksdorf (1977)
describes an arrangement which can be used for the gradient transport
growth of solid solution crystals using ACRT.

The period of the cycle used in ACRT is typically about 1 minute, and
changes in melt temperature at any location do occur with the same period.
Corresponding changes in crystal composition presumably occur, but on the
scale of 1 μm or less which would be difficult to detect. A novel
alternative to ACRT, called coupled vibrational stirring (CVS) was
recently investigated (Liu et al., 1987). The vertical axis supporting
the crucible is subjected to a Lissajous figure, the simplest case being a
circle in a horizontal plane. This exerts a very powerful stirring action
in which liquid is swept downwards near the crucible wall, inwards at the
base and upwards in a central spiral column. To our knowledge this
technique has not been applied to flux growth and it would be of interest
to compare the size and quality of crystals with those produced using
ACRT. The period of the oscillations is typically 5-10 Hz, so the amount
of material grown per cycle would be only around 30 Å.

It must be emphasized that a successful stirring action must be
smooth, since excessive stirring can induce a shower of nuclei and so
would have a very negative effect.

5.   CRYSTAL DEFECTS

Space does not allow a detailed discussion of the defects likely to
be found in flux-grown crystals, but a summary of the most important
defects is given here:

(a) _Inclusions_. Solvent inclusions are likely to be trapped by the crystal whenever there is growth instability.

(b) _Chemical impurities_ can be introduced via impurities in the original chemicals or can be major constituents of the flux itself. In addition, traces of platinum or other container materials can often be detected by sensitive techniques.

(c) _Dislocations_ are always present and normally exert a strong influence on the growth kinetics and so on the crystal habit. However, dislocation densities are normally lower than those in equivalent melt-grown crystals except in those few cases where dislocation-free crystals can be grown from the melt. Dislocation densities of 10 cm^{-2} are typical when reasonable care is taken to achieve stable growth.

(d) _Sector boundaries_ are seen in internal regions which separated adjacent habit faces during growth. These may be seen by X-ray topography which is the most powerful general tool for investigating physical defects.

(e) _Striations_, which are periodic changes in composition, can often be related to periodic temperature changes. There is some evidence for an intrinsic striation-inducing mechanism associated with the growth process (see Gornert and Hergt, 1973).

(f) _Strain_ is always present and surface strains of the order of 10^{-5} have been measured by X-ray reflection topography. These may be associated with the layer growth mechanism.

It should also be noted that studies of the defect structure can be used to follow the growth history of the crystal.

6. CONCLUSIONS

Flux growth of crystals is now a mature subject and is used mainly for studies of novel materials and for the commercial growth of a small number of materials, particularly oxides, which cannot be grown from a pure melt. There is a steady development of important new materials which require the flux method, for example the non-linear optical material β-BaB_2O_4 (Jiang et al., 1986). Crystal growth of the new high temperature superconductors with transition temperatures above 50 K will be an important challenge for flux growers, and other materials in the same general family may emerge from studies of related systems.

Experience has shown that large, high quality crystals can be grown by the flux method given sufficient research to find the right melt composition, right technique and appropriate temperature-time program. Research into the nature of a "good" flux and solution will continue and it is to be hoped that some research groups will be able to find the resources to apply new measurement techniques and to develop improved models of the liquid structure. The development of techniques to allow direct visual observation of crystals growing in fluxed melts (Tsukamoto et al., 1983) is another important advance which should be valuable for studies of systems of low volatility. Finally the desirability of optimizing fluid flows and developing new approaches to stirring even enclosed melts has been stressed.

REFERENCES

Chernov, A. A., 1972, Soviet Phys. Crystallog., 16:734.

Elwell, D., 1975, in: "Crystal Growth and Characterization", R. Ueda and J. B. Mullin, eds., N. Holland, Amsterdam, p. 155.

Elwell, D., 1975, in: "Crystal Growth", E. Kaldis and H. J. Scheel, eds., N. Holland, Amsterdam, p. 605.

Elwell, D., 1980, in: "Crystal Growth", B. R. Pamplin, ed., Pergamon, Oxford, Ch. 12.

Elwell, D., and Coe, I. M., 1978, J. Crystal Growth, 44:553.

Elwell, D., and Morris, A. W., 1979, J. Mater. Sci., 14:2139.

Elwell, D., and Scheel, H. J., 1975, "Crystal Growth from High-temperature Solutions", Academic Press, London-New York.

Elwell, D., Capper, P., and D'Agostino, M., 1975, J. Crystal Growth, 29:321.

Gornert, P., and Sinn, E., 1985, in: "Crystal Growth of Electronic Materials", E. Kaldis, ed., Elsevier, Amsterdam, p. 23.

Gornert, P., and Wende, E., 1976, Phys. Stat. Sol., (a) 37:505.

Jiang, A., Chang, F., Lin, Q., Cheng, Z., and Zheng, Y., 1986, J. Crystal Growth, 79:970.

Liu, W.-S., Wolf, M. F., Feigelson, R. S., and Elwell, D., 1987, J. Crystal Growth, 81:359.

Roberts, K. J., and Elwell, D., 1981, J. Crystal Growth, 63:249.

Scheel, H. J., 1982, Prog. Cryst. Growth Charact., 5:277.

Scheel, H. J., and Elwell, D., 1973, J. Electrochem. Soc., 120:818.

Timofeeva, V. A., 1978, "Rost Kristallov iz Rastorov-Rasplavov", Nauka, Moscow.

Tolksdorf, W., 1977, in: "Crystal Growth and Materials", E. Kaldis and H. J. Scheel, eds., N. Holland, Amsterdam, p. 640.

Tolksdorf, W., 1985, in: "Crystal Growth of Electronic Materials", E. Kaldis, ed., Elsevier, Amsterdam, p. 175.

Tsukamoto, K., Abe, T., and Sunagawa, I., 1983, J. Crystal Growth., 62:215.

van Erk, W., 1979, J. Crystal Growth, 46:539.

Wanklyn, B. M., 1977, J. Crystal Growth, 37:334.

Wanklyn, B. M., 1983, J. Crystal Growth, 65:523.

Wanklyn, B. M., 1987, private communication.

Wanklyn, B. M., Smith, S. H., and Garrard, B. J., 1983, J. Crystal Growth, 63:77.

Wanklyn, B. M., Watts, B. E., and Fenin, V. V., 1984, J. Crystal Growth, 70:459.

FUNDAMENTALS OF EPITAXY

R. Kern

CRMC2-CNRS
Campus de Luminy, Case 913
F-13288 Marseille Cedex 09, France

This contribution introducing us to the fundamental knowledge of epitaxy is divided into three parts. In Part I the historical basis is reviewed and the geometrical aspects of epitaxy are treated. In Part II these aspects are examined from the point of view of equilibrium thermodynamics and some examples are given. In Part III references are given.

I   HISTORICAL BACKGROUND: GEOMETRICAL LAWS

I.1. For a very long time it has been known that two different mineral species are able to overgrow in a regular way [1]. This phenomenon was called regular overgrowth. In Figures I.1 and I.2 two such examples are illustrated. These are, respectively, the overgrowth of triclinic cyanite (disthen) $Al_2O[SiO_4]$ on the orthorhombic staurolith $\{Al_2O[SiO_4]\}Fe(OH)_2$ and the case of hematite, α-Fe_2O_3 trigonal, supporting on its basal plane tetragonal rutile, TiO_2 crystals.

From the beginning it was clear that regular overgrowth has neither to do with chemistry in the sense of stoichiometry or isomorphism, nor with the symmetry of both crystals in contact. However, there is an invariant in all known examples: the host and the guest crystals have two of their own crystal faces and one or more crystal edges, located in these planes, which have parallel orientation in common. The geometrical nature of regular overgrowth was, therefore, secured. At that time the lattice theory of crystals was not yet properly established and it was assumed that crystal faces are parallel to dense lattice planes and crystal edges are parallel to dense lattice rows. Therefore the adjective "regular" preceding the noun "overgrowth" meant, in terms of lattice theory, that the regular association of both minerals prefers, and there is no exception to it, that one or two dense lattice rows of each plane want to exhibit strictly parallel orientations [2] for both crystal species.

At that time X-ray diffraction by crystals was not known so the lattice theory had no direct experimental support.

I.2. In 1928 L. Royer [3] could confirm the geometrical laws when X-ray diffraction was available. He was able to specify these laws since now the absolute lattice parameters of many minerals were known. Taking

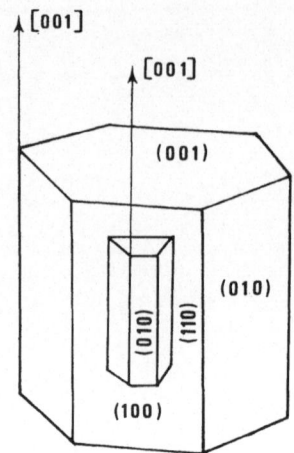

Figure I.1

the example $TiO_2/\alpha Fe_2O_3$ we notice, in Figure I.3a, that the basal
hexagonal lattice plane (0001) of αFe_2O_3 has a parameter of 5.04 Å. In
Figure I.3b we see in the (101) lattice plane of TiO_2 rows with [001]
(2.95 Å) and [100] (4.58 Å). These lattice planes (0001) and (101) are
precisely those which are in contact (Figure I.2). Superimposing these
two plane lattices of Figure I.3a and Figure I.3b it is readily seen that
both have, for a given precise azimuthal orientation, a rigorous
parallelism of the rows $[10.0]_h//[100]_r$ and $[001]_r//[11.0]_h$. Three such
orientations, each from the other at $2/3$, are equivalent exactly as
observed in nature (Figure I.2).

More can be said since both lattice planes, in these orientations,
have a common planar mesh:

- the 1 x 3 mesh of hematite and the 1 x 3 mesh of rutile are in
close metric coincidence;

- this metric coincidence is only an approximate one, that means
that in two orthogonal directions there is either a positive or a negative
metric misfit.

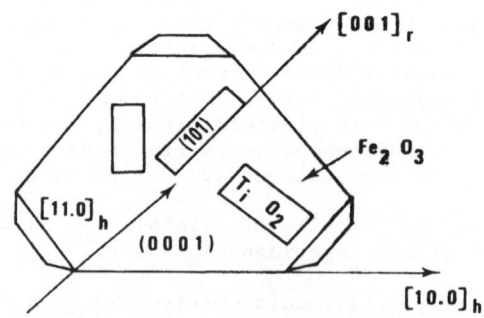

Figure I.2

(0001) Fe₂O₃

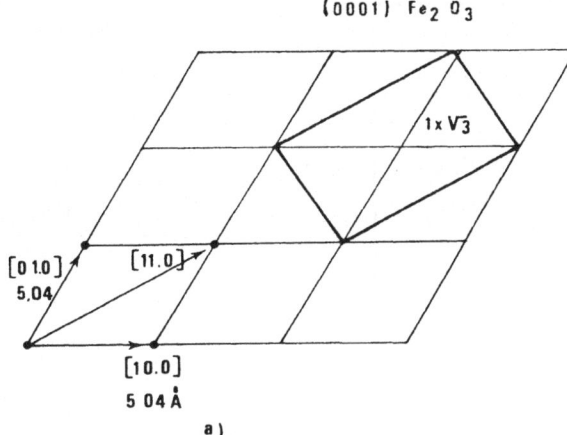

(101) TiO₂

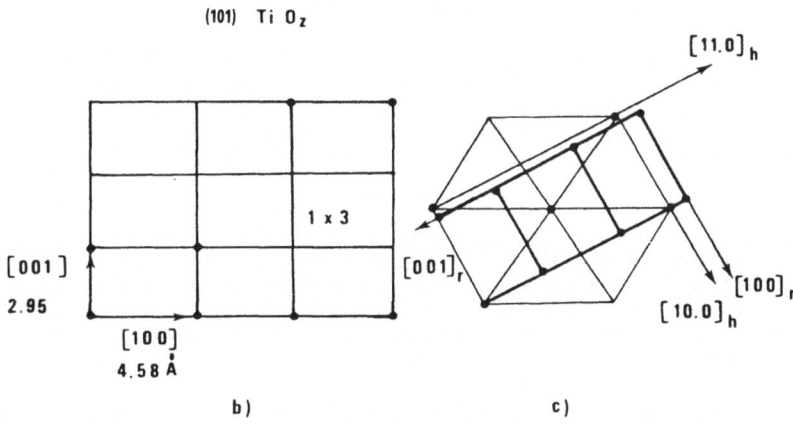

b)   c)

Figure I.3

Species	Rows	Å	Misfit %
Fe₂O₃	[11.0] x 1	5.04 x √3	-3
	[10.0] x 1	5.04	
TiO₂	[001] x 3	2.95 x 3	+10
	[100]	4.58	

    Many other examples are possible. Very often the common mesh is a
simple mesh, the chosen example is one where this common mesh is a
multiple one in both species: 1 x √3 in the contact plane of hematite,
1 x 3 in the contact plane of rutile. In general the misfit m is smaller
than |m| < 10%.

    I.3. Royer [3] however wanted to assess his statement also on an
experimental basis. We mention two types of illuminating experiments:

    - On a fresh cleavage plane (001) of an alkali halide crystal A(a)
of parameter a(cubic) he deposited a droplet of a solution containing an

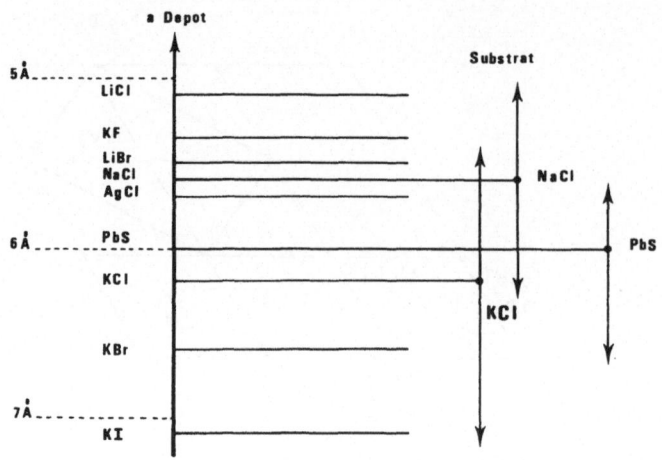

Figure I.4

alkali halide B(b). After a short evaporation time of the water, B(b) crystallized on A. Substrate A and deposit B showed a regular overgrowth: the (001) planes are parallel and the lattice rows <010> are in rigorous coincidence. The coincidence cell is a square mesh, the misfit m is defined as m = (a-b)/a.

Figure I.4 gives the result. The arrows indicate the domain where regular overgrowth is observed for a given couple A/B. For ten different deposits on three different substrates it was found that there is a limiting misfit of approximately m ≃ ±10%. Outside the arrows of this limiting misfit, the deposited crystals are no more in a regular orientation but are randomly oriented.

- Another series belonging to the calcite, $CaCO_3$ structure type, also gave Royer some information about the angular tolerances (angular misfit). These crystals are trigonal, parameters a and α, and the cleavage plane (100) has a lozangic planar mesh containing the rows [001] and [010] and making the angle α, both having the parameter a. He used the substrates $MgCO_3$, $MnCO_3$, $FeCO_3$ and $CaCO_3$ and deposited from solution sodium nitrate on the corresponding cleavage planes. The a,α characteristics are the following ones:

Species	$MgCO_3$	$MnCO_3$	$FeCO_3$	$CaCO_3$	$NaNO_3$
a Å	5.84	6.01	6.02	6.41	6.48
α	$103°24'$	$102°5'$	$103°3'$	$104°55'$	$102°47'$

$NaNO_3$ gives a regular overgrowth only on $CaCO_3$, but not on the other substrates. Clearly this couple has the smallest metric misfit of m = +1.1% and an angular misfit $\Delta\alpha \simeq +2°8'$. The other substrates do not have a prohibiting metric misfit since for them m ∿ 7% (except $MgCO_3$ with m ∿ +12%). This means that the angular misfit $\Delta\alpha$ helps severely to restrict the limiting metric misfit.

Notice, moreover, that for these couples there is not an unique regular orientation. There are two distinct orientations rotated each by an angle $\Delta\alpha$, that means that the rows <001> and <010> may be superimposed separately.

146

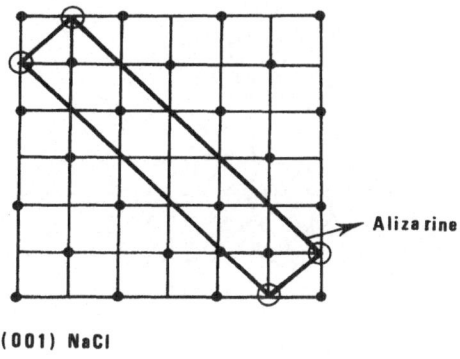

(001) NaCl

Figure I.5

From that time, 1928, Royer could give a short name to this experimentally confirmed phenomenon and he called it "epitaxy"*. This name says exactly what happens. There is an arrangement, an order (taxis) which takes place upon (epi) something. Today technicians in crystal growth use the word epitaxy or epi in a much more restrictive sense, thinking on the growth of silicon on silicon. Webster's Dictionary's definition is also misleading: "the growth on a crystalline substrate of a crystalline substance that mimics the orientation of the substrate". We have seen in Figure I.2 that there is no mimic at all in some cases.

Neuhaus (a review in [4]) confirmed with many other experimental examples Royer's laws either by crystallization from solution, or by sublimation. The orthorhombic alizarine crystals, e.g., deposited with their (010) plane on the (100) plane of NaCl, KCl, KBr and KI (see Figure I.5) have their respective rows [001] and [110] in parallel orientation. There is a multiple common mesh. The misfit is smaller than 10% in the direction of the great superposition period. Along the orthogonal direction no misfit limitation seems to intervene. Such epitaxies have been called monoperiodic epitaxies by Monier [5] and Hocart in contrast to the preceding ones, called biperiodic.

At that time technology came into the field; one of the first applications had been the preparation of flat light polarizers (polaroids) by depositing huge, but very thin, crystals of iodoquinine on polyacetate single crystal films.

A very exhaustive revue of epitaxy was given by Pashley [6] but special emphasis was given for metals on ionic crystals as well as for metals on metals. In general these epitaxial systems are produced by fast vacuum evaporation of the metal. Such a process is characterized by far from equilibrium conditions. Then secondary phenomena occur, as solid-solid recrystallization, coalescence and as a consequence Royer's laws, especially the limiting misfit law, is not obeyed.

I.4. The epitactic contact between A and B crystals, where there is a geometrical misfit, was a very intricate problem from the point of view of energetics at molecular level. Kantorova and Frank were the first in this field and since 1949 van der Merve [7] got involved in this problem. If two rows of different parameters a and b are superimposed one

---

* The name epitaxy was suggested to Royer by Father Derville S. J., who prepared at that time his thesis in Petrology at the same Institute at Strasbourg University.

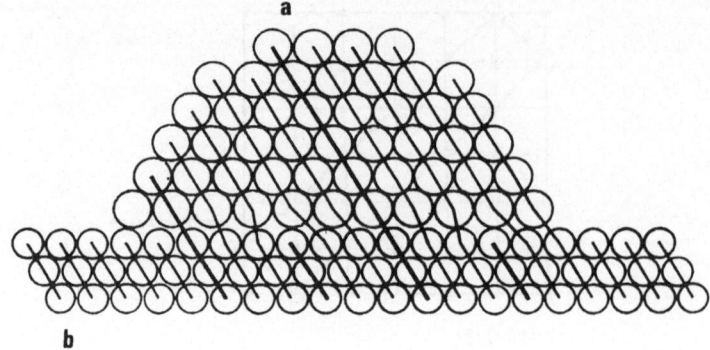

Figure I.6

on the substrate and the other one on the deposit, only nodes distant at p are in exact coincidence, according to the Vermier-Nonius relation:

$$p = Na = (N + 1)b \equiv \frac{ba}{(a - b)} \; ; \; N \text{ integers.}$$

If rigid atoms are located on the nodes of these two rows, only a fraction a/p of the atoms are located in the "potential valleys" of the substrate (Figure I.6). The potential energy of this system is probably not in its minimum and the corresponding "adhesion energy" is not at its maximum.

An opposite situation may be imagined when both rows are either in compression or in dilatation in order to have (a < a') = (b' < a), where a' and b' are the new common parameters. Both solids in contact are now coherent at their interface but are homogenously deformed so that the elastic energy which is stored is $\alpha \; Vm^2$, V being the volume of the two layers and m the original misfit. Far from the interface however, each crystal wants to recover by relaxation of its own original parameter a and b.

A simple elastic and static analysis shows (substrate is a rigid row, the deposit atoms are in the periodic potential of this substrate and are connected by springs) that none of the preceding extreme solutions give the minimal energy. Neither the juxtaposition nor the coherent interface are valuable solutions.

At the interface a relaxation process takes place periodically with a period approximately Na. Around each coincidence node of some Vernier-Nonius, the atoms go into potential wells; these large zones are separated by narrow zones where the greatest part of inhomogeneous deformation concentrates (interfacial dislocations). Figure I.7 gives a schematic illustration of this. Crossing the interface and beneath the coincidence

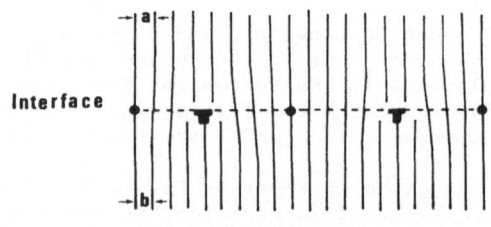

Figure I.7

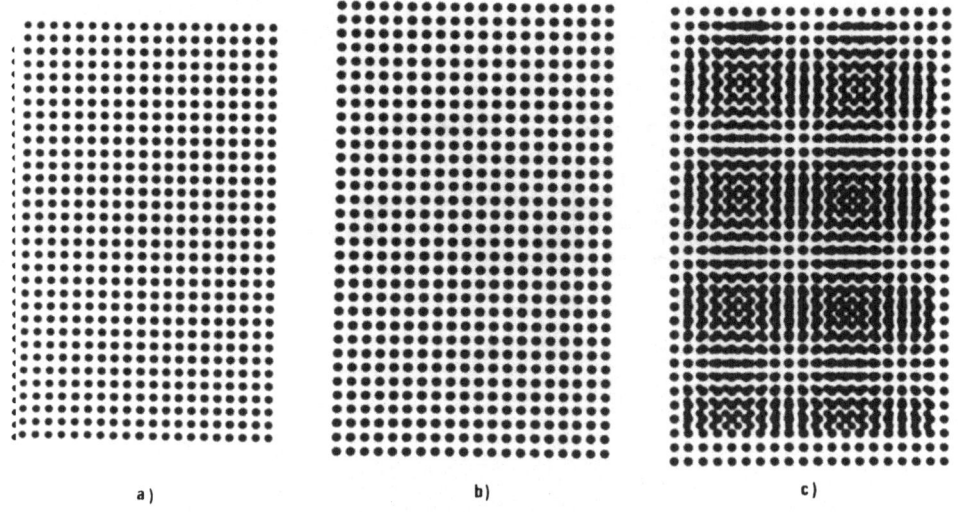

a)   b)   c)

Figure I.8

nodes the lattice planes of A and B are much less distorted, but between
such nodes there is a lack of coherence. The narrow zones are the cores
of the interfacial dislocations, they may be characterized by the
introduction of supplementary lattice planes. In this case the interface
may be described by a one-dimensional lattice of edge dislocations; the
interface is the glide plane and the Burgers vector is in this plane and
is normal to the dislocation lines.

In a two-dimensional case, with metric misfits only, m and m', a two-
dimensional coincidence lattice is defined and two crossing edge
dislocation periodic arrays are formed. Figures I.8a and I.8b show the
atoms of the interface A and B. The nets are square with Na = (N + 1)
b = 10. By superposing both nets in a parallel orientation a Moiré
pattern appears (Figure I.8c). The superperiods of the coincidence
lattice are seen. Dark areas are separated by lighter ones, each area
simulating regions of better or worse coincidence. By a relaxation
process in the plane of the atoms A and B there would appear periodic
zones of small extension with bad coincidence (high strain) and large ones
of good coincidence (very low strain).

When the two lattices A and B have only an angular misfit then the
energetic description is given by an array of screw dislocations (Figure
I.9). See for these interfacial dislocation descriptions Ref. [8].

In the case of $NaNO_3(100)$ on $(100)CaCO_3$ mentioned above, there is not
only an angular misfit but also a metric one, the full description is
given in terms of a mixed dislocation array.

I.5. The direct observation of interfacial dislocations was made
possible in 1965 by the school of Kuhlmann-Wilsdorf and Mathews [9]; see
also the two volumes of Ref. [10]. Epitaxies have been produced on very
thin monocrystalline films and observed in dark field transmission
electron microscopy (image with only one diffraction spot); the Burgers
vector was, therefore, identified [11]. Not only the interfacial
dislocations are seen but also the volume dislocations in A and B, as well
as stacking faults when present.

149

Figure I.9

Let us consider in more detail the epitaxial growth of iron on a monocrystalline thin film of gold [12]: Fe(110)/Au(111); [001]Fe/Au[110]. Iron monolayers are deposited on gold. At the beginning only defects of the substrate are seen in the microscope until a critical thickness (∿ 15 Å) of iron is obtained. If the iron layer is now thicker, linear contrast lines are observed when using the Fourier components 111 or 131 for the image formation. To demonstrate that these lines are located in the interface, an elegant method is to dissolve either the thin substrate (Au) or the deposit (Fe). When using a solution of KCN gold dissolves, diluted HC1 dissolves iron. Doing one or the other, in both cases the linear defects are no longer seen, but the bulk defects in the substrate or the deposit are still seen after these processings.

The theoretical periodicity of the interfacial dislocations may be calculated but there is in fact no strict periodicity. Probably, during the very irreversible film formation by vacuum evaporation, their configurational equilibrium could not be obtained.

In this example the deposit of iron on gold proceeds by layer by layer growth (other methods such as Auger spectroscopy have been used more recently to demonstrate this fact). However, when the critical thickness of ∿ 15 Å is obtained the interfacial dislocations appear. Therefore, the film of less than 15 Å was homogeneously deformed. Just when the thickness of ∿ 15 Å is surpassed there are three-dimensional crystals of iron on the layers with a slightly different lattice constant from that of the pseudomorphic layers previously formed.

I.6. This very consistent geometrical treatment was followed, but with some overlap, by a surface physics period. The real initiators have been Finch [13], Farmsworth [14], Rhodin [15] and Bauer [16] using electron diffraction methods very sensitive to the surface layers. Bauer defined in 1958 three mechanisms of epitaxial growth as will be outlined in Part II.

150

From the 1960s surface physics has expanded exponentially because
behind such activities there are areas of applications such as
semiconductors, catalysis, friction, and so on. In particular, new
experimental methods have been introduced: low energy electron diffraction
(LEED), reflection high energy electron diffraction (RHEED), light
ellipsometry, Auger spectroscopy, and many others. Often, such methods
could be used in situ that means during epitaxy. There are tutorial
papers about the methods and the way of thinking [17] about fundamentals
of epitaxy.

More practical aspects of epitaxial film preparation, chemistry,
molecular beam techniques, ion implantation, hydrodynamics, and so on, and
technological problems have been considered periodically since 1970 at
International Symposia, the Proceedings of which are published by North-
Holland every three years.

II   SOME EQUILIBRIUM PHENOMENA (FROM ADSORPTION TO EPITAXY)

Nothing has been said up to now about the physico-chemical conditions
which induce epitaxy. Thermodynamics can give an unique answer for near
equilibrium conditions.

In the following pages we proceed gradually, considering first the
macroscopic description of the formation of a crystal without substrate
(Section II.1) then on a substrate without structure (II.2). Looking at
the molecular level, a model crystal, isomorphous with the substrate, is
considered with only first neighbor interactions (II.3), then further
interactions are introduced (II.4), and finally in the general case where
the isomorphism condition is released the effect of misfit is inspected
(II.5). In (II.6) the consequences are summarized concluding to the
existence of three epitaxial possible growth modes. Examples are then
given.

II.1.   Consider a saturation pressure p_∞ at a temperature T that
means the pressure which is given by a very big crystal made up of the
same molecules. If in an actual vapor at T the pressure is p, p/p_∞ is
called the degree of supersaturation when $p/p_\infty > 1$, of undersaturation
when $p/p_\infty < 1$. $\Delta\mu = kTLnp/p_\infty$ is the thermodynamical super(under)
saturation according to $\Delta\mu <> 0$ for the corresponding perfect vapor and
represents the work per molecule to be done to bring this vapor from state
p_∞ to p at T.

In this vapor a crystal of i faces of surface tension σ_i of area S_i
exists or is formed. At constant T and total pressure P, the variation of
free enthalpy of formation is at constant $\Delta\mu$:

$$\Delta G = -n\Delta\mu + \sum_i \sigma_i S_i \qquad (1)$$

where n is the number of molecules in the crystal; n = V/v, v being the
volume of a molecule in the crystal, V the volume of the actual crystal.
The crystal being a polyhedron of volume

$$dV = \frac{1}{3} \sum_i h_i dS_i \qquad (2)$$

since it is composed of as many pyramids i of height h_i as there are faces
of area S_i. Differentiating (2), $dV = (1/3) \sum_i (h_i dS_i + S_i dh_i)$ or as some
approximation, $dV = \sum_i h_i dS_i$ so that

$$dV = \frac{1}{2} \sum_i h_i dS_i \qquad (3)$$

and from (1) at constant $\Delta\mu$:

$$d\Delta G = \sum_i [-\frac{\Delta\mu}{2v} h_i + \sigma_i] dS_i \equiv \sum_i \frac{\partial \Delta G}{\partial S_i} dS_i. \qquad (4)$$

For an arbitrary variation dS_i, at equilibrium $(\partial\Delta G/\partial S_i) = 0$ for all i, so that

$$\sigma_i/h_i = \Delta\mu_{eq}/2v, \qquad (5)$$

$\Delta\mu_{eq}$ being the supersaturation at equilibrium which can exist only in a closed volume. Relation (5) is the Wulf theorem telling that the central distances h_i of the equilibrium shape crystal are proportional to the surface tensions σ_i. The equilibrium shape is constructed by drawing from a point along the normal to the faces i, segments of lengths $h_i \alpha \sigma_i$ and normal planes at the end of these segments. The smallest polyhedron among all so defined concentric ones is the equilibrium shape crystal. Its absolute size is determined by the equilibrium supersaturation $\Delta\mu_{eq}$ of the closed system.

II.2. Consider now a substrate B of surface tension σ^B. We suppose it to be inert which means that it does not mix with the deposit A, neither in the solid state nor in the vapor phase. This substrate is also considered to be structureless. The crystal A is brought in contact with this substrate B with one of its faces i = j, all the others $i \neq j$ remaining free, the area of contact is $S_{i=j}$.

The contact is characterized by an adhesion free energy $\beta_{i=j} = W/S_{i=j}$ where W is the work to be spent for separating the deposit A from substrate B.

Dupré's relation defines an auxiliary excess quantity $\sigma^x_{i=j}$:

$$\sigma^x_{i=j} = \sigma^A_{i=j} + \sigma^B - \beta_{i=j} \qquad (6)$$

called interfacial tension coming naturally from the fact that two crystals A and B are each separated in two parts and then brought mutually in contact so that two bicrystals A/B in the right orientation are produced and a new interface is created.

Now when the analogue of (1) has to be written for the formation of the deposit A on B there is:

$$\Delta G = -n\Delta\mu + \sum_{i \neq j} (\sigma^A_i S_i) + (\sigma^x_{i=j} - \sigma^B) S_{i=j} \qquad (7)$$

since, at the contact area, there is created the interface of interfacial tension $\sigma^x_{i=j}$ but some free surface $S_{i=j}$ of the substrate of surface tension $\sigma^B_{i=j}$ then disappears.

Calling h_i the central distance to the faces, but distinguishing those $h_{i \neq j}$ pointing to the free faces and one $h_{i=j}$ measuring the distance of the substrate, the truncated volume of the crystal is:

$$V = \frac{1}{3} \{ \sum_{i \neq j} h_i S_i + h_{i=j} S_{i=j} \} \qquad (8)$$

provided that the $h_{i \neq j} > 0$ but $h_{i=j}$ may be positive or negative according to whether the central point lies outside or inside the substrate respectively, so that as before

$$dV = \frac{1}{2}\left\{\sum_{i \neq j} (h_i dS_i) + h_{i=j} dS_{i=j}\right\}$$ (9)

and from (7) and n = V/v the equilibrium conditions split into two types

$$\frac{\partial \Delta G}{\partial S_{i \neq j}} = 0 \text{ and } \frac{\partial \Delta G}{\partial S_{i=j}} = 0$$

and there is the Wulf-Kaischew theorem which states:

$$\sigma_i^A/h_i = \Delta\mu_{eq}/2v \text{ for all } i \neq j,$$ (10)

that means the free faces of the deposit and

$$(\sigma_{i=j}^A - \beta_{i=j})/h_{i=j} = \Delta\mu_{eq}/2v$$ (10')

for the contact equilibrium. For (10') we used again (6) so that only the basic quantities $\sigma_{i=j}^A$ and the adhesion free energy $\beta_{i=j}$ appear.

In the absence of substrate, $\beta_{i=j} = 0$, (10') looks as (10) but for all i and the equilibrium shape it is again one of a free crystal (5).

However, when $\beta_{i=j} \neq 0$ and since $\beta_{i=j}$ is a positive quantity, there results $h_{i=j} < h_{i=j}^0$ where $h_{i=j}^0$ is the central distance for $\beta_{i=j} = 0$. Increasing adhesion energy flattens the equilibrium shape.

For $\sigma_{i=j}^A = \beta_{i=j}$ the shape is half truncated ($h_{i=j} = 0$), for $\beta_{i=j} > \sigma_{i=j}^A$, $h_{i=j}$ becomes negative which means the central point goes into the substrate. Finally when $\beta_{i=j} \rightarrow 2\sigma_{i=j}^A$ the equilibrium shape becomes of molecular thickness. This point and $\beta_{i=j} > 2\sigma_{i=j}^A$ will be understood from II.3.

Let us look now at another point by following the change of ΔG as a function of the number n of molecules contained in the crystal. When for different n values the polyhedra are similar we can write $S_{i \neq j} = n^{2/3} \cdot C_{i \neq j}$ and $S_{i=j} = n^{2/3} \cdot C_{i=j}$ where the C_i are geometrical constants, the polyhedra containing n molecules, the surface $n^{2/3}$. Then ΔG can be written:

$$\Delta G(n) = -n\Delta\mu + n^{2/3} \sum_{i \neq j} [\sigma_i^A C_i] + [\sigma_{i=j}^A - \beta_{i=j}]C_{i=j}$$

At constant $\Delta\mu$ then an extremum of $\Delta G(n)$, $(d\Delta G/dn) = 0$ is possible, provided that $\Delta\mu > 0$. This extremum is a maximum since $(d^2\Delta G/dn^2) < 0$. When $\Delta\mu < 0$ the deposit cannot form since $\Delta G(n)$ is forever increasing. However at supersaturation $\Delta\mu > 0$ the system has to overcome an activation barrier:

$$\Delta G_S^x = \frac{4}{27} [\sum_{i \neq j} \sigma_i C_i + (\sigma_{i=j}^A - \beta_{i=j})C_{i=j}]/\Delta\mu^2$$

which is clearly smaller when there is a subtrate present ($\beta_{i=j} > 0$). Figure II.1 illustrates these situations: for increasing adhesion energies the nucleation barrier becomes smaller. In every case a substrate has some nucleation power. This nucleation is called heterogenous three-dimensional (3D) nucleation in contrast to the case $\beta_{i=j} = 0$ called homogeneous nucleation.

II.3. Let us reformulate the results in another way by looking at the problem at a molecular level. For this we consider now the substrate as a crystal and no more as a structureless body. First, and for

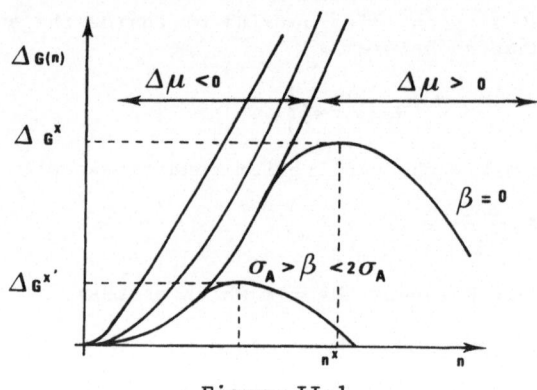

Figure II.1

simplicity, the substrate B is a Kossel crystal (cubic, simple with first neighbor interactions), as well as the deposit A, both having the same lattice parameters, have their contact with the (001) planes in parallel orientation. Then the surface tensions are given from the broken bond energies ϕ_{AA}, ϕ_{BB}, ϕ_{AB}: $\sigma^A = \sigma^A_{001} = \phi_{AA}/2a^2$, $\sigma^B = \sigma^B_{001} = \phi_{AA}/2a^2$ and the adhesion $\beta_{i=j} = \beta_{001} = \phi_{AB}/a^2$, the entropy terms being omitted. The condition of three-dimensional nucleation $0 < \beta_{001} < 2\sigma^A$ is written:

$$0 < \phi_{AB} < \phi_{AA}, \text{ for } \Delta\mu > 0. \tag{11}$$

It is easy to remember with this model that three-dimensional growth occurs when the bonding between A and B is less strong than the bonding in the deposit crystal. It takes place at supersaturation of A.

In principle the bond energy ϕ_{AB} may be also stronger than ϕ_{AA} so that we can look now at the meaning of the equivalent condition $\beta_{001} > 2\sigma^A_{001}$.

ϕ_{AB} may be the bond energy (or the reversible work of separation) of an isolated molecule A adsorbed on a substrate site. Let n^A be the number of such molecules, n_s the maximum number of sites accessible to such molecules (a full monolayer) so that $0 < (\theta = n^A/n_s) < 1$, where θ is called the degree of coverage. It represents also the probability that a mean site is covered. A given A molecule has therefore in this mean field approximation an energy $-u = \phi_{AB} + (1/2)(4\phi_{AA}\theta)$ the molecule having four potential neighbors of bond probability θ. Among the n available surface sites there are a number of ways W to realize such a layer so that the entropy of configuration per molecule $s = k/n_s \times LnW \equiv k/n_s \times Ln[n_s!/ (n_s - n^A)!.n^A.s \simeq -k[(1 - \theta)Ln(1 - \theta) + Ln\theta]$, that means the mixing entropy after having used Stirling's formula ($Lnn! = nLnn - n$). The surface being exposed to a vapor of actual vapor pressure p^A has a chemical potential per molecule $\mu^V = \mu^{OV} + kTLnp^A$ and the change of free energy Δf per molecule is:

$$\Delta f(\theta) = u - sT - \mu^V. \tag{12}$$

The condition $(\partial\Delta f/\partial\theta)_T = 0$ gives the equilibrium coverage, the adsorption isotherm $\theta(p^A)$ represented in Figure II.2 for this mean field approximation.

This family of isotherms at T = Cte for different ϕ_{AA} parameters all have (as every adsorption isotherm on a flat surface) an initial slope of $k_1(T)exp(\phi_{AB}/kT)$ (Henry's law) depending only on the normal adsorption

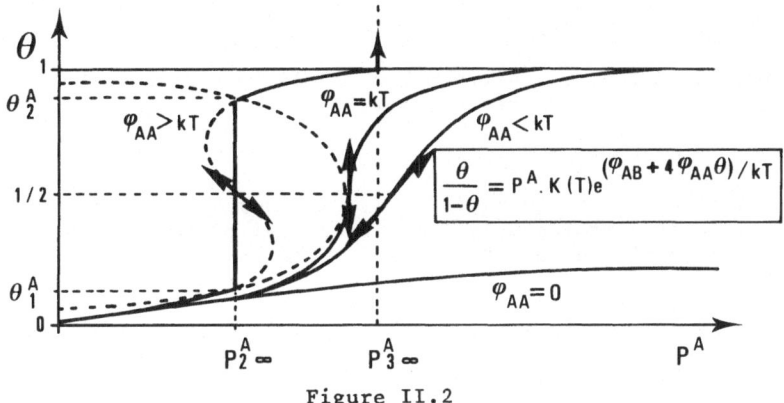

Figure II.2

bond ϕ_{AB}. For greater coverages their nature depends on the ϕ_{AA} parameter. When $\phi_{AA} = 0$, which means no lateral interactions, the coverage saturates monotonously to $\theta \to 1$ for $p^A \to \infty$ (Langmuir isotherm). However if $\phi_{AA} > 0$, there appears an inflexion point at $\theta = 1/2$. Its slope is positive as long as $\phi_{AA} < kT$ and $\theta(p^A)$ is a single valued function.

For $\phi_{AA} > kT$ its slope becomes negative after a discontinuity at $\phi_{AA} = kT$ where it passes from $+\infty$ to $-\infty$. The isotherm has the form of a loop suggesting an instability. For $\phi_{AA} > kT$ there are now two minima in the course of $\Delta f(\theta)$, one for a small coverage θ_1^A and another for a high value θ_2^A but giving the same Δf_{min} value. There is at equilibrium a diluted two-dimensional (2D) phase covering the surface by θ_1^A and a dense one covering θ_2^A at the temperature T and at a constant pressure $p^A(\theta = (1/2)) \equiv p_{2\infty}^A$. The stable isotherm at this pressure is the vertical full line in Figure II.2 and not the loop (van der Waals loop).

The constant pressure at this phase transition is determined using the equation in Figure II.2 for $\theta = 1/2$:

$$p_{2\infty}^A = k_2(T)^{-1} exp[(\phi_{AB} + 2\phi_{AA})/kT]$$ (13)

Figure II.3 shows that the bonding energy in the exponential is that one of an A molecule in a 2D crystallized phase.

This vapor pressure has to be compared to that one of a tri-dimensional (3D) crystal of A whose vapor pressure is written:

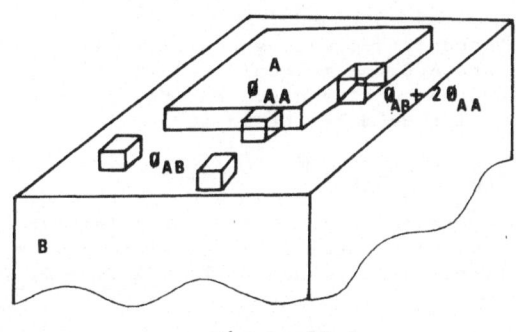

Figure II.3

155

$$p_{3\infty}^{A} = k_3(T)^{-1}\exp(-3\phi_{AA}/kT), \tag{14}$$

$-3\phi_{AA}$ being the potential energy per molecule in this Kossel crystal. Inserting (14) in (13) we obtain the relation:

$$p_{2\infty}^{A}/p_{3\infty}^{A} = \exp[-(\phi_{AB} - \phi_{AA})/kT]. \tag{15}$$

In this relation we omitted $k_3(T)/k_2(T)$, an inessential term for our purposes since it contains only the vibrational properties of a molecule in the bulk crystal and the 2D crystal. From this relation (15) it is seen that as long as $kT < \phi_{AA} < \phi_{AB}$, the 2D phase on the substrate is the most stable one, a condition which is equivalent to $\beta_{001} > 2\sigma_{001}^{A}$. This 2D phase forms at $p_{2\infty}^{A}/p_{3\infty}^{A} < 1$ that means, $\Delta\mu < 0$ or at undersaturation, in contrast to the 3D phase which forms only on the substrate or as a free crystal, as we have seen in II.2 at $\Delta\mu > 0$ that means oversaturation. This is typical for two-dimensional phase formation.

This point is not trivial since when bringing in a same closed vessel a 3D crystal A and the substrate B, at a temperature T, there is a stable 2D crystal A on the substrate, provided

$$kT < \phi_{AA} < \phi_{AB} \tag{16}$$

is satisfied. Outside this condition the molecules A adsorb only as a very diluted layer on the substrate but do not form a stable 2D phase.

In Figure II.2 we put at the pressure scale the value $p_{3\infty}^{A}$ for one of the isotherms and it is seen that this isotherm has at this pressure a vertical part which means that there is stability for a further 2D layers on the still formed first 2D layer. In this first neighbor model the second layer has still a $\phi_{AB}^{(2)} = \phi_{AA}$ so that relation (15) gives $p_{2\infty}^{(2)} = p_{3\infty}^{A}$ and also for all further n layers.

Notice that the first layer as well as all the others do not form spontaneously at their equilibrium pressure respectively $p_{2D}^{(1)A}$ and $p_{2\infty}^{(n)A} \equiv p_{3\infty}^{A}$. As for the 3D phase formation on the substrate there is a need of supersaturation of the vapor in respect to these equilibrium pressures, in order to overcome the corresponding surface creation energy. For these 2D islands the created surfaces are the edges limiting the 2D island. The corresponding activation barriers for nucleation are those found for 2D nucleation in crystal growth [18].

II.4. The model can now be extended to two crystals A and B having higher molecular interactions than the first neighbor ones.

Let us have monotonously decreasing additive pair interactions between molecules A-A and A-B. A molecule (A) in n^{th} layer has (n-1) underlying A layers and the substrate B. For the potential energy of a molecule in a nth layer let us make the ansatz:

$$-\varepsilon(n) = 3\phi_{AA} + [\phi_{AB}(1) - \phi_{AA}]n^{-\alpha} \tag{17}$$

where α is a positive exponent, e.g., $\alpha = 3$ for molecular crystals. The energy in a layer of a big A crystal is $-3\phi_{AA}$ since $n(\infty) = -3\phi_{AA}$, $-\phi_{AB}(1)$ is the energy in the first layer, since $-n(1) = 2\phi_{AA} + \phi_{AB}(1)$, as we had in the former section.

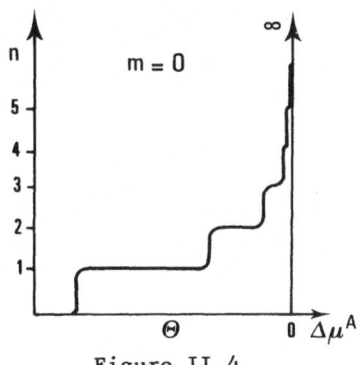

Figure II.4

The adsorption isotherm has now instabilities and phase transitions at pressures similar to (15), but:

$$p_{2\infty}^A{}^n/p_{3\infty}^A = \exp[-(\phi_{AB}(1) - \phi_{AA})n^{-\alpha}/kT].\tag{18}$$

The vapor pressure of the A crystal was again given by (14) but with the new meaning of ϕ_{AA} where other than first neighbor interactions are also contained. Written in supersaturation $\Delta\mu_{2\infty}(n) = kTLnp_{2\infty}^A{}^n/p_{3\infty}^A$ there is

$$\Delta\mu_{2\infty}(n) = -[\phi_{AB}(1) - \phi_{AA}]n^{-\alpha}.\tag{19}$$

Since this relation is valid for $\phi_{AB}(1) > \phi_{AA}$ there is an infinite number n of A layers formed on the B surface which are stable at $\Delta\mu_{2\infty}^A(n) < 0$ that means at undersaturation in respect to a bulk A crystal. The isotherm shows an infinite number of vertical steps (2D phase transitions) at $p^A < p_{3\infty}^A$ coming very closely spaced when $p^A \rightarrow p_{3\infty}^A$. The isotherm has $p_{3\infty}^A$ as an asymptote (see Figure II.4).

II.5. Now, let us go to the last step with the model, releasing the condition of isomorphism of the deposit with the substrate. The parameter a and b of the A and B crystal may differ by a natural misfit m = (a-b)/a. A layer of A having a thickness Z being in parallel orientation with the substrate parameters is formed, the substrate is considered as a very thick one.

The film to be deposited has to be strained equally in the two orthogonal directions in the contact plane so that the parameter a

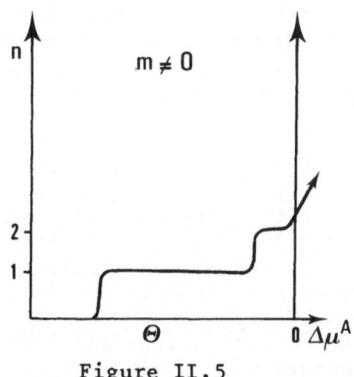

Figure II.5

approaches b and some contraction of the deposit normally to the interface occurs. There are two contributions to the free energy by doing this: a bulk contribution which is the elastic homogeneous free strain energy written shortly:

$$\Delta F^H = KZm^2 \tag{20}$$

of the film having now a thickness Z and work is spent against surface forces $\int_0^m d[\sigma_A S_A + (\sigma^* - \sigma_B)S_A]$. The first term is that one of the free surface of the film, the second corresponds to the exchange energy at the interface (see again (7)). From (6) we obtain also $\int_0^m d[2\sigma_A \sigma_A - \beta_{AB}S_A]$.

Since we use model crystals obeying additive pair potentials the surface tension and the adhesion energy can be written (neglecting entropy), after summation over the pairs $\sigma_A(Z) = k_A E_A/2S_A^2$ and $\beta_{AB}(Z) = k_{AB}E_{AB}/S_A$. E_A, E_{AB} are bulk energies and k_A, k_{AB} numerical factors coming from the summations depending only on the orientation of the interfaces in the crystals. A change dS_A due to strain, since the change of the bulk energies are still accounted in (20), is only

$$d\sigma_A = -\frac{k_A E_A}{2S_A} \frac{dS_A}{S_A} = -\sigma_A \frac{dS_A}{S_A}$$

and similarly for $d\beta_{AB}$ so that each term in the integral becomes zero* and

$$\int d[\sigma_A \sigma_A + \sigma^* - \sigma_B)S_A] = 0. \tag{21}$$

Therefore the only contribution is the homogeneous strain energy (20)**. In the outermost layer the contribution per unit area of substrate is $b^2 \Delta F^H/n$, a quantity to be added to (19) for having the new stability criterion for 2D layer growth.

For cubic crystals K in (20) as well as Z(n) can be expressed in a not too cumbersome way in terms of the three elastic constants C_{11}, C_{12}, C_{44}. When these crystals are supposed to be isotropic then

$$K = 2G_A \left(\frac{1+\nu}{1-\nu} \right) \text{ and } Z = na(1 - \nu m) \tag{22}$$

where ν is the Poisson ratio, the factor indicating vertical contraction of the film due to the lateral extension; G_A is the shear modulus of the deposit parallel to the interface. The new criterion for stability of 2D layer growth can be written as a function of the number of layers n and for a small misfit m:

$$\Delta\mu_{2\infty}(n) = -[\phi_{AB}(1) - \phi_{AA}]n^{-\alpha} + 2G_{AA}b^3 \left(\frac{1+\nu}{1-\nu} \right)m^2 \leqslant 0, \tag{23}$$

the second term being the elastic energy stored in the last layer per atom. Provided the adsorption energy of an A atom on the substrate is greater than on its own surface, $\phi_{AB}(1) > \phi_{AA}$, G_{AA} being a positive quantity there is necessarily a limited number n_1 of 2D layers able to grow:

---

* In fact the definition of σ(or β) in the case of pair potential does not contain the contribution due to surface (interface) relaxation due to the fact that when creating the surface (interface) there is some reorganization of the surface atoms. Such a contribution W_{rel}/S which is a negative quantity is, however, only a small fraction of σ (less than several percents) so that the effect of strain on $\Delta W_{rel}/S$ is a very weak second order effect.

** This is no more true when dislocations arise in the interface. The second term in (20) becomes non zero.

$$n_1 = [[\frac{\phi_{AB}(1) - \phi_{AA}}{2G_A b^3 (1 + \nu)(1 - \nu)}] \frac{1}{m^2}]^{1/\alpha}. \tag{24}$$

The number n_1 of layers able to grow is thus more limited: the natural misfit m is high; the deposit is elastically stiff; the heterobonding A-B is close to the homobonding A-A; and these different behaviors are thus moderated, the interactions are of short range (α great).

The adsorption isotherm is now no more asymptotic to the vertical pressure line $p^A = p^A_{3\infty}$ of A but cuts this line with a finite slope

$$(\frac{d\Delta m_{2\infty}}{dx})_{n_1} = \alpha [\phi_{AB}(1) - \phi_{AA}] n_1^{-(1 + \alpha)} \tag{25}$$

depending on the number of limiting layers (Figure II.4).

Before this limiting number n_1 of layers is attained the homogeneously strained film may suffer some collapse. From experiments it is known that interfacial dislocations may be introduced spontaneously and reduce the stored elastic energy in some conditions. The homogeneous strain energy drops from $F^H(m)$ to $F^H(m')$, $m' < m$ being the residual misfit, the dislocations storing $F^d(m - m')$, the total free energy being

$$F^T(m') - F^H(m') + F^d(m - m'). \tag{26}$$

There may be a value of the residual strain $m' = m'_{eq} < m$ which minimizes the total free energy (26). When this is realized then from (24) where we have to introduce instead of m the value m'_{eq}, as a consequence the number of limiting layers n'_1 becomes greater than n_1.

But in (26) the free energies depend also on the thickness Z of the film. For the homogeneous strain free energy there is from (20) per unit area of substrate $F^H(m') = KZm^2$, with K,Z given by (22).

For the free energy of the dislocations, when non-interacting, it is proportional to their number N per unit area. The period of each array being $p = b/(m - m')$, where $(m - m')$ is the misfit accommodated by the dislocations ($m' < m$), their number density is then per unit area $N_d = 2/p = 2(m - m')/b$, taking into account both arrays. The value b is also the modulus of their Burgers vector. The free energy of an edge dislocation in an interface AB writes $\simeq b^2 K'[1 + LnR/b]$ with

$$K' \simeq G_A G_B / 2\pi (G_A + G_B)(1 - \nu) \tag{27}$$

where R is a cut-off radius of the strain field. In the periodic array and along the interface the strain field falls to zero halfway between the dislocations, $R = 1/2p$. Normal to the interface this happens also. Approximately the surface of the film is the cut-off distance $Z = R$, provided the film is thinner than the half period of the dislocations

$$Z < \frac{1}{2} p \tag{28}$$

which can be checked later experimentally.

Now the total free energy (26) can be written for the two non-interacting arrays:

$$F^T(m',Z) = km'^2 . Z + 2k'b[1 + Ln^Z/b](m - m')$$

provided $m' < m$ and for $\partial F^T/\partial m' = 0$ the residual equilibrium strain is:

$$m'_{eq} = \frac{K'b}{K} (1 + LnZ/b) \frac{1}{Z} , \text{ provided } m'_{eq} < m. \tag{29}$$

Starting growth, increasing Z, as long as the criterion (29) says that $m'_{eq} > m$, there are no dislocations but only a homogeneous strain, the stored energy being $\Delta F^H(m)$. When (29) says m' = m, dislocations are introduced, the natural misfit m starts to be shared by the dislocations and by the homogeneous residual strain of energy $\Delta F^H(m'_{eq})$. The critical thickness Z_C where just this happens, or the number of critical layers n_C upon which dislocations are introduced, is from (29) with $m'_{eq} = m$ the solution of the equation

$$\frac{1}{n_c} (1 + Ln\ n_c) - \frac{K}{K'} m = 0. \tag{30}$$

In Figure II.6 the relation (29) is represented, i.e., the residual misfit m'_{eq} as a function of number of layers $n \simeq Z/b$ for small misfits using (22), (27) and the typical value $\nu = 1/3$. The two curves represent couples of extreme materials, at the left for fairly equally stiff partners $G_A \simeq G_B$, at the right for very stiff substrates in respect to the deposit ($G_B/G_A \to \infty$). For a couple of natural misfit m the critical number of layers n_C according to (30) is read by making $m'_{eq} = m$ in Figure II.6. As an example, for $m = 10^{-2}$, $n_c \simeq 10$ for equally stiff materials, respectively $n_c \simeq 27$ for a stiff substrate and soft deposit. These layers grow up to n_c without dislocations, then when dislocations start, the homogeneous strain m'_{eq} decreases progressively becoming zero for a very thick film.*

Now there may be two scenarios depending on the criterion (24) about the limiting number n_1 of layers. When (24) says $n_1 > n_c$, the layers remain fully strained at the value m up to the limiting number of layers. This happens preferentially for soft deposits on stiff substrates.

But there may be also $n_c < n_1$ from both criteria (24)(29) so that dislocations are introduced during the layer growth. This happens preferentially for equally stiff partners A-B. The limiting layer n_1 has then a strain $m'_{eq}(n_1)$ smaller than in the former case where it was $m > m'_{eq}$. What happens when dislocations are introduced at $n = n_c$. From

---

* In order to have simple formulae we considered isotropic cubic crystals. None of the cubic crystals are really isotropic except Ni!. However, among the cubic crystals the shear moduli vary by a factor of ten, but their anisotropy is much smaller (factor two) when different orientations are considered. Therefore the formulae are good guides for estimating an order of magnitude.

This also remains true when considering crystals of lower symmetry provided their crystal structure is homodesmic. That means that the crystals are made up by only one kind of bond type (either metallic, covalent, ionic, of hydrogen type, or van der Waals bonds).

However, when hetrodesmic structures are considered (for example, layer structures such as graphite, BN, MoS_2, and so on, where there are very dissimilar bonds in the same structure, very strong covalent bonds inside the layers and very weak van der Waals bonds between them), then clearly the used formula gives a wrong order of magnitude. In such cases, even the macroscopic elastic constants and the appropriate energy formulae are fully misleading for the cases we are considering here. For example, for graphite taken as a substrate the only correct estimation of the stiffness of the substrate parallel in its plane has to come from spectroscopic data where from the stiffness of covalent C-C bonds, G_B is calculated. For such substrates the curve on the right of Figure II.6 clearly holds.

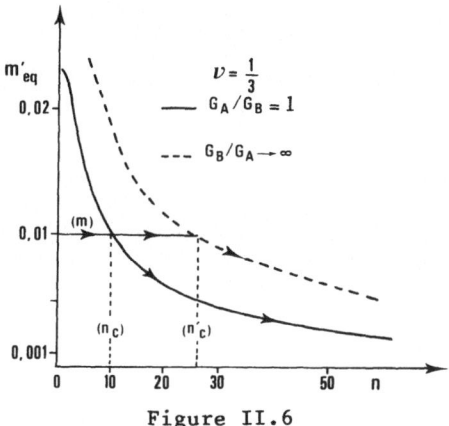

Figure II.6

(23)(29)(30) we have to evaluate the changes δ of free enthalpy ΔG just when $n = n_c$ passes to $n = n_c^+$. There is:

$$\delta \Delta \mu_{2\infty}(n_c) = \Delta \mu_{2\infty}(n_c^+) - \Delta \mu_{2\infty}(n_c^-) = \delta \left(\frac{\partial \Delta G}{\partial n} \right)_{n_c} = 0$$

and

$$\delta \left(\frac{d\Delta \mu}{dn} \right)_{n_c} = \left(\frac{d\Delta \mu}{dn_c} \right)_{n_c^+} - \left(\frac{d\Delta \mu}{dn_c} \right)_{n_c^-} = \delta \left(\frac{\partial^2 \Delta G}{\partial n^2} \right)_{n_c} = \frac{b^3 G_{AA}}{2\pi(1-\nu)} \frac{1}{n_c}$$

for equally stiff partners. This is exactly what is called a second order transition.

For both scenarios, when the limiting number n_1 of layers is obtained 2D growth is relayed by a 3D growth of A on A(n_1) since now $\Delta \mu^A > 0$ and there is a misfit between them, either m or $m'_{eq}(n_1)$, depending on both scenarios. The formulation of the problem becomes more difficult. When 3D crystals of A form on the ultimate layer numbered n_1, there is an inhomogeneous strain in the deposited 3D A crystal and in the substrate. The deposited 3D crystals do not remain forever fully strained when growing. The system is able also to minimize the total strain energy by introduction of new interfacial dislocations in its own interface with the underlaying layers. These underlaying layers may or may not develop, depending on the two scenarios described previously, interfacial dislocations in the substrate interface. Such situations have been considered by Kuhlmann-Wilsdorf [18] and Mathews [9]. When dislocations are introduced in the interface of a growing 3D crystal this happens in a discrete way and very curious but well explained effects, have been observed by transmission electron microscopy. Due to the misfit some Moire patterns show discontinuous ample rotations when a new dislocation enters. This does not, however, mean that the crystal rotates slightly.

What we described qualitatively may clearly be true also for couples A-B which start to grow on the substrate directly as 3D crystals, i.e., when $\phi_{AB}(1) < \phi_{AA}$.

II.6. The foregoing explanation makes clear, and surface physics experiments, confirm that there are two extreme growth modes of epitaxy:

- mode I called Frank-van der Merwe mode, where an infinite number of layers are stable or when there is a natural misfit between the couples there is a limitation to several layers, 3D growth following them;

- mode III called Volmer-Weber mode where inside a very diluted $(\theta < 10^{-2})$ adsorption layer 3D crystals form;

- an intermediary mode II, called Stranski-Krastanov mode, which has its own characteristics, the formation of only one layer followed by 3D crystal formation.

Mode I and II have the similarity that in both cases the bonding energy $\phi_{AB}(1)$ is greater or equal to ϕ_{AA} or $\beta_{AB} \geqslant 2\sigma_A$, the near equality being the domain of mode II and the definite inequality that of mode I. In fact, the consideration of the structure of deposit and substrate reveals much more drastic differences so that mode II no longer appears as a limiting case as stated in the above criterion. Surface experiments during epitaxial growth show:

- In mode I the A layers have the sme structure as the corresponding layers in the stable bulk structure of A. The plane of A coming in contact with the substrate has periodicities (simple or multiple) which can be brought in coincidence with those of the substrate by only a slight homogeneous strain in the contact plane (Royer's misfit laws) so that the molecules of the deposit come into the potential wells (or a fraction of them) of the substrate.

- In mode II the notion of misfit makes no more sense in general. The molecules of A, however, go into the potential wells of the substrate but by doing this they take a 2D structure which is either so distorted that the layer does not reassemble to a layer of the bulk A crystal, or it takes a structure which is much more similar to the substrate, or it takes a structure resembling neither of the above. The 2D layer in the mode II has in every case a new structure in respect of the stable bulk phase A. Sometimes such a layer has the structure of a non-stable bulk polymorph of A.

Some examples illustrating these general considerations will bring about a better understanding.

Mode III

Starting with mode III things are simpler since less details have to be known except the mutual crystallographic orientations, the morphology and the dislocation content. Transmission electron microscopy (TEM), besides other methods, is the most informative tool for the study of this mode. Actually there are no methods available which provide structural information about the interface in the mode III, except that very recently high resolution TEM methods are proving useful for studying such interfaces.

Mode III is expected and demonstrated by the proper methods (RHEED, Auger spectroscopy) for all deposits having stronger bonds than the substrate:

- Metals on alkali-halides as Ag, Cu, Au, Ni, and so on, (CFC); W, Mo (CC) on the whole series LiF, NaF, NaCl, KCl, KI, and so on, CaF_2 etc.

- Metals, as Mo, W on other metals, as Au, Ag, Zn, Cd.

For the reciprocal systems, exchanging substrate and deposit, these facts are not true.

A simple guide for the mode III is, therefore, to compare the heats of sublimation of the A, B partners when such partners do not react chemically (no formation of compounds). The real criteria we used in the preceding subsections $2\beta/\sigma_A$, $\phi_{AB}(1)/\phi_{AA}$ cannot be used since the quantities characterizing the interface β, or $\phi_{AB}(1)$ are in general not known. When the simple guidelines do not give a decisive answer then only careful experiments can help.

When the deposits are done under nearly equilibrium conditions (high substrate temperatures so that the incoming and outcoming flux of A are of the same order) Royer's laws predicts well the mutual orientations when the misfit is not too high.

In non-equilibrium conditions many other things may happen. A 2D-condensation can take place which by annealing transforms into 3D-crystals. This is the case of gold deposited on graphite at ordinary temperature. By annealing at 150°C the film breaks into 3D-islands which become thicker and thicker. Finally the 3D-equilibrium shape is obtained at 1000°K. Even 3D-crystals may be deposited having a non-stable structure. This is illustrated by iron deposited on (001) Cu where γ-Fe crystals form. By annealing they transform back to the stable CC' structure. See [19] for metastable phase formation.

Mode I

The epitaxial mode I is probably the most studied one, especially on systems of academic interest, but is very important for a fundamental understanding. This is the case for van der Waals crystals as rare gases, CH_4, CF_4, ethane especially on lamellar crystals as graphite, NB, MX_2 compounds but also on MgO crystals. The reason is firstly because the deposit interactions can be modeled more easily, secondly because these crystals have very homogeneous surfaces either by cleavage or when produced as highly dispersed crystals (high specific surface area of ~ 20 m^2/g) since then their surface can be explored by "bulk methods" such as X-ray and neutron scattering bringing clear structure and mobility information or by calorimetry for information about the scaling laws of a second order transition around critical points.

On such substrates of hexagonal symmetry (except MgO) the layers are close packed as in the (111) plane of the bulk CFC phases, the natural misfits of all the series of molecules remaining in a very limited range of a few percent. The coincidence lattice is of the $\sqrt{3} \times \sqrt{3}$ type. Many layers n may be formed if the misfit is small. In all cases $\phi_{AB}(1)/$ $/\phi_{AA} \simeq 2$ as measured, the interaction law $n^{-\alpha}$ has been verified with $\alpha = 3$ as it is expected for van der Waals interactions. Even the strain energy localized in the latter can be estimated in some cases. The ease when the number of layers in a given system can be brought from an infinite number to a finite one by a temperature $-\Delta\mu$ change has been called "wetting transition" (see [20]). The introduction of edge dislocations in the first layer has been measured by the analysis of diffraction intensities in the case of Xe or Kr on (0001) graphite. Changing the temperature or $\Delta\mu$ can lead to such a layer going progressively out of register and becoming an incommensurate 2D layer. This happens especially for this type of substrate where the potential wells are not deep at all for some of the molecules of that size.

But there are also studies on metallic couples such as Na on (001)W. Layers grow up to a great number. The CFC Na has its (001) plane in contact in parallel orientation, the coincidence mesh is 2 x 2 with a natural misfit of 4%. The strain energy in the layers is small because the shear modulus of Na is small. The sublimation energy of the first layer has been measured to 2.15 eV per Na atom, that of the following layers is practically constant and equal to the sublimation energy of bulk

Na, 1.09 eV. So a Na atom in the second layer does not feel the underlying substrate.

Cs on (110) GaAs is of more practical interest. In a first layer there are successive 2D three phases when the coverage is increased. The full layer is a (110)Cs CC plane having a C(4 x 4) coincidence mesh with the substrate but with the high misfit of 14%. The following layers, however, form in a very great number. The study of the adsorption isobars shows, however, that these layers are probably 2D liquids having similar energy and entropy as bulk liquid Cs.

Mode II

In order to stress the fundamental difference between mode II and I, let us give first an example of a limiting case of growth mode I.

Ag(CFC) forms on (001)Mo(CC) only one layer which is followed by 3D Ag(CFC) crystals, both being observed by Auger imaging. The layer (as well as the 3D crystals) have in the contact plane their (001) plane in a 45° orientation with the substrate, the natural misfit being 9%. The first layer is, therefore, only a strongly strained (001) layer of bulk silver having all its atoms in the sites of the Mo substrate. Due to the high strain energy no other layer is stable. This is clearly a situation of a limiting case of mode I. This layer does not fulfill the requirement of the mode II where this layer should have a "foreign" structure.

Just the opposite will be the case with the example of Pd/(110)W. One monolayer of Pd only forms but this layer has a structure of the underlying (110)W plane, it is called "pseudomorphic". Pd which in the bulk is CFC has adopted the structure of the underlying structure of the (110)W plane which is of the CC structure type. This is a "foreign" structure for Pd. When after this layer the 3D crystals of Pd(CFC) form they put in contact their (111) plane and two equivalent orientations <110>Pd//<1$\bar{1}$1>W result.

Other examples illustrate that the 2D layer in mode II may have such a strange structure that it resembles neither the substrate nor the bulk A phase. The simplest of these examples are Ag or Pb on (111)Si or (111)Ge.

In all combinations the first monolayer has the composition 2/3 Ag(Pb) : 1 Si(Ge). The diffraction pattern shows a $\sqrt{3}$ x $\sqrt{3}$ coincidence mesh so that the most probable structure of this layer is a honeycomb as given in Figure II.7. The metal atoms are probably in the triangular potential wells of Si(111), two-thirds of them being occupied, one-third empty. These metal atoms have enough space when considering the atomic diameter of the bulk metal they do not touch each other. The in-plane bonds are very weak $\sim$ kT at temperatures T $\sim$ 1000°K, the energy of $\phi_{AB}(1)$ bonds being a bit smaller than the heat of vaporization of the bulk metals $\sim$ 30(20).10^3k. Instead of a second layer above this exotic one, 3D crystals of CFC Ag(Pb) are formed in an epitaxial parallel orientation. In respect to the substrate period there is a Vernier 4<1$\bar{1}$0>$_{Metal}$ $\simeq$ $\simeq$ 3<1$\bar{1}$0>$_{Substrate}$ of only several percent.

By discussing the three modes on examples or with a simple theory we have avoided those cases where interfacial or bulk mixing of AB, i.e., compound formation, also occurs. In practice such couples do exist and are of great interest but very few examples are known about in detail.

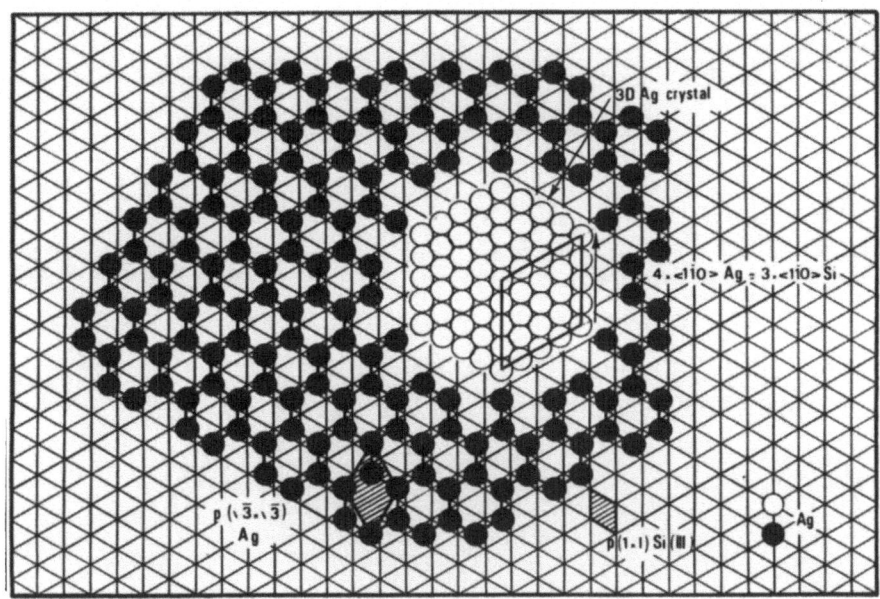

Figure II.7

III REFERENCES

1. F. Wallerant, Bull. Soc. Fr. Min., 25:180 (1902); O. Mugge, Neues Jahrb. Min., 16:335 (1903).
2. G. Friedel, "Lecons de Cristallographie (1926)"; Berger-Levrault, new edition, Blanchard, Paris (1966).
3. L. Royer, Bull. Soc. Fr. Min., 51:7 (1928).
4. A. Neuhaus, Fortsch. Min., 29:136 (1950).
5. J. C. Monier, Bull. Soc. Fr. Min. Crist., 77:1029 (1954).
6. D. W. Pashley, Adv. Phys., 5:173 (1956); Adv. Phys., 14:183 (1965).
7. G. H. van der Merve, Surf. Sc., 31:1988 (1972).
8. J. Friedel, "Dislocations", Pergamon (1964).
9. "Grain Boundaries" in Surf. Sci., 31 (1972).
10. J. W. Mathews, "Epitaxial Growth A and B", Material Science Series, Academic Press, NY (1975).
11. P. B. Hirsch, "Electron Microscopy of Thin Crystals", Butterworth (1965).
12. P. Guegen, M. Cahoreau and M. Gillet, Thin Solid Films, 16:27 (1973).
13. G. I. Finch and A. G. Quarrel, Proc. Roy. Soc., A14:398 (1933).
14. H. E. Farnworth, Phys. Rev., 43:900 (1933).
15. T. N. Rhodin, in: "Crystal Growth", Disc. Far. Soc., 5:215 (1949).
16. E. Bauer, Z. Krist., 110:372 (1958).
17. E. Bauer, in: "Interfacial Aspects of Phase Transitions", Nato Advanced Study Institute Series, Series C, pp 1-32, 411-432, Reidel Co. (1982); R. Kern, ibidem, pp 287-314; R. Kern, in: "Synthesis, Crystal Growth and Characterization", pp 119-134, North Holland (1982); R. Kern, G. Le Lay and J. J. Métois, in: "Current Topics in Materials Science", Vol. 3, pp 178-419, North Holland (1979).
18. W. K. Burton, N. Cabrera and F. C. Frank, Trans. Roy. Soc., A243:299 (1951).
19. R. Kern, in: "Current Topics in Materials Science", 12:81, North Holland (1985).
20. M. Bienfait, Surface Sci., 162:411 (1985); J. Suzanne, Annales de Chimie, 11:37 (1986).

FUNDAMENTALS OF DENDRITIC GROWTH

M. E. Glicksman

Materials Engineering Department
Rensselaer Polytechnic Institute
Troy, NY 12180, USA

I. INTRODUCTION

I.1 Importance of Dendritic Crystal Growth

Dendritic growth is perhaps the most common form of solidification especially in metals and other systems that freeze with relatively low entropies of transformation. Dendritic or branched growth in alloys generates microsegregation as well as other internal defects in castings, ingots, and weldments. More subtle effects introduced by the complex dendritic microstructure in solidified materials include crystallographic texturing, hot cracking, suboptimal toughness, and reduced corrosion resistance. Moreover, the dendritic microstructure and its effects may be modified by subsequent heat treatments, but they are seldom fully "erased". As such, the understanding and control of dendritic growth in solidification processing is crucial in order to achieve specific material properties in final products.

Dendritic growth is a coupling of two seemingly independent growth processes: the steady-state propagation of the dendrite main stem and the non-steady-state evolution of dendrite branches. Until recently, the time-dependent features of dendrites were completely ignored and theoretical models of dendrites were limited to the mathematical description of a branchless, paraboloidal needle, which grew at a constant rate in a shape-preserving manner. Furthermore, theoretical studies of dendritic growth have concentrated on the steady-state development of a one-component needle-dendrite growing in a pure melt, wherein the major transport process is heat conduction.

I.2 Steady-state Dendritic Growth

Despite the fact that dendrite formation seems to involve both steady-state attributes (near the tip) and time dependent features (side branches) the earliest models employed steady-state descriptions of needle-like, branchless dendrites growing in a shape preserving manner [1-3]. The dendrite was assumed to grow at a constant axial rate, V, into a melt of uniform supercooling, ΔT, such that the surrounding thermal field would appear to be stationary in a coordinate frame traveling with the dendrite tip. The steady-state shape was chosen, ab initio, such that the prescribed solid-liquid interface remained at its bulk thermodynamic

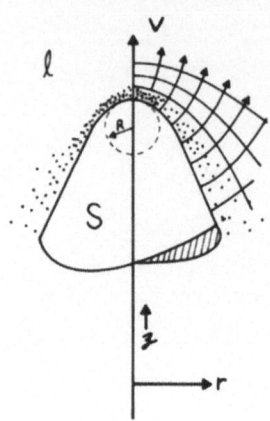

Fig. 1. Sketch of Ivantsov's paraboloid of reolution, having tip radius R
and moving at constant velocity V in the z-direction. The arrows
represent lines of heat flow from the interface in the moving
frame which are orthogonal to the paraboloidal isotherms in the
liquid phase.

melting temperature, T_m. Imposition of an isothermal boundary condition
retained linearity of the heat flow solution and led to a large class of
steady-state "dendrite" solutions which depended on the arbitrarily chosen
shape. Ivantsov's solution for the paraboloid of revolution is typical of
these linear shape-preserving solutions.

I.3 Ivantsov's Transport Solution

    An exact solution to the non-dimensional temperature distribution was
first developed by G. P. Ivantsov in 1947, for the case of a paraboloidal
"needle" dendrite, the shape for which was first suggested by Papapetrou
in 1935.

    Ivantsov (1947) modeled dendritic growth in the form of an isothermal
paraboloid of revolution (see Figure 1) with a steady-state tip radius R,
growing at a uniform rate of V along the z-axis. (z,r) is a coordinate
moving at the same speed V as the tip. Invantsov set the temperature of
the paraboloidal s/ℓ interface to T_m, and the temperature far from the
dendrite at T_∞, so $\Delta T = T_m - T_\infty$.

    Ivantsov correctly suggested that a steady-state solution to the heat
flow equations could be obtained in the moving (z,r) frame. He solved in
the melt phase (ℓ) $\nabla^2 T_\ell + (v/\alpha)\nabla T = (1\partial T_\ell/\alpha\partial t) = 0$, and realized that the
heat flow solution for the solid (isothermal paraboloid) was trivial,
$T_s(z,r) = T_m$.

    If α, C, and ΔH are chosen as constants, then the temperature
equation is conveniently non-dimensionalized as $\nabla^2\Theta + 2Pe\nabla\Theta = 0$, where
operators ∇^2 and ∇ are taken with respect to non-dimensional distances z/R
and r/R.

    The following dimensionless parameters prove useful in developing the
heat flow solution first given by Ivantsov

(dimensionless temperature scale) $\Theta = \dfrac{T - T_\infty}{-\Delta H/C}$ (I.1)

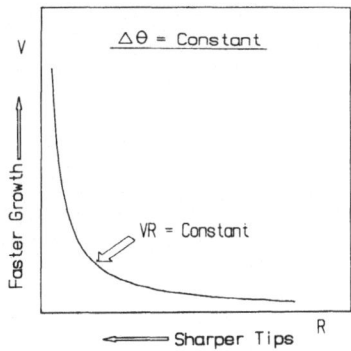

Fig. 2. Velocity vs. tip radius for a fixed value of $\Delta\Theta$, obtained from Ivantsov's heat-flow solution for a paraboloid of revolution.

(Peclet number, $\alpha \equiv$ diffusivity for heat in the melt ($i = \ell$) or solid ($i = s$)) $\qquad$ $Pe = \dfrac{VR}{2\alpha_i}$ $\qquad$ (I.2)

Ivantsov used the <u>approximations</u> of an isothermal s/ℓ interface and a uniform far-field temperature in the supercooled liquid, viz.

<u>At the s/ℓ interface</u> $\Theta = \hat{\Theta}$ $\qquad$ (I.3)

<u>Far from the interface:</u> $\Theta = \Theta_\infty$. $\qquad$ (I.4)

Ivantsov's solution $\Theta(z/R,\ r/R)$ may be used to relate supercooling, $\Delta\Theta$, to the Peclet number, Pe, namely

$$\Delta\Theta = Pe\ \exp(Pe)E_1(Pe), \qquad (I.5)$$
where
$$\Delta\Theta = \hat{\Theta} - \Theta_\infty, \qquad (I.6)$$

i.e., $\Delta\Theta$ is the dimensionless supercooling responsible for the free energy change "driving" the dendritic crystal growth process.

$E_1(Pe)$ is the 1st exponential integral – a tabulated function, defined here as

$$E_1(Pe)\int_{Pe}^{\infty} \frac{e^{-s}ds}{s}\ . \qquad (I.7)$$

Figure 2 shows that for a given value of $\Delta\Theta$, a unique value of Pe occurs which provides the hyperbolic relationship between V and R. This is also referred to as the "point effect", well-known in the theory of diffusion.

Note also that as $\Delta\Theta \rightarrow 1$, $Pe \rightarrow \infty$ and that at small supercooling the behavior is almost linear*. It is also important to understand that only the product $V{\cdot}R$ is specified by Ivantsov's solution, and that a manifold of growth states is possible. Thus, although $V = V(R)$ is known, <u>the unique, or operating state of the dendrite is not specified</u> by transport theory.

---

* This "linearity" is with respect to log-log coordinates. Actually the slope of log $\Delta\Theta$ versus log Pe approaches 2 as $Pe \rightarrow 0$ (see Figure 3).

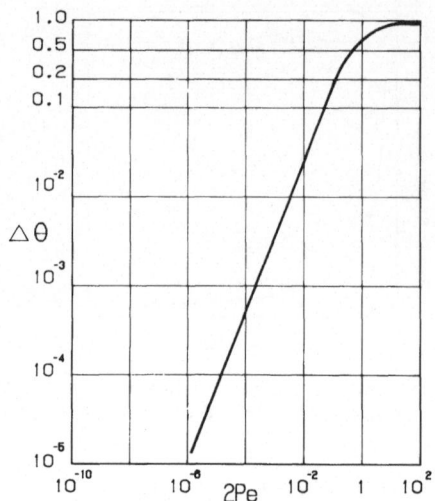

Fig. 3. Log-log plot of $\Delta\Theta$ vs. 2Pe from Ivantsov's solution. The slope of this plot approaches 2 for small values of Pe and $\Delta\Theta$, while Pe $\rightarrow \infty$ as $\Delta\Theta \rightarrow 1$.

This is perhaps understandable based on the fact that the only length scale appearing in the transport problem, per se, is the diffusion length, α/V. A second independent length scale is required to go further and determine the unique growth state $V(\Delta\Theta)$, $R(\Delta\Theta)$ for a dendritic crystal growth system.

II. INFLUENCE OF THE SOLID-LIQUID INTERFACE

II.1 Need for Capillarity Effects

Equation (I.5) relates $\Delta\Theta$ to the Peclet number, Pe = $VR/2\alpha$, where R is the radius of curvature of the dendrite tip, and α is the thermal diffusivity of the melt. The inverse of Eq. (I.5), although not expressible in terms of known functions, does establish that $V \cdot R = f(\Delta\Theta)$, which provides an infinite range of hyperbolic solutions, i.e., unbounded values for V and R for a given value of $\Delta\Theta$. Clearly, the transport solutions for the steady-state dendrite lack uniqueness when the only physical length scale introduced into the problem is the characteristic, but unknown, diffusion distance, α/V, or, alternatively, the equally unknown dendrite tip radius of curvature, R. This limitation was recognized over twenty years ago by Bolling and Tiller [4], who then introduced a non-linear boundary condition into the problem which effectively places an upper bound on V and a lower bound on R. Bolling and Tiller suggested that local thermodynamic equilibrium along the dendrite surface requires that the melting temperature, T_e, is a function of the mean interfacial curvature, κ, namely, $T_e = T_m - \Gamma\kappa$, which is the well-known Gibbs-Thomson equation. Here $\Gamma = \gamma\Omega/\Delta S$; γ is the solid-liquid surface energy; Ω is the molar volume of the solid; and ΔS is the molar entropy of fusion. The Gibbs-Thomson equation requires, therefore, that the dendrite grows with a non-isothermal interface.

II.2 Non-Isothermal Dendritic Growth

Introduction of the non-isothermal temperature boundary condition raises a severe difficulty, inasmuch as the steady-state shapes which were

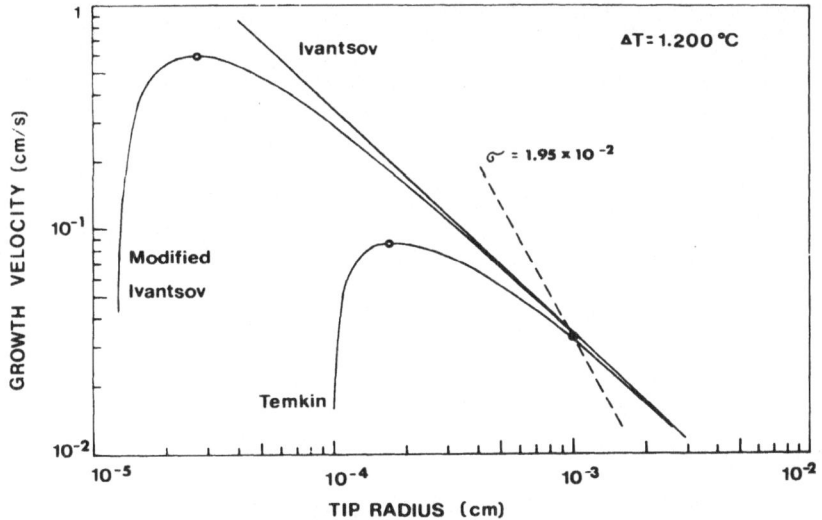

Fig. 4.   Velocity (logarithmic scale) vs. dendrite tip radius (logarithmic
scale) at a fixed supercooling, ΔT, of 1.20 K. The isothermal
(Ivantsov) solution is characterized by VR = constant, whereas
the nonisothermal (Temkin and modified Ivantsov) solutions show
progressive departures from the line VR = constant as the tip
radius decreases. VR^2 = constant (---) represents the condition
of marginal morphological stability with a separation constant σ^*
of 1.95 x 10^{-2}, close to those discussed in the text.

treated as a class of shape preserving solutions by Horvay and Cahn [2] no
longer simultaneously satisfy both energy conservation and the non-
isothermal equilibrium temperature boundary condition. Several
approximate theories were developed in which the interface shape was
chosen to satisfy either energy conservation [5-7] or the non-isothermal
condition [8,9], and a decade ago, a self-consistent theory was finally
developed [10] which determined the dendrite shape as part of the solution
and satisfied both physical requirements. All of these non-isothermal
theories shared the common result that the values of V and R, which
constitute possible operating states, lie along curves with a maximum in
the value of V. Figure 4 shows two typical V versus R relationships at a
fixed value of the supercooling.

II.3 Maximum Growth Rate Hypothesis

The value of V at the maximum was selected as the probable operating
state over the manifold of the possible operating states. Note that there
were never compelling reasons used to incorporate within these theories
the hypothesis that the maximum velocity is the unique velocity
characterizing axial dendritic growth. As shown in Figure 4 the states of
maximum velcoity for non-isothermal models have tip radii such that the
Peclet number at $V = V_{max}$ is one-half that of the isothermal Ivantsov
model. The tip radius for the Modified Ivantsov model is just twice the
critical radius R*, where R* is the radius which depresses the dendrite
tip temperature by an amount equal to the supercooling. Without a
temperature difference between the interface and the supercooled melt,
latent heat cannot be transferred and V = 0. R* is easily estimated from
the Gibbs-Thomson relationship by finding the value of $\kappa^* = 2/R^*$ which
reduces the tip temperature, T_m, to T_∞. Figure 4 corresponds to a

dimensionless supercooling $\Delta\Theta$ = 0.05, which is relatively small, yet as seen in this figure R* $\simeq$ 10^{-5} cm. Even Temkin's analysis predicts a tip radius of only 10^{-4} cm, which is still much smaller than the observed scale of dendrites at such small supercoolings. Of course, estimates from theory for the size scale R depend sensitively on the parameters chosen for the critical radius, namely

$$R* = 2\Gamma c/\Delta\Theta L, \tag{II.8}$$

among which Γ is least well-known, insofar as it contains the solid-liquid interfacial energy, γ. Thus, even qualitative observations of dendritic structures led some investigators to suspect that the hypothesis of maximum velocity was incorrect, long before quantitative kinetic data were available to challenge its validity directly.

III MORPHOLOGICAL STABILITY AND TIME DEPENDENCE

III.1 Perspective Commentary

As outlined in the previous Section, purely steady-state theories of shape preserving dendritic growth all failed, the level of their sophistication notwithstanding. Indeed, even solving the steady-state growth problem with non-linear boundary conditions in an exact self-consistent form [10] served to show mainly that the maximum steady-state velocity was a relatively poor upperbound to the true operating state. Two disparate viewpoints arose on this issue: (1) that the steady-state optimized dendrite was correct to first order but needed inclusion of nonsteady-state features such as side branching [17,18], and (2) that dendritic growth was intrinsically time-dependent and unstable [19,20-21]. Analyses based on the first point of view showed that the steady-state needle dendrites were unstable when tested for morphological stability using linear perturbation methods fashioned after those first used by Mullins and Sekerka [22] and by Voronkov [23]. An unfortunate aspect of these approaches was that the steady-state dendrite shapes themselves were only approximations, and therefore were intrinsically unstable without perturbation. Consequently, viewpoint (1) only served to emphasize the deficiency of steady-state approaches, and did not lead to new insights into the problem. The second viewpoint was originally proposed a decade ago by Oldfield [19] who was the first investigator to stress that the size of a dendrite tip might be selected through a balance of destabilizing forces arising from diffusion by stabilizing forces arising from capillarity. He found by simple logic and some numerical analysis that
$$VR^2 \simeq 100\alpha\Gamma c/L, \tag{III.1}$$

which, as we shall show, is remarkably close to the results obtained later by linear perturbation analysis. Oldfield also demonstrated through computer generated cinematography that such a dendrite was actually a fully time-dependent object, with branches emanating as waves from a nearly steady-state tip. The numerical character of Oldfield's work, unfortunately, prevented its wide acceptance at that time.

III.2 Dynamical Theory of Dendrites

The proper estimation of size scales for morphologically unstable systems begins with Mullins and Sekerka's ideas that any Fourier component of a perturbed planar interface represented as $\delta = \delta_0\exp(i\omega x)$ is subject initially to a time dependence described by $\delta(t) = \delta\exp(\sigma t)$, where σ, in general, is a complex eigenvalue of the linearized dynamical equations of the interface motion [21]. The quantity ω is the Fourier component's wave number, t is time, and δ_0 denotes the initial (small) amplitude of this component at t = 0. If the real part of σ is negative, then the perturbation decays to zero amplitude, whereas if the real part of σ is positive then δ grows exponentially. If σ is purely imaginary, then on average the amplitude of δ remains equal to δ_0 and the interface is deemed to be marginally stable. The condition of marginal stability for a pure material growing from its supercooled melt may be shown to be [21]

$$\text{Re}(\sigma) = 0 = -\Gamma\omega^{*2} - \bar{G}, \tag{III.2}$$

where $\bar{G}$ is the average thermal gradient (weighted by the thermal conductivities of each phase) and ω^* is the wave number of the marginal perturbation. We now adopt the remarkable suggestion originally made by Langer and Müller-Krumbhaar [22] that the wavelength, $\lambda^* = 2\pi\omega^{*-1}$, of the marginal perturbation sets the scale of the dendrite tip radius, R. Moreover, the average temperature gradient acting on the tip may be found from the transport solution of the underlined, steady-state, shape preserving dendrite. For example, the paraboloid of revolution, as described by Ivantsov's solution [2], is an isothermal shape with a zero gradient within the solid phase and with a dimensionless temperature gradient in the melt phase ahead of the tip $G_\ell = -2\text{Pe}$. Again, Pe, the Peclet number, can be related to the supercooling through Eq. (I.5), and thus the average dimensionless gradient G = -Pe. The gradient G can be dimensionalized to $\bar{G}$ by rescaling by the characteristic temperature L/C divided by the dendrite tip radius, R. Thus, the average temperature gradient at the dendrite tip is

$$\bar{G} = \frac{-\text{Pe}}{R}\frac{L}{C} = \frac{-VL}{2\alpha C} . \tag{III.3}$$

If Eq. (III.3) is substituted into Eq. (III.2) we find that the marginally stable state Re(σ) = 0 occurs when R = λ^*, which after some rearrangement yields the condition for growth

$$VR^2 = 8\pi^2\alpha C\Gamma/L, \tag{III.4}$$

which, except for a slight difference in the numerical coefficient, agrees with Oldfield's expression, Eq. (III.1).

Equation (III.4) is the central result obtained from dynamical analysis of dendrite tip motion. If we recall the definition Pe = VR/2α, then Eq. (III.4) may be recast in an especially convenient form, namely

$$\text{Pe} = \frac{4\pi^2}{R}\frac{C\Gamma}{L} . \tag{III.5}$$

Now, Ivantsov's transport solution, Eq. (I.5), may be written in operator form as $\Delta\Theta = \text{Iv}[\text{Pe}]$, where Iv[] represents the series of operations carried out on the right-hand side of Eq. (I.5). We can formally invert Eq. (I.5) to stress that Pe is some function of $\Delta\Theta$, viz.,

$$\text{Pe} = \text{Iv}^{-1}[\Delta\Theta], \tag{III.6}$$

although the inversion operator $\text{Iv}^{-1}[]$ cannot be represented exactly by any known series of algebraic or transcendental operations. Nonetheless the inversion operator Iv^{-1} exists, if only as a graph or an asymptotic expansion. Equations (III.5) and (III.6) can now be combined, eliminating explicit dependences on Pe, to yield the operating state of the dendrite under marginally stable dynamical conditions. Specifically, we find that

$$R = \frac{4\pi^2}{L} \frac{C\Gamma}{Iv^{-1}[\Delta\Theta]} ,$$

(III.7)

and

$$V = \frac{1}{4\pi^2} \left(\frac{2\alpha\Delta SL}{\gamma\Omega C} \right) \{Iv^{-1}[\Delta\Theta]\}^2 .$$

(III.8)

IV   SCALING LAWS AND OBSERVATIONS

IV.1 Analysis

The coefficient $4\pi^2$ appearing in Eqs. (III.7) and (III.8) is often defined as $(\sigma*)^{-1}$. Also the parameter grouping

$$\frac{C\Gamma}{L} = \frac{C\gamma\Omega}{L\Delta S} \equiv d_o ,$$

(IV.1)

where d_o is called the capillary length scale. d_o is about 1 nm in size for many common materials, and represents the second physical length scale introduced.

Thus, we can rewrite Eq. (III.7) and (III.8) in a scaled form:

$$\frac{d_o}{R} = \sigma*Iv^{-1}[\Delta\Theta],$$

(IV.2)

and

$$\frac{V}{V_o} = \sigma*\{Iv^{-1}[\Delta\Theta]\}^2,$$

(IV.3)

where V_o is the "characteristic" velocity of the dendrite forming system given by $V_o = (2\alpha\Delta SL/\gamma\Omega C)$. Note that the scaled radius only depends on the dimensionless supercooling, $\Delta\Theta$, and $\sigma*$, as does the scaled velocity. All the materials parameters appear either in d_o or V_o.

The predicted operating point (V,R) of dendrites can be expressed in a form especially convenient for checking against experiments, namely

$$V \cong 0.018 \frac{-\Delta S\alpha\Delta H}{\Omega\gamma C} \Delta\Theta^{2.5},$$

(IV.4)

$$R \cong 55 \frac{C\gamma\Omega}{(-\Delta H)\Delta S} \Delta\Theta^{-1.25}.$$

(IV.5)

These expressions are accurate for moderate values of $\Delta\Theta$. In typical experiments, $0.01 < \Delta\Theta < 0.3$. At extremely small values of $\Delta\Theta$, $V_o \sim \Delta\Theta^2$, and $R_o \sim \Delta\Theta^{-1}$. At large values of $\Delta\Theta$, the local equilibrium assumption breaks down and the theory, as presented here, fails.

Several careful experiments (see Figure 5) conducted by Glicksman, Schaefer and Ayers [11], Huang and Glicksman [24], and by Fujioka [15], confirm this result to within experimental error and the current knowledge of the thermophysical parameters. To date, three systems have been checked:

a)   succinonitrile
b)   ice/water
c)   pivalic acid

Here $g = 0.018 (-\Delta S\alpha\Delta H/\Omega\gamma C)$.

Figure 6 shows that in a transparent model system such as succinonitrile (SCN) the images of the dendrite tip do indeed scale with

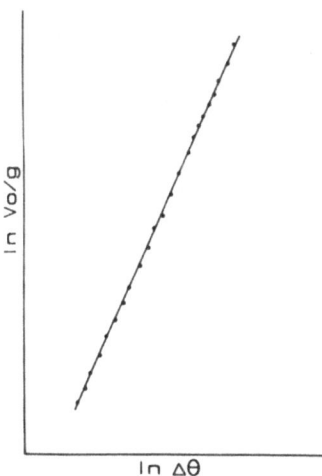

Fig. 5. Experimental values of growth velocity, V_O (normalized by the
characteristic velocity of the material, $g = 0.018$ ($\Delta S \alpha L / \Omega \gamma C$))
vs. dimensionless undercooling, $\Delta \Theta$. The slope of this log-log
plot is about 2.5, which agrees with the result obtained from a
marginal stability criterion at the tip (Eq. IV.4 in the text).

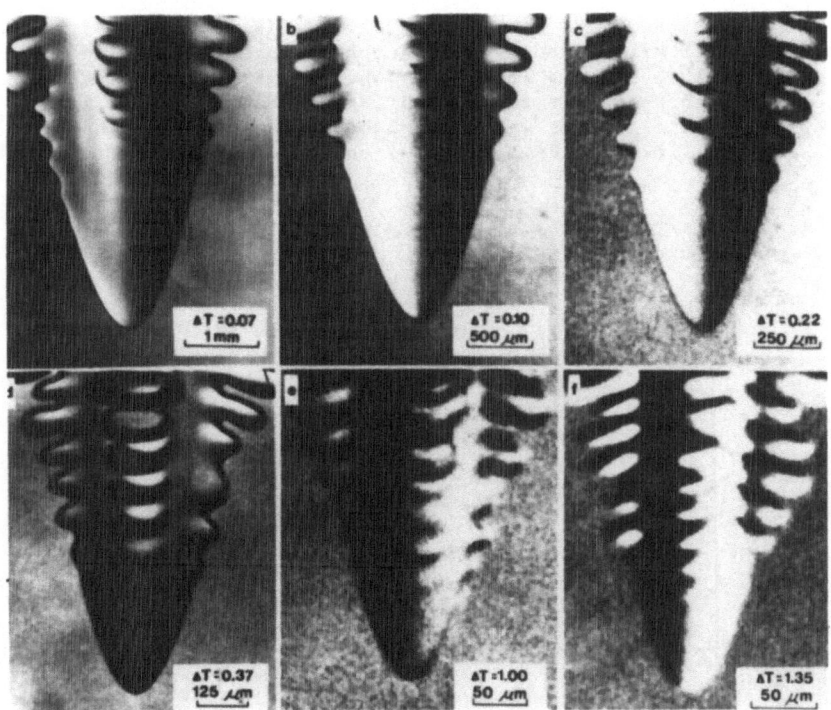

Fig. 6. Tip morphologies of succinonitrile dendrites growing at various
supercoolings. Photographic magnifications have been adjusted to
produce constant apparent tip radii. It should be noted that at
increasing supercoolings the side branches amplify more rapidly
and encroach on the steady state region near the tip.

175

the supercooling. Also, Figure 8 shows that the dendrite tip radius, R, scaled to the critical wavelength, $\lambda*$, is of unit order, where

$$\lambda* = \frac{d_o}{\sigma* Iv^{-1}[\Delta\Theta]} \qquad (IV.6)$$

That is, $R \overset{\sim}{=} \lambda*$, which is the fundamental scaling arising from the dynamical theory.

Another interesting scaling relationship for dendrites can be obtained from the stability analysis by inserting Eq. (II.1) into Eq. (III.7) and then solving for the ratio of the tip radius to the nucleation or critical radius, $R/R*$. We find

$$\frac{R}{R*} = \frac{2\pi^2\Delta\Theta}{Iv^{-1}[\Delta\Theta]} \qquad (IV.7)$$

which by virtue of Eqs. (I.5) and (III.6) and the definition of $\sigma* = 1/4\pi^2$ may be rewritten in the forms

$$\frac{R}{R*} = \frac{Iv[Pe]}{2\sigma* Pe} = \frac{\exp(Pe)E_1[Pe]}{2\sigma*} . \qquad (IV.8)$$

In the range of small supercoolings ($\Delta\Theta \ll 1$), specifically where the value of Pe is sufficiently small that $E_1(Pe) \to -1N\ Pe$, Eq. (IV.8) becomes

$$\frac{R}{R*} (\Delta\Theta \ll 1) = \frac{-\ln Pe}{2\sigma*} . \qquad (IV.9)$$

The value of $R/R*$ predicted from Eqs. (IV.7) - (IV.9) over the typical range of experimentally useful "small supercoolings" ($10^{-3} < \Delta\Theta < 0.1$) is of the order of 100, clearly indicating that marginally stable dendrites ought to grow with their tip radii much larger than $R*$, which is a morphological scaling law at variance with the steady-state theories that predict small multiples of $R*$. Figure 7 shows measurements of SCN

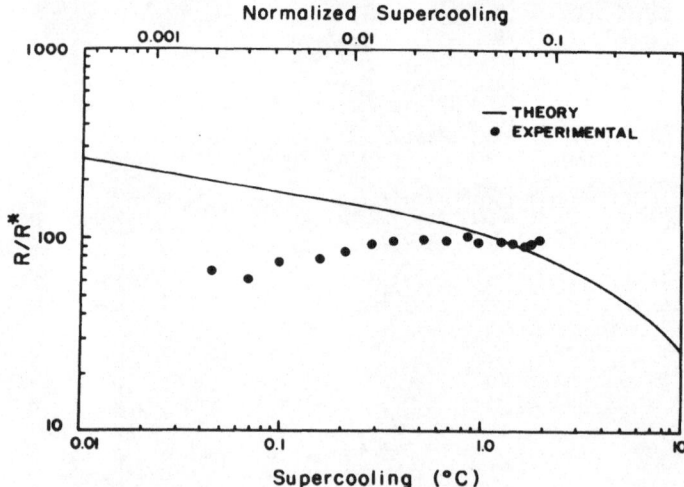

Fig. 7. Dendrite tip radius R scaled to the critical radius $R*$ (logarithmic scale) vs. supercooling (logarithmic scale): ——, theory; ·, experimental. Data, based on measurements performed on succinonitrile, show that the operating states of dendrites occur at large multiples of $R*$.

dendrite tip radii, obtained by Huang and the author [24], scaled to the critical radii, R*. These data shows that the value of R/R* is well approximated by Eqs. (IV.7) - (IV.9). The precise fit of the data in Figure 7 to the scaling law is limited at very small supercoolings by the presence of convection. Convection effects, relative to transport by diffusion, become negligible in SCN above a supercooling of about 1 K [29], and, as shown in Figure 7, a rather close correspondence occurs near and above $\Delta T = 1$ K. It is unfortunate that obtaining reliable morphological data above a supercooling of 2 K is, at present, not technically feasible with SCN. This is due simply to the difficulty in obtaining adequate photographic resolution of the dendrite tip structure as the tip radius decreases to 1 μm and its speed exceeds 1 mm/s. Figure 6 also shows that the series of SCN dendrite images growing at increasing levels of supercooling decline in optical quality as the supercooling increases. Lappe [30] has confirmed many of the morphological and kinetic measurements originally reported by Huang [31], and has further demonstrated that at a supercooling of 2 K, or beyond, a growing SCN dendrite cannot be resolved optically to permit an accurate measurement of the tip radius. Figure 6 also shows another effect which might be explained by the scaling law shown in Figure 7 and described in Eqs. (IV.4) and (IV. 5), namely, that the dendrites are not self-similar, except for the tip itself. Inspection of the micrographs in Figure 6 show that the side branches intrude on the tip as the supercooling increases, due primarily to the faster amplification rate of the marginal eigenmode, $\omega*$. It would be of extreme interest to follow this trend into the regime of large supercoolings and rapid solidifcation where a great deal of current research interest is focused. For the present, we can accept that the ratio R/R* should decrease at large supercoolings, with the dendrite becoming commensurately less stable. Eventually, the interfacial molecular attachment rate will become rate limiting, causing the interface to depart radically from local equilibrium and from the scaling laws based on local equilibrium. This remains as an interesting topic of research.

Finally, the fundamental assertion of Langer and Müller-Krumbhaar is proved by the data in Figure 8 where the scaling law $\lambda* = R$ is shown to hold for SCN over two decades of supercooling. Specifically, $R = 1.2 \lambda*$ in Figure 8 which is equivalent to an error of 20% in the value of $\sigma* = 1/4\pi^2$. As shown in Table 1, the analysis of tip stability based on spherical harmonics almost provides the $\sigma*$ value needed, viz., $\sigma* = 0.021$.

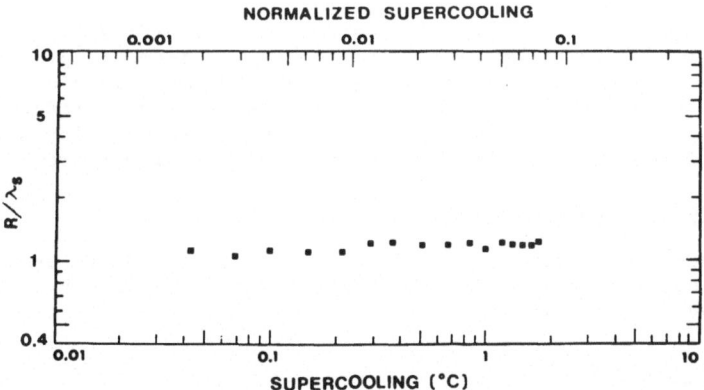

Fig. 8. Dendrite tip radius R scaled to the wavelength of the marginal eigenmode $\lambda*$ (logarithmic scale) vs. supercooling (logarithmic scale). The scaling law demonstrated here remains valid over at least two decades in supercooling, i.e., $R/\lambda* = 1.2$.

Table 1. Values of the Parameter σ* for Freely Growing Dendrites

σ*	Stability Model	Ref.
0.02	Oldfield's "force balance"	[19]
0.0253	Planar front	[21,24]
0.025	Parabolic eigenstate	[22,32,33]
0.0192	Spherical harmonic (1 = 6)	[25-27,20]
σ* exp.	System	Ref.
0.0195	SCN	[24]
0.022	PVA	[28]

Although the differences among the σ* values appearing in Table 1 seem modest, the detailed physical assumptions employed in each stability analysis are markedly different. For example, the planar front model bears little geometrical similarity to a dendrite tip and cannot include factors such as crystal symmetry or anisotropy. For the spherical harmonic model, by comparison, only the BCC system (SCN) and an fcc high anisotropy system (PVA) have been shown to display kinetics and tip morphologies which are in quantitative agreement with the dynamical theory of dendritic growth at low to moderate levels of supercooling. Much work remains to establish such correspondences over broader classes of materials such as metals, semiconductors and ceramic crystals.

V SOLUTE DENDRITES

V.1 Importance of Alloy Solidification

Most materials are solidified as alloys, so that solute diffusion and constitutional supercooling are encountered, in addition to thermal supercooling as already introduced for pure substances. Virtually all castings are formed from alloy dendrites, and, indeed, a typical cast microstructure is delineated by strong microsegregation of alloying additions surrounding the dendritic crystals.

A brief outline is presented covering the solidification fundamentals of binary alloy dendritic growth. The theory, as formally presented, covers the general situation of coupled heat amd mass transfer at the solid-liquid interface. Both thermal and constitutional effects are included, so the case described is that of an equi-axed alloy dendrite, of the sort often encountered in the central zones of slowly frozen castings.

V.2 Combined Transport Model

The basic transport model of Ivantsov Eq. (I.5) can be used to describe solute transport as well as heat transport. We merely define a dimensionless supersaturation,

$$U = \frac{c_\ell^* - C_o}{c_\ell^*(1 - k_o)} , \qquad (V.1)$$

where c_ℓ^* is the concentration in the liquid phase just ahead of the dendrite tip; C_o is the nominal alloy concentration established far from

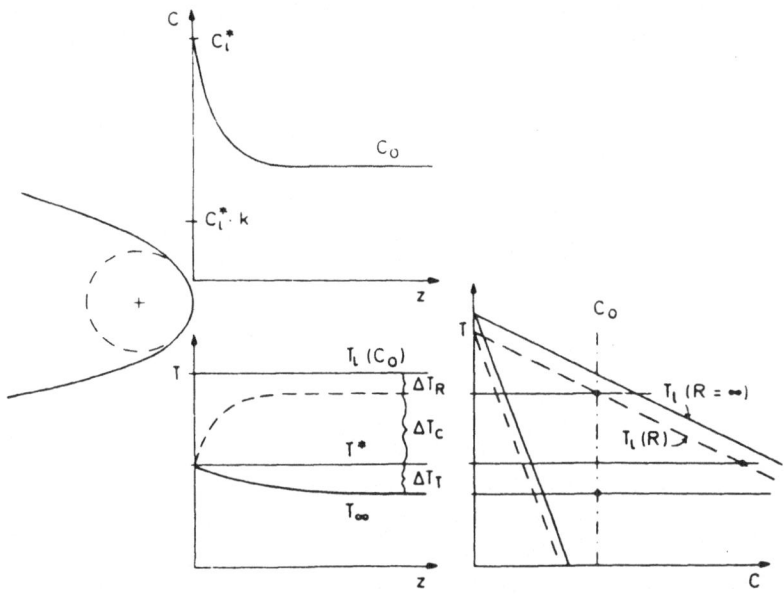

Fig. 9. Solute concentration profile and thermal fields ahead of a
growing alloy dendrite. The portion of the phase diagram at the
right shows the capillary shift in equilibrium concentrations due
to the interfacial radius of curvature, R. The temperature -
distance plot illustrates the capillary, solutal and thermal
contributions to the effective undercooling in the liquid phase.

the dendrite; k_o is the equilibrium distribution coefficient. Thus we may
rewrite Eq. (I.5) for alloys as

$$U = Pc \exp(Pc)E_1(Pc),\qquad\qquad (V.2)$$

where Pc is the chemical Peclet number (cf. Pe, which is the thermal
Peclet number). By analogy with $Pe \equiv VR/2\alpha$, we define $Pc \equiv VR/2D$, and D
is the chemical diffusivity of the solute in the molten phase.

Figure 9 shows the various relationships concerning the distribution
of solute, C(Z), and temperature, T(Z), ahead of a dendrite. The key
assumptions required in arriving at these distributions are:

a) The dendrite grows steadily into an undercooled melt of constant
undercooling.
b) The shape of the dendrite tip is described by a paraboloid of
revolution (isothermal and isoconcentrate tip).
c) The heat and mass transport in the solid and liquid is controlled by
diffusion only (i.e., convection plays a negligible role).

The melt undercooling, ΔT, determines the driving force for the
dendrite growth. In pure materials ΔT can be set approximately equal to
the thermal undercooling. In the solute case the situation becomes more
complicated as shown in Figure 9. The total undercooling ($\Delta T = T_\ell - T_\infty$)
is subdivided into three parts: ΔT_T, the thermal undercooling ($T^* - T_\infty$);
ΔT_c, the solutal undercooling ($T_\ell(R)-T^*$); ΔT_R, the curvature undercooling
($T_\ell(C_o)-T_\ell(R)$).

Two of these undercoolings can be expressed through the Ivantsov transport solution, viz.

$$\Delta T_T = (\Delta H/C_p) I_v(Pe),$$ (V.3)

and

$$\Delta T_c = mC_o \{1 - \frac{1}{1 - (1 - k_o) I_v(P_c)} \}.$$ (V.4)

Equation (V.4) can be obtained simply from Eq. (V.2) and the definition of supersaturation, knowing that $T_\ell - \Delta T_c = m(C_o - C_\ell^*)$ [43].

The third contribution to the total undercooling of an equi-axed alloy dendrite is the capillarity term determined by the Gibbs-Thomson equation, as already described in Section II.1, again given by

$$\Delta T_R = 2 \Gamma/R.$$ (V.5)

If the kinetic (interfacial) undercooling is negligible, then the sum of the 3 contributions to the undercooling gives the total supercooling. (See also the temperture coupling diagram, Figure 9.) Thus,

$$\Delta T = \Delta T_T + \Delta T_c + \Delta T_R.$$ (V.6)

The second equation needed to describe equi-axed alloy dendrites can be found from the condition of marginally stable tip growth [22,32]. This criterion (the same one as used for pure materials) postulates that a dendrite tip grows at the margin of stability, i.e., the radius at the tip is equal to the wavelength of the fastest growing instabilities to be formed under the local growth conditions. To simplify the problem we use the lower limit of the perturbation wavelength of a plane front [44]. This condition has been shown to correspond closely to the growth conditions encountered in dendrites of pure SCN (pure thermal dendrites) [24].

At low Peclet numbers (corresponding to low undercoolings, as encountered in castings) the instability wavelength $\lambda*$ is given by [44]

$$\lambda* = R = (\frac{\Gamma}{\sigma*(mG_c - \bar{G})})^{1/2}.$$ (V.7)

Here $\sigma*$ is a stability constant on the order of $1/(4\pi^2) \simeq 0.025$; m is the liquidus slope, assumed to be constant; G_c and $\bar{G}$ are the concentration gradient and conductivity-weighted temperature gradient at the interface, respectively.

The concentration and temperature gradients at the tip of the dendrite are determined from the transport solutions (E. (V.2)) to be

$$G_c = \frac{2P_c C_\ell^*(1 - k_o)}{R},$$ (V.8)

and

$$G_\ell = - \frac{2Pe \Delta H}{C_p R}.$$ (V.9)

The conductivity-weighted average temperature gradient is $\bar{G} = (\kappa_s G_s + \kappa_\ell G_\ell)/(\kappa_s + \kappa_\ell)$ where κ_s and κ_ℓ are the thermal conductivities of solid and liquid, and G_ℓ and G_s are the thermal gradients in liquid and solid. With equal thermal conductivities in liquid and solid, and with the isothermal Ivantsov dendrite ($G_s = 0$) one obtains for the mean temperature gradient ahead of the tip

$$\bar{G} = - \frac{Pe \, \Delta H}{C_p R} .$$ (V.10)

Substituting Eqs. (V.8) and (V.10) in (V.7) yields an expression for the tip radius

$$R = \frac{\Gamma}{\sigma^*} \left[\frac{-\Delta HPe}{C_p} - \frac{2P_c \dot{m} C_o (1 - k_o)}{1 - (1 - k_o) Iv(P_c)} \right]^{-1},$$ (V.11)

and one for the capillarity undercooling at the tip,

$$\Delta T_R = 2\sigma^* \left[\frac{-\Delta HPe}{C_p} - \frac{2P_c m C_o (1 - k_o)}{1 - (1 - k_o) Iv(P_c)} \right].$$ (V.12)

The solutal Peclet number is simply related to the thermal Peclet number by $P_c = Pe(\alpha/D)$ (with α the thermal and D the solutal diffusivity), so the only unknown appearing in Eq. (V.6) is the product $(R \cdot V)$, which appears in the Peclet numbers.

Finally, the growth rate can be calculated from the definition of the Peclet number

$$V = 2\alpha Pe/R.$$ (V.13)

Note that Eq. (V.6) with (V.3), (V.4), (V.12) and Eq. (V.11) define completely the growth problem at low Peclet numbers. For example, for a given composition, C_o, and undercooling, ΔT, the radius, R, and the growth rate, V, can be calculated <u>uniquely</u> by simultaneously satisfying the transport equations for heat and solute and the condition of marginal stability. These equations constitute a coupled non-linear set which must be solved by iterative methods on a computer.

V.3 Comparison with Experiment

Chopra [42] has carried out careful measurements of dendrite growth speed, V, and dendritic tip radius, R, as functions of solute concentration C_o. Figures 10 and 11 compare Chopra's observations on acetone-succinonitrile alloys with the predictions of equations (V.11) and (V.12).

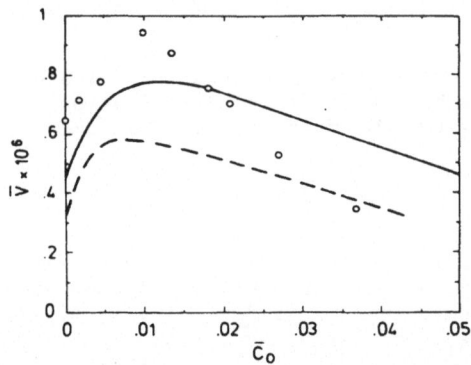

Fig. 10. Variation of dimensionless tip velocity with dimensionless concentration for succinonitrile - acetone mixtures with an undercooling of 0.5 K. Points: experimental values; solid line: Lipton-Glicksman-Kurz (LGK) model [36]; dashed line: Karma-Langer (KL) [37].

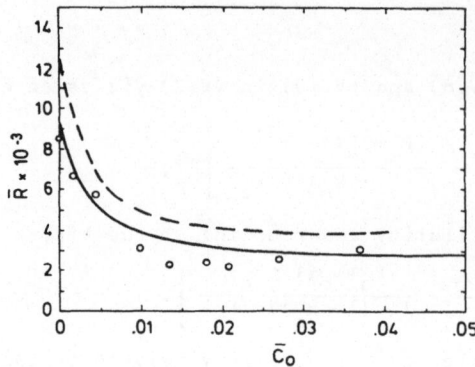

Fig. 11. Relationship between dendrite tip radius and dimensionless concentration for the system described in Figure 10.

The general trends appear correct, with a maximum appearing in V at small concentrations and a sudden drop in R occurring at small solute concentrations. Figure 12 shows what the model predicts for acetone-succinonitrile alloys assuming different values for k_O. Other details have been discussed regarding this model of alloy dendrite growth and the interested reader is referred to Ref. [45].

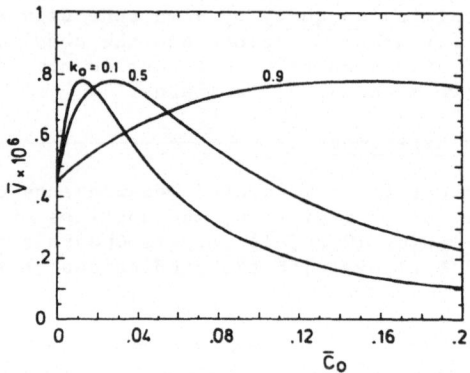

Fig. 12. Velocity versus solute concentration for soccinonitrile alloys with solutes having different distribution coefficients k_O, as predicted by the LGK model [36]. Mean undercooling is 0.5 K.

VI. REFERENCES

1. J. C. Fisher, as referenced by Bruce Chalmers, "Principles of Solidification", p. 105, John Wiley and Sons, New York (1964).
2. G. P. Ivantsov, Dokl. Akad. Nauk SSR., 58:567 (1947).
3. G. Horvay and J. W. Cahn, Acta Met., 9:695 (1961).
4. G. F. Bolling and W. A. Tiller, J. Appl. Phys., 32:2587 (1961).
5. R. F. Sekerka, R. G. Seidensticker, D. R. Hamilton and J. D. Harrison, Investigation of Desalination by Freezing, Westinghouse Res. Lab. Rep., Ch. 3 (1967).
6. M. E. Glicksman and R. J. Schaefer, J. Crystal Growth, 1:297 (1967).
7. M. E. Glicksman and R. J. Schaefer, J. Crystal Growth, 2:239 (1968).
8. D. E. Temkin, Dokl. Akad. Nauk SSR., 132:1307 (1960).

9. R. Trevedi, Acta Met., 18:287 (1970).
10. G. E. Nash and M. E. Glicksman, Acta Met., 22:1283 (1974).
11. M. E. Glicksman, R. J. Schaefer and J. D. Ayers, Met. Trans., A7:1747 (1976).
12. R. J. Schaefer, M. E. Glicksman and J. D. Ayers, Phil. Mag., 32:725 (1975).
13. C. Zener, Trans. AIME, 167:550 (1964).
14. I. Jin and G. R. Purdy, J. Crystal Growth, 23:25 (1974).
15. T. Fujioka, PhD Thesis, Carnegie-Mellon University (1978).
16. S. C. Hardy, Phil. Mag., 35:471 (1977).
17. G. R. Kotler and W. A. Tiller, J. Crystal Growth, 2:287 (1968).
18. R. Trivedi and W. A. Tiller, Acta Met., 26:67 (1979).
19. W. Oldfield, Mat. Sci. Engr., 11:211 (1973).
20. R. D. Doherty, B. Cantor and S. Fairs, Met. Trans., A9:621 (1978).
21. W. W. Mullins and R. F. Sekerka, J. Appl. Phys., 34:323 (1963).
22. J. S. Langer and H. Muller-Krumbhaar, Acta Met., 26:1681;1689;1697 (1978).
23. V. V. Voronkov, Sov. Phys. Solid St., 6:2378 (1964).
24. S. C. Huang and M. E. Glicksman, Acta Met., 29:701 (1981).
25. S. R. Coriell and R. L. Parker, J. Appl. Phys., 36:632 (1965).
26. S. R. Coriell and R. L. Parker, Proc. ICCG, Boston, Mass., 1966, Suppl. to J. Phys. Chem. Solids, H. Steffen Peiser, ed., J-3:703 (1967).
27. R. Trivedi, H. Franke and R. Lacmann, J. Crystal Growth, 47:389 (1979).
28. Narsingh Bahadur Singh, private communication.
29. M. E. Glicksman and S. C. Huang, Adv. Space Res., 1:25 (1981).
30. U. Lappe, KFA Report, Kernforschungsanlage Julich, FRG (1980).
31. S. C. Huang, PhD Thesis, Rensselaer Polytechnic Institute (1979).
32. J. S. Langer and H. Müller-Krumbhaar, J. Crystal Growth, 42:11 (1977).
33. J. S. Langer, Rve. Mod. Phys., 52, No. 1:1 (1980).
34. M. H. Burden and J. D. Hunt, J. Crystal Growth, 22:99 (1974).
35. S. Witzke, J. P. Riquet and F. Durand, Acta Met., 29:365 (1981).
36. Hasse Fredricksson, in: "Materials Processing in the Reduced Gravity Environment of Space", G. E. Rindone, ed., p. 619, Elsevier, Amsterdam (1982).
37. R. Trivedi and W. A. Tiller, Acta Met., 26:679 (1978).
38. J. S. Langer, Phys. Chem. Hydrodyn., 1:41 (1980).
39. C. Lindenmeyer, PhD Thesis, Harvard University (1959).
40. M. E. Glicksman, Narsingh Bahadur Singh and M. Chopra, in: "Materials Processing in the Reduced Gravity Environment of Space", G. E. Rindone, ed., p. 461, Elsevier, Amsterdam (1982).
41. W. Kurz, J. Lipton and M. E. Glicksman, unpublished work (1983).
42. M. Chopra, PhD Thesis, Rensselaer Polytechnic Institute (1983).
43. J. Lipton, M. E. Glicksman and W. Kurz, Mater. Sci. Eng., 65:57-63 (1984).
44. W. Kurz and D. J. Fisher, Acta Met., 29:11-20 (1981).
45. J. Lipton, M. E. Glicksman and W. Kurz, Met. Trans., 18A:341-345 (1987).

TWINNING IN CRYSTALS

B. Březina and J. Fousek

Institute of Physics
Czechoslovak Acad. Science
Na Slovance 2, 18040 Prague

1. INTRODUCTION

A crystal is said to be twinned when it contains regions whose atomic structures, being uniform in each region, are mutually related by some well defined law and this law is repeatedly observed in many samples of the crystalline species. These requirements distinguish twins from polycrystalline aggregates in which the crystallites are randomly oriented. Since the beginning of crystallography as a science, twinned crystals provided objects interesting to describe. Their investigation was rather academic, with useful aspects in the determination of minerals. New impulses for the investigation of twinning have been connected with practical uses of crystalline materials (metals, quartz).

When the as-grown crystals are twinned, we speak about growth twinning. When twins are produced by mechanical deformation, they are often referred to as mechanical twins. In a large category of materials important for today's optoelectronic devices, the existence of twins is intimately related to the fact that the crystal undergoes a structural phase transition; these are transformation twins (domains). It is unfortunate that the languages describing twinning properties have been to a considerable degree developed independently for all three mentioned kinds of twins. The more so that this classification is not unambiguous. Indeed, transformation twins often occur during the growth process, mechanical twins may be identical with domains. Still, we shall use this classification as a key in this presentation. In Section 2 we recall the basic concepts used for the description of twins. Sections 3 and 4 deal with the growth and mechanical twinning, respectively. In recent decades, the understanding of transformation twinning made considerable progress and we shall treat the involved phenomena in more detail in Section 5. Depending on a particular purpose, the presence of domains may be required (to obtain, for example modulated structures with desired properties) or suppressed (to reach maximum figures of merit); paragraphs 5.2 and 5.3 are devoted to these aspects.

The present notes are to be considered as a condensed guide to the subject in which the problems will be illustrated by practical examples, and as a source of references.

2. DESCRIPTION OF TWINS

Crystallography of twins is not a subject to be explained on a few pages. In fact a full treatment covering all presently known aspects of twinning and related phenomena is still not available. Here we shall confine ourselves to introduce basic concepts used in describing and classifying twinned crystals.

The homogeneous crystalline regions constituting a twin are called twin components (TC). In the interface between two TC's the crystal structure is disturbed; its lateral dimension is always orders of magnitude smaller than that of TC's. Often the interface is necessarily planar and referred to as composition plane; the two coexisting TC's then represent a contact twin. When the interface is irregular a penetration twin results.

A pair of identical crystal individuals in any mutual orientation can be brought into exact coincidence by a combination of a rotation and translation. The pair represents a twin if the involved rotation is one of the proper or improper rotations encountered in crystallography. Then this rotation (we now neglect translations) is called twinning operation (TO) denoted by t. Pairs of TC's of a single substance can be related by different TO's; one speaks about different twinning laws for the substance in question. For a particular twinning law, the orientation of possible composition planes is usually well defined.

In a large number of crystals, the TO is a reflection (m) or a rotation by $180°$ (2). Such twins are referred to as reflection or rotation twins, respectively. In the former case, m is often called the twin plane. Older textbooks on crystallography claim that the majority of twins are of these types. When the TC's are related by operation $\bar{1}$ (this occurs also when m and 2 both are TO's), we have an inversion twin. All these three kinds of twins have the following property: denoting the two TC's as A,B, it holds A = tB as well as B = tA. If m, 2 or $\bar{1}$ is the only present TO, we have two-component twins. The possibility of multi-component twins is obvious.

A useful classification of twins has been proposed by Friedel [1] who distinguishes four types of twinning. I - Twinning by merohedry: Bravais lattices of the two TC's coincide. II - Twinning by reticular merohedry. Here the lattices are not parallel but there is a coincidence site lattice containing only some points common to the A,B lattices. The coincidence lattice has some symmetry elements extra to that of the crystal lattice. In these categories I and II the TO is a true crystallographic operation. III - Twinning by pseudo-merohedry and IV - Twinning by reticular pseudo-merohedry; these types of twinning are derived from the above categories I and II by allowing a small but exactly defined and repeatedly encountered deformation of the lattices of the TC's such that after the deformation they are brought into full (III) or partial (IV) coincidence. Then new TO's appear which are of crystallographic character.

It is obvious that grains in a polycrystalline sample or in general bicrystals do not constitute twins, the requirements of TO being crystallographic and/or of deformation being always the same are not met.

Let us finally mention concepts connected with the shape of TC's. When TC's in the form of thin slabs alternate, we speak about lamellar twinning. The occurrence of very fine lamellae filling a relatively large volume is known as polysynthetic twinning. The external symmetry of such a complex is higher than that of a single TC; this phenomenon is generally called mimetic twinning.

In more detail, geometry of twinning is dealt with in Refs. [1 to 4].
Many minerals may serve as examples of these concepts; quartz provides a
very illuminating one [3].

3.   GROWTH TWINS

Here we wish to mention growth conditions that may influence the
occurrence of twinned crystals, supposing the growth process proceeds
peacefully with no external forces (mechanical or electrical) applied to
the growing individual; artificial twinning will be discussed in Section
5.   In the literature, a large number of twin laws for different compounds
can be found [5] which we do not reproduce here.   However, the amount of
well established data about the mechanisms of the twin origin is small.

A nucleus arising by fluctuation in a supersaturated solution, may be
already twinned, giving rise to a twinned crystal.   Alternatively, the
twin may develop when the growing crystal makes an error in depositing the
next layer: a cluster of atoms arrives, already formed, and due to its
instantaneous position, attaches itself in a twin orientation.   A more
detailed consideration connects this picture with Friedel's definition of
twins [2].   In order that enough completed clusters be available the
supersaturation must be high.   Experience corroborates this expectation.
Conversely, in a barely saturated solution untwinned nuclei grow
preferably; this may be explained by a preferential redissolving of embryo
twins, due to a large surface to volume ratio [6].

The generation of a stacking fault on the growing lattice plane leads
to the occurrence of a reflection twin [7]; repeated events constitute a
polysynthetic twin.

Specific impurities may greatly enhance the production of twins.   The
involved mechanism seems to be connected with the ionic radii [2].

Refs. [8,9] may serve as examples of the analysis of growth twinning
for a particular substance: $BaTiO_3$.

Not all minerals or artificially grown crystals show a tendency for
twinning.   It can be assumed that two conditions when fulfilled enhance
the probability of twinning.   First, a lattice plane must exist capable of
serving as a composition plane.   We expect that the structure disturbance
at the interface after the twin is formed is small: most of the atoms
within the composition plane should have almost the same positions as in
the normal structure.   Then the surface energy density of this plane is
low.   Often the structure of the interface region may be close to the one
of a possible polymorphic modification of the given substance.   Such is
the case of transformation twins where the internal structure of
composition planes resembles that of the parent phase.   These
considerations about the geometry of interfaces tend to indicate that
isomorphous substances would twin with the same frequency.   This is often
but not always so and this brings us to the second condition.   It requires
that even the long range broken bonds, such as second-nearest neighbor
interactions, do contribute only moderately to the interface energy.
Twinning in isostructural crystals of NaCl and PbS provides an outstanding
example [2].   It is very frequent in PbS where nearest neighbor covalent
bonds carrying the energy are almost unaffected by the composition plane.
In NaCl the ionic bonds to second-nearest neighbors contribute
considerably to the crystal internal energy and their disturbance at the
interface increases its energy density.

4. MECHANICAL TWINNING

Under suitable geometry, many metals subjected to a shear stress undergo plastic shear deformation due to the formation of a twin whose TC's differ in shape. It is a well defined process with technical impact and as such it was studied in detail and analyzed in monographic literature [10,11]. Here we only make a few remarks. Mechanical twins comply with Friedel's definitions. However, Friedel's categories I and II are obviously ruled out. In the geometrical considerations on mechanical twinning, two lattice planes K_1, K_2 are introduced in which vectors between all lattice points remain unchanged during the shear; in addition, shear direction η_1 and another vector η_2 characterize the twin. Tables exist of <u>twinning elements</u> K_1, K_2, η_1, η_2 characterizing mechanical twins in different materials. Depending on whether the pairs (K_1, η_2) and (K_2, η_1) are rational or not, two classes of mechanical twins are defined whose TO's are either reflection or rotation by 180°. It should be mentioned that any ferroelastic transformation twin can be considered as a mechanical twin since it can be created by a shear deformation. However, we shall see in the next section that there the TO's can be more general than those connected with symmetry elements m or 2.

Usually twins produced by external mechanical forces are taken as identical with twins changing the external crystal shape. But this is not necessarily so. An applied stress creates elastic energy in a crystal individual. This can be relaxed in two ways. First by adjusting the crystal shape which is the case considered above. Second by introducing a twin component whose elastic compliance, described in a fixed laboratory frame, is larger than that of the original material. It may not change the crystal shape. This is a twinning process known from quartz and we shall come to it once more in Section 5 (ferrobielastic domains). It is mechanical in nature and can be classified as such.

5. TRANSFORMATION TWINNING

5.1 General Laws

Most of the distortive structural phase transitions (SPT) in crystals are marked by the reduction of point symmetry from G (parent phase) to F⊂G (distorted or ordered phase). Numbers of molecules Z_G, Z_F in the primitive unit cells of the two phases are equal for the so-called <u>equitranslational</u> SPT's. In the alternative case $Z_F > Z_G$ the translational symmetry changes. Because of the symmetry reduction, in the ordered phase the crystal may exist in several degenerate states which may be distinguished when viewed from a fixed laboratory frame; these are <u>domain states</u> D_i, i = 1,...,q. A sample in which these states coexist contains <u>domains</u>, i.e., it is twinned. Symmetry aspects of SPT's relevant to domain phenomena are explained in Refs. [12 to 14]. A symmetry approach to studying domains has been developed by Janovec [15,16]. Here we shall only mention a few basic concepts.

Let us select the symmetry group F_1 which leaves invariant D_1; it may or may not leave invariant other D's as well. Now the group G can be written as the sum of the left cosets of the subgroup F_1:

$$G = F_1 + t_{12}F_1 + \ldots + t_{1q}F_1;$$

the coset $t_{1k}F_1$ encompasses all twinning operations transforming D_1 into D_k. If we operate in the realm of point groups, D_i's are <u>orientational domain states</u>. It holds that q = (order of G) / (order of F)

and its maximum value is 48 (for G = m3m, F = 1). Note that the coexistence of D_i's has not yet been mentioned.

If $Z_F > Z_G$, the volume of the primitive cell in F is roughly a q_t-multiple of that in G. In each orientational state we may have q_t different $\underline{\text{translational states}}$,

$$q_t = Z_F/Z_G.$$

They differ in the position of the (larger) cell of the distorted phase with respect to the origin defined in G. When coexisting, we call them $\underline{\text{translational}}$ or $\underline{\text{antiphase domains}}$. The total number of domain states is qq_t.

Orientational domain states differ in macroscopic tensorial properties referred to the laboratory frame. We may connect this frame with the crystallographic axes of the parent phase; then transformation matrices of the operations t_{jk} determine how property A changes going from D_j to D_k. Domain states or even SPT's may be classified on the basis of A's in which the domain states differ. The most significant property A from the point of view of twinning is spontaneous strain u_{mn}. If two domain states D_j, D_k differ in some component of strain, $u_{mn}^{(j)} \neq u_{mn}^{(k)}$, this pair of states is said to be $\underline{\text{ferroelastic}}$ and the SPT can be given the same attribute. For the number q_u of ferroelastic domains we may have $q_u = q$ or $q_u < q$. In the latter case, ferroelastic domains contain "subdomains" with different values of the order parameter responsible for the SPT.

Domain states may differ, in addition to u_{mn}, in other macroscopic properties; some are shown in Table 1, together with the designation of the phase [17,18,19]. Electric field E or mechanical stress T or their combinations can be applied in which the degeneracy of states is removed; relevant energy terms are also given in the Table. From them, switching forces can be derived. It is a common case that two given domains fall into two or more of these categories simultaneously; switching forces are then competitive. For ferroelastic crystals, the switching force is mechanical stress and we come back to mechanical twinning. But the stress removes also the degeneracy of ferrobielastic domain states. What used to be a rarity (piezocrescence [2] in quartz) is a simple consequence of this more general approach.

Table 1.   Some Ferroic Phases

Quantity distinguishing the domain states	Phase designation	Energy term removing domain degeneracy
Polarization P_i	ferroelectric	$(P_i^{(1)} - P_i^{(2)})E_i$
Strain u_{ij}	ferroelastic	$(u_{ij}^{(1)} - u_{ij}^{(2)})T_{ij}$
Piezoelectric coefficients d_{ijk}	ferroelastoelectric	$(d_{ijk}^{(1)} - d_{ijk}^{(2)})E_i T_{jk}$
Elastic compliance s_{ijkl}	ferrobielastic	$(s_{ijkl}^{(1)} - s_{ijkl}^{(2)})T_{ij} T_{kl}$

Considering two coexisting states in a sample we have first to determine possible orientations of the composition planes (<u>domain walls</u>). A plane (hkl) constitutes a <u>permissible wall</u> [20] if the strains u_{mn} of the two neighboring domains are equal at this plane, i.e., the plane is stress-free. For non-ferroelastic domains any plane is permissible. For ferroelastic domains usually two mutually perpendicular planes are permissible. In some cases, however, no permissible wall exists. Note that ferroelasticity plays a key role in determining the composition planes. Parameters in Table 1 other than strain do not put any constraint on the wall orientation with the exception of polarization in ferroelectrics; however, electric fields can be compensated by free charges while elastic ones cannot. Tables of permissible walls are available [21,22].

The existence and orientation of permissible walls can be derived purely on the basis of symmetry considerations, from the knowledge of TO's between domain states. Table 2 shows examples [16].

In praxis, walls sometimes slightly deviate from permissible orientations (wedge domains). These departures are made possible by twinning dislocations [3].

In most of these considerations the strain is taken into account but at the same time it is considered to be zero. Only then the TO's (t_{ij}) remain effective and stay in positions they had in the parent phase. In reality, the shear strain required to transform one ferroelastic domain into the neighboring one is finite [23]. Therefore only few TO's from the left coset in question are operative; such are the reflections in the domain wall or 180°-rotations. Here we come back to Friedel's categories of twins III and IV: deformations of the lattices are required (bringing them back to the parent-phase) to restore all crystallographic TO's.

At the end let us consider a multi-domain sample containing a statistical mixture of all possible domains. As for macroscopic properties of such a sample, it can be shown that any tensorial property which arose due to the symmetry breaking averages to zero. How many domains there exist in such a sample? When asking this, we have not in

Table 2. Orientation of Permissible Walls in Transformation Twins as given by Twinning Operations. (Examples only, after Ref. [16]. The Term "Diad" denotes a Two-fold Rotation or Inversion Axis.)

TO's included in the Left Coset	Permissible Walls
$\bar{1}$ or two non-perpendicular diads or more than two diads of different directions	any plane
Two perpendicular diads	any of two planes perpendicular to these diads
One diad but no rotations about an axis parallel to this diad	any of two planes, one of which is perpendicular to the diad
Only rotations of higher order than two about the same direction; at least two of them are not related by $\bar{1}$	two planes, both with non-crystallographical orientation

mind previously mentioned domain states but real coexisting domains, whose orientations refer to a common coordinate system. Two ferroelastic states D_j, D_k differ in spontaneous strain $\omega_{mn}^{(j)} \neq u_{mn}^{(k)}$ measured with respect to the parent axes. If the two domains meet along a permissible wall, each of them must be subject to some rotation, $\omega^{(j)} = -\omega^{(k)}$. Up to now two domain states result in two domains. However, for a pair of ferroelastic domains there are two (if any) permissible walls which are mutually perpendicular (cf. Table 2) when indexed in the parent phase. It is an experimental fact [26] that the two walls remain perpendicular even in a real sample in the distorted phase. Obviously, while strains u occurring for the two domains in question are independent of the composition plane, the rotations necessary to maintain physical contact between domains may differ in sign for the two permissible walls. Therefore the number of domains (lattices) in a given sample may exceed that of domain states. These phenomena have been indeed observed [24,25]; e.g., in KH_2PO_4 four domains are shown [26] by X-rays to exist while there are only two domain states.

Symmetry of transformation twins, orientation of walls and other mentioned topics have been analyzed for a number of crystals with SPT's. To mention just a few: WO_3 [15], $Gd_2(MoO_4)_3$ [16,27], $(NH_4)_2SO_4$ [28], $TlNO_3$ [29], $Cd_2Tl_2(SO_4)_3$ and $Cd_2(NH_4)_2(SO_4)_3$ [30], $LiKSO_4$ [31]. Domains occurring in a number of ferroelastics have been reviewed by Boulesteix [32].

5.2 Producing Transformation Twins

Twinning in crystals undergoing SPT's can proceed either by itself during the crystal growth or during the cooling of the sample into the distorted phase (natural twinning) or may be due to an aimed effort in the laboratory (artificial twinning). Twinning during the growth occurs under conditions discussed in Section 3. Here, however, the symmetry features of defects or dopants serving as nuclei of domains are accentuated; e.g., the role of polar defects in ferroelectrics is obvious. On cooling the crystal through the transition temperature T_{tr}, other mechanisms become effective. The SPT's are often of the 1st kind and at T_{tr} the cell dimensions change discontinuously. Then it holds that when coexisting the two phases are never mechanically compatible along any plane. If the SPT is ferroelastic, a planar phase boundary is, nevertheless, often realized so that on the side of the distorted phase the crystal is densely twinned; in this way the elastic energy is reduced. These twins may stay in the sample even after the transition is completed [33]. Another mechanism consists in a chaotic nucleation of the distorted phase in a sample cooled through T_{tr}, different nuclei corresponding to different domain states. Another case of natural twinning occurs in ferroelectric samples where domains differing in polarization directions lead to the reduction of the electrostatic energy produced by bound electric charges [34]. In either of these cases the resulting pattern of twins is usually irregular and cannot be compared, e.g., to highly regular ferromagnetic domains.

There are several factors backing the attempts to produce twins artificially. First, regular twin patterns are arrays of regions with alternating macroscopic properties and as such they represent super-lattices with new features. For instance, when the period of a slab-like domain structure is tuned to the coherence length, we obtain a highly effective nonlinear-optical element. A similar texture of ferroelastic domains may serve as an acoustic filter and optical grating. Second, if the twinning texture copies an optical pattern projected into a crystal plate, we have an optically written and read storage medium.

Any of the switching forces (cf. Table 1) corresponding to a given SPT can be used to produce twins artificially. In the ferroelastic case, the simplest way to produce a regular domain pattern is to subject a properly oriented plate to a bend deformation [35]. It was also shown that in materials with low coercive stress, pulses of properly oriented uniaxial stress result in a highly regular slab-type texture of ferroelastic domains [36]; mechanisms involved are yet to be understood. In the ferroelectric case, a well-functioning procedure to produce regular domain patterns is to apply positive and negative electric fields to the growing crystal, for required periods of time. This has been very successful for the Czochralski technique operating with melts in electrically conductive crucibles [37,38], less effective for crystals growing from water solutions [39].

As for the information storage by domains, much effort was devoted to hybrid elements in which a ferroelectric plate is covered on one side by a photoconductive layer and then sandwiched between two transparent electrodes. The formation of domains proceeds only where illumination exceeds some intensity. If these domains are also ferroelastic, the information is written in as birefringence modulation and can be read by optical means [40]. $Bi_4Ti_3O_{12}$ is a practical material.

In ferroelastoelectrics, a combination of electric field and mechanical stress provides the switching force. This mechanism may compete (e.g., in quartz [41,42]) with ferrobielastic switching, both being simultaneously allowed by symmetry. The latter can be achieved, e.g., by using laser beams to produce local stress.

Whatever the mechanism by which domains are produced, the twinned crystal is usually in a metastable state since the composition planes have an extra positive energy, though it may be a long-living state. There are, however, exceptions like domains stabilized by a special distribution of defects or domains in incommensurate phases.

5.3 Detwinning

The existence of domains in a sample reduces the magnitude of any morphic property (one due to the symmetry breaking). Further, it adds to the imaginary part of any property connected with wall shifts and thus it increases the noise of the sample. These two reasons, plus the stability considerations, have resulted in efforts to make samples of crystals undergoing SPT's single-domain.

One obvious way is to apply a sufficiently high switching force to a properly oriented crystal cut (poling process). Since the coercive values of switching forces diminish near T_{tr}, one often proceeds by cooling the sample from the parent phase in the presence of the switching force. The most common case is the poling of ferroelectrics [43] and ferroelastics [44]: however, the method has its setbacks since the sample may tend to crack at T_{tr} when the relaxation of stress (cf. Section 5.2) is suppressed by the poling field. Elaborate modifications are possible, such as using thermal gradients to relieve the stress field or photo-inducing charges to relieve electric fields [45,46].

These are methods suitable for the laboratory. In a larger scale production they would be too cumbersome and processes are looked for ensuring the whole as-grown crystal or its major portions to be untwinned. In addition to what was said in Section 3, especially about means of limiting supersaturation and its fluctuations near the growing crystal, two methods are employed: either the application of switching forces

Table 3. Ferroelectric Transitions (G → F) in which One
Orientational Domain State can in Principle be
Fixed by Chiral Dopants

G	6/m	$\bar{6}$	$\bar{3}$	4/m	$\bar{4}$	2/m	m	$\bar{1}$
F	6,3,2,1	3,1	3,1	4,2,1	2,1	2,1	1	1

during crystal growth or the addition of special dopants. The former
approach has already been mentioned in Section 5.2.

As far as the dopants are concerned, it has been known since the
early studies of ferroelectrics that some crystal individuals have
"biased" hysteresis loops, meaning a tendency of a larger crystal part to
exist in a given domain state. This has been explained by the presence of
impurities. Long-range (electrostatic) interactions between polar
impurities ensure that their moments are parallel and also the lattice
between them chooses the domain state accordingly [43]. In this model the
resulting domain state is purely accidental. Indeed, different growth
pyramids may have a different biasing tendency or may even be biased in
opposite directions [47].

The most interesting case of dopants is that a certain admixture
makes only one domain state possible. When this happens, the SPT is,
strictly speaking, suppressed but most of the macroscopic properties not
connected with the behavior of domains remain in effect. The only known
case is the fixing of one polarity of ferroelectric TGS-type crystals by
admixture of alanine molecules [48,49]. Its analysis leads to the
following generalization [50,51]. Suppose the ordered phase is chiral
(i.e., F contains no improper rotations) and that a symmetry operation in
G exists which is the twinning operation and changes at the same time the
chirality. In some species, mostly in molecular crystals, a well defined
microscopic unit will be responsible for this change of chirality, by
small atomic shifts. Now if part of these units are substituted by
dopants with a fixed chirality, the resulting crystal as a whole will
acquire this chirality and the corresponding domain state as well. Table
3 shows ferroelectric transitions for which in principle this mechanism
could be used. A similar analysis has been also performed for
ferroelastic SPT's. Though the number of known materials fulfilling these
symmetry conditions is at present small, this method seems to provide the
most efficient way of detwinning.

ACKNOWLEDGEMENTS

Discussions with Dr V. Janovec on various aspects of twinning are
gratefully appreciated.

REFERENCES

1. G. Friedel, Bull. Soc. Franc. Minéralogie, 56:262 (1933).
2. R. W. Cahn, Adv. Phys., 3:363 (1954).
3. B. K. Vainshtein, V. M. Fridkin and V. L. Indenbom, "Modern
   Crystallography", vol. II, Springer-Verlag, Berlin (1981).
4. G. Donnay and J. D. H. Donnay, Canad. Mineral., 12:422 (1974).
5. P. Ramdohr and H. Strunz, "Klockmanns Lehrbuch der Mineralogie",
   Ferdinand Enke Verlag (1967).

6. M. J. Buerger, in: "Phase Transformations in Solids", R. Smoluchowski, ed., p. 183, Wiley, New York (1951).
7. H. Tabata and E. Ishii, J. Crystal Growth, 49:753 (1980).
8. R. C. De Vries, J. Am. Ceram. Soc., 42:547 (1959).
9. J. W. Nielsen, R. C. Linares and S. E. Koonce, J. Am. Ceram. Soc., 45:12 (1962).
10. J. W. Christian, "The Theory of Transformations in Metals and Alloys", Pergamon Press, Oxford (1965).
11. L. A. Shuvalov, A. A. Urusovskaya et al., "Modern Crystallography", vol. IV, Springer-Verlag, Berlin (1981).
12. L. A. Shuvalov, J. Phys. Soc. Japan, 28 Suppl., 38 (1970).
13. K. Aizu, Phys. Rev., B2:754 (1970).
14. V. Janovec, V. Dvořák and J. Petzelt, Czech. J. Phys., B25:1362 (1975).
15. V. Janovec, Czech. J. Phys., B22:974 (1972).
16. V. Janovec, Ferroelectrics, 12:43 (1976).
17. R. E. Newnham and L. E. Cross, Mat. Res. Bull., 9:927,1021 (1974).
18. K. Aizu, J. Phys. Soc. Japan, 34:121 (1973).
19. R. E. Newnham, American Mineralogist, 59:906 (1974).
20. J. Fousek and V. Janovec, J. Appl. Phys., 40:135 (1969).
21. J. Fousek, Czech. J. Phys., B21:955 (1971).
22. J. Sapriel, Phys. Rev., B12:5128 (1975).
23. L. A. Shuvalov, E. F. Dudnik and S. V. Wagin, Ferroelectrics, 65:143 (1985).
24. L. G. Shabelnikov, V. Sh. Shekhtman and O. M. Tsarev, Fizika Tverdogo Tela, 18:1529 (1976).
25. N. Ziegler, M. Rosenfeld, W. Känzig and P. Fischer, Helv. Phys. Acta, 49:57 (1976).
26. A. M. Balagurov, I. D. Datt, B. N. Savenko and L. A. Shuvalov, Fizika Tverdogo Tela, 22:2735 (1980).
27. J. R. Barkley and W. Jeitschko, J. Appl. Phys., 44:938 (1973).
28. J. Tomek, V. Janovec, J. Fousek and Z. Zikmund, Ferroelectrics, 20:253 (1978).
29. V. K. Wadhawan and M. S. Somayazulu, Phase Transitions, 7:59 (1986).
30. M. Glogarová and J. Fousek, Phys. Stat. Sol., (a) 15:579 (1973).
31. A. M. Balagurov, N. C. Popa and B. N. Savenko, Phys. Stat. Sol., (b) 134:457 (1986).
32. C. Boulesteix, Phys. Stat. Sol., (a) 86:11 (1984).
33. S. Mendelson, Ferroelectrics 37:519 (1981).
34. J. Fousek and M. Glogarová, Jap. J. Appl. Phys., 4:403 (1965).
35. J. Fousek and M. Glogarová, Ferroelectrics, 11:469 (1976).
36. S. W. Meeks and B. A. Auld, Appl. Phys. Lett., 47:102 (1985).
37. A. Feisst and P. Koidl, Appl. Phys. Lett., 47:1125 (1985).
38. W. Wang, Q. Zhon, Z. Geng and D. Feng, J. Crystal Growth, 79:706 (1986).
39. W. Wang and M. Qi, J. Crystal Growth, 79:758 (1986).
40. J. C. Burfoot and G. W. Taylor, "Polar Dielectrics and Their Applications", Macmillan, London (1979).
41. J. W. Laughner, R. E. Newnham and L. E. Cross, Phys. Stat. Sol., (a) 56:K83 (1979).
42. E. Bertagnolli, E. Kittinger and J. Tichý, J. Appl. Phys., 50:6267 (1979).
43. K. Nassau, H. J. Levinstein and G. M. Loicono, J. Phys. Chem. Solids, 27:989 (1966).
44. S. C. Abrahams, J. L. Bernstein, J. P. Chaminade and J. Ravez, J. Appl. Cryst., 16:96 (1983).
45. P. W. Haycock and P. D. Townsend, Appl. Phys. Lett., 48:698 (1986).
46. G. Metratand A. Deguin, Ferroelectrics, 13:527 (1976).
47. B. Březina and M. Havránková, Ferroelectrics Letters, 4:81 (1985).
48. P. J. Lock, Appl. Phys. Letters 19:390 (1971).

49. B. Březina and M. Havránková, <u>Crystal Res. and Technology</u>, 20:781,787 (1985).
50. Z. Zikmund and J. Fousek, "Proc. Int. Symp. on Applications of Ferroelectrics", V. Wood, ed., Lehígh University, Bethlehem, Pa (1987).
51. Z. Zikmund and J. Fousek, <u>Ferroelectrics</u>, 79:73 (1988).

POLYTYPISM AND CRYSTAL GROWTH OF INORGANIC CRYSTALS

Alain Baronnet

CRMC2-CNRS
Campus de Luminy, case 913
F-13288 Marseille cedex 09, France

1. INTRODUCTION

Polytypism is a crystallographic property of now a large number of crystallized, organic as well as inorganic, substances. Among inorganic substances, this property is displayed by crystals of some metals, metallic alloys, semiconducting compounds of the III-V, II-VI, I-VII and ternary types as well. It is also found in silicate minerals like chain-, ribbon-, and sheet-silicates but also among non-silicate minerals like diamond, graphite, molybdenite, pyrrhotite, perovskites, for instance.

In a general way, polytypism affects some of the crystals the structures of which may be described by close-packing of atoms, ions or molecules (close-packed crystals) and/or by the stacking of almost structurally invariant layers or modules (layered crystals) [1].

Polytypes of one substance are found as an almost unlimited number of crystallographically distinct structures with constant, or nearly constant, chemical composition. These structures differ only in the way their component layers are stacked along one single crystallographic direction. Structurally speaking, polytypism may be regarded as a kind of one-dimensional polymorphism [2]. However, contrary to polymorphs which are limited in number and have a thermodynamic stability range each, there is yet no convincing experimental evidence showing that polytypes are really stable phases [3,4], except perhaps for the cubic (3C)-sphalerite - and hexagonal (2H)-wurtzite - close packed modifications of zinc sulphide [5]. The probable lack of visible stability domains for polytypes is suggested by the expected exceedingly small internal energy differences between polytypic modifications [1]. It is so because polytypes appear in some crystals in which the first coordination polyhedron of definite atoms may be satisfied in different manners, the second - or even higher - coordinations being only modified.

However, equilibrium theories of polytypism have been suggested [6,7] which take care of the role of entropy differences, even if they are expected to be small [8], upon the stabilization of polytype structures. The Disorder Theory [6] suggested that metastable equilibrium given by maxima of the total entropy reached by polytype structures containing a given amount of disorder in their stacking sequences could account for polytypism. Unfortunately, this model is unable to explain the usual

observation of fully ordered polytypes with long repeat distances. The
axial next nearest neighbor Ising (ANNNI) models of statistical mechanics
have been recently developed up to third nearest neighbors to draw phase
diagrams of polytypes, polytypoids and polysomes [7,9-11]. Its success
rests in the prediction, in a unique scheme, of stability domains for
short-repeat polytypes and some long-repeat polytypes as well, and to show
that the latter possibility could result from short-range interactions
between structure-building modules. Nevertheless, ANNNI models predict a
rather limited number of polytype structures when compared to those
actually observed.

On the other hand, many investigators of this phenomenon believe that
polytypism is mostly of kinetic origin [e.g., 4,12,13]. Primary arguments
for kinetic influence are based on the following facts: 1) most of natural
and synthetic polytypic crystals are made of a mosaic of various polytype
structures, more or less affected by stacking faults; 2) it is exceedingly
difficult, if not impossible for the moment, to control the ab initio
production of a desired polytype even under pretty well controlled and
stabilized external growth conditions. As a matter of fact, the existence
of numerous structures (or phases) under fixed P,T, composition conditions
is an obvious violation of the Gibbs phase rule.

Instead, the variety of polytype structures in a "single" crystal may
be regarded as responses to the non-unique processes of the crystal
formation including nucleation, crystal growth, and eventually solid-state
modification of the as-grown crystal.

The following sections will briefly summarize these expected and/or
experimentally demonstrated connections.

2. BASIC STRUCTURES AND COMPLEX POLYTYPES

If equilibrium theories try now to give a unique answer for the
existence of all kinds of polytypes, kinetic models clearly distinguish
two kinds of polytypes:

The short-period polytypes, also called basic structures. They are
by far the most frequent ones, e.g., 2H, 3C, 4H and 6H in ZnS [14,15], 1M,
$2M_1$, $2M_2$ and 3T in micas [18]. Some of these structures may be considered
as nearly stable phases (e.g., 3C and 2H for ZnS, 1M, $2M_1$ for specific
micas) while the others may be considered as metastable. For the latter,
gradations in their metastability and related significant energy-
differences between them may be inferred from the existence of structural
transformations among some of these short-period modifications [19]. Such
basic structures are frequently more or less faulted by randomly or semi-
randomly distributed stacking faults.

The long-period polytypes or complex polytypes. Although much less
frequent than basic structures in any substance, they are most interesting
in that they may have very widely scattered repeat distances, with
periodicities sometimes beyond more than one hundred modules. Such
polytypes are often extremely well ordered and their detailed stacking
sequences of modules are made of periodically faulted substructures which
correspond to some of the above basic structures [4]. In absence of any
better explanation, it is yet agreed that no ordering forces over such
large distances may be conceived and that such ordering can only be due
to a "mechanical" process provided by dislocations.

3.   FACTORS ACTING ON THE FORMATION OF BASIC STRUCTURES

The role of factors like temperature, impurity content, departure from stoichiometry, rate of crystallization on the stabilization of specific basic structures has been experimentally investigated. Results are often questionable because the variation of one factor could not be clearly separated from the others in the relevant experiments.

Temperature

SiC crystals are grown in the range 1300-2800°C.   2H forms preferentially between 1300 and 1600°C, whereas 6H dominates 4H and 15R between 1800 and 2800°C [20].

For ZnS, sphalerite(3C) which is stable below 1020 C converts to wurtzite(2H) above this temperature.   Both phases are stable and are related by a reversible transformation [21].

Among the micas, muscovite $KAl_2AlSi_3O_{10}(OH)_2$ exhibits a thermal trend, disordered $\rightarrow$ 1M $\rightarrow$ $2M_1$, up to its breakdown [22]. $2M_1$ seems to be the only stable form of muscovite.

2H-CdI transforms irreversibly to 4H on thermal annealing [23,24]. Thus 4H seems to be the only high temperature stable phase of CdI .

Impurity Content

There are conflicting results about the role of Al as an impurity on the changes of structure types of SiC at constant temperature.   Either no effect [20,25] or preferential stabilization of 4H occurs [26].   4H was found also stabilized by Sc and Tb [27].   It has also been shown that impurities are essential to stabilize 2H-SiC under conditions where 3C-SiC should have grown in the pure system [28].   On the other hand, 3C-SiC is stabilized by nitrogen at high temperature at which normally -SiC should have been obtained [29].

Cu and Ag, probably under the form of copper and silver sulphides, triggers the 2H to 3C transformation of polycrystalline ZnS [30].   These works remain largely qualitative and not amenable to detailed interpretation of the mechanisms involved because "impurities" may disturb the stoichiometry of the compounds [13], may change the growth rates through absorption phenomena and/or if they form clusters may even originate dislocations as well.

Departure from Stoichiometry

The departure from stoichiometry in the growth zone of the two components of binary compounds may have a significant effect on growing basic structures.

For SiC, Si enrichment contributes to growth of 3C crystals with a small amount of stacking regions of hexagonal type, while Si depletion (C enrichment) promotes the hexagonality [31].

For mixed crystals around the ZnS end-member grown from the melt by the Bridgman method, it has been shown that for $Zn_{1-x}Cd_xS$, $ZnS_{1-x}Se_x$ and $ZnS_{1-x}Te_x$, the structure hexagonality of the as-grown 3C modification increases with x [32,33].   The solid-state transformation behavior of such solid solutions also depends on x.   For instance, $2H-Zn_{1-x}Cd_xS$ crystals were annealed in the range 300 to 1100°C [34,35]. For $0 \leqslant x \leqslant 0.02$, 2H transforms to a disordered twinned 3C structure like pure ZnS, while for $0.02 \leqslant x \leqslant 0.05$, an intermediate disordered 6H form is obtained at 600°C which itself converts to the disordered twinned 3C on further annealing at 800°C. When $0.05 \leqslant x \leqslant 0.07$, final conversion of disordered 6H is to 2H instead of 3C at T = 1050°C. Intermediate stabilization of 6H with x also takes places for $Zn_{1-x}Mn_xS$ [35,36].

Micas exhibit extended solid solutions involving many end-members: basic structures of lepidolites, for instance, close to muscovite $KAl_2Si_3O_{10}(OH)_2$ are mainly $2M_1$, but rather 1M, $2M_2$ or 3T when their composition plots closer to polylithionite $KAlLi_2Si_4O_{10}(F,OH)_2$ and trilithionite $KAl_{1.5}Li_{1.5}AlSi_3O_{10}(OH,F)_2$ [37].

The role of departure from stoichiometry on the polytypic behavior is thus strong but yet poorly understood, especially for very small concentrations of the minor end-members where the average atomic structure is not significantly modified. To find the reasons for that would contribute to understand why an end-member (e.g., ZnS) shows extensive polytypism while the other (e.g., CdS), although perfectly isostructural, does not.

Rate of Crystallization

Although closely related to temperature, supersaturation and adsorption phenomena, the rate of crystallization has been considered as a whole and shown to influence the basic structure type and its disorder degree. For instance, 3C SiC largely dominates the equilibrium hexagonal forms at high growth rates [38]. The higher the temperature, the longer the equilibrium forms persist as a function of growth rate. The role played by supersaturation on the polytype structure of CdI_2 crystals grown from solution by the evaporation technique is to increase the amount of disorder structure [39,40]. The same effect holds for self-nucleated muscovite in a closed hydrothermal system at constant temperature [41].

4. GROWTH MECHANISMS OF BASIC STRUCTURES

For as-grown polytypic crystals, there are now experimental proofs for the generation of basic structures during the successive 3D-nucleation and layer-by-layer growth of the faces parallel to the layers or modules. This has been recently shown directly to be the case for synthetic muscovite crystallites, devoid of linear defects, nucleated at constant temperature, and examined by HRTEM techniques [41]. The disordered and more or less randomly faulted 1M and $2M_1$ structures proved to originate during the successive early nucleation and layer-growth stages.

While 3D-homogeneous nucleation (at very high supersaturation) or 3D-heterogeneous nucleation (at moderate supersaturation) takes place, embryonic stacking sequences of faulted basic structures form. The stacking scheme of the nuclei may or may not be inherited during the subsequent layer-by-layer growth (2D-nucleation process): it seems to be so if both thermodynamic and kinetic factors do not change too much as expected between heterogeneous 3D-nucleation and 2D-nucleation. In the case of homogeneous 3D-nucleation followed by 2D-nucleation, the structure memory is often lost between the two growth processes, thereby indicating that environmental factors dominate the structural controls.

Especially on the modular (or basal) faces of rather large crystals, the different types of 2D-nucleation are expected to affect the lateral polytype homogeneity of the crystals. For the mononuclear process for which the layer-spreading rate is high as compared to the nucleation rate, lateral polytype homogeneity should be preserved. It is not so for the "birth-and-spread" and polynuclear processes where these two rates compete [3]. Resulting lateral discontinuities of stacking of single layers at the surface of synthetic micas have been recently reported [42].

On the other hand, coalescence due to face-to-face autoepitaxial settlement of already formed 3D-nuclei is expected to contribute to inhomogeneity along the normal to the layers and may partly account for syntaxic coalescence of polytype domains and/or to twinning of a basic structure [3].

During solid-state transformations between basic structures of ZnS or SiC, 3D-nucleation of the new form takes place more easily on stacking faults of the parent matrix which are an embryonic prefiguration of this new form. Increase of thickness of such nuclei statistically takes place by nucleation of stacking faults separated by two (2H to 3C trans-formation) or three (2H to 6H transformation) layers [43]. Such nuclei are almost free to expand laterally because the leading partial dislocations are pushed out by the attached stacking faults because of their negative stacking-fault energies.

5. GROWTH MECHANISMS OF COMPLEX POLYTYPES

If probably not the only one possible, the spiral growth of real crystals [44] as related to the outcrop of screw dislocation(s) on modular faces provides the most accepted mechanism of generation and development of complex polytypes.

Perfect crystallographic periodicities of several hundred angstroms as shown by many complex polytypes may be explained by the piling up over itself of a stacking sequence slab, as resulting from the winding-up of the exposed ledge around the dislocation core. According to this screw dislocation mechanism, a "mechanical memory" of the structure propagates along the dislocation line, as long as the dislocation survives. Good correlations between the pitch of the screw dislocation as measured from the height of the spiral step and the periodicity of the polytype as shown by X-ray diffraction were reported for SiC [45,46] and CdI_2 [47].

The original screw dislocation theory for the generation of polytypes [48] assumed that the crystal nucleated first as a perfect platelet, then grew by 2D-nucleation. At this stage, the crystallite adopts any basic structure according to environmental conditions. Finally, after sufficient drop of supersaturation, the basal face should have developed further on by spiral growth. These successive growth stages were actually found for self-nucleated synthetic micas evolving in a closed system [49,50].

The spiral growth model considered first two cases: 1) the Burgers vector is an integral multiple of the basic structure repeat distance: the basic structure, supposed perfect, is further developed by spiral growth; 2) the Burgers vector is a non-integral multiple of this distance: a new polytype, with a repeat distance equal to the pitch of the screw and which admits the basic structure as a substructure, is generated. According to this model, referred to as the "perfect matrix model", all possible polytypes were generated for SiC, CdI_2, micas, etc. [51-53].

After comparing the predicted structures with those actually observed, it came out that: 1) some predicted structures were not observed, and 2) some observed structures were not predicted. Case 1) could be partially resolved considering their depressed probability of occurrence due to important lattice mismatch or structure disturbance in the cut plane of the dislocation [51,54]. The explanation of case 2) required further sophistications of the original model. One attempt has been to consider the interplay of more than one screw dislocations on the basal face and the manners according to which they were originated (e.g., [55,3]). But the most convincing model, because both simple and efficient, was to consider a single dislocation the exposed ledge of which could contain a unique low-energy stacking fault. It has been first developed for close-packed substances like SiC [55,56], CdI_2 [56] and PbI_2 [57] and then extended to micas [58]. Most of the already known polytype structures of these substances were thus found to match the predictions of this "faulted-matrix model".

6. COMPLEX POLYTYPES FORMED BY DISLOCATION-GUIDED SOLID-STATE TRANSFORMATIONS

The complex polytype formation in synthetic ZnS has been explained in terms of a periodic slip mechanism [5,59-62]. At high temperature, 2H-ZnS crystals grew first as needles by spiral growth around a unique screw dislocation axially located along the z axis. Such needles thicken and grow further as platelets. The Burgers vector of the screw dislocation may be any integral multiple of c(2H). During the cooling down of the crystals, stacking faults nucleate and expand up and down along the spiral ramp offered by distorted (0001) lattice planes around the dislocation line. Since any set of stacking faults will be recovered at distances along the dislocation line corresponding to the Burgers vector strength, almost perfectly ordered polytypes will result. This repeat period will be that of the pitch of the screw dislocation. This mechanism explains why all presently known synthetic ZnS polytypes involve an even number of single layers in their unit cell [61].

Similar dislocation-assisted phase transformations are known for $TiS_{1.7}$ [63].

This mechanism illustrates the post-growth effect of as-grown dislocations on the ordering of phase transformations.

7. ATTEMPTS FOR CONTROLLED PRODUCTION OF POLYTYPES

Some of either close-packed or layered polytypic substances display remarkable changes of their physical and electrical properties as a function of the polytype concerned and of its degree of order [13,64,65]. It is especially the case for semiconducting, photovoltaic and dielectric properties. Therefore, controlled production of some polytype of a substance should result in obtaining desired properties for technological applications. Significant results have been obtained recently for SiC [13] by controlling temperature, supersaturation and impurities in the growth zone. However, the best way to obtain a desired polytype is yet to produce its lateral overgrowth onto a seed crystal with the same structure. Furthermore, this technique is most promising because lateral epitaxic overgrowth of a non-polytypic but isostructural substance will force it to acquire the polytype structure of the seed. By doing so, new properties of spontaneously non-polytypic substances will be explored for the benefit of technology.

REFERENCES

1. A. R. Verma and P. Krishna, "Polymorphism and Polytypism in Crystals", Wiley, New York (1966).
2. C. J. Schneer, Acta Cryst., 8:279 (1955).
3. A. Baronnet, Prog. Cryst. Growth Charact., 1:151 (1978).
4. D. Pandey and P. Krishna, in: "Prog. Cryst. Growth Charact.", 7:213 (1983).
5. I. T. Steinberger, in: "Prog. Cryst. Growth Charact.", 7:7 (1983).
6. H. Jagodzinski, Neues Jb. Miner., Abh., 84:49 (1954).
7. J. M. Yeomans and G. D. Price, Bull. Mineral., 109:3 (1986).
8. P. Krishna, in: "Prog. Cryst. Growth Charact.", 7:3 (1983).
9. R. Angel, G. D. Price and J. M. Yeomans, Acta Cryst., B41:310 (1985).
10. G. D. Price and J. M. Yeomans, Acta Cryst., B40:448 (1984).
11. G. D. Price and J. M. Yeomans, Min. Mag., in press (1985).
12. A. Baronnet, in: "Current Topics in Materials Science", 5:447 (1980).
13. Y. M. Tairov and V. F. Tsvetkov, in: "Prog. Cryst. Growth Charact.", 7:111 (1983).
14. W. L. Bragg, Phil. Mag., 39:647 (1920).
15. C. Frondel and C. Palache, Science, 107:602 (1948).
16. H. Baumhauer, Z. Krist., 55:249 (1915).
17. Z. G. Pinsker, Acta Physicochim. URSS., 14:503 (1941).
18. J. V. Smith and H. S. Yoder, Min. Mag., 31:209 (1956).
19. D. Pandey, Proc. Indian Natn. Sci. Acad., A47:78 (1981).
20. W. F. Knippenberg, Philips Res. Rep., 18:161 (1963).
21. E. T. Allen and J. L. Crenshaw, Am. J. Sci., 34:341 (1912).
22. H. S. Yoder and H. P. Eugster, Geoch. Cosmoch. Acta, 8:225 (1955).
23. G. Lal and G. C. Trigunayat, J. Solid State Chem., 9:132 (1974).
24. A. K. Rai, R. S. Tiwari and O. N. Srivastava, Phys. Stat. Sol., 35:719 (1976).
25. H. Jagodzinski and H. Arnold, in: "Silicon Carbide – A High Temperature Semiconductor", p. 136, Pergamon, London (1960).
26. J. Ruska, L. J. Gauckler and G. Petzow, Sci. Ceram., 9:332 (1977).
27. M. Tairov, I. I. Khlenikov and V. F. Tsetkov, Phys. Stat. Sol., (a), 25:349 (1974).
28. W. F. Knippenberg and G. Verspui, Mat. Res. Bull., 4:S.33 (1969).
29. A. Addamiano and L. S. Staikoff, J. Phys. Chem. Solids, 26:669 (1965).
30. M. Aven and J. A. Parodi, J. Phys. Chem. Solids, 13:56 (1960).
31. Yu. A. Vodakov, E. N. Mokhov, A. D. Roenkov and M. M. Anikin, Pisma v Jurnal Teknich Phisiki, 5:367 (1979).
32. M. J. Kozielski, J. Crystal Growth, 30:86 (1975).
33. M. J. Kozielski, Bull. Acad. Polon. Sci., 26:193 (1978).
34. M. T. Sebastian and P. Krishna, Phys. Stat. Sol., 79A:271 (1983).
35. M. T. Sebastian and P. Krishna, Bull. Mat. Sci., 6:369 (1984).
36. M. T. Sebastian and P. Krishna, J. Crystal Growth, 66:586 (1984).
37. M. D. Foster, US Geol. Surv. Prof. Pap., 354B:11 (1960).
38. W. Spielman, Z. Angew. Phys., 19:93 (1965).
39. T. Minagawa, Acta Cryst., A34:243 (1978).
40. G. C. Trigunayat and A. R. Verma, Acta Cryst., 15:499 (1962).
41. M. Amouric and A. Baronnet, Phys. Chem. Minerals, 9:146 (1983).
42. M. Amouric, Acta Cryst., B43:57 (1987).
43. P. Krishna and M. T. Sebastian, Bull. Min., 109:99 (1986).
44. W. K. Burton, N. Cabrera and F. C. Frank, Phil. Trans. Roy. Soc. London, Ser. A, 243:299 (1951).
45. A. R. Verma, Proc. Roy. Soc. London, A240:462 (1957).
46. T. Nishida, Mineral. J., 6:216 (1971).
47. M. R. Tubbs, Acta Cryst., B27:857 (1971).
48. F. C. Frank, Phil. Mag., 42:1014 (1951).

49. A. Baronnet, Amer. Mineral., 57:1272 (1972).
50. A. Baronnet, Fortschr. Mineral., 52:203 (1975).
51. R. S. Mitchell, Z. Kristal., 109:1 (1957).
52. P. Krishna and A. R. Verma, Z. Kristal., 121:36 (1965).
53. A. Baronnet, Acta Cryst., A31:345 (1975).
54. A. Baronnet and M. Amouric, Bull. Min., 109:489 (1986).
55. D. Pandey and P. Krishna, J. Crystal Growth, 31:66 (1975).
56. D. Pandey and P. Krishna, Phil. Mag., 31:1113; ibid., 1133 (1975).
57. D. Pandey and P. Krishna, Phys. Lett., 51A:209 (1975).
58. D. Pandey, A. Baronnet and P. Krishna, Phys. Chem. Minerals, 8:268 (1982).
59. B. K. Daniels, Phil. Mag., 14:487 (1966).
60. E. Alexander, Z. H. Kalman, S. Mardix and I. T. Steinberger, Phil. Mag., 21:1237 (1970).
61. S. Mardix, Bull. Min., 109:131 (1986).
62. I. T. Steinberger, in: "Prog. Cryst. Growth Charact.", 7:7 (1983).
63. J. J. Legendre, R. Moret, E. Tronc and M. Huber, in: "Prog. Cryst. Growth Charact.", 7:309 (1983).
64. A. M. Fernandez Samuel, M. Rao and O. N. Srivastava, in: "Prog. Cryst. Growth Charact.", 7:391 (1983).
65. M. Rao, S. Acharya, A. M. Samuel and O. N. Srivastava, Bull. Min., 109:469 (1986).

CRYSTAL GROWTH FROM TRANSPARENT SOLUTION - TWO METHODS TO STUDY

CRYSTAL GROWTH AND HYDRODYNAMICS OF THE SOLUTION AT A TIME

F. Bedarida

Instituto di Mineralogia
Università di Genova
26 C. Europa 16100 Genova, Italy

INTRODUCTION

The subject of this paper deals with applications of holography in crystal growth from transparent solutions. The method may be easily applied also to crystal growth from transparent melt, but no experiment has been done up to now in this field.

Holography had been discovered by Dennis Gabor in 1948 and is an interference phenomenon between waves. The waves may be in theory of any type: sound, X-rays, corpuscular waves, light, but the most valuable results have been achieved with coherent light.

In this paper the results obtained in the application to crystal growth from solution will be discussed along with the perspectives of this type of work.

The word holography has the greek root "olos" that means the whole (same root!) and probably Gabor who invented the word, intended that both information on amplitude and phase are given.

The normal photograph is a two-dimensional projection of the object, described by the distribution of the square of the amplitude $A^2 \sim 1$. The hologram on the contrary reconstructs the wave field reflected by an object.

But even if the discovery of Gabor was done in 1948, the first appreciable holograms were obtained only in 1963 by E. Leith and J. Upatnieks using the light of the laser [1,2] that was discovered in 1960. At that time Dennis Gabor sustained that he had almost forgotten holography [3]. Holography is then a way to reconstruct the wave field diffused by an object.

A step forward is holographic interferometry, a method to check the interference between two registered wavefronts, or one registered wavefront and the wavefront from an object: the two wavefronts must be slightly different. One practical point, that is an advantage, has to be kept in mind: the quality of the optical tools has not to be of such a high standard as in conventional interferometry. The only point is that they must not have any change during the experiment. Today this method is largely used to measure:

1. surface displacements of opaque objects;
2. refractive index variations of transparent objects.

Both these informations may be used for crystal growth, but this paper will deal mainly with the second one.

The first applications of holographic interferometric methods in this laboratory were applied to check microtopography of surfaces of crystals as grown: their effectiveness is good and their application is easy [4,5,6]. Two advantages are that no matching surface and in many instances no silvering are needed.

FIRST METHOD

One Beam Holographic Interferometry

The first experiments on crystal growth from solution with holographic interferometry were started in the Mineralogical Institute of the University of Genova, many years ago [7]. The scheme of the experimental apparatus is given in Figure 1. Two collimated laser beams obtained through the variable beam splitter are reflected by the mirrors and are spread by the objective and the lenses. The first beam, the object beam, crosses the solution where nothing is going on and interferes with the second one, the reference beam, on the holographic plate. The plate after exposure is developed and then replaced exactly in the same position and enlightened with the reference beam. The reconstructed wavefront registered by the hologram can interfere with the wavefront that has crossed the solution. If the plate is in the exact position and nothing has happened to the solution, no fringes will be seen. If the concentration of the solution has changed with some irregularity, interference fringes are formed which show how much the optical path is changing.

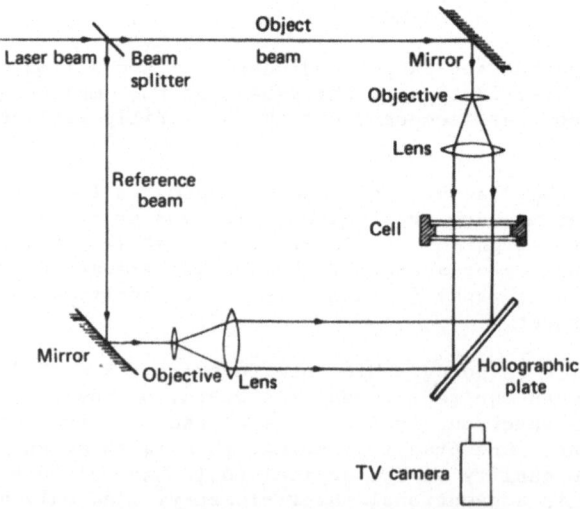

Fig. 1. Schematic diagram of the experimental apparatus.

An apparatus based on the same optical schema and qualified for space, has been built by DFVLR in Germany and has been used in flight: the program of research in microgravity previews crystal growth as one of the most important points. Avoiding buoyancy and sedimentation, crystals are likely to grow with a higher perfection. The progress of an experiment on crystal growth (on ground or even in space) may be followed in real time and simultaneously registered using a TV camera, a monitor and a TV recorder. Sodium chlorate was chosen in the beginning, since in aqueous solution it has a broad metastable region under supersaturated conditions and well shaped single crystals may grow from a seed without simultaneous precipitation of many tiny crystals. A thermostated cell has been used and the schema is shown in Figure 2. Bigger vessels and crystals are allowed by the use of coherent light, as compared to the classical methods. The supersaturation desired was achieved by lowering the temperature. From the fringes shown in Figure 3 it is possible to deduce the movements that take place in the solution during the process of growth. At present polar crystals are under study (saccarose and epsomite).

In saccharose crystals the (110) face grows faster than the ($1\bar{1}0$) face as it is shown in the interferograms of Figures 4a-4d by the different number of the fringes near the faces. The rates of growth of the faces were carefully measured. It has been possible to obtain isotherms that show a peculiar behavior. An example is given in Figure 5. The growth rate has been plotted versus supersaturation at a temperature of 30°C. The mechanism sequence resulting for the (110) face is as follows: (1) parabolic laws; (2) exponential law; (3) linear law; (4) exponential law; and (5) linear law, while for the ($1\bar{1}0$) the sequence stops when the linear law is reached.

A paper will be published in the Journal of Crystal Growth where it has been checked with "one beam" interferometry how the diffusion boundary layer varies from face to face of a single sucrose crystal and how the distribution of captured impurities on a given crystal face can

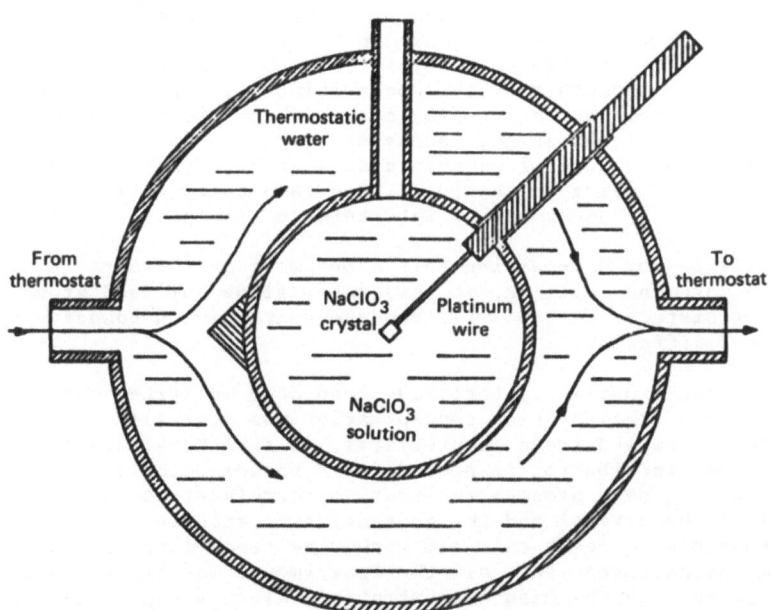

Fig. 2. Schematic diagram of the thermostatic cell.

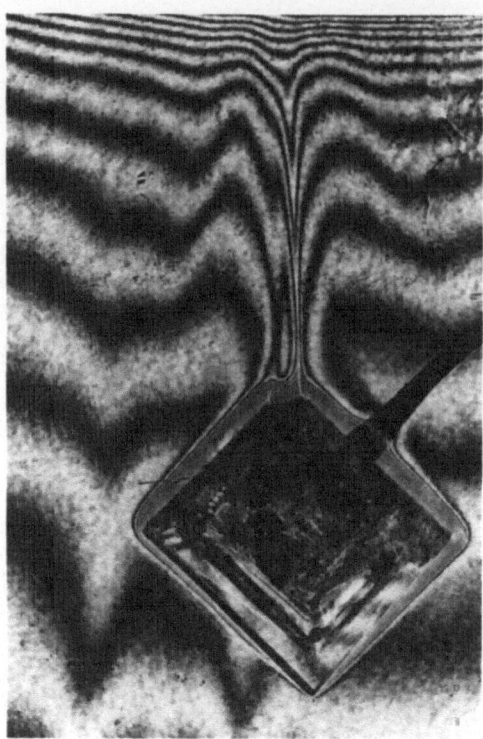

Fig. 3. Interferogram of a NaClO$_3$ crystal growing in water solution.

influence the iso-concentration curves and hence the variation of the boundary layer in time and position on the face.

The effect of the introduction of a doping substance (raffinose in this case) is shown in good evidence in Figures 6a-6d. The upper interferograms are related to a crystal growth from the pure solution, the lower ones to the growth from the doped solution; the two interferograms on the left are related to an early stage of the growth process, the two on the right to a later stage, when a steady state of growth is reached. The concentration change of supersaturated solution near the different crystal faces belonging to the [001] zone, and their corresponding growth, can be compared by the number of interference fringes.

It will be worth repeating this experiment in microgravity because convective movements will be, in this case, almost nil and growth will be depleted of disturbances and will be dependent only on boundary (volume and surface) diffusion.

In all the experiments that have been done on earth, the spatial configuration of the solution concentration was generally rather complex inside the plumes and around the crystal. A real knowledge of the concentration distribution is needed for a phenomenological description of the stationary growth process in order to correlate the growth rates of the faces of the crystal and the concentration gradient in the solution. It is impossible to reach this aim with "one beam" interferometry. Even if it has the advantage that all the experiments may be followed in real time, it gives only the mean concentration value averaged over the depth of the solution crossed by the wavefront.

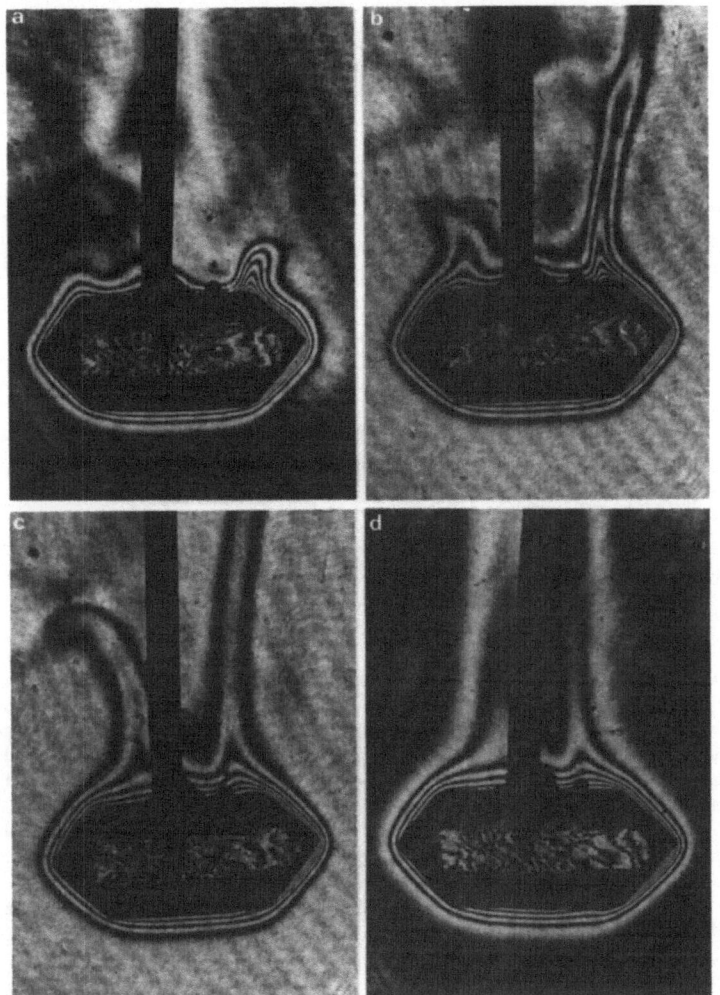

Fig. 4. a-d: Interferograms at different times of a sucrose crystal growing in water solution: (110) grows faster than (1$\bar{1}$0).

SECOND METHOD

Multi-directional Holographic Interferometry

A premise has to be made before talking in detail of multi-directional holographic interferometry. The problem that deals with this experimental technique is a very general one in physics and is the problem of image reconstruction from projections. With the word "image" the distribution of a physical property in the three dimensions of space is meant. In this meaning, "images" are used to represent the distribution of some property of an object or physical system: for instance the brightness of a star, the distribution of drops in a volume or conversely of bubbles in a liquid, the measurement of a variable concentration or a non-constant absorption coefficient, etc. [9].

The set of line or strip integrals corresponding to a particular angle of view is said to be a projection of the object, where the linear

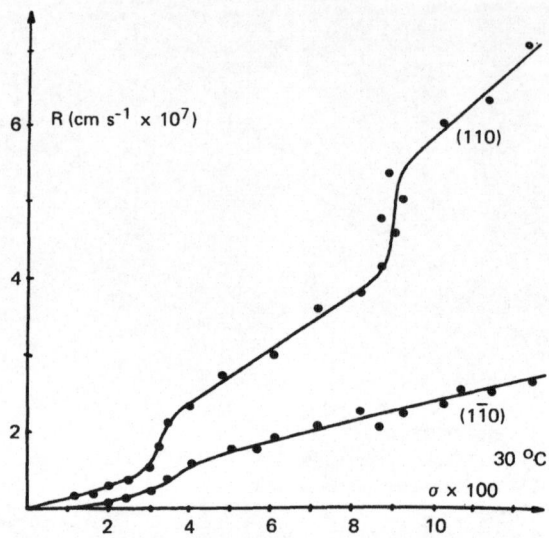

Fig. 5. Diagram of the growth rate versus supersaturation of a sucrose
crystal growing in aqueous solution.

integrals are actually integrals of functions related to the physical
definition of the system. Given a number of such projections at different
angles of view, the estimation of the corresponding scalar or vectorial
field within the object is the basic problem of image reconstruction from
projections. A scalar field is more common: for instance a density, an
index of refraction, an absorption. A vectorial field could be the
electrical polarization of an anisotropic substance, the internal stress
of a birefringent object and so on.

It is well-known that in medicine computed X-ray tomography and more
recently NMR (nuclear magnetic resonance) are very probably the most
significant applications up to this moment of image reconstruction from
projections.

One of the outstanding points of these techniques is that no invasive
examination is needed. Multi-directional holographic interferometry being
based on the same principles, is also not invasive.

An idealized projection measurement process, according to G. T.
Herman and R. M. Lewitt [9] may be described in the following way. The
function g(x,y) represents the distribution of the physical property of
interest within the object, which we take to be "two-dimensional" for the
purpose of illustration, see Figure 7. Projection data are estimates of
line integrals of "g" along lines of known location. In two dimensions
each line is specified by two parameters "s" and "θ". Then the line
integral of "g" along the line specified by (s,θ) can be denoted by
$[\mathcal{R}g](s,\theta)$ where we use $\mathcal{R}$ for the operator in honor of J. Radon who
apparently was the first to study the transformation which maps function
"g" into the function $\mathcal{R}g$. Clearly

$$[\mathcal{R}g](s,\theta) = \int_{-\infty}^{\infty} g(s\cos\theta - u\sin\theta,\ s\sin\theta + u\cos\theta)du.$$

The reconstruction of images from projections is based on the
development of techniques for solving the system of the above integral
equations. No single technique has been found up to now capable of

210

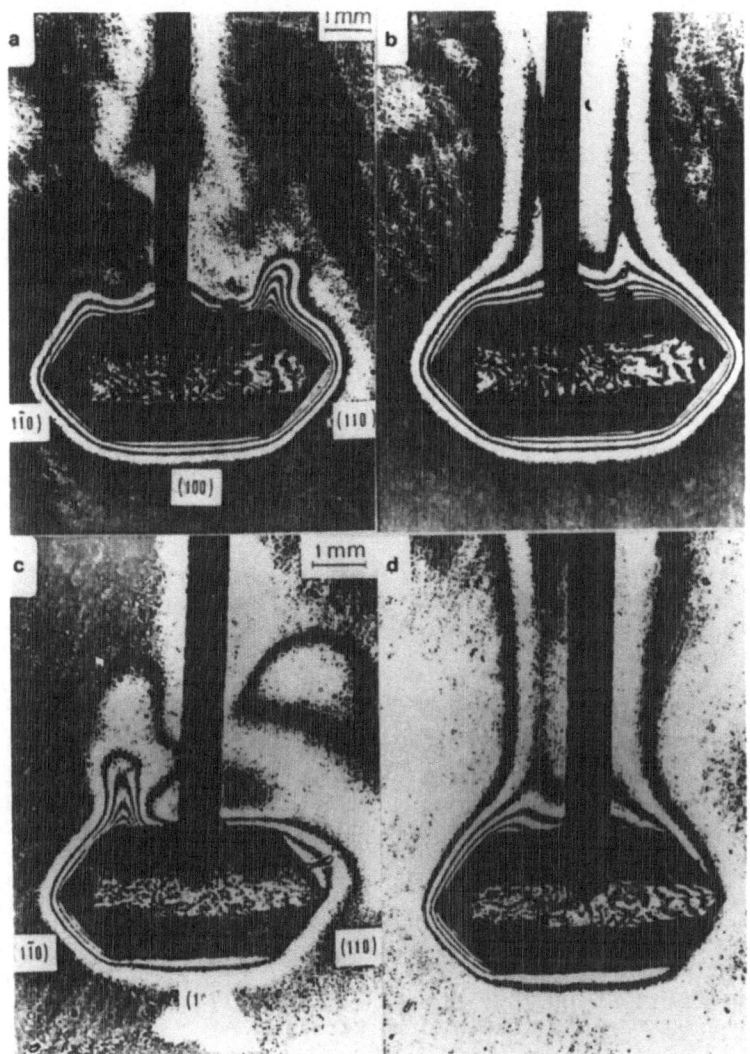

Fig. 6. a-d: Interferometric images of different stages of growth of two
sucrose crystals respectively from a pure solution (a and b) and
from a solution doped with raffinose (c and d).

processing satisfactorily the wide variety of projection-measurement
geometry and quantity and precision of data which occur in practical
applications. When operating experimentally at the present stage of
research, in order to trust the results, it is very useful in practice to
do mathematical models at the same time, and control the results in
parallel with the experimental ones.

        In multi-directional holographic interferometry we make use of
parallel beams, discrete in number, each with a different direction.
These directions are the ones along which the integrals will be
calculated. They have been obtained through a suitable diffraction
grating. The angle of divergence is very important in the measurements.
The aim of the experiment is to reconstruct a concentration map inside the
solution in every plane parallel to the upper face of the growing crystal.

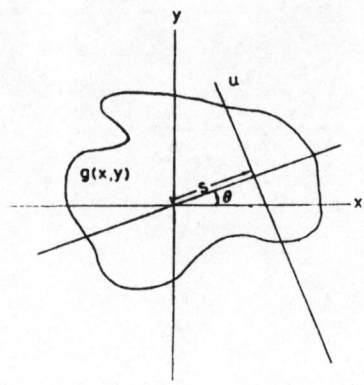

Fig. 7. Projection measurement geometry. Each projection measurement is
an estimate of a particular line integral of g(x,y) where the
line of integration is specified by the parameters "s" and "θ"
(after G. T. Herman and R. M. Lewitt).

The precision of the method seems in some way to be linked to the angle of
aperture of the rays along which the integrals are calculated. At present
a divergence of 50° has been achieved. A 100° divergence would be a good
result.

The essential experimental procedure is summarized in the following
points [10]:

1.   A double exposure multi-directional hologram of the growing process
     is recorded; it stores on a single plate N different holograms
     corresponding to as many projections along N directions of the
     refractive index variation field.
2.   The N images obtained are processed by an image elaboration system
     that gives the grey levels corresponding to the intensity values on
     each point of the image.
3.   These data are elaborated (filtered, smoothed, etc.) and are the
     input of a computer program which performs the reconstruction of a
     tridimensional field from its projections. For this reason this
     method could be called optical tomography.

Figure 8 shows the scheme of the experimental apparatus; the multi-
directional double exposure hologram records at the same time on a single
plate the phase variations of several (13-15) plane waves passing through
the growth cell in different directions. The plane waves are obtained by
placing the phase grating (P in Figure 8) in front of the growth cell and
the phase variations are due to a change of the refractive index as
consequence of the growth process. In Figure 9 an interferogram out of
the 13 obtained on the same plate is given. It is only a particular one,
but meant to show how to carry on the experimental procedure. The growing
crystal is seen at the top of the photograph surrounded by interference
fringes that appear also in the convective plume starting from the upper
face of the crystal. Two different stages of the digitation procedure are
given in the diagrams under the hologram. The diagram below is the
intensity evaluated along a horizontal line at a fixed height above the
crystal (the height is marked with arrows in the upper photogram). It is
very easy to see by the diagram below that the data need an anti-noise
filtering. Observing that the fringes have a spatial frequency lower than
the noise the proper algorithm is a Fourier filtering. This is done in a
computer with iterative operations that take off the high functions in an

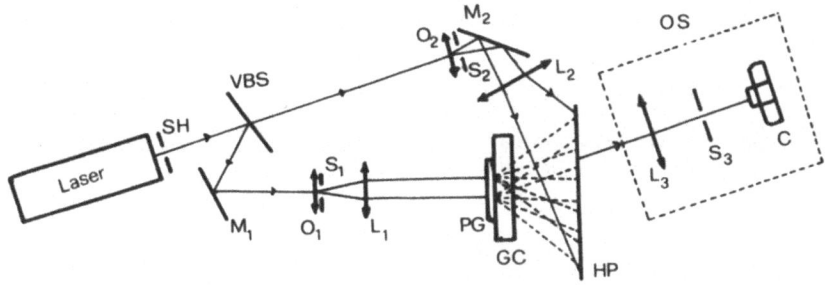

Fig. 8.   Schematic diagram of the experimental apparatus for multi-
          directional holographic interferometry: SH = shutter, VBS =
          variable beam splitter, M = mirror, O = objective, S = spatial
          filter, L = lens, PG = phase grating, GC = growth cell, HP =
          holographic plate, C = camera, OS = optical system.

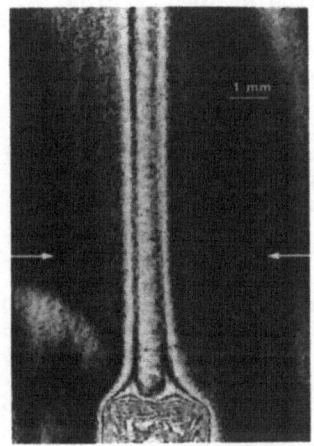

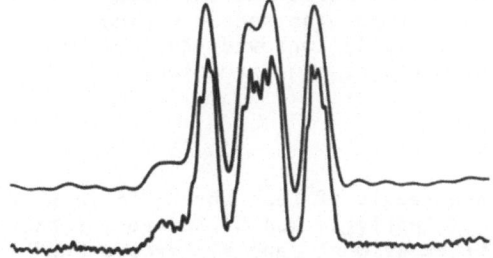

Fig. 9.   One out of the 13 holograms taken on the same plate. The two
          diagrams below are: the lower one without filtering, the higher
          one after filtering.

arbitrary mode, because at this stage of the research those saw steps are considered to be without any meaning. The result is shown in the upper part of the diagram.

The light intensity evaluated in each point of an interferometric image is:

$$I(s,\theta) = C[1 + \cos\Delta\phi(s,\theta)].$$

Where "s" and "θ" are the coordinates shown before, C is a constant and $\Delta\phi$ is the phase variation between the two interfering wavefronts at a fixed height. Therefore the phase variation is obtained by the equation:

$$\Delta\phi(s,\theta) = [2a \; arcos \sqrt{\frac{I(s,\theta) - I_1}{I_2 - I_1}} + b].$$

Where "a" and "b" are suitable parameters, I_1 and I_2 are respectively the minimum and the maximum values of the digitized intensity. This process is repeated using each one of the N orders obtained by the phase grating. A system of equations may be written for each level z:

$$\Delta\phi_k = \frac{2\pi}{\lambda} \int \{n(s,\theta) - n_o\} du_k \qquad k = 1,2,\ldots,N$$

where u_k is the kth direction of observation, $\Delta\phi_k$ is the phase variation relative to the direction u_k and the integrals are calculated along straight lines in the refractionless approximation that is good since the concentration gradients are small.

By the technique of reconstruction from projections the set of these equations may be solved, obtaining the distribution n(x,y) of the refractive index at the chosen level z above the crystal and hence the concentration. Applying two different numerical methods suitable in this particular case called "sinc" and "grid" algorithms the reconstruction of the concentration at different levels above the crystal has been obtained.

In an experiment with sodium chlorate ($NaClO_3$) the difference in concentration between two consecutive contour lines was checked as 0.125%, the initial concentration of the supersaturated solution was 50.2%. After a forced stop, this research is again in progress on both sides: the experimental apparatus and the mathematical way of implementation.

The techniques just described have many advantages to study crystal growth from solution or from transparent melts on earth and may be very suitable also in space. The first method, "one beam" interferometry allows a control in real time, the second method, "multiple beam" interferometry, experimentally improved and implemented, will give good results on the measurements of concentration.

ACKNOWLEDGEMENTS

This paper is the result of team work. It is my pleasure here to thank my colleagues C. Pontiggia and L. Zefiro for their cooperation on the laser side of the experiment, and M. Bertero and P. Boccacci for the implementation of the data.

REFERENCES

1. E. Leith and J. Upatnieks, J. Opt. Soc. Amer., 53:1377 (1963).
2. E. Leith and J. Upatnieks, J. Opt. Soc. Amer., 54:1295 (1964).

214

3. D. Gabor, Opt. Spectra, 32, Oct. (1970).
4. F. Bedarida and C. Pontiggia, Acta Cryst., A24:614 (1968).
5. F. Bedarida, C. Pontiggia and L. Zefiro, J. Crystal Growth, 24/25:327 (1974).
6. F. Bedarida, C. Pontiggia and L. Zefiro, J. Crystal Growth, 34:79 (1976).
7. F. Bedarida, C. Pontiggia and L. Zefiro, in: "Applications of Holography and Optical Data Processing", E. Marom and A. A. Friesem, eds., Pergamon Press, 249 (1977).
8. D. Aquilano, F. Bedarida, P. Boccacci, G. Mantovani, M. Rubbo, G. Vaccari and L. Zefiro, in: "Materials under Extreme Conditions", H. Ahlborn, H. Fredrikson and E. Luscher, eds., Les editions de physique, 33 (1985).
9. G. T. Herman and R. M. Lewitt, in: "Topics in Appl. Physics", G. T. Herman, ed., Springer, Berlin, 1 (1979).
10. F. Bedarida, P. Boccacci, L. Zefiro and C. Pontiggia, Phys. Chem. Hydrodynam., 2:327 (1982).
11. F. Bedarida, J. Crystal Growth, 79:43 (1986).
12. F. Bedarida, L. Zefiro, P. Boccacci and C. Pontiggia, in: "Proc. 4th European Symp. on Materials under Microgravity", Madrid, 139 (1983).

INDUSTRIAL CRYSTALLIZATION

R. J. Davey

Imperial Chemical Industries plc
Chemicals and Polymers Group
Research and Technology, The Heath
Runcorn, Cheshire, UK

1. INTRODUCTION

Crystallization and precipitation are widely used by the chemical industry as a means of preparing chemicals in solid form. Many materials, including pharmaceuticals, dyestuffs, agrochemicals and polymers are synthesized by reactions which occur in the liquid phase to yield either solutions or melts containing the required products. Crystallization from these reaction media allows the separation of products in convenient solid form. These solids may then be further processed to yield dispersions, tablets, pastes, powders etc. for sale to customers.

The important properties of these crystallized products will vary but include purity, phase, composition, crystal shape and size distribution.

The overall objectives of a crystallization process are threefold:

1. To produce solid particles.
2. To produce these particles with specified properties.
3. To achieve this at minimum operating costs and optimum profit margins.

In the economic context it is important to recognize two types of product. The first is a commodity material. These are produced in large bulk quantities and sell at prices which are only marginally above production costs. Only by selling vast amounts manufactured by a sophisticated process to a tight specification can a satisfactory profit margin be attained.

Chemicals such as salt, ammonium nitrate, PVC, terephthalic acid and chlorine would fall into this category.

The second class of product is termed a speciality or effect chemical because it is sold for the effect it produces. Such materials sell at prices well in excess of production costs so that profit margins are high. Pharmaceuticals, dyestuffs and pesticides are examples of this category of product as would be many polymer based materials used increasingly in electronic components and as advanced construction materials. The price

reflects the development and proving of the product, and the identification and maintenance of exacting product performance.

In terms of specific crystallization techniques it is common for bulk
commodity materials to be crystallized in continuously operating plants
whilst specialities are produced batch-wise. This allows product costs to
be minimized for commodity products.

2. CONTINUOUS CRYSTALLIZATION

Continuous crystallizers are normally designed specifically for the
production of one particular chemical. They are not flexible pieces of
equipment and there is no such item as a typical continuous crystallizer.
In volume they vary from 40 to 100 m^3 and are capable of production rates
of 100's of tonnes per day. Normally a solution containing product is fed
to the crystallizer and supersaturation is produced either by cooling,
evaporation or a combination of the two depending on the solubility
characteristics of the system. For example, sodium chloride is
crystallized by evaporation because its solubility is very insensitive to
temperature. For a detailed description of continuous crystallizers the
reader is referred to the book of Mullin [1].

Much work has been devoted to the modelling of continuous
crystallizers in order to elucidate the factors which control the crystal
size of the product [2]. Because a continuous system operates at steady
state a mathematical description is readily achieved. Temperature and
supersaturation are fixed and the size distribution will result from some
combination of nucleation, crystal growth, crystal breakage and crystal
agglomeration occurring within the crystallizer. A convenient means of
achieving a mathematical description is to consider the number of crystals
of each size which exist in the crystallizer. A density function is first
defined as

$$\lim \Delta N/\Delta L = dN/dL = n \qquad \Delta L \to 0$$

in which n is called the population density and ΔN is the number of
crystals in the size range ΔL per unit volume. Note that n is a function
of L, and n(L) is thus the size distribution and n has the dimensions
number per unit size per unit volume. If a size distribution is
determined by sieving the product crystals then the mass, w, retained on a
sieve size L_{n+1} represents the mass having sizes between L_{n+1} and L_n. The
mean size of these crystals is

$$\bar{L} = \frac{1}{2} (L_n + L_{n+1})$$

Thus the number of particles in this interval, ΔL

$$= \frac{\text{mass of particles}}{\text{mass of one particle}} = w/fv\rho\bar{L}^3$$

f_v is a volume shape factor ($= 4/3 \, \Pi$ for a sphere), ρ the density, and

$$n = \frac{W}{fv\rho\bar{L}^3 \cdot \Delta L \cdot V} \quad \text{in the size range } \Delta L$$

For a continuous crystallizer these concepts may now be used in order
to elucidate the parameters having most influence on the crystal size
distribution. This crystallizer is referred to universally as the Mixed
Suspension Mixed Product Removal Crystallizer (MSMPR) and is characterized
as follows:

(a) Steady state operation is achieved (i.e., supersaturation, nucleation rate, crystal growth rate and temperature are constant);
(b) no crystals in feed stream;
(c) no crystal breakage or agglomeration;
(d) all crystals grow at the same rate;
(e) the product size distribution is the same as that in the crystallizer; and
(f) well mixed contents.

With these constraints a <u>number balance</u> for a given size range ΔL can be written as

$$
\begin{array}{l}
\text{number} \\
\text{entering the} \\
\text{size range} \\
\text{due to growth}
\end{array}
=
\begin{array}{l}
\text{number leaving} \\
\text{due to growth}
\end{array}
+
\begin{array}{l}
\text{number leaving as} \\
\text{product}
\end{array}
$$

Let the total crystallizer volume be V and the crystal growth rate, G and consider the number balance over a time interval Δt. This may be written as

$$n_1 VG\Delta t = n_2 VG\Delta t + Q\Delta t\ \bar{n}\ \Delta L$$

n_1, n_2 and n are population densities for sizes L_1, L_2 and averaged. Thus:

$$VG\ \frac{(n_2 - n_1)}{\Delta L} + Qn = 0$$

and defining V/Q as τ the residence time of slurry in the crystallizer this becomes, in the limit of $\Delta L \to 0$,

$$\frac{dn}{dL} + \frac{n}{G\tau} = 0$$

This equation can be integrated using the boundary conditions at L = 0, n = n^0, this being the nuclei population density. Thus

$$\int_{n^0}^{n} \frac{dn}{n} = \int_0^L - \frac{dL}{G\tau}$$

or n = $n^0 \exp(-L/G\tau)$.

Actually the nucleation rate, B, is the number growing through the nuclei size, i.e., B = $n^0 G$, so that

$$n = \frac{B}{G} \exp(-L/G\tau) \tag{1}$$

This relation shows that the size distribution is determined solely by the mean residence time, τ, and the rates of nucleation, B and crystal growth G.

It also follows that, taking logs of Eq. 1

$$\ln(n) = \ln(B/G) - L/G\tau$$

so that if the size distribution in an MSMR crystallizer is measured, it is possible to evaluate directly both G (slope) and B/G (intercept) from a lot of $\ln(n)$ versus L.

A further result of these analyses is that the dominant size in the distribution is linked to the residence time and growth rate according to:

$$L_D \approx 3G\tau$$

A consequence of this is that increasing τ would be considered an obvious way of increasing L_D. Unfortunately in a continuous system increasing τ also decreases the supersaturation. This decreases G and in fact the value of L_D is rather insensitive to τ.

3.   BATCH CRYSTALLIZERS

Batch Crystallizers are essentially enlarged versions of laboratory experiments. What has been successful in a 1 liter beaker is transferred to a 40 m^3 stirred tank. The tank is often made of stainless steel and may be glass or rubber lined depending on the chemicals to be used. It will be provided with a means of heating and cooling (often a steam/water jacket) and entry ports for addition of chemicals.

Supersaturation in such a system may be created in numerous ways:

(i) Chemical Reaction

In some instances the reaction product may be extremely insoluble in the medium and precipitate out as the reaction proceeds. This occurs in the production of azo pigments and disperse dyes from coupling reactions and in many inorganic reactions, e.g.,

$$Ca(OH)_2 + CO_2 \rightarrow CaCO_3 \downarrow + H_2O.$$

(ii) Salting Out

The basis of this method is the common ion effect in which a reaction product, present as an anion, is forced out of solution by addition of excess quantities of a counter cation.

One widely used example of this is to be found in sulphonic acids which are often precipitated as sodium salts.

(iii) Drowning Out

A further commonly used technique is to add to a reaction medium a second liquid component which is miscible with the medium but which reduces the solubility of the reaction product. Pentaerythritol tetranitrate is crystallized in this way by addition of acetone to an aqueous solution of the solute. Sodium sulphate can be recovered from waste streams by addition of methanol.

(iv) Cooling

For materials whose solubilities depend strongly on temperature, a straightforward cooling of the batch reaction vessel will be sufficient to crystallize the product. Many pharmaceutical products are prepared in this way.

(v) Evaporation

It is used for materials with flat solubility/temperature curves.

To illustrate the concepts a cooling batch crystallizer will be considered which has a volume V (m^3), a total mass of crystals M (kg) and a solution concentration C (kg/m^3). A solute mass balance can be written since the rate of increase of crystal mass = rate of decrease of solute concentration in solution:

$$\frac{d\overline{M}}{dt} = - \frac{d(Vc)}{dt}$$

Since V is constant this reduces to

$$\frac{Vdc}{dt} + \frac{d\overline{M}}{dt} = 0$$

or if Θ is the temperature of the batch and the concentration reaches its saturation value C*, at each temperature

$$\frac{Vdc^*}{d\Theta} \cdot \frac{d\Theta}{dt} + \frac{d\overline{M}}{dt} = 0$$

Since V is a constant and dc*/dΘ is the slope of the solubility curve it follows that the crystallization rate, d$\overline{M}$/dt, is determined only by the cooling rate, dΘ/dt.

4.   PHYSICAL CHEMISTRY

The physico-chemical aspect of crystallization in industrial situations are identical to those covered elsewhere in this text. Thus nucleation in batch crystallizers is essentially by a primary mechanism and crystal growth is governed, both in batch and continuous crystallizers by various factors associated with both mass transfer and surface reaction.

There are, however, three important features of industrial crystallization which are not covered in other chapters and these are discussed here.

(a) Secondary Nucleation

Continuous crystallizers, as we have seen previously, operate at fixed values of supersaturation generally well within the metastable zone, where, the rate of primary nucleation, $J \approx 0$. Crystals are always present in a continuous crystallizer and the necessary energy to create new surface comes not from the supersaturation but from mechanical energy and crystal fracture.

Crystals are moving around within the crystallizer due to forced convection brought about by the agitator. Crystals inevitably collide with each other, with the vessel walls and with the impeller.

Two situations may then develop:

(a) the stirrer may act as an indenter producing plastic deformation in the crystal surface and microcrystalline debris, or

(b) the stirrer will impart strain energy to the crystals which may result in the gross fracture. In order for a crystal to store sufficient strain energy to initiate a crack it must be larger than a critical size given by

$$d_c = \frac{\propto ER}{Y^2}$$

in which E is Young's Modulus, R the fracture surface energy and Y the plastic yield stress. For KCℓ in aqueous solutions this critical size is $\sim$30 μm, it increases with increasing material plasticity.

Both these collision events lead to the creation of new crystal surfaces. This is secondary nucleation and it produces crystal nuclei in the size 1 to 50 μm.

It follows that the secondary nucleation rate is a function of many poorly defined variables:

B ∝ (Collision Energy) (Collision Frequency) (Nature of Surface)
          ↓   ↓   ↓
B ∝ (Stirrer Speed)   (Stirrer Speed +   (Supersaturation)
                         Slurry Density)

Which gives an expression:

$$B = KN^b \cdot M^r \cdot \sigma^m$$

in which the secondary nucleation rate, B, is related to the stirrer speed, N, the slurry density, M, and the supersaturation σ, through the empirical exponents b, r and m. Which is indeed borne out in practice.

For further details of this mode of nucleation the reader is referred to Ref. [3].

(b) Polymorphic Phase Transitions

Many industrial systems demonstrate the property of polymorphism in the form of true polymorphs (same molecule, various structures.), solvates (solvent molecules form part of the lattice) compounds, gels, amorphous phases etc. Table 1 is a list of common examples.

Because a continuous crystallizer operates with fixed temperature and composition it is possible to operate it so that even in a multiphase system one particular phase is crystallized.

Table 1. Common Chemical Products

Chemical Product	Uses	Number of Polymorphs or Hydrates
Ammonium nitrate	Explosives, Fertilzers	5
Calcium carbonate	Filler for rubber and plastics	3
TiO	Pigment	4
Copper phthalocyanine	Pigment	4
Titanium trichloride	Catalyst	4
Indigo	Disperse dye	2
Aspirin	Pharmaceutical	4
Sorbitol	Confectionery sweetener	4
Lead chromate	Pigment	2
Triglycerides	Fat and oils	Numerous
Sucrose	Sweetener	4 hydrates
Tetraethyl lead	Petrol additive	6
Phenobarbitone	Pharmaceutical	13
Sodium carbonate	Glass manufacture	3 hydrates
Pentaerythritol tetranitrate	Explosive	2
Lead azide	Explosive	2

In batch processes this is not the case and metastable phases are often formed, for kinetic reasons and these subsequently undergo phase transitions to stable phases [4]. This is well-known in the pharmaceutical industry.

Because the crystals are in contact with solution such phase transitions are driven by solubility differences between phases and take place by solution mediation in which the metastable phase dissolves and the stable phase grows. Three basic expressions are relevant:

growth of the stable phase II: $\quad dr_{II}/dt = k_G\sigma_{II}$

dissolution of the metastable phase I: $dr_I/dt = K_d\sigma_D \quad \sigma_D = (x - x_I)/x_I$

and the solute mass balance: $\quad \sigma_i = \sigma_{II} + {}^{(x_{Is} + x_{IIs})}/x_{II}$

These expressions may be solved numerically to yield the phase volumes, V_I and V_{II} and the solution supersaturation as a function of time. Transformation rates dominated by growth of the stable or dissolution of metastable phases are both possible [5].

In terms of a batch crystallization it is important to know the time scale of such a transformation process so as to avoid the isolation of mixtures of crystalline phases as well as insuring consistency of the products.

(c) Impurities

Industrial processes are often "dirty" due to the presence of by-products in the crystallizing system. These impurities may be difficult to remove by crystallization or they may have a detrimental effect on the crystallization kinetics and crystal morphology.

In other cases trace quantities of selected impurities may be added specifically to control the crystallization process.

For example, in the production of urea, the by-product dimer, biuret, is an unwanted impurity and it influences the morphology of urea crystals [6]. In the crystallization of sodium chloride, on the other hand, ferricyanide is added specifically to produce a dendritic form of salt. For more details of impurity effects the reader is referred to Ref. [1].

5. PRODUCT CHARACTERISTICS

Having defined a crystallization process it is important to have an understanding of the key product qualities by which the degree of success of the process can be judged. These are listed below and are often referred to collectively as the "Physical Form":

Crystal size
Crystal shape
State of Aggregation
Crystalline Phase
Crystal Purity

Crystal size and shape are important in processing of solids since they influence filtration, drying and rheologies of slurries. Generally large crystals > 100 μm will be preferred. On the product size small crystals may lead to undesirable caking and dustiness of materials whilst in some applications, such as powders for ceramics or pigments, crystals smaller than 1 μm may be mandatory. Crystal shape can have a major influence on the bulk density, mixing and tableting of powders, needles in particular are to be avoided.

One method of achieving a good balance of size and shape is to crystallize aggregates of crystals. These tend to be spherical and ideal for many applications.

Crystalline phase is an important consideration since phase transitions give problems both during processing and as products. This is well-known in the pharmaceutical industry.

Crystal purity is of most concern in the product where impurity may lead to toxicological problems or may even cause deterioration of physical properties of materials. One example of this effect would be in the crystallization of urea which is done to remove biuret, a toxic by-product. Urea is sold as a fertilizer, biuret is phyto-toxic particularly toward citrus fruits.

In considering a process which produces solids it is important to take the process as a whole and to optimize the physical form at each processing stage. An understanding of the trade off between process design and product characteristics is vitally important in the establishment of a solids producing process.

REFERENCES

1. J. W. Mullin, "Crystallization", 2nd Edition, Butterworths (1972).
2. A. D. Randolph and M. A. Larson, "The Theory of Particulate Processes", Academic Press, New York (1971).
3. J. Garside and R. J. Davey, Chem. Eng. Commun., 4:393 (1980).
4. R. J. Davey, P. T. Cardew, D. McEwan and D. E. Sadler, J. Crystal Growth, 79:648 (1986).
5. P. T. Cardew and R. J. Davey, Proc. R. Soc. London, A398:415 (1985).
6. R. J. Davey, W. Fila and J. Garside, J. Crystal Growth, 79:607 (1986).

CRYSTAL GROWTH AND CRYSTAL CHEMISTRY

H. Arend

Laboratory of Solid State Physics
Swiss Federal Institute of Technology
CH-8093 Zurich, Switzerland

INTRODUCTION

The primary aim of crystal chemistry was for many years the study of the relationship between chemical composition and crystal structure.

In order to obtain and classify systematically data, as they were compiled, for example, for the inorganic world of chemistry in Refs. [1-3] a huge amount of single crystals, used primarily for structure determinations, had to be grown by means of a variety of crystal growth methods. Much less systematic knowledge exists presently - inspite of big progress in computational techniques used for molecular structure determination - in the field of organic crystals (see, for example, Ref. [4]), where it is still very difficult to foresee from molecular data resulting crystal structures.

A second major aim of crystal chemistry is to establish structure-property relations which are so urgently needed in modern materials' research. This is again very difficult in the field of organic materials, whereas several successful attempts were already made in the inorganic world [5,6]. Here crystal chemistry has played a dominant role in crystal engineering, the systematic tayloring of crystal properties. The importance of crystal chemistry in materials' science was recently reviewed by Th. Hahn within the program of an International Summer School on Crystal Growth [7].

Thus materials' research is often based on a "closed loop circuit" which is shown in Figure 1.

Several turns in this circuit are sometimes necessary, before materials with optimized properties are found for device applications. Let us mention a few examples of important electronic materials where crystal engineering based on the cooperation of crystal growth, crystal chemistry, crystallography and solid state physics were successfully used in technology. They are:

1.   semiconductor materials of the $A^{III}B^{V}$-family,
2.   magnetic oxide-materials within the spinel and garnet families,
3.   ferroelectric oxide materials of the perovskite family, and
4.   high T_c superconducting oxides.

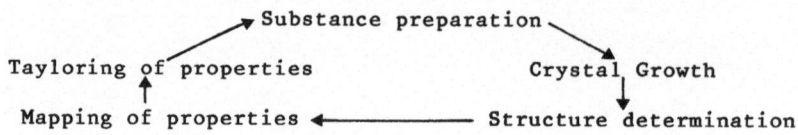

Fig. 1. Materials' research of crystalline materials.

Having demonstrated the important links of crystal growth and crystal chemistry in materials' research let us now consider some problems arising in the search of new materials. Crystal growth is necessarily a tool for obtaining samples for a first mapping of structural and related physical and chemical properties. The difficulty lies, however. often in the lack of relevant physical or chemical data facilitating the selection of suitable crystal growth techniques for new material families.

Bearing in mind that a first assessment of relevant properties should be done as fast as possible and in an economic way crystals which are only as large and as good as absolutely necessary have to be prepared. So-called exploratory techniques of crystal growth are very valuable which are based on simple readily available growth equipment and on the knowledge and use of self-regulating mechanisms inherent to different crystal growth techniques which may help to overcome the difficulties of work with new materials.

In the following part we shall first propose a classification of methods encountered in the crystal grower's arsenal and shall then outline some selected techniques giving also examples of their use.

The classification of crystal growth methods in Table 1 is based on ideas expressed by Laudise in Ref. [8] and uses the division into a single or a multi-component mother phase as proper growth medium and the presence of secondary phases added to the growth medium during the growth process.

Examples described there-upon in some detail are:

1. solution growth based on temperature difference procedures (II-1-b' of classification in Table 1),
2. zone fusion in a lamp heated elliptical mirror oven (I-2-a'),
3. gel growth procedures (II-1-b'), and
4. reactive sintering in closed ampoules (II-2-b').

The self-regulating mechanisms active in each case will be written in bold letters.

SOLUTION GROWTH BASED ON TEMPERATURE DIFFERENCE PROCEDURES

Before any solution growth technique can be applied, a solvent with a suitable solubility has to be found, its temperature dependence has to be established and we have, moreover, to be sure that the primary phase of crystallization really is the desired compound. Complications can arise either in the case of solvate formation, of incongruent solubility or, if solid solution crystals have to be grown and the distribution coefficient differs strongly from unity, in the solvent used. For solving these problems and for finding the temperature dependence of solubility an apparatus, the use of which was illustrated in Refs. [9,10], has been helpful. A direct visual observation of a stirred suspension in a heating cycle made it possible to determine liquidus points with an accuracy better than ± 1 K and the precipitate after cooling back can be used for

Table 1. Classification of Crystal Growth Methods for Preparing Bulk Crystals

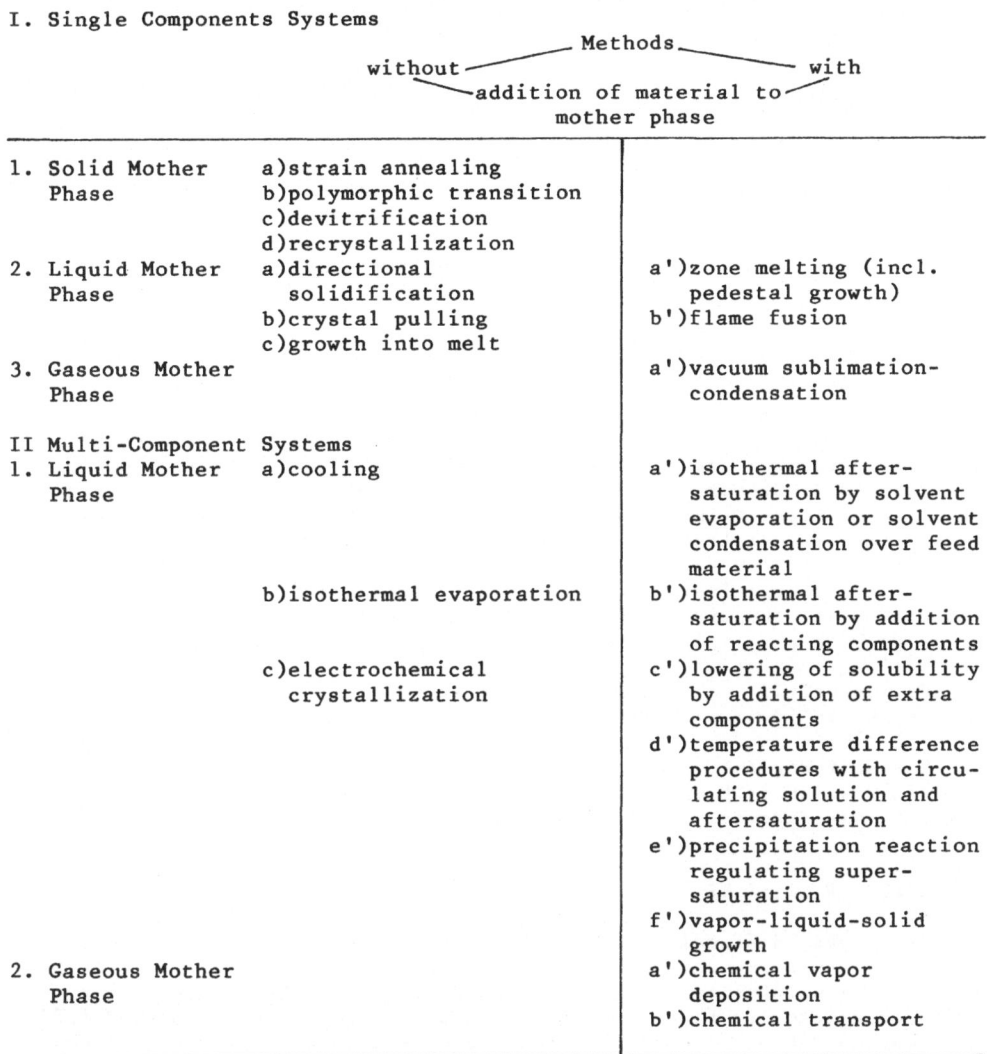

I. Single Components Systems

Methods — without / with — addition of material to mother phase

	without addition of material to mother phase	with addition of material to mother phase
1. Solid Mother Phase	a) strain annealing b) polymorphic transition c) devitrification d) recrystallization	
2. Liquid Mother Phase	a) directional solidification b) crystal pulling c) growth into melt	a') zone melting (incl. pedestal growth) b') flame fusion
3. Gaseous Mother Phase		a') vacuum sublimation-condensation

II Multi-Component Systems

	without addition of material to mother phase	with addition of material to mother phase
1. Liquid Mother Phase	a) cooling b) isothermal evaporation c) electrochemical crystallization	a') isothermal after-saturation by solvent evaporation or solvent condensation over feed material b') isothermal after-saturation by addition of reacting components c') lowering of solubility by addition of extra components d') temperature difference procedures with circulating solution and aftersaturation e') precipitation reaction regulating super-saturation f') vapor-liquid-solid growth
2. Gaseous Mother Phase		a') chemical vapor deposition b') chemical transport

identifying the primary crystallization product or for a rapid estimation of distribution coefficients in the case of solid solution crystals.

Temperature difference procedures, in which **feed material** is dissolved at a temperature T_2 and is **transported**, preferably **by thermal convection**, onto a seed crystal at a temperature T_1 and is resaturated there-upon by passing again over the seed material, maintain automatically (as long as there is a sufficient supply of feed material) the **supersaturation at a constant level**. In case solid solution crystals have to be grown the use of an **equilibrated solid solution feed material** as described in Ref. [10] **ensures** thus **automatically** also a **constant composition** of the **liquid growth solution** and leads to the growth of **homogeneous solid solution single crystals**. In Ref. [10] procedures based either on spontaneous nucleation (mostly sufficient for purposes of crystal chemistry) or seed growth (if crystals of cm^3 size are needed) are described.

A typical example of the application of such techniques is shown in [10a]. In the case of organic-inorganic double halides a new field of application is, for example, the search of new materials to be used in electro-optics [10b]. Looking at the many promising substances mentioned, e.g., in Ref. [11], and being aware that in the majority of the reported compounds structure determinations were not yet performed and accordingly many possibilities of crystal engineering were not exhausted, we should never forget that besides melt growth in capillaries the described techniques of temperature difference procedures can be really very useful as demonstrated, e.g., in Ref. [12] for the crystallization of a purely organic material where melt growth is often difficult.

ZONE FUSIOON AND ZONE LEVELLING CRYSTAL GROWTH IN ELLIPTICAL LAMP HEATED MIRROR CAVITY

Lamp heated elliptical mirror cavity pulling apparatus was described in the past, e.g., by Nitsche and his co-workers [13], for materials' research under microgravity. A very simple modification of such equipment was proposed by Bednorz et al. [14]. Crystals of 20-25 mm in length and 5-6 mm in diameter with melting temperatures up to 2100°C can be grown in air. A simple modification, i.e., placing the samples into quartz-bulbs, makes it possible to use controlled atmospheres in this oven up to $\sim$ 1700 K [15]. The self-regulating effect of zone levelling can be used to produce homogeneous crystals of solid solution systems as shown by Bednorz in his Ph.D thesis [16] for the systems:

$SrTiO_3$-$CaTiO_3$ $SrTiO_3$-$LaAlO_3$
$SrTiO_3$-$BaTiO_3$ $SrTiO_3$-Nb_2O_5

The **self-regulating mechanism** in such a growth procedure, **zone levelling,** which is **keeping** the **composition of solid solution crystals constant,** was discussed already by Pfann [17].

GEL GROWTH OF CRYSTALS

It is very difficult to grow single crystals of materials with a low solubility and a low thermal stability. In such a situation gel growth with several helpful self-regulating mechanisms will be very useful as mentioned already in the first review on this crystal growth method [18]. These mechanisms are due to the presence of a more or less rigid three-dimensional lattice of a polymerized or polycondensated gel-forming agent. The pore size is usually of the order of $\sim$ 100 Å and, if possible, the gel lattice should give way to the growing crystal and avoid thus its incorporation.

The self-regulating mechanisms encountered are:

a. **Material transport** is **controlled** essentially **by diffusion or** under special conditions by a **slow laminar percolation under hydrostatic overpressure.** Such a "gel micro-pump" can deliver over years amounts of solutions down to 0.1 cm^3/day [19].
b. Spontaneously nucleated crystals are fixed at their nucleation sites (gravity is blocked) and a **crystal protects itself by its own growth process against competitive nucleation in its vicinity.**

c. In initial stages of gel growth dendritic crystallization is often
    encountered. At a later stage the decreasing supersaturation leads
    to an **automatic change from dendritic to stable growth.**
d. In unknown, complicated systems where several phases can be formed,
    time and **site dependent nucleation and growth** and **lead to the
    consecutive growth of individual phases** [20].

Gel growth has been successfully used for a preliminary simulation of
space growth experiments [21], and it can be used for the scanning of
whole families of low solubility compounds, such as for example of
"monetites" belonging to the $PbHPO_4$ family [22].

REACTIVE SINTERING IN CLOSED AMPOULES

Reacting in a sealed ampoule the starting components A and B to form
a new compound AB - after having previously found a non-reactive, non-
volatile material for the ampoule - effects of evaporation or interactions
with the surrounding atmosphere can be avoided in the reaction taking
place within the ampoule. Thus, for example, well defined non-
stoichiometric oxide materials or new members of different oxide fluoride
families can be synthesized. If a **single phase composition** is reached in
the ampoule the chemical composition of the compound formed is **known from
a weight control** of the reacted ampoule and **bigger crystal individuals**
encountered **can be used directly for a structure determination** without the
necessity of a chemical analysis. The advantages of this technique were
used over the years by Magnéli and his co-workers [23] for the study of
non-stoichiometric binary and ternary oxides, e.g., of V, Ti, Nb, Ta, Mo,
W, where the composition was regulated by reacting mixtures containing
known amounts of metal and/or well defined metal oxides. Another example
are oxide-fluoride materials where volatility problems make it difficult
to maintain a stable composition of reaction products. Reactions
occurring within the ampoule - which is usually heated isothermally over
the whole volume - may be very ocmplicated [24]. The usefulness and
efficiency of the method was demonstrated, e.g., in the search of new
ferroelectric oxide fluoride materials, performed over the years in the
group of Hagenmüller [25].

CONCLUSION

The four crystal methods described are certainly only examples of
exploratory techniques to be used in materials' research. Others,
suitable for different materials and using also some other self-regulating
mechanisms should be at least mentioned here. They are, for example:

1. The "classical" hydrothermal technique as discussed recently, e.g.,
    by Rabenau [26].
2. The hydrothermal technique in acid solutions in quartz ampoules as
    described in the past by Rabenau in Ref. [27].
3. Fiber growth based on the pedestal technique [28].
4. Chemical transport reactions in sealed ampoules [29].

REFERENCES

1. A. F. Wells, "Structural Inorganic Chemistry", 5th ed., Clarendon
    Press, Oxford (1984).
2. J. Donohue, "The Structures of the Elements", Wiley, New York (1974).
3. W. B. Pearson, "The Crystal Chemistry and Physics of Metals and
    Alloys", Wiley, New York (1972).

4. A. I. Kitaigorodskij, "Organic Chemical Crystallography", Consultants Bureau, New York (1961).
5. O. Muller and R. Roy, "The Major Ternary Structural Families", Springer, Berlin-New York (1974).
6. R. E. Newnham, "Structure-Property Relations", Springer, Berlin-New York (1974).
7. Th. Hahn, in: "Crystal Growth of Electronic Materials", E. Kaldis, ed., pp 299-307, Elsevier Science Publ. BV. (1985).
8. R. A. Laudise, "The Growth of Single Crystals", Prentice Hall, New Jersey (1970).
9. H. Arend et al., J. Crystal Growth, 74:321 (1986).
10a. H. Arend, in: "Crystal Growth of Electronic Materials", E. Kaldis, ed., pp 307-311, Elsevier Science Publ. BV. (1985).
10b. W. S. Wang, J. Hulliger and H. Arend, to be published in Proc. 1st European Conf. on Appl. of Polar Dielectrics, 1st Symp. on Appl. of Ferroelectrics 1988; to be published as special issue of Ferroelectrics.
11. R. J. Twieg and K. Jain, in: "Nonlinear Optical Properties of Organic and Polymeric Materials", J. Williams, ed., pp 57-80, ACS Symposium Series 233, Washington DC (1983).
12. P. Günter et al., Appl. Phys. Letters, 50:486 (1987).
13. A. Eyer et al., J. Crystal Growth, 47:219 (1979).
14. J. G. Bednorz and H. Arend, J. Crystal Growth, 67:660 (1984).
15. E. Dieguez, unpublished results.
16. J. G. Bednorz, Ph.D. Thesis, ETH Zürich (1982).
17. W. G. Pfann, "Zone Melting", Wiley, New York (1958).
18. K. H. Henisch, "The Gel Growth of Crystals", Penn State University (1970).
19. H. Arend and W. Huber, J. Crystal Growth, 12:179 (1972).
20. H. Arend and J. Perrison, Mat. Res. Bull., 6:977 (1972).
21. F. Lefaucheux et al., J. Crystal Growth, 47:313 (1979).
22. R. Blinc et al., Phys. Stat. Sol., (b), 74:425 (1976).
23. See A. Magnéli, S. Andersson and L. Kihlborg et al., in Annual Reports of S. Arrhenius Laboratory, Univ. Stockholm (1960-1980).
24. H. Arend et al., Mat. Res. Bull., 5:753 (1970).
25. A. Simon and J. Ravez, Ferroelectrics, 24:305 (1980).
26. A. Rabenau, Angewandte Chemie, Internat. Ed., 24:1026 (1985).
27. A. Rabenau, in: "Crystal Growth: An Introduction", P. Hartman, ed., pp 198-209, North Holland, Amsterdam (1973).
28. R. Feigelson, J. Crystal Growth, 79:669 (1986).
29. H. Schäfer, "Chemische Transportreaktionen", Chemie-Weinheim (1962).

CRYSTAL GROWTH IN SOLID STATE PHYSICS

E. Kaldis

Laboratium für Festkörperphysik ETH
CH-8093 Zürich
Switzerland

1. SOLID STATE CHEMISTRY AND SOLID STATE PHYSICS

Although it is now generally acknowledged that crystal growth is very important in solid state physics, there are still many opinions how this interaction should take place. The present author believes that it should be based on fundamental studies of the solid state chemistry of the materials whose single crystals are ultimately wanted. Systematic thermodynamic and structural studies are necessary in order to decide on the best strategy for crystal growth. Another supporting argument is the sensitivity of many physical properties on the chemical and thermochemical conditions of the crystal growth process. For example, as we demonstrate further down, the control of the non-stoichiometry of the crystals is very important in order to optimize certain physical phenomena. Growing crystals by trial and error or concentrating only on nucleation and growth kinetics aspects, will give us crystals with unknown stoichiometries or defect concentrations and therefore with unreproducible physical properties.

Unfortunately, this widening of the crystal growth horizon by fundamental studies in solid state chemistry is exercised very seldom, and in most cases crystals for physical measurements are grown by trial and error experiments. The reasons for that are many and can be divided into two groups: (a) fundamental investigations are expensive and time consuming, needing appreciable manpower and equipment even if only a single group of compounds is investigated. Average physics institutes at universities do not have funds for large solid state chemistry laboratories. But even more important is that (b) traditionally educated physicists and chemists have difficulties to understand, or interact with each other in the strongly interdisciplinary field of solid state research. This may seem almost unbelievable in view of the fact that there are scarcely universities left without materials science departments. However, if one analyzes the degree of scientific interaction between physics and chemistry, only very few such departments are worth their names.

1.1 Crystal Growth in the Frame of Solid State Research

The branch of solid state chemistry suitable for such collaboration with solid state physics has its origin in the very important inorganic preparative chemistry of the first half of this century. However, now the goal is not only the synthesis of a new compound but also its crystallization and in particular its characterization of crystals and in a backfeeding process (from solid state physics to solid state chemistry) the optimization of their properties. The latter can be only achieved if the dependence of the physical properties on the most relevant conditions of chemical synthesis and crystallization is systematically studied. It is very important and often difficult and time consuming to find the conditions which can be used then as optimization parameters for the material. In a very general way, one can say that these conditions depend on temperature, pressure and composition. Whereas the importance of temperature has been recognized very early in crystal growth, the pressure conditions have been neglected. The historical reason for that is that the crystal growth of electronic materials was from the early 50's in the hands of metallurgists who traditionally were not involved with the pressure parameter. An additional reason is that the main electronic material up to now, silicon, does not have appreciable vapor pressure under the conditions of crystal growth from the melt. It was, therefore, only in the recent years that the oxygen-in-silicon problem introduced the partial pressure control parameter. In the early crystal growth work, the pressure parameter became important in chemical transport, dissociative sublimation etc. [1], particularly for the II-VI compounds. In the real world of applications and mass production of electronic materials, however, the problem of partial pressure control was introduced particularly via the $Hg_{1-x}Sn_xTe$ family of IR active materials and in the last years via GaAs and all other III-V's. As a matter of fact one of the main difficulties in the production of high quality GaAs crystals and layers is the partial pressure control.

The third parameter for the control of physical properties is composition, in the sense either of impurities (e.g., in silicon) or non-stoichiometry (ratio of metal/nonmetal) like, for example, in GaAs and the rest of the III-V. Partial pressure and non-stoichiometry are intimately related via the thermodynamics, which makes the quality control of electronic materials for practical purposes much more complicated. In view of the extreme importance of non-stoichiometry for the variation of physical properties, as discussed in the case studies of the following sections, we may very briefly consider the question how non-stoichiometry influences the physical properties? Although many books were written on this subject, in most cases the exact mechanism is not known. In general, a variation of the ratio metal/nonmetal leads to the formation of either statistically distributed point defect, or ordered extended defects. In the latter case the ordering introduces changes of symmetry and triggers structural phase transitions. In addition, carrier concentration changes result with non-stoichiometry and lead alone or with the symmetry changes to changes of the electronic band structure. Non-stoichiometry can be neither predicted nor treated quantitatively from the solid state theory, because the most important parameter, the interaction between the resulting defects, cannot be even estimated at present. Therefore, progress in this field depends substantially on the quality of the experimental investigations performed by solid state chemists. Only, after a dependence of a certain physical property on a certain range of non-stoichiometry has been discovered, the main work of solid state physics can start.

Our work was concentrating for many years in the field of the chalcogenides, pnictides, and hydrides of the rare earths. Discovering and/or using large non-stoichiometric ranges in these compounds triggered very interesting physical effects. Some case studies from this work are briefly presented here with adequate literature references for further studies.

2. CHEMICALLY TRIGGERED METAL-SEMICONDUCTOR TRANSITIONS IN MIXED-CRYSTAL SERIES

The problem of homogeneity is a very important one for single phase electronic materials and particularly for solid solutions. An interesting question is if the homogeneity of a solid solution depends on the physical properties of the end compositions. Into this category belongs also the very interesting problem of the mixed crystal series between a metal and a semiconductor. An appreciable number of experiments in our laboratory with various isostructural solid solution series of two binary compounds (one metallic and one semiconducting) with a common nonmetal element has clearly shown that large miscibility gaps exist. According to the theory of thermodynamic solutions this shows that appreciable repulsive interactions are existing. In most cases the repulsive forces can be attributed to strain due to different ionic radii. In some cases, however, purely electronic effects, due to the different bandstructures of the end members must be also considered. Thus one could possibly expect a continuous solid solution in the case of a semiconductor alloying with a metallic compound to form an impurity band. With increasing concentration this band would lead to degeneration and then disappearance of the band gap. Unfortunately, to our knowledge no such systems have been investigated with enough chemical expertise to be really sure that small miscibility gaps are not existing.

Of course in many cases, quenching of crystals and layers may suppress the phase dissociation at least in the range of sensitivity of X-ray structural investigations (2-3%). This is also the case in solid solutions of the III-V compounds. However, one can be rather sure that more sensitive methods (e.g., electron microscopy) would detect at least a partial phase decomposition (high concentration of defects).

2.1 Metal-Semiconductor Transitions in Solid Solutions of Rare Earth Hydrides LnH_2- LnH_3

The metal-semiconductor transition in the light rare earth (RE)-hydrides was discovered in 1957 [2]. It was found that by changing the stoichiometry between REH_2 (RE = La-Nd) and REH_3 a compositionally triggered metal-semiconductor transition takes place. As both compounds have the cubic (CaF_2) structure, it was generally accepted that this is an isostructural transition and therefore, the mixed crystal series REH_2-REH_3 belongs to a homogeneous solid solution.

In 1969 a peritectic phase diagram based on a small number of measurements was proposed and a crystal growth method by precipitation of the hydride from the metallic melt was introduced [3]. Using these single crystals it was clearly shown by electrical resistivity measurements that the above mentioned transition was existing [4]. Since that time many physical properties of the solid solutions CeH_2-CeH_3 have been measured, but the work of various laboratories was partly in disagreement with each other and a clear picture of the physical properties could not be obtained under the assumption of complete homogeneity. It was first in 1982 [5] that a tetragonal distortion was discovered only for one composition by X-ray and later confirmed by neutron diffraction.

Our interest in the RE-hydrides was due to the wide non=stoichiometry range which could allow the manipulation of the physical properties and also due to the compositionally driven metal-semiconductor transition. We wanted to investigate the chemical parameters influencing this transition, based on two experimental advantages: extreme purity of the samples and direct measurement of the non-stoichiometry by chemical methods.

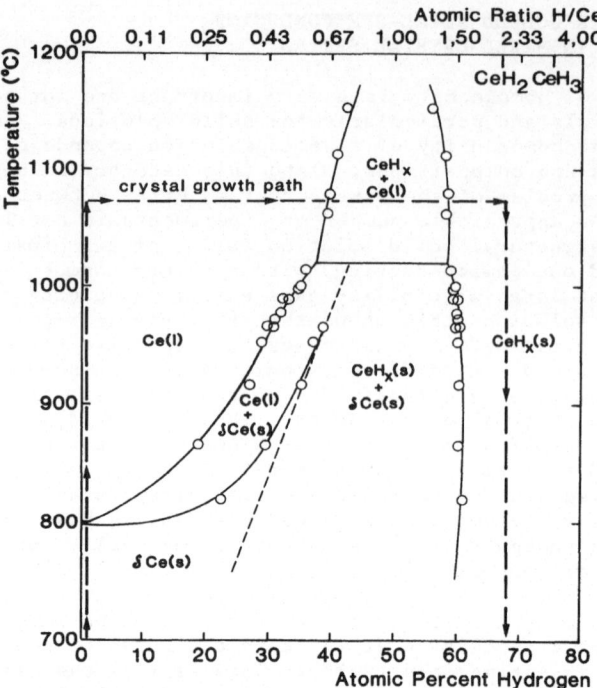

Fig. 1. The new Ce-H$_2$ phase diagram after [7]. The peritectic character allows the precipitation of the hydride from the hydrogen saturated Ce melt. The path of the crystal growth process, after [3] is also shown in the Figure.

To understand the phase relationships in these systems we have measured first the phase diagrams Ce-H$_2$ and La-H$_2$. Both were found to differ appreciably from those described in the literature [6]. However, in the Ce-H$_2$ system we could reconfirm the peritectic melting proposed earlier [3]. Figure 1 shows the new Ce-H$_2$ phase diagram and the path used for crystal growth. A first scouting of the solid solution range CeH$_2$-CeH$_3$ was performed by measuring at room temperature the lattice constants as a function of composition [7]. It was found that for pure samples the tetragonal distortion is existing not only for one composition but for the whole range CeH$_{2.15}$-CeH$_{2.60}$. (Revised hydrogen content due to a recent refinement of the H-determination [8].) After this finding the investigation of the lattice constants as a function of temperature and composition became imperative in order to find the regions of stability of the tetragonal distortion. Figure 2 shows the results achieved up to now [9]. Although there is still some doubt about the dashed phase boundaries due to missing measurements, a cubic range is seen by X-rays between CeH$_{2.60}$-CeH$_{2.75}$ dividing the stability range of the tetragonal phase into two fields. Three aspects of Figure 2 should be stressed in this very brief discussion:

(a) At room temperature the metal-semiconductor phase transition at approximately CeH$_{2.57}$ seems to overlap with the phase boundary of the tetragonal-cubic transition, indicating that the lattice distortion is important. This is the first direct proof of some previous theoretical work [10].

234

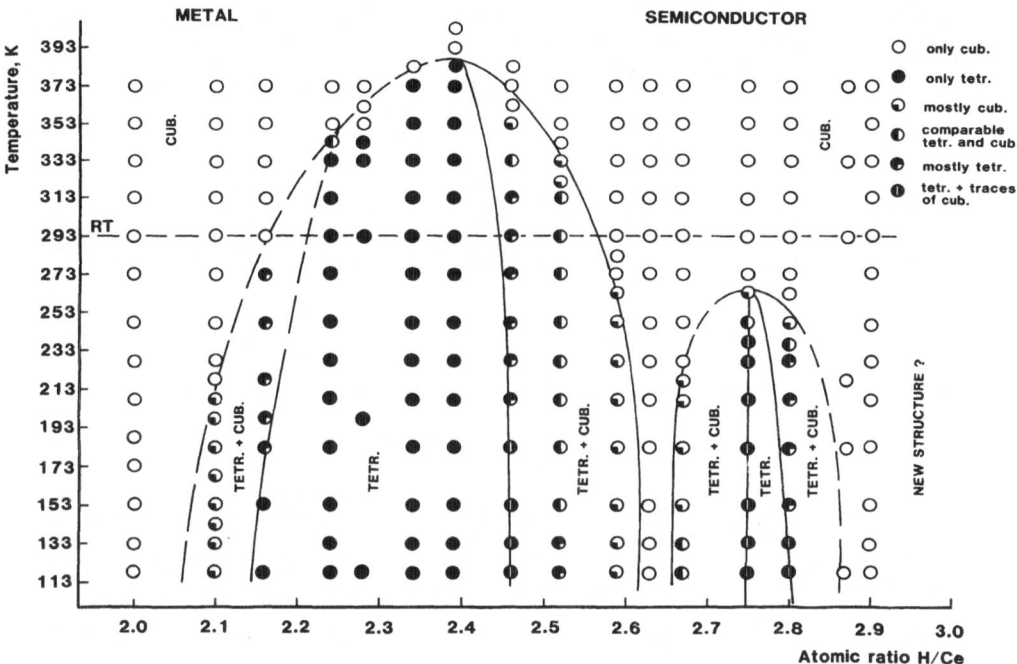

Fig. 2. The phase relationships in the solid solution range CeH$_2$ (metal)-
CeH$_3$ (semiconductor) after [9] according to X-ray investigations.
Clearly the metal-semiconductor transition leads to an
inhomogeneous phase field.

(b) In the tetragonal fields neutron diffraction [12] shows the
existence of superstructures different from those reported in the past
[11].

(c) Last but not least it was found that new reflections appear both
in X-ray and neutron diffraction in the range CeH$_{2.9}$-CeH$_{3.0}$ and LaH$_{2.9}$-
-LaH$_{3.0}$ [12]. The reproducibility of various carefully prepared samples
shows that this is not the result of contamination but of a new structure
appearing together with an appreciable amount of an X-ray amorphous phase
(due to high concentration of defects). In view of the fact that for
heavier RE the trihydride has another structure (hexagonal) the above
result seems very important and can be understood by the increasing
repulsive interactions as the last octahedral interstitial sites are
filled.

In conclusion it can be said that solid state chemistry gives us for
the first time a clearer picture of the phenomena involved in this metal-
semiconductor transition.

3. CONTROL OF VALENCE FLUCTUATIONS BY VARYING
    THE NON-STOICHIOMETRY OF SINGLE CRYSTALS

From the point of view of applications in electronics a material
which could be switched from metal to semiconductor at will seems very

desirable. The materials with valence fluctuations have brought us nearer to this goal, because roughly speaking they can combine under certain conditions properties characteristic of metals and semiconductors. This is the result of proximity effects between the localized 4f-level and the delocalized states in the 5d-band. Several reviews have been written about this group of materials, so that a comprehensive introduction concerning the materials aspects can be found in [13] and [14] and concerning the physical phenomena in [15].

The mechanism to influence the physical properties in the direction of semiconductor or metal is the variation of the crystal-field splitting. This is achieved either by external (mechanical) pressure or by internal chemical pressure (i.e., chemical substitution with a large ion) [13-15]. In this way the overlap of the t_{2g} orbitals increases, the crystal field splitting increases and the 5d-band expands, decreasing the gap between 5d and 4f-levels. Finally, this can lead to a slight overlap and a hybridization of the delocalized, light (small effective mass), 5d-electrons with the atom-like, heavy (large effective mass), strongly localized 4f-electrons. f-d hybridization in the same atom is not possible because, for example, the $4f^6$-configuration in SmS has a total angular momentum $J = 0$. Thus, hybridization is achieved by a parallel promotion of a $4f^6$-electron to the 5d-band in some atoms and transfer of one of these 5d-electrons to the 4f-state of another neighboring atom (valence fluctuation) [16]. As a consequence of it a non-integral occupation of the 4f-states results and, therefore, a non-intregal mixed valence. The degree of mixed valence depends on the degree of this hybridization and, therefore, on the energy difference between $4f^n$ and $4f^{n-1}5d^1$. The frequency of the valence fluctuation is very high, in the 10^9-10^{15} Hz range and it can be experimentally detected only with very few, fast methods (e.g., Mössbauer spectroscopy). Our goal has been to vary the internal, chemical pressure in TmSe, by a controlled change of non-stoichiometry or in the case of mixed crystals by substituting Se by Te or Tm by Eu.

From the point of view of crystal growth the possibility to grow single crystals with various controlled non-stoichiometries, at temperatures up to 2200°C and equilibrium between the solid crystal and the liquid and vapor nutrient phases, was of utmost importance. This was only possible by using the electron beam sealed tungsten crucible technique [1] and using charges with various metal/nonmetal ratios. In this way it was possible to measure the physical properties on single crystals grown from the melt - as a function of non-stoichiometry.

Figure 3 shows a beautiful example of what can be achieved if the non-stoichiometry of the Tm_xSe crystals is varied in a reproducible way. At the Se-rich side ($Tm_{0.87}Se$) the electrical resistivity has metallic character. But already for the stoichiometry $Tm_{1.0}Se$ an increase of resistivity takes place with decreasing temperature, which for the Tm-rich phase boundary ($Tm_{1.05}Se$) becomes several orders of magnitude higher due to onset of valence fluctuations. Most interesting: due to the magnetic moment of the Tm-ion, by switching a magnetic field the maximum resistivity collapses and the material becomes metallic [17,18]. In principle, therefore, we have here a crystal which can be switched between a semiconducting and a metallic state by pulsing a magnetic field of a few KOe.

The controlled variation of the non-stoichiometry of the crystals of Tm_xSe, was the result of a systematic thermodynamic analysis: the phase diagram was measured up to 2200°C (see, for example, [13,14] and the references therein); the thermodynamic stability (enthalpy of formation)

236

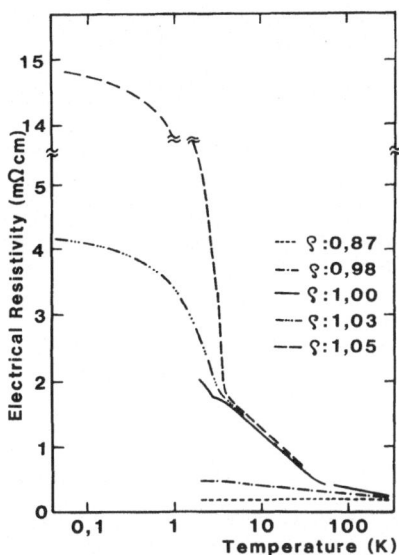

Fig. 3. Resistivity vs. temperature as a function of non-stoichiometry
(: metal/nonmetal) for Tm_xSe single crystals. After [17,18].

was measured as a function of non-stoichiometry using calorimetric methods
(e.g., for fluor-combustion calorimetry see [19]).

Although the discussion of this thermodynamic characterization is too
lengthy to be included here, the most important conclusion can be briefly
mentioned: in the middle of the homogeneity range of the Tm_xSe phase
(NaCl-structure) a miscibility gap (0.94 < x < 0.98) appears which is
separating the normal valence range (Se-rich) from the valence fluctuating
range (Tm-rich). No superstructure or other structural anomalies could be
found by electron diffraction. We can, therefore, conclude that the
repulsive interactions leading to the miscibility gap are due to the
different band structures of the two homogeneity subranges.

4. CONCLUSION

Crystal growth plays an important role in solid state physics.
However, trial and error growth methods - although some times very
important to open a new field - are not enough. For the selection of the
strategy of crystal growth a full thermodynamic analysis (phase diagram,
non-stoichiometry, thermodynamic stability measurements etc.) should be
performed followed by the thorough characterization of the crystals and
the optimization of their properties.

It would not be serious to expect that this work could be done in
less than 3-4 years by a group of 3 experienced people.

REFERENCES

1. E. Kaldis, Principles of the vapor growth of single crystals, in:
   "Crystal Growth, Theory and Techniques", C. H. L. Goodman, ed.,
   pp 54-191, Plenum Press (1974).
2. B. Stalinski, Bull. Acad. Pols. Sci., 3,5:997 (1957).

3. G. G. Libowitz and J. G. Pack, J. Chem. Phys., 50:3557 (1969).
4. G. G. Libowitz, J. G. Pack and W. P. Binnie, Phys. Rev., B6:4540
    (1972).
5. P. Knappe, H. Muller and H. W. Mayer, J. Less-Common Met., 95:323
    (1982).
6. R. Bischof, M. Tellefsen and E. Kaldis, J. Less-Common Met., 110:99
    (1985).
7. M. Tellesfsen, E. Kaldis and E. Jilek, J. Less-Common Met., 110:107
    (1985).
8. K. Conder and E. Kaldis, J. Less-Common Met., 146:205 (1989).
9. E. Boroch and E. Kaldis, Inorganica Chimica Acta, 140:89 (1987);
    Zeitschrift für Physik. Chemie, (1989), in press.
10. Khulikov, J. Less-Common Met., 107:111 (1985).
11. See, for example, V. K. Feodotov, V. G. Feodotov, M. E. Kost and E.
    G. Ponyatovskii, Sov. Phys. Solid State, 24:1253 (1982); and
    references therein.
12. K. Conder, J. Scheffer and E. Kaldis, Zeitschrift für Physik. Chemie,
    (1989), in press.
13. E. Kaldis and B. Fritzler, Progr. in Solid State Chemistry, 14:95
    (1982).
14. E. Kaldis, B. Fritzler and H. Spychiger, in: "Solid State Chemistry",
    R. Metselaar et al., eds., p. 89, Elsevier (1983).
15. J. M. Lawrence, P. S. Riseborough and R. D. Parks, Rep. Progr. Phys.,
    44:1 (1981).
16. C. M. Varma, Rev. Modern Physics, 48:219 (1976).
17. B. Batlogg, H. R. Ott, E. Kaldis, W. Thoni and P. Wachter, Phys.
    Rev., B19:247 (1979).
18. P. Haen, F. Holtzberg, F. Lapierre, J. Mignot and R. Tournier, Phys.
    Rev. Letters, 43:304 (1979).
19. H. Spychiger, E. Kaldis and B. Fritzler, J. Less-Common Metals,
    110:61 (1985).

CRYSTALLIZATION OF PROTEINS AND OTHER BIOLOGICAL MOLECULES

Charles E. Bugg

Center for Macromolecular Crystallography
University Station
University of Alabama
Birmingham, Alabama 35294, USA

Major advances in the technology involved in determining protein crystal structures have facilitated several new and existing applications for protein crystallography. Protein crystal growth, the one major bottleneck in this field, has recently received much attention and several new developments hold promise for the future.

Since completion of the first crystallographic study of a protein structure (myoglobin) in 1960 [1], crystallography has become a valuable tool for determining the three-dimensional structures of complicated biomolecules. Crystallographic studies of proteins and nucleic acids have played key roles in establishing the structural foundations of molecular biology and biochemistry, and for revealing structure/function relationships that are of major importance in understanding how macromolecules operate in biological systems. Recently, crystallographic studies of proteins have become of interest to the pharmaceutical, biotechnology, and chemical industries, as promising tools in protein engineering, drug design and other applications to biological systems.

The information that can be gleaned from the three-dimensional structure of a protein is utilized in several ways. Structural information has been of paramount importance in understanding the fundamental mechanisms by which enzymes, immunoglobulins, hormones, etc. function in biological systems. The detailed three-dimensional structural information can be useful for designing novel drugs which selectively interact with proteins to augment or inhibit their action [2-6]. Several major pharmaceutical companies have now established protein crystallography groups to pursue this approach of rational drug design.

An equally exciting application of protein crystallography has emerged in the field of protein engineering [7-16]. Present day techniques of molecular biology enable investigators to specifically alter protein molecules by site-directed mutagenesis. Detailed crystallographic results for the protein of interest are often invaluable in suggesting amino acid changes that may alter biochemical and physical properties in specific ways. Although potential applications of protein engineering are under intense development at this stage, it is generally accepted that these methods will prove to be of tremendous practical use for the design of modified enzymes, and for the development of proteins that have

239

carefully engineered physical and biological properties. The interest in and potential applications of site-directed mutagenesis is continually expanding, and several protein crystallography groups in academic institutions and in biotechnology-oriented companies are broadening their programs in order to exploit the potential of this technology.

Another use of protein crystallography is in the design of synthetic vaccines [17-20]. Several recent studies have indicated that effective vaccines might be made from synthetic peptides that are representative of protein segments found on the surfaces of target proteins. Protein crystallography provides one of the most effective techniques for locating these peptides.

RECENT ADVANCES IN PROTEIN CRYSTALLOGRAPHY

"Protein crystallography" is a generic term that is widely used to describe crystallographic studies of biological macromolecules, such as proteins and nucleic acids. Excellent reviews of protein crystallography are available [21,22], and the general field will not be covered in this paper. However, it is worthwhile to briefly summarize the major steps that are involved in protein crystallography, along with some of the recent advances in the field, in order to place the importance of protein crystal growth in proper perspective.

To determine the structure of a protein by crystallography, it is generally necessary to grow discrete single crystals for detailed X-ray diffraction studies. These crystals usually must be about 0.3-1.0 mm on a side; considerably larger crystals are required for neutron diffraction studies. High-resolution data sets require accurate measurements for the intensities of thousands of diffraction profiles from crystals of the protein, typically referred to as "native" protein crystals. Generally, heavy-atom derivatives (e.g., metal complexes or complexes that contain multi-electron atoms such as iodine) must be prepared, either by diffusing these complexes into the native protein crystals, or by crystallizing the macromolecule in the presence of the heavy-atom complexes. The techniques of multiple-isomorphous replacement [21], which have been widely used for solving protein structures, require that a series of different heavy-atom complexes must bind to the protein without disrupting the native crystal-packing scheme. Once potential derivatives are identified and diffraction data sets recollected for each of the heavy-atom complexes, an electron density map is calculated. Finally, a protein model that matches the map is constructed and refined, by adjusting atomic parameters so that the calculated diffraction patterns agree with the measured data.

In practice, all of the major steps that are involved in determining a protein structure by crystallographic techniques are subject to a number of experimental difficulties. Most of the proteins that have been studied during the past three decades required many years of intense effort before the complete three-dimensional structure was known. Consequently, until recently, there was limited interest in using protein crystallography as a general tool in biological research, and protein crystallography programs were limited to a few laboratories, primarily in academic institutions.

Several recent advances in the technologies required for protein crystallographic studies have made it much easier to determine the crystal structure of a protein or other macromolecule. Recombinant DNA techniques have made it possible to obtain proteins that would have been impossible to isolate in sufficient quantities for X-ray diffraction studies several years ago [7-14]. Synchrotron radiation sources, which provide X-ray beams with intensities that are several orders of magnitude greater than those available from laboratory sources, permit data sets to be collected rapidly, and from relatively few crystals [23,28]. Synchrotrons also facilitate unique wavelength-dependent experiments. The availability of electronic area detector systems permits data to be measured much faster than was possible with the single-counter detector systems widely used in the past [29,30]. Recent applications of anomalous dispersion measurements have enabled protein structures to be determined without the addition of heavy-atom derivatives [31-33]. Intense efforts are in progress in several laboratories to develop direct methods for determining protein phases, which are analogous to the powerful statistical techniques that have essentially eliminated the need for heavy-atoms in "small molecule" crystallography [34-40]. Computer graphics methods have revolutionized protein model construction from electron density maps [41,42]. Computer graphics techniques have also made it possible to understand and to interpret structures with ease [42,43], and the widespread use of computer graphics in the pharmaceutical, chemical and biotechnology industries has served to further stimulate interest in applications of protein crystallography. A variety of approaches and software systems have been developed for refining protein structures, once an initial model is constructed from the electron density maps [44-47]. Refinements are now being pursued routinely to high resolutions, and the increasing availability of super computers will make refinements of large proteins relatively routine [48].

Because of the widespread fundamental and practical importance of knowing the structures of biological materials, the overall interest in protein crystallography has increased rapidly during the past few years. Solution techniques, such as two-dimensional NMR spectroscopy, are likely to become more and more useful during the next few years, but it is unlikely that any of these techniques will be competitive with crystallography in the near future for routinely determining three-dimensional structures of large proteins and other complicated biological macromolecules. Unfortunately, protein crystallography has the unique requirement that relatively large, high-quality single crystals must be obtained before a structural study can be pursued. Therefore, protein crystal growth has become a topic of considerable importance.

It is noteworthy that the major advances in protein crystallography involve those experimental steps that are of importance after suitable crystals have been obtained. For the most part, the general procedures that are used for growing protein crystals have not changed appreciably during the past few years. Most protein crystals are still grown by brute-force, trial-and-error methods, which require investigations of large numbers of experimental conditions in hopes of identifying those that will produce usable crystals. However, once good crystals of a particular protein have been obtained, the crystallography often moves along at a rapid rate.

CURRENT TECHNIQUES FOR PROTEIN CRYSTAL GROWTH

Various techniques have been utilized to grow protein crystals [49]. In all of these methods, crystallization is dependent upon control of solubility, and by the kinetics of nucleation and subsequent growth. Solubility is influenced by inherent properties of the macromolecule itself as well as by its environment (i.e., solvent, temperature, pH, etc.). A basic principle followed is to slowly approach a condition of supersaturation while trying to maximize favorable intermolecular interactions [49]. It is generally believed that the protein must be homogeneous and free of contaminants to optimize the change of producing large high-quality crystals [49,50]. Advances in purification techniques (i.e., HPLC, affinity chromatography, isoelectric focusing, etc.) have been of importance in providing homogeneous samples for crystal growth studies. However, for many proteins of interest, it remains difficult to produce large quantities of purified material.

A major achievement in protein crystal growth technology has been the development of novel micromethods that permit crystal growth studies using microliter quantities of sample. The most widely used microtechniques for crystal growth are dialysis, liquid-liquid diffusion and vapor-diffusion. In the dialysis technique, the protein solution is held within a small chamber by a semipermeable membrane. The protein solution is then slowly brought toward a state of supersaturation by dialysis against an external solution containing a precipitating agent [49,51]. One clear advantage of this technique is that crystal growth conditions can be easily adjusted by simply controlling the composition of the external solution (i.e. precipitant concentration, pH, buffer type, etc.). Microliter quantities of sample in specially constructed dialysis cells [49,51], can be used for screening conditions and unsuccessful experiments (precipitated samples) can often be reused by modification of the external solution. Another inherent advantage of the dialysis method is that as the difference in concentrations inside and outside the dialysis cell decreases, so too does the rate of equilibration, thereby permitting a gradual asymptotic approach to equilibrium.

Crystallization by free interface liquid-liquid diffusion is similar to dialysis except that the protein solution is not separated from the precipitant by a semipermeable membrane. The two solutions are layered atop each other (typically the more dense solution is placed at the bottom) and allowed to diffuse until equilibrium is reached. Consequently, the high protein and precipitant concentrations initially present at the interface create conditions which are suitable for crystal nucleation. As diffusion continues, this condition changes to one of lower protein and precipitant concentrations which favors optimal crystal growth from pre-existing nuclei [52,53].

The technique that is now most widely used for protein crystal growth involves equilibrium by vapor diffusion [49]. Droplets of protein solution (as small as 0.5 microliter) containing concentrations of precipitant below that needed to render the protein insoluble, are placed over reservoirs containing higher precipitant concentrations. The droplet then slowly equilibrates with the reservoir through the vapor phase. The process is typically used for screening large numbers of conditions by placing the protein droplets on siliconized coverslips, which are subsequently inverted and placed over small (1 ml) reservoirs in multi-chamber plates. Several hundred different conditions including multiple combinations of pH, buffer type, precipitant type, precipitant concentration, temperature, etc. can be examined quickly by this approach using only a few milligrams of protein.

MAJOR PROBLEMS IN PROTEIN CRYSTAL GROWTH

Despite the development of standard micromethods for crystallizing biological materials, protein crystal growth has continued to be much more of an art than a science. Generally, suitable crystals are obtained only by screening procedures that involve examination of hundreds or thousands of experimental conditions. These trial-and-error techniques have permitted a number of proteins to be crystallized during the past few years.

However, there is a growing list of important proteins that have not yet been crystallized despite intensive efforts. Some proteins merely form amorphous precipitates that display no evidence of crystallinity. In other cases, small microcrystals may be obtained readily but these crystals cannot be induced to grow large enough for structural studies. There are also numerous examples of proteins that produce relatively large crystals which display such severe internal disorder that high-resolution structural analyses are not possible.

RECENT ADVANCES IN PROTEIN CRYSTAL GROWTH

Although protein crystal growth continues to be a major bottleneck in widespread applications of protein crystallography, several promising new approaches are emerging. One of the most exciting developments is the progress that has been made in crystallization of membrane proteins. Three-dimensional crystals of integral membrane proteins of sufficient quality to produce X-ray diffraction patterns were first reported for the protein bacteriorhodopsin, isolated from the purple membrane of halobacteria [54] and for the protein porin, an outer membrane protein from Escherichia coli [55]. A remarkable recent achievement was the crystallization and structural analysis of the photoreaction center from "Rhodopseudomonas viridis" [56,57]. Crystallization of these membrane proteins was achieved using short chain surfactants in the crystallization solution. In some cases, the quality of the crystals (as evidenced by both appearance and resolution limits of the X-ray diffraction patterns) could be improved using small amphiphilic molecules such as heptane 1,2,3 triol [58-60]. To date, several membrane proteins have been crystallized with some yielding crystals of sufficient quality for high-resolution crystallographic analysis [56,58,60-65]. Experience with new synthetic detergents and amphiphilic molecules will most probably increase the chances of obtaining useful crystals of many other membrane proteins. It has also been found that detergents such as β-octylglucoside are also useful for crystallization experiments with hydrophilic proteins [66], possibly by interfering with nonspecific interactions that compete with those involved in crystal growth.

Especially encouraging progress is now being made in experimental and theoretical studies of the principles of protein crystal growth. Investigations regarding the mechanisms of protein crystal nucleation, growth and growth cessation have recently been reported [67-76], and considerable new activity in this area is underway. These studies have indicated that optimum protein crystal growth requires much more careful monitoring and control of major parameters than is possible with current techniques. Several prototype systems that permit dynamic control of equilibration rates during protein crystal growth processes have been developed recently [77,78], and efforts to automate these systems are in progress. Substantial progress has also been realized in applications of robotics to protein crystal growth experiments [79-85]. Dynamic light-scattering methods have been used to monitor nucleation events during protein crystal growth, thus permitting nucleation and growth to be separated and controlled as different processes [86,87]. The considerable body of theoretical and experimental data relating to the growth of organic and inorganic crystals is now being applied to protein crystal growth. It is likely that the multi-disciplinary fundamental protein crystal growth studies now in progress will lead to a much better understanding of the basic processes involved, and will result in much more powerful experimental systems for dynamically monitoring and controlling protein crystal growth.

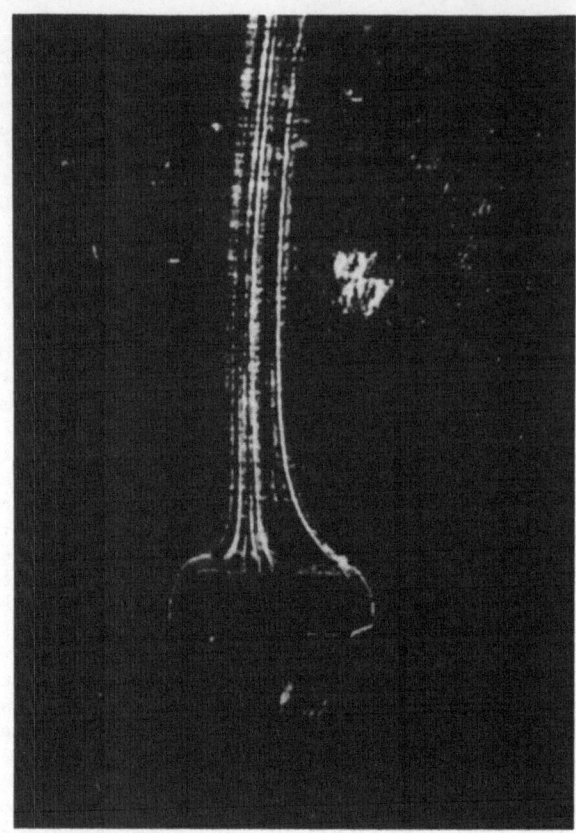

Fig. 1. A Schlieren photograph of a triglycine sulfate crystal growing
under unit gravity at isothermal conditions. The crystal has
approximate dimensions of 1.5 x 0.4 cm and is attached to a glass
rod, which is vertical in the photograph. The lighter regions
show those portions of the solution that have lower densities.
This photograph depicts the type of density-dependent convective
flow patterns that occur when a crystal grows from solution under
gravitational influence. The photograph was obtained and kindly
furnished by Drs. Roger Kroes and William Witherow of the
Marshall Space Flight Center in Huntsville, Alabama.

SPACE EXPERIMENTS IN PROTEIN CRYSTAL GROWTH

    One interesting new development in protein crystal growth involves
studies of crystal growth in the microgravity environment obtainable in
space. Crystal growth has been of considerable interest to the National
Aeronautics and Space Administration (NASA) and to other space-oriented
researchers for a number of years [88], and several fundamental studies of
crystal growth in space are in progress. The major motivation behind
these space experiments is to examine the effects that density-driven
convective flow has on crystal growth. The nature of convective flow can
be seen in Figure 1, which is a photograph taken by Schlieren techniques
(Schlieren photography highlights variations in refractive index) of a
triglycine sulfate crystal growing from solution under isothermal
conditions on earth. As material is incorporated on to the surface of the
growing crystal, density variations occur in the surrounding solution.
Under gravitational influence, the low density solution around the crystal
rises, thereby producing convective flow patterns. The turbulence created

244

by these flow patterns may influence crystal growth by creating non-uniform growth conditions across the crystal surface; by forcing solution to flow by the crystal at such a rate that steady-state diffusion ceases to be a rate-limiting step in growth; and by generating convective stirring effects within the solution. Under microgravity conditions, these convective flow patterns are eliminated. The absence of gravity therefore provides a direct way to examine the role of convection on crystal growth. In addition, microgravity conditions may serve to minimize sedimentation, which can interfere with uniform growth of protein crystals. Since protein crystals are extremely fragile, being stabilized by relatively weak crystalline interactions, one might expect that protein crystal formation would be especially affected by fluctuations in the growth environment, including those caused by gravitational effects.

Several laboratories around the world are involved in efforts to investigate gravitational effects on protein crystal growth. The first reported space experiments are those of Littke and John [89], which describe the growth of lysozyme and β-galactosidase crystals on Spacelab I. These preliminary studies indicated that the space-grown protein crystals are larger than crystals of these proteins obtained under the same experimental conditions on earth.

One of the long-range efforts to evaluate gravitational effects on protein crystal growth is being coordinated through NASA's Marshall Space Flight Center in Huntsville, Alabama. This interdisciplinary effort now involves a large group of co-investigators from a number of academic, industrial and government laboratories. Along with the space experiments, various fundamental ground-based studies of protein crystal growth are being pursued. The overall objectives of this effort are to better understand the fundamentals of protein crystal growth; to develop new improved hardware and techniques for growing protein crystals; and to evaluate the effects that gravitational fields exert on solution growth of crystals of proteins, nucleic acids and other biological materials. The space experiments, which are still in the initial stage, have involved prototype hardware development, and crystal growth on four different shuttle flights in 1985 and 1986. Figure 2 shows the latest hardware version which was flown in January 1986 on shuttle flight STS-61C. The apparatus consists of twenty-four vapor-diffusion chambers, with syringe mechanisms for activating and deactivating the experiments. Once in orbit, the experiments are activated by extruding protein solutions from the syringes forming droplets on the syringe tips. These droplets are then permitted to equilibrate with reservoir solutions in the chambers.

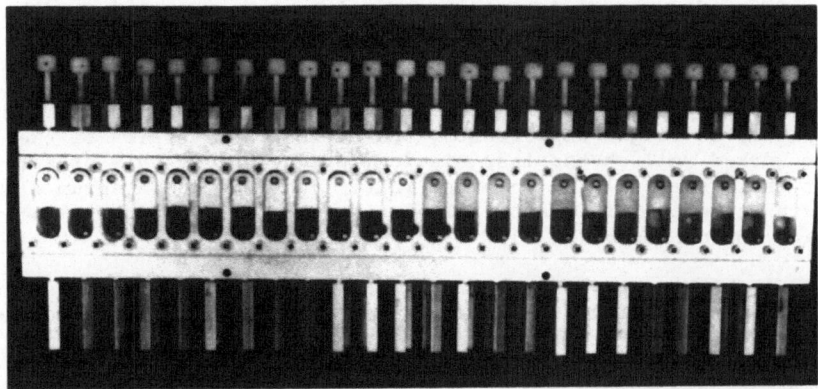

Fig. 2. Vapor diffusion crystal growth hardware. One complete unit consists of twenty-four syringes and opposing plungers.

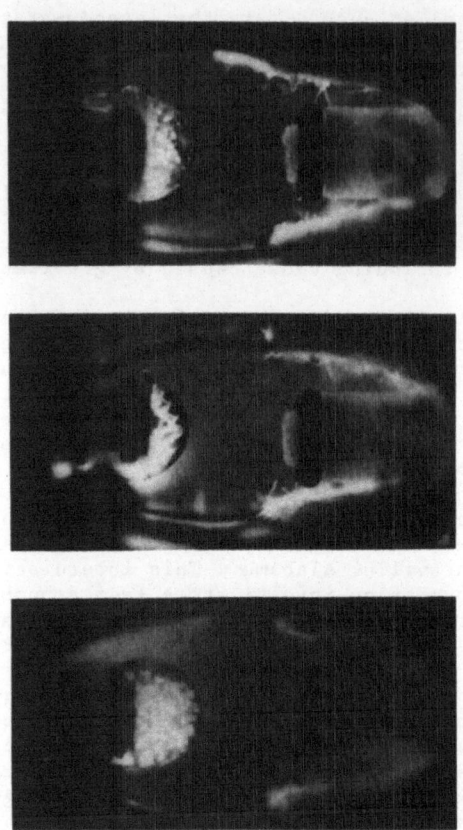

Fig. 3. Photograph obtained in orbit of crystals growing in 40 µl droplet on STS-61C.

Figure 3 is a photograph taken in space on STS-61C of protein crystals growing in 40 µl droplets. Figure 4 shows payload specialist, Congressman Bill Nelson, performing protein crystal growth experiments in space on STS-61C. The NASA protein crystal growth project is in the initial stages, and the space experiments have been primarily used to develop reliable vapor-diffusion techniques that may be used in the future for systematically evaluating gravitational effects on protein crystal growth. The prototype system shown in Figure 2 evolved over the course of the four shuttle flight experiments, and the current version of the hardware worked well on the last mission, producing crystals of all proteins that were included (hen egg white lysozyme, human serum albumin, human C-reactive protein, bacterial purine nucleoside phosphorylase, canavalin, and concanavalin B).

During this hardware development stage, it was not possible to control major parameters such as temperature, and quantitative conclusions about protein crystal growth cannot be drawn from these experiments. However, it is clear that microgravity conditions can have remarkable effects on crystal morphology in some cases, as evidenced particularly by the crystal growth studies with canavalin. Canavalin crystals grown on earth quickly sediment to the bottom of the crystallizing solutions and

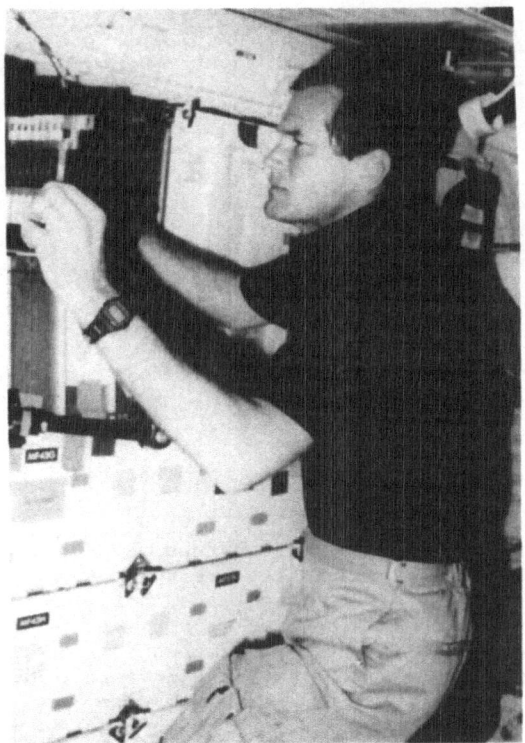

Fig. 4.   Payload specialist, Congressman Bill Nelson, performing protein
          crystal growth experiments in space on STS-61C.

grow as fused aggregates of poorly formed crystals (Figure 5).   Under
microgravity conditions, canavalin crystals grow dispersed through the
droplets (Figure 3) resulting in uniform morphologies for nearly all of
the canavalin crystals (Figure 6).   Interesting morphology changes were
also noted for crystals of human C-reactive protein and for human serum
albumin.

Fig. 5.   Crystals of canavalin grown in flight hardware under controlled
          conditions on earth.

Fig. 6.   Crystals of canavalin grown on shuttle flight STS-61C.

Although the space experiments to date have been useful for evaluating possible techniques and hardware that can be used for protein crystal growth under microgravity conditions, detailed studies of microgravity effects on protein crystal growth must await future shuttle flights. The prototype hardware that has been developed will form the basis for the design of automated systems that permit dynamic monitoring and control of key variables including temperature, protein concentration, ionic strength, pH, and precipitating agent concentration. This hardware should prove useful for both ground-based and space experiments that are directed at better understanding protein crystal growth process.

ACKNOWLEDGEMENTS

This work has been supported by NASA grant #NAGW-813 and NASA contract #NAS8-36611.

REFERENCES

1. J. C. Kendrew, R. E. Dickerson, B. E. Strandberg, R. G. Hart, D. R. Davies, D. C. Phillips and V. C. Shore, Nature, 185:442 (1960).
2. P. J. Goodford, J. Med. Chem., 27:555 (1984).
3. T. Blundell, B. L. Sibanda and L. Pearl, Nature, 304:273 (1983).
4. R. A. Johanson and J. Henkin, J. Biol. Chem., 260:1465 (1985).
5. T. Wilson and A. Klausner, Biotechnology, 511 (1984).
6. W. Goddard, Science, 227:917 (1985).
7. K. P. Nambiar, J. Stackhouse, D. M. Stauffer, W. D. Kennedy, J. K. Eldredge and S. A. Benner, Science, 223:1299 (1984).
8. E. T. Kaiser and D. S. Lawrence, Science, 226:505 (1984).
9. L. J. Perry and R. Wetzel, Science, 226:555 (1984).
10. A. R. Fersht, J.-P. Shi, J. Knill-Jones, D. M. Lowe, A. J. Wilkinson, D. M. Blow, P. Brick, P. Carter, M. M. Y. Waye and G. Winter, Nature, 314:235 (1985).
11. C. S. Craik, C. Largman, T. Fletcher, S. Roczniak, P. J. Barr, R. Fletterick and W. J. Rutter, Science, 228:291 (1985).
12. J. Galloway, Nature, 314:228 (1985).
13. R. P. Wharton and M. Ptashne, Nature, 316:601 (1985).
14. D. A. Estell, T. P. Graycar and J. A. Wells, J. Biol. Chem., 260:6518 (1985).
15. T. H. Maugh, Science, 223:269 (1984).

16. T. N. C. Wells and A. R. Fersht, Nature, 316:656 (1985).
17. E. Westhof, D. Altschuh, D. Moras, A. C. Bloomer, A. Mondragon, A. Klug and M. H. V. Van Regenmortal, Nature, 311:123 (1984).
18. R. J. P. Williams and G. R. Moore, Trends Biochem. Sci., 10:96 (1985).
19. J. L. Marx, Science, 226:819 (1984).
20. J. A. Berzofsky, Science, 229:932 (1985).
21. T. L. Blundell and L. M. Johnson, "Protein Crystallography", Academic Press, New York (1976).
22. A. A. Kossiakoff, Ann. Rev. Biochem., 54:1195 (1985).
23. T. J. Greenhough and J. R. Helliwell, Progr. Biophys. Mol. Biol., 41:67 (1983).
24. H. D. Bartunik, P. N. Clout and B. Robrahn, J. Appl. Cryst., 14:134 (1981).
25. R. Usha, J. E. Johnson, D. Moras, J. C. Thierry, R. Fourme and R. Kahn, J. Appl. Cryst., 17:147 (1984).
26. K. S. Wilson, E. A. Stura, D. L. Wild, R. J. Todd, D. J. Stuart, Y. S. Babu, J. A. Jenkins, T. S. Standing, L. N. Johnson, R. Fourme, R. Kahn, A. Gadget, K. S. Bartels and H. D. Bartunik, J. Appl. Cryst., 16:28 (1983).
27. U. W. Arndt, T. J. Greenhough, J. R. Helliwell, J. A. K. Howards, S. A. Rule and A. W. Thompson, Nature, 298:835 (1982).
28. K. Moffat, D. Szebenyi and D. Bilderback, Science, 223:1423 (1984).
29. R. Hamlin, C. Cork, A. Howard, C. Nielson, W. Vernon, D. Matthews, N. G. H. Xuong and V. Perez-Mendiz, J. Appl. Cryst., 14:85 (1981).
30. S. C. Harrison, Nature, 309:408 (1984).
31. W. A. Hendrickson and M. Teeter, Nature, 290:107 (1981).
32. W. E. Furey, A. H. Robbins, L. L. Clancy, D. R. Winge, B. C. Wang and C. D. Stout, Science, 231:704 (1986).
33. J. L. Smith, W. A. Hendrickson and A. W. Addison, Nature, 303:86 (1983).
34. H. Hauptman, S. Potter, C. Weeks, J. Karle, R. Srinivasan, S. Parthasarathy and G. N. Watson, Acta Cryst., A38:289 (1982).
35. H. Hauptman, S. Potter and C. M. Weeks, Acta Cryst., A38:294 (1982).
36. M. M. Woolfson, Acta Cryst., A40:32 (1984).
37. J. Karle, Acta Cryst., A40:1 (1984).
38. J. Karle, Acta Cryst., A40:4 (1984).
39. J. Karle, Acta Cryst., A40:526 (1984).
40. G. Bricogne, Acta Cryst., A40:410 (1984).
41. A. Jones, J. Appl. Cryst., 11:268 (1978).
42. A. J. Morffew and D. S. Moss, Acta Cryst., A39:196 (1983).
43. P. A. Bash, N. Pattabiraman, C. Huang, T. E. Ferrin and R. Langridge, Science, 222:1325 (1983).
44. J. H. Konnert and W. A. Hendrickson, Acta Cryst., A36:344 (1980).
45. J. L. Sussman and A. D. Podjarny, Acta Cryst., B39:495 (1983).
46. A. G. W. Leslie, Acta Cryst., A40:451 (1984).
47. O. Herzberg and J. L. Sussman, J. Appl. Cryst., 16:144 (1983).
48. A. M. Silva and M. G. Rossmann, Acta Cryst., in press.
49. A. McPherson, "Preparation and Analysis of Protein Crystals", John Wiley & Sons, New York (1982).
50. R. Giege, A. C. Dock, D. Kern, B. Lorber, J. C. Thierry and D. Moras, J. Crystal Growth, 76:554-561 (1986).
51. M. Zeppenzauer, H. Eklund and E. Zeppenzauer, Arch. Biochem. Biophys., 126:564 (1968).
52. F. R. Salemme, Arch. Biochem. Biophys., 151:533 (1972).
53. Methods in Enzymology, 114:140 (1985).
54. H. Michel and D. Oesterhelt, Proc. Natl. Acad. Sic., USA., 77:1283 (1980).
55. R. M. Garavito and J. P. Rosenbusch, J. Cell Biol., 86:327 (1980).
56. J. Deisenhofer, O. Epp, L. Miki, R. Huber and H. Michel, J. Mol. Biol., 180:385 (1984).

57. J. Deisenhofer, O. Epp, K. Miki, R. Huber and H. Michel, Nature, 318:618 (1985).
58. H. Michel, J. Mol. Biol., 158:567 (1982).
59. H. Michel, Trends Biochem. Sci., 8:56 (1983).
60. H. Michel, EMBO J., 1:1267 (1982).
61. R. M. Garavito, J. A. Jenkins, J. N. Jansonius, R. Karlsson and J. P. Rosenbusch, J. Mol. Biol., 164:313 (1983).
62. R. M. Garavito, U. Hinz and J. M. Neuhaus, J. Biol. Chem., 259:4254 (1984).
63. J. P. Allen and G. Feher, Proc. Natl. Acad. Sci., USA., 81:4795 (1984).
64. W. Welte, T. Wacher, M. Leis, W. Kreutz, J. Shinozawa, N. Gad'on and G. Drews, FEBS Letters, 182:260 (1985).
65. C.-H. Chang, M. Schiffer, D. Tiede, U. Smith and J. Norris, J. Mol. Biol., 186:201 (1985).
66. A. McPherson, S. Koszelak, H. Axelrod, J. Day, R. Williams, L. Robinson, M. McGrath and D. Cascid, J. Biol. Chem., 261:1969-1975 (1986).
67. J. Schlichtkrull, Acta Chem. Scand., 11:439 (1957).
68. R. W. Fiddis, R. A. Longman and P. D. Calvert, Trans. Faraday Soc., 75:2753 (1979).
69. Z. Kam, H. B. Shore and G. Feher, J. Mol. Biol., 123:539 (1978).
70. M. L. Pusey, R. S. Snyder and R. Naumann, J. Biol. Chem., 261:6524-6529 (1986).
71. G. Feher and Z. Kam, Methods in Enzymology, 114:78-110 (1985).
72. F. Rosenberger, J. Crystal Growth, 76:618-638 (1986).
73. R. J. Davey, J. Crystal Growth, 76:637-644 (1986).
74. W. A. Tiller, J. Crystal Growth, 76:607-617 (1986).
75. M. Pusey and R. Naumann, J. Crystal Growth, 76:593-599 (1986).
76. S. D. Durbin and G. Feher, J. Crystal Growth, 76:583-592 (1986).
77. Dan Carter, NASA Protein Crystal Growth Meeting, December (1986).
78. F. L. Suddath, L. J. Wilson and J. Bertrand, Dynamic Control of Protein Crystallization, American Crystallographic Association Meeting, March 15-20, Austin, Texas, Abst. H-2 (1987).
79. N. D. Jones, J. B. Decter, J. K. Swartzendernber and P. L. Landis, Apocalypse - An Automated Protein Crystallization System, American Crystallographic Association Meeting, March 15-20, Austin, Texas, Abst. H-4 (1987).
80. K. B. Ward, M. A. Perozzo and W. M. Zuk, Automated Preparation of Protein Crystals Using Laboratory Robotics and Automated Visual Inspection, American Crystallographic Association Meeting, March 15-20, Austin, Texas, Abst. H-3 (1987).
81. P. C. Weber and M. J. Cox, Experiments in Automated Protein Crystal Growth, American Crystallographic Association Meeting, March 15-20, Austin, Texas, Abst. H-5 (1987).
82. K. B. Ward, M. A. Perozzo and J. R. Deschamps, Automation of protein crystallization experiments, in: "Automation of Protein Crystallization Experiments", J. R. Strimaitis and G. L. Hawk, eds., pp 413-433, Zymark Corporation, Hopkinton, Massachusetts.
83. N. D. Jones and K. B. Ward, A New Protein Crystallization Plate, Abstract of American Crystallographic Association Annual Meeting, June 22-27, PA-26 (1986).
84. W. M. Zuk and K. B. Ward, Robotics in the Industrial Laboratory, Preparation of Protein Crystals Using Robotics and Automated Visual Inspection, 93rd ACS National Meeting, Denver, Section B, paper 49.
85. W. M. Zuk and K. B. Ward, Detection and Monitoring of Crystal Growth in Protein Crystallography Experiments Using Automated Image Analysis, American Crystallographic Association Meeting, Austin, Texas.

86. C. W. Carter, Jr. and C. W. Carter, <u>J. Biol. Chem</u>., 254:12219-12223 (1979).
87. E. T. Baldwin, K. V. Crumley and C. W. Carter, Jr., <u>Biophys. J</u>., 49:47-48 (1986).
88. G. E. Rindome, ed., "Materials Processing in the Reduced Gravity Environment of Space", North-Holland, New York (1982).
89. W. Littke and C. John, <u>Science</u>, 225:203 (1984).

126. ... ...

127. Huntington, H.B. Champ. and O.A. Correction. Physical ...

128. ... Eighnes, P.D. ... ... Anzuschung in the Rate, an enzyme ... Problems in Atoms, Solids & ... Model Analysis ...

129. ... ... Acta Mecanica, 42 247 (1982).

CRYSTAL GROWTH AND GEOSCIENCES

Ichiro Sunagawa

Institute of Mineralogy, Petrology and
Economic Geology, Faculty of Science
Tohoku University, Sendai, 980 Japan

1.   INTRODUCTION

    In any geological processes, irrespective of the places where they
occur, nucleation, growth, dissolution, transformation or replacement of
crystals are involved. Through these processes, weathering, superficial
precipitation, diagenesis, metamorphism, metasomatism, and magmatic
crystallization proceed, and sedimentary, metamorphic and igneous rocks or
ore deposits are formed. Depending on how and under what conditions these
processes proceed, different mineral assemblages, different morphologies
and perfections of crystals, and different textures of rocks or ores
appear. Therefore, it is essential to understand the fundamentals of
crystal growth mechanisms so as to properly analyze the genesis of
minerals, rocks and ores, which is one of the main subject of geosciences.

    The basic approach in geosciences to analyze mineral genesis was at
first descriptive, and then has been based on equilibrium thermodynamic
considerations and phase studies. Temperature and pressure conditions of
mineral formation have mainly been evaluated on the basis of phase
relations and mineral assemblages. However, as compared to the efforts,
having been paid to analyze macroscopic initial and final products, not
enough attention except for a few serious efforts, such as seen in the
list of selected books and review papers, has hitherto been paid to
analyze microscopic kinetic problems, i.e, how mineral crystals grew and
what sort of thermal or stress histories they have experienced in
geological circumstances. It is clear that both kinetic and equilibrium
thermodynamic approaches are required to understand mineral genesis
properly.

    In understanding the process and mechanism of mineral formation,
the knowledge obtained through theoretical and experimental studies of
crystal growth mechanisms in simple systems is applicable and useful.
However, it is necessary to modify the knowledge by taking the
characteristics of natural crystallization into consideration.

    It is the purpose of this contribution to outline what are the
essential factors to be taken into consideration when the science of
crystal growth is applied to geosciences.

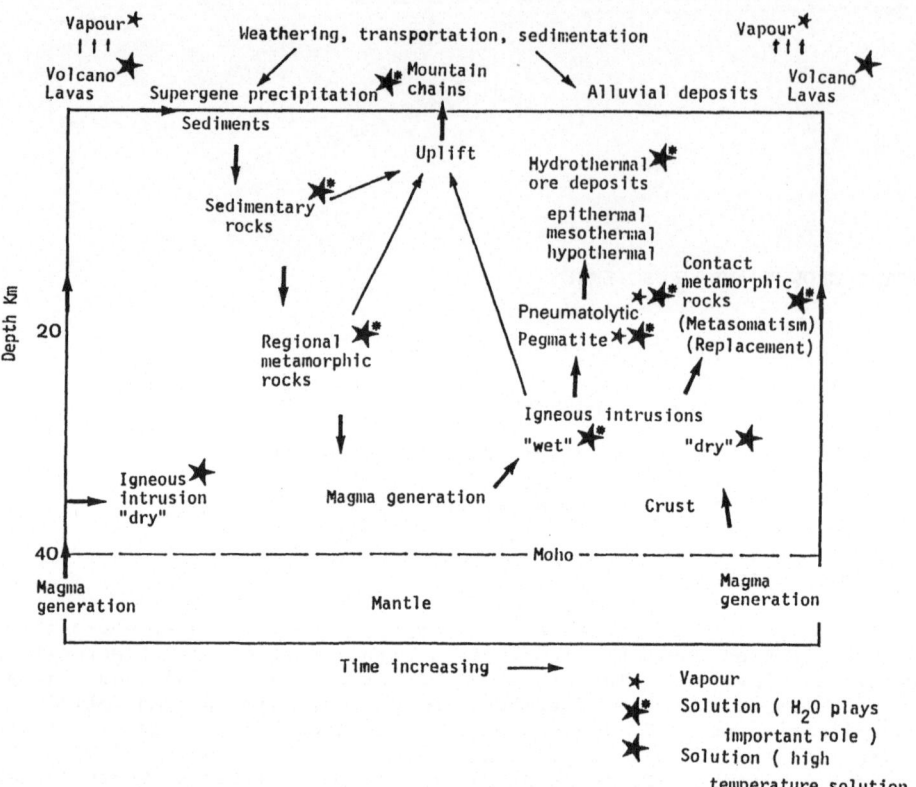

Fig. 1. A schematic diagram to show characteristics of crystal growth in various geological processes [3].

2.   CRYSTALLIZATION IN GEOLOGICAL PROCESSES

Crystallization taking place in most geological processes, including metamorphism and metasomatism, is essentially growth of crystals from aqueous or high temperature solution phases [1,2,3]. This is schematically illustrated in Figure 1. Melt growth and pure vapor growth are absent, and recrystallization is limited to, for example, the formation of marble from limestone or of quartzite from sandstone, since natural crystallization is the growth of crystals from impure, multi-component systems. Since natural vapor growth always involves chemical reactions, it is not a physical vapor deposition (PVD) but a chemical vapor deposition (CVD) process, in which solid-fluid (solute-solvent) interaction is involved. In this respect, natural vapor growth has a close similarity to solution growth.

The characteristics of solution growth as compared to melt growth and pure vapor growth are:

1)   It is a growth from dilute phase, and thus transportation and condensation processes are essential. Contribution of mass transfer is major while that of heat transfer will at most be 5%.
2)   The growth temperature is greatly lowered as compared to melt growth.
3)   Solid-liquid (solute-solvent) interaction is always involved, which is absent in melt and pure vapor growth. Therefore, desolvation processes are essential. This will slow down the growth rate, and

also results in the formation of buoyancy driven convection when
crystals grow under the influence of the Earth's gravity. The
behavior of buoyancy driven convection gives a strong effect upon the
growth, morphology and perfection of crystals.
4)   Gradients in diffusion boundary layers, and interfacial phenomena are
     important.
5)   The Berg effect [4] can be more enhanced in solution growth than in
     vapor or melt growth.

     In Figure 2, the characteristics of solution growth are schematically
illustrated.

     As seen in Figure 1, mineral crystals grow under a wide range of
conditions, from high temperature, high pressure magmatic conditions to
low temperature, atmospheric superficial conditions. The crystallization
never occurs from pure, simple systems, but takes place in impure, multi-
component systems. The growth conditions may fluctuate or change
drastically during growth. Also, in natural crystallization, macroscopic

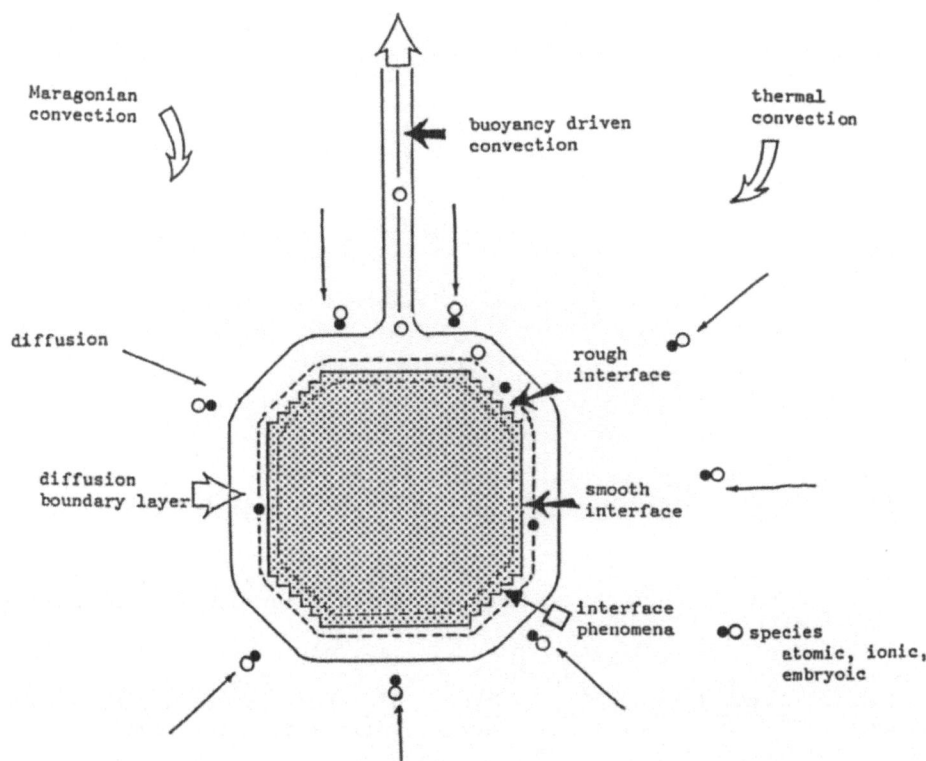

Fig. 2.   Generalized feature of solution growth in real systems. Growth
          units coupled with solvent molecules are transported from the
          bulk solution by diffusion, thermal or Marangonian convections,
          and brought into the diffusion boundary layer. Through the
          process of desolvation and incorporation of growth units at the
          interface, the solution around a growing crystal becomes lighter,
          creating buoyancy driven convection. The buoyancy driven
          convection behaves differently depending upon the bulk
          supersaturation. The solid-liquid interface is assumed to be not
          bare. Various interface phenomena are expected to occur.

and microscopic flow of solution often plays an important role. So in applying the knowledge obtained through the studies of simple systems to the analysis of mineral genesis, care should be taken in respect to the following characteristics:

1) impure, multi-component systems,
2) wide range of growth conditions,
3) changing or fluctuating conditions,
4) processes under the effect of flow,
5) possibility of the presence of growth units of varying sizes and states, like polymerized growth unit.

3. RESULTS OBTAINED THROUGH CHARACTERIZATION OF CRYSTALS

Since it is not possible to observe in-situ the process of mineral formation, except in special cases, it has been the orthodox way to analyze mineral genesis or growth and post-growth histories of mineral crystals based on their characterization, supplemented by the information on growth temperatures, pressures, chemical environments evaluated through thermodynamic analyses and mineral assemblages. The following properties have mainly been investigated for such purposes.

1) External morphology of crystals, i.e., habit changes.
2) Surface microtopography of crystal faces.
3) Internal morphology, such as growth banding, growth sectors, etc., and element partitioning in different growth sectors.
4) Primary inclusions of solid, liquid or other foreign phases.
5) Distribution patterns and nature of dislocations.
6) Precipitation, exsolution and other microtextures.
7) Modes of aggregation of crystals.
8) Textures of polycrystalline and polymineral aggregations.
9) Chemical properties, particularly impurities and stoichiometry.

In short, physical and chemical properties which deviate from the ideal state have been used to analyze growth and post-growth histories of mineral crystals.

A representative successful example of this type of investigation is the case of natural diamond crystals. It has been clarified that:

1) Natural diamond crystals are formed under high temperature and pressure conditions, i.e., in its stable field, at a depth greater than 120 Km, and brought up near the Earth's surface with a speed of approximately 100 Km/hr [5].
2) Two distinct chemical environments, i.e., ultra mafic and eclogitic suites, are postulated for their growth environments [6].
3) Natural diamond crystals grew from a high temperature solution phase and neither from a melt phase nor by solid state recrystallization process [5,7].
4) Three types, single crystalline, polycrystalline and combined types, are distinguished from their morphological characteristics, which reflect different growth conditions and environemnts [7].
5) Diamond crystals of single crystalline type are formed by the spiral growth mechanism [8].
6) Polycrystalline type diamonds are likely formed in eclogitic suite in the subduction zone [7]
7) At least a part of the carbon source is biogenetic, and derived from organic sediments, through the subduction zone [5].

8) Natural diamond crystals experienced annealing, weak plastic
deformation and slight dissolution while they were transported to the
Earth's surface, where they were quenched [5,9].

Another example is quartz. SiO_2 crystallizes in a form of high
temperature quartz, low temperature quartz, chalcedony and amorphous opal,
in addition to other polymorphs. High temperature quartz grows in silica
rich acidic to intermediate magmas by the spiral growth mechanism, and

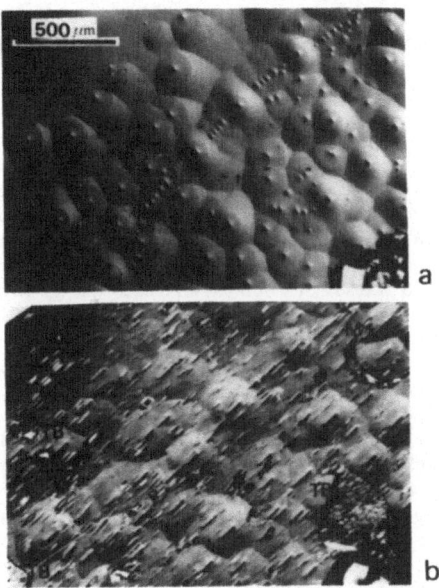

Fig. 3. Corresponding natural (a) and etched (b) {10Ī1} surfaces of a
high quartz crystal. Both circular (in (a)) and elongated
polygonal (in (b)) patterns are depressions, i.e., etch pits.
Circular etch pits were formed above the phase transition
temperature, 573°C, while the crystal holds hexagonal symmetries,
whereas elongated polygonal etch pits appear by laboratory
etching and reflects the symmetries of low quartz. M: percussion
mark, TB: Dauphiné twin boundary, P-P: array of dislocations
formed by plastic deformation. Note the increment of etch pit
density and the formation of Dauphiné twinning in (b) [10].

usually experiences disequilibrium conditions and weakly dissolves to form
etch pits above the high-low transition temperature, 573°C, and is
quenched and transforms to the low temperature polymorph by an adiabatic
expansion of the magmas. These events are well documented on the surface
microtopography, particularly etch figures, of hexagonal bi-pyramidal
faces, distribution and type of dislocations, and distribution of Dauphiné
twin boundaries [10]. In Figure 3 an example of surface microtopographs
is shown.

Fig. 4. An example of quartz crystals from epithermal vein deposits. Polarization photomicrograph.

Low temperature quartz, chalcedony and opal precipitate in a wide range of conditions from pegmatitic, pneumatolytic to hydrothermal veins, geode to supergene and superficial conditions. The growth temperatures range from ca. 500°C to ordinary temperature. The crystal perfection of quartz crystals shows a tendency to diminish as growth temperature decreases. Quartz crystals with habit faces formed in, for example, pegmatite at elevated temperatures show higher perfection, and thinner growth layers on their faces, yielding good X-ray topographs. Those formed at lower temperature, like in epithermal vein or in geodes, are poorer in the perfection than those formed at higher temperatures and usually consist of split domains. Figure 4 shows such an example. Further down the growth temperature, cryptocrystalline chalcedony or its varieties are formed, and at the lowest temperature, SiO_2 precipitates as opal, consisting of amorphous silica spheres. Such a change implies that at decreasing temperature, polymerization of silica occurs, and the entities to incorporate into a growing crystal become larger. This demonstrates the necessity of taking a variety of growth units into consideration in analyzing mineral genesis.

Diamond and quartz crystals are examples of mineral crystals grown in a non-confined environment, i.e., the crystallization takes place in liquid state hydrothermal or high temperature solutions in which crystals can grow freely.

When crystals grow through metasomatic or metamorphic processes, crystallization takes place in confined environments, i.e., in the presence of foreign solid materials. How crystallization proceeds in such environments, namely whether it is solid-state crystallization or through a dissolution/precipitation process, has been a question. To answer this problem, Sunagawa and Koshino [11] and Tomura et al. [12] compared the surface microtopographic characteristics of the basal plane of kaolin group minerals, sericite and chlorite of various origins, by appying the decoration technique of electron microscopy. They demonstrated that:

1) These crystals usually exhibit growth spirals of mono-molecular step height when they occur in hydrothermal veins and metasomatic deposits, whereas growth spirals are not observed on those grown in regional metamorphic rocks.

2) Crystals occurring in hydrothermal veins are characterized by a common occurrence of coalesced crystals, whereas those occurring in metasomatic deposits are usually isolated and independent crystals. This

indicates that in hydrothermal veins, crystals grow and coalesce in moving solutions, whereas in hydrothermal metasomatic deposits, crystallization takes place in a thin film of aqueous solution through a dissolution/precipitation process.

3) Regional metamorphic crystals exhibit both smooth and rugged steps of mono-molecular height, which vary depending on the metamorphic grade and the positions within one metamorphic zone. From the analysis of the latter observation [12], it was concluded that Ostwald ripening played an important role in the crystallization of regional metamorphism. As to Ostwald ripening applied to geological materials, the readers may refer to a review paper by Baronnet [13].

4. DYNAMIC CRYSTALLIZATION EXPERIMENTS

In the equilibrium phase studies of silicate systems, isothermal crystallization experiments followed by quenching and identification are the principal techniques. When lunar samples and pillow lavas solidified on the oceanic floor were collected, feather-like or dendritic morphologies exhibited by the constituent rock-forming minerals in them stimulated the interest of geoscientists.

Dynamic crystallization experiments were initiated with the hope of reproducing such igneous rock textures and of understanding the morphological changes of rock-forming crystals depending on supercooling and composition. Also nucleation problems are investigated [14,15]. In these experiments, the degree of supercooling and also the cooling rate are varied.

For morphological changes of crystals in relation to the composition and the degree of supercooling, dynamic crystallization experiments on various systems have offered essentially similar results as those obtained and analyzed in the field of crystal growth. As the driving force increases, the morphology of crystals changes from polyhedral bounded by flat surfaces, via hopper growth, to dendritic and further to spherulitic aggregates. Figure 5 indicates morphological changes in relation to driving force vs. growth rate, to three different growth mechanisms, to smooth and rough interfaces and thus to the α-factor.

The positions of two transitional driving forces, $\Delta\mu/kT^*$ and $\Delta\mu/kT^{**}$, in Figure 5, may vary depending on the phases, solvents, or materials, all related to the generalized α-factor [2,3,16,17]. Based on Figure 5, one may also evaluate how internal morphologies in a single crystal may change, as growth conditions change. Figure 6 shows various internal morphologies expected in relation to the changes in growth conditions. Three morphological types exhibited by natural diamond crystals described in the preceding section are demonstrated in these figures [7].

Dynamic crystallization experiments have been successful in reproducing various igneous textures, from volcanic to plutonic rocks. However, it failed to reproduce the hyalopilitic texture of basalt, namely a texture consisting of phenocrysts and microphenocrysts in the groundmass, Figure 7a, which is commonly exhibited by natural basalt. Two stage nucleation experiments produced a texture consisting of phenocrysts and groundmass with radiating fibers, Figure 7b.

By applying forced flow to a system containing phenocrysts in a basaltic melt, Kouchi et al. [18] were successful in reproducing the hyalopilitic texture. The texture was reproduced because (1) the nucleation rate increased sharply due to the presence of phenocrysts and

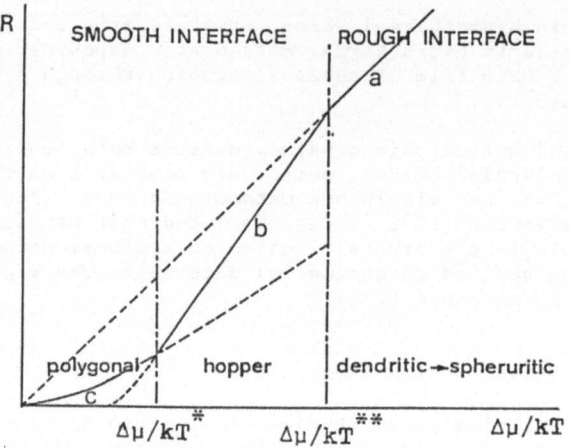

Fig. 5. Schematic drawing to show variation of morphologies of crystals depending on growth rate (R) versus driving force ($\Delta\mu/kT$) relations. Also three growth mechanisms, and smooth and rough interfaces, are shown [3]. a: adhesive type growth; b: two-dimensional nucleation growth; c: spiral growth.

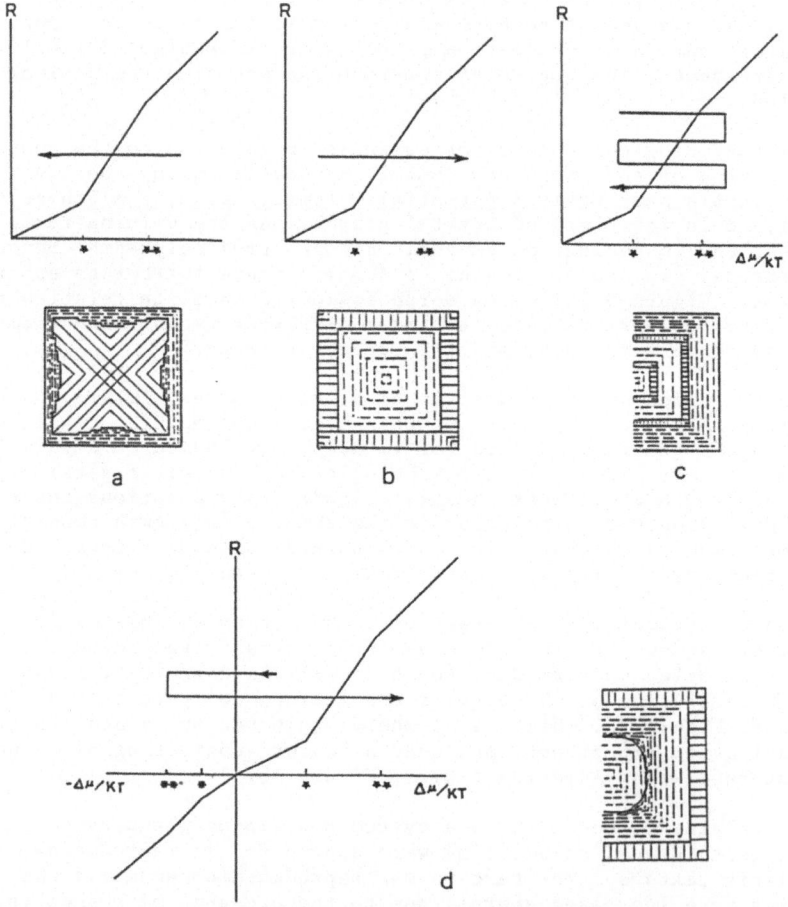

Fig. 6. See next page for Legend.

a   b

Fig. 7.   Sketches of basalt textures reproduced by the dynamic
          crystallization experiment (without flow) (a) and by the
          experiments under the effect of forced flow (b).   Based on Kouchi
          et al. [18].

flow in the system, and (2) morphological changes depending on the
supercooling changed drastically by the presence of flow.   The former can
be related to secondary nucleation phenomena, and the latter to the
decrement of the thickness of the diffusion boundary layer due to the
presence of flow.

     This simple experiment indicates the importance of flow in natural
crystallization, and of the fundamental knowledge of crystal growth
mechanisms in the studies of mineral genesis.   Kouchi and Sunagawa [19,20]
also demonstrated that andesite and banded dacite (corresponding to banded
pumice) can be easily produced in a matter of hours simply by mixing
basaltic and dacitic lavas.   This provided the first direct experimental
verification for the model of magma mixing as a possible origin of
adnesite.   Since flow on a larger or small scale, laminar or turbulent, is
always involved in natural crystallization, the investigation on this
problem, which has not been treated extensively in the field of crystal
growth, will be important for future studies.   The readers may also refer
to a review paper by Huppert [21] on this subject.

     The above experiments demonstrate the importance of the secondary
nucleation phenomenon in natural crystallization.   This phenomenon has
been known for many years in the field of industrial crystallization,
although still much more should be understood with respect to its
mechanism.   The previous investigations on secondary nucleation phenomena
have suggested three possible mechanisms: initial breeding, contact
breeding and fluid shear breeding.   All the three mechanisms assume the
destruction of and the detachment of fine particles from the seed crystals
as possible sources for secondary nucleation.   In this respect, recent
observations on the appearance of high density layers adjacent to the
interface of a growing crystal, when the supersaturation is increased
above a certain critical value, is worthwhile to be cited here [22,23].
By means of an in-situ observation technique coupled with the Schlieren
technique and Mach-Zehnder interferometry, they could visualize the
density changes around a growing $Ba(NO_3)_2$ crystal in an aqueous solution.
They demonstrated that buoyancy driven convection due to desolvation
behaves differently depending on the bulk supersaturation, and such

Fig. 6.   A schematic illustration of expected internal morphologies of
          crystals in relation to the changes in growth conditions.   Based
          on Sunagawa [3].

behaviors have a decisive effect upon the growth rate versus super-saturation relation.

An exciting observation in their experiment was that above 8% of bulk supersaturation, jelly-like high density layers appeared on the growing interface. When the system was stirred, the jelly-like layer was carried away into the bulk solution, and after a few tens of seconds, an avalanche of tiny crystals appeared. This implies that embryonic particles were already present in the jelly-like layer, which acted as nucleation centers. In supergene or superficial precipitation of crystals at lower temperature, like in the precipitation of authigenic minerals during diagenesis, or in the solidification of high temperature magmas in a form of lava flow or sheet, similar phenomena may be expected to occur universally.

5. NUCLEATION

In natural crystallization, homogenous nucleation is not expected. Nucleation in geological processes is principally heterogeneous nucleation around impurities or foreign particles. This decides the size of crystals and rock textures. A large difference of crystal sizes of the same mineral depending on geological environments may be accounted for on this basis. Muscovite crystals $KAl_2(AlSi_3O_{10})(OH,F)_2$, attain a few tens cm across, when they grow in a pegmatite, but their sizes diminish down to less than a micron when they crystallize in metamorphic or metasomatic rocks.

An additional important problem related to nucleation in natural crystallization is metastable nucleation, namely nucleation of thermodynamically unstable phases prior to the nucleation of the stable phase. This is commonly encountered in biological activities. A representative example is the formation of aragonite, a high pressure phase of $CaCO_3$, in shells of bivalves and gastropoda. Metastable nucleation also takes place commonly in high temperature silicate systems, which has been ascertained through the in-situ observation technique, which will be discussed in the following section.

6. IN-SITU OBSERVATION

Principal approaches to investigate the molecular mechanisms of crystal growth have been either (1) through characterization of already grown crystals (surface microtopography and internal topography, etc.), or (2) measurements and analyses of growth rates versus driving force relations. Both approaches are essentially in-direct ones. If growth or dissolution processes of crystals could be observed on the molecular scale under precisely controlled growth conditions, and mass and heat flows around growing crystals could be visualized, together with precise measurements of growth rates and conditions, this will form the most direct method to investigate growth mechanisms.

We at first developed such an in-situ observation method for ordinary temperature aqueous solution growth [24], and further developed the method to be applicable to elevated temperatures [25]. Also a method to visualize mass flow around a growing crystal using the Schlieren technique and Mach-Zehnder interferometry, and a method to measure clustering and the states of species using light and Rayleigh scattering, laser-Raman spectroscopy, etc., have been developed.

In the in-situ observation techniques, it is essential to improve the visibility, to control the conditions as precisely as possible, and to measure the growth rates and conditions precisely. Various improvements have been made to meet these requirements, including a production of newly designed objective lenses whose optical aberration due to the presence of a liquid layer or glass plate is corrected [26].

Through these improvements, it is now possible to carry out in-situ observations of the movements of growth or dissolution steps having the height in the order of 10 Å, of growth processes of crystals of rock-forming minerals in silicate solutions at $1600 \pm 0.3°C$ under an optical microscope. Mass and heat flows around a growing crystal, formation of embryoic clusters are now visualized and quantitatively measurable. We are now able to prepare a magma and observe magmatic crystallization under the optical microscope. Various problems related to magmatic crystallization, such as partial melting, liquid immiscibility, reaction between crystals and a magma, nucleation, growth or dissolutioon of rock-forming crystals, formation of zoning and sector structures in single crystals, etc. may be directly observed and investigated with the in-situ observation method developed in our laboratory. It was shown by a couple of films how rock-forming minerals nucleate, grow, move or dissolve in a thin film of high temperature silicate solution. Although there are various topics to be discussed, we shall select only one topic, i.e., metastable nucleation, investigated by the in-situ observation technique at elevated temperatures.

The activation energy for nucleation can be expressed in the following simple formula

$$A_N = \frac{1}{\eta} \frac{\gamma^3}{\Delta\mu^2}$$

where η is the viscosity, γ is the interfacial free energy between the crystal and the liquid, $\Delta\mu$ is the chemical potential difference between solid and liquid phases. Therefore, due to the competition between γ and $\Delta\mu$, it is possible that phases which are not expected in the equilibrium phase diagram can nucleate prior to the nucleation of the stable phase, when a certain degree of supercooling is realized. We shall call this phenomenon metastable nucleation.

Metastable nucleation phenomena in silicate systems are far more commonly encountered than one may expect. Even starting from a pure diopside melt, the initial phase to nucleate is not diopside but forsterite when ΔT is larger than ca. 70°C. Forsterite is not expected to appear on the basis of equilibrium thermodynamics. The metastably nucleated forsterite crystals can grow steadily at this temperature. The growth rate is, however, one order lower than that of diopside, the stable phase, which nucleates later. As a result, metastably nucleated forsterite is shadowed by the later nucleated diopside. In Table 1, metastable nucleation phenomena we have observed by the in-situ observation methods are tabulated [27].

If temperature is elevated after a metastable nucleation takes place, the crystal of the metastable phase takes a rounded form above a certain temperature, namely it starts to dissolve. If the temperature is decreased, the crystal becomes polyhedral, i.e., starts to grow again. The whole process is reversible and can be traced under the microscope. The melting point of the metastable phase, thus, can be determined precisely. It is on the extension of the liquidus line of the phase in its stable region. This indicates the importance of using metastable phase diagrams, and not equilibrium phase diagrams, in analyzing natural crystallization.

Table 1. Metastable Nucleation Observed by the
In-situ Method (After Nakamura [27])

Liquid Composition	Expected Phase	Observed Phase
An_{100}	An	An
$Di_{40}An_{60}$	An	Di
$Di_{50}An_{50}$	An	Di
$Di_{60}An_{40}$	An	Di
Di_{100}	Di	Fo
$Di_{80}Tp_{10}Ak_{10}$	Di	Fo
$Di_{65}Tp_{25}Ak_{10}$	Fo	Pv, Di
$Di_{55}Tp_{35}Ak_{10}$	Pv	(Fo)
$Di_{85}En_{15}$	Di	Fo

An: Anorthite	$CaAl_2Si_2O_8$	
Di: Diopside	$CaMgSi_2O_6$	
Fo: Forsterite	Mg_2SiO_4	
Pv: Perovskite	$CaTiO_3$	
Tp: Ti-pyroxene	$CaTiAl_2O_6$	
Ak: Akermanite	$Ca_2Mg_2SiO_7$	
En: Enstatite	$MgSiO_3$	

7. CONCLUDING REMARKS

Although geoscientists gradually begin to realize the importance of kinetics and to understand the fundamentals of crystal growth mechanism, there is still a big discrepancy when the knowledge obtained through theoretical and experimental studies on crystal growth mechanisms of simple systems is applied to mineral genesis occurring in complicated systems. This is because the interests of geoscientists are focused on more macroscopic phenomena, whereas those of crystal growers moved to microscopic details, and the discrepancy appears to be too big for geoscientists to overcome. As a result, geoscientists are still rather reluctant to learn from the science of crystal growth. Efforts should be made from both sides to fill up this gap.

REFERENCES

1. I. Sunagawa, Natural crystallization, J. Crystal Growth, 42:214-223 (1977); Vapour growth and epitaxy of minerals and synthetic crystals, J. Crystal Growth, 45:3-12 (1978).
2. I. Sunagawa, Characteristics of crystal growth in nature as seen from the morphology of mineral crystals, Bull. Mineral., France, 104:81-87 (1981).
3. I. Sunagawa, Growth of crystls in nature, in: "Materials Science of The Earth's Interior", I. Sunagawa, ed., pp 63-105, Terra Sci. Publ. Co., Tokyo/D. Reidel Pub. Co., Dordrecht (1984a).
4. W. F. Berg, Crystal growth from solutions, Proc. Roy. Soc., London, A164:79-95 (1938).
5. F. C. Frank, Defects in diamond, in: "Proc. Intern. Indust. Diamond Conf., Oxford", R. Berman, ed., pp 119-135, Industrial Diamond Information Bureau, London (1966).
6. J. W. Harris and J. J. Gurney, Inclusions in diamond, in: "Physical Properties of Diamond", R. Berman, ed., pp 555-591, Academic Press, Oxford (1979).
7. I. Sunagawa, Morphology of natural and synthetic diamond crystals, in: "Materials Science of The Earth's Interior", I. Sunagawa, ed.,

pp 303-330, Terra Sic. Pub. Co., Tokyo/D. Reidel Pub. Co.,
Dordrecht (1984b).

8. I. Sunagawa, K. Tsukamoto and T. Yasuda, Surface microtopographic and
    X-ray topographic study of octahedral crystals of natural diamond
    from Siberia, in: "Materials Science of The Earth's Interior", I.
    Sunagawa, ed., pp 331-349, Terra Sci. Pub. Co., Tokyo/D. Reidel
    Pub. Co., Dordrecht (1984).

9. A. R. Lang, Internal structure, in: "The Properties of Diamond", J.
    E. Field, ed., pp 425-469, Academic Press, London (1979).

10. I. Sunagawa and A. Sugibuchi, Growth and post-growth histories of
    high quartz as revealed by natural and laboratorical etching, J.
    Japan. Assoc. Min. Petr. Econ. Geol., 81:348-358 (1986).

11. I. Sunagawa and Y. Koshino, Growth spirals on kaolin group minerals,
    Amer. Mineral., 60:407-412 (1975).

12. S. Tomura, M. Kitamura and I. Sunagawa, Surface microtopography of
    metamorphic white mica, Phys. Chem. Minerals, 4:1-17 (1979).

13. A. Baronnet, Ostwald ripening in solution, the case of calcite and
    mica, Estudios Geol., 38:185-198 (1982).

14. A. W. Hofmann, B. J. Giletti, H. S. Yoder, Jr. and R. A. Yund, eds.,
    Geochemical Transport and Kinetics, Carnegie Institute of
    Washington, Washington, pp 353 (1974).

15. R. B. Hargraves, ed., Physics of Magmatic Processes, Princeton Univ.
    Press, N.J., pp 585 (1980).

16. I. Sunagawa and P. Bennema, Morphology of growth spirals, theoretical
    and experimental, in: "Preparation and Properties of Solid State
    Materials", W. A. Wilcox, ed., Vol. 7, pp 1-129, Marcel Dekker
    Inc., New York (1982).

17. I. Sunagawa, Morphology of crystals in relation to growth conditions,
    Estudios Geol., 38:127-134 (1982).

18. A. Kouchi, A. Tsuchiyama and I. Sunagawa, Effect of stirring on
    crystallization kinetics of basalt: texture and element
    partitioning, Contrib. Mineral. Petrol., 93:429-438 (1986).

19. A. Kouchi and I. Sunagawa, Mixing basaltic and dacitic magmas by
    forced convection, Nature, 304:527-528 (1983).

20. A. Kouchi and I. Sunagawa, A model for mixing basaltic and dacitic
    magmas as deduced from experimental data, Contrib. Mineral.
    Petrol., 89:17-23 (1985).

21. H. E. Huppert, Multicomponent crystallization and convection beneath
    volcanoes, J. Crystal Growth, 79:12-18 (1986).

22. K. Onuma, Role of mass flow in aqueous solution growth, MSc Thesis,
    Tohoku Univ. (1986).

23. K. Tsukamoto, In-situ direct observation of a crystal surface and its
    surroundings, in: Morphology and Growth Unit of Crystals,
    Proceedings of the Oji International Seminar", I. Sunagawa, ed.,
    Terra Sci. Pub. Co., Tokyo/D. Reidel Pub. Co., Dordrecht, to be
    published.

24. K. Tsukamoto, In-situ observation of mono-molecular growth steps on
    crystals growing in aqueous solution, I., J. Crystal Growth,
    61:199-209 (1983).

25. K. Tsukamoto, T. Abe and I. Sunagawa, In-situ observation of crystals
    growing in high temperature melts and solutions, J. Crystal
    Growth, 63:215-218 (1983).

26. K. Tsukamoto and I. Sunagawa, In-situ observation of mono-molecular
    growth steps on crystals growing in aqueous solution, II.
    Specially designed objective lens and Nomarski prism for in-situ
    observation by reflected light, J. Crystal Growth, 71:183-190
    (1985).

27. H. Nakamura, Observation and analysis of metastable nucleation in
    some silicate systems, MSc Thesis, Tohuku Univ. (1986).

Suggested Books and Review Papers

The following books and review papers deal with natural crystallization:

Books:

R. A. Berner, "Early Diagenesis, A Theoretical Approach", Princeton Univ. Press, N.J., pp 241 (1980).

D. P. Grigoriev, "Ontogeny of Minerals", Israel Programme for Scientific Translations, Jerusalem, pp 250 (1965).

D. P. Grigoriev and A. G. Jeabin, "Ontogeny of Minerals", Nauka,Moscow, pp 339, in Russian (1975).

G. G. Lemmlein, "Morfologiya i genezis kristallov ("Crystal Morphology and Genesis"), Nauka, Moscow, in Russian (1973).

S. K. Saxena, "Kinetics and Equilibrium in Mineral Reactions", Springer-Verlag, New York, pp 273 (1983).

A. Spry, "Metamorphic Textures", Pergamon Press, Oxford, pp 350 (1979).

R. L. Stanton, "Ore Petrology", McGraw-Hill Book Co., New York, pp 713 (1972).

I. Sunagawa, ed., "Materials Science of The Earth's Interior", pp 653, Terra Sci. Pub. Co., Tokyo/D. Reidel Pub. Co., Dordrecht (1984).

I. Sunagawa, ed., "Morphology of Crystals, Part A, B", pp 743, Terra Sci. Pub. Co. Tokyo/D. Reidel Pub. Co., Dordrecht (1988).

R. H. Vernon, "Metamorphic Processes, Reactions and Microstructure Development", George Allen &Unwin Ltd., London, pp 247 (1976).

Review paper:

W. A. Tiller, On the cross-pollenation of crystallization ideas between metallurgy and geology, Phys. Chem. Minerals, 2:125-151 (1977).

CRYSTAL GROWTH AND THE PROPERTIES OF POLYMERS

D.C. Bassett

J.J. Thomson Physical Laboratory
University of Reading, Whiteknights
PO Box 220   Reading RG6 2AF, UK

Crystal growth is especially important in polymers because it exerts such a strong influence on properties. At the same time the legacy of crystal growth processes is a rich record of microstructure whose study has developed into a field of its own: Polymer Morphology[1]. This can be read not only to reveal the history of a sample and its molecular characteristics but also to provide insight into crystal growth mechanisms themselves. This contribution seeks to introduce the principal and salient facts of polymeric growth, relate these to properties, both known and potential and to outline the current state of understanding.

Firstly, what is a polymer? It is a substance whose molecules are large — hence the alternative name, macromolecule — and built by repetition of many parts. Those polymers which can crystallize form a sub-set based on long chain molecules. We shall restrict discussion to synthetic, carbon polymers — materials such as polyethylene, polypropylene, nylons (polyamides) etc. — although there is much that carries over both to natural polymers and to other chain chemistries. These are materials which have assumed large economic importance in the latter half of this century. Polyethylene is by far the most important economically. It also has the simplest chemical structure and has been the most studied. It is probably true that the crystal growth of polyethylene has been investigated more than that of any other material. The reasons for this are the variability of its properties linked to variations in its microstructure and the novel phenomena displayed by the crystals of this and other crystallizable polymers. Especially because of recent developments this remains a very active topic.

Our present understanding stems from findings discovered thirty years ago in 1957. Prior to that it had been appreciated that certain polymers gave crystalline X-ray reflections but that, in general, they also showed diffuse scattering akin to that shown by liquids. Moreover, the crystalline reflections were comparatively broad i.e. the crystallites were small, with estimated dimensions of, say, 20 nm. These features were combined in the fringed micelle model of polymer structure according to which the solid contained two principal types of order, small crystallites and amorphous regions, with individual molecular chains passing through both[2]. The properties of such a microstructure would depend on the degree of crystallinity: the more crystalline a sample the higher its density, modulus and strength, for example. The fringed micelle model served to rationalize polymeric behaviour for many years during which there was no direct means of observing texture of the supposed crystallite dimensions. It has now been superseded: real textures contain lamellae, often 20nm thick, but sometimes much more, and up to μm or more wide. Separate lamellar crystals were discovered in 1957 by crystallizing polyethylene from solution and examining the product in the new electron microscopes[3,4,5]. (It was inherent in the fringed micelle model that crystallites were inextricably linked to amorphous regions i.e. separate crystals were not expected). These lamellae had similar shapes to those of short chains of the same chemical structure i.e. n paraffins in the case of polyethylene in which chains are normal or nearly so to lamellar. Electron diffraction confirmed that this was also true for polymers even though chains were, say, fifty times longer than the crystal thickness. This situation can be resolved if molecules are folded back on themselves alternately at each crystal surface. This was proposed by Keller[3] and termed chainfolding: ideally regular chainfolding is depicted in Fig. 1.

Lamellar crystallization is widespread, general and basic in crystalline polymers. For a long while its relevance to melt-as opposed to solution — crystallized systems was questioned but new techniques for electron microscopy have shown that lamellae are characteristic even of poorly crystalline polymers such as poly(ethylene terephthalate), PET. What has been controversial is the extent to which there is regular chainfolding[6]. Extensive neutron scattering data on polyethylene have shown that, under conditions where folding is effectively minimized, (very rapid growth) the probability of adjacent re-entry folding by the same chain crystallizing from the melt is 0.65[7]. This is effectively a lower limit: slower growth gives greater regularity. The essential point is that chainfolding accommodates the necessary change in density between crystal and surroundings in a way the fringed-micelle model cannot.

The consequences of lamellar crystallization for properties are great. In particular it can account for the familiar large discrepancy between molecular properties and those of the macroscopic sample. Values of Young's modulus and tensile strength are typically lower by at least two orders of magnitude than molecular values. This is because of chainfolding which reduces stress-bearing covalent bands across fold surfaces. At the same time it suggests how to overcome this restriction and this has now been

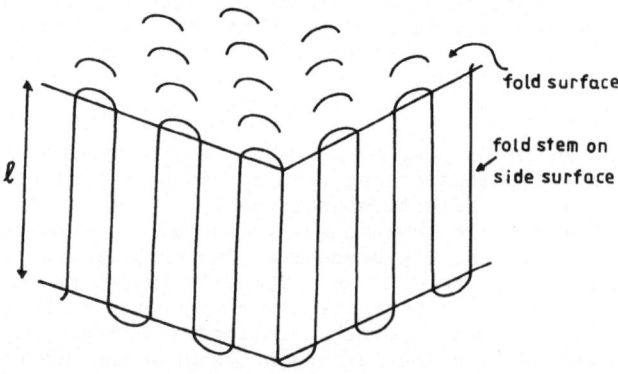

Fig. 1. Sketch of regular molecular chainfolding in a polymer lamella. Chainfolding and its terminology.

done. High modulus polyethylenes are now available commercially which approach theoretical values for the crystal lattice. What is important is the arrangement of lamellae in the microstructure. Series arrangements of lamellae are weak because of the intervening van der Waals bondings parallel ones with continuous covalent linkages much stronger. With lamellar textures, it is not the amount of crystallinity which is important, but its relative disposition i.e. the morphology of the sample.

Two major questions arise accordingly concerning crystal growth in polymers. One concerns the understanding of lamellar growth itself, the other how to manipulate lamellar morphologies to achieve textures with appropriate properties. Regarding the growth of lamellae there has been considerable theoretical discussion of two aspects, firstly, the lamellar thickness i.e. the interval at which chains fold and, secondly, the kinetics of growth. We shall consider these in turn.

The thickness of polymer lamellae is primarily a physical rather than a chemical quantity. It is not constant but generally increases with crystallization temperature; approximately as ΔT^{-1} where ΔT is the supercooling. This is behaviour shown by classical nucleation and, until recently, viable theories of chainfolding have all been of this kind.

Consider, as a guide, the formation of a crystalline nucleus of thickness ι containing ν chain segments arranged on a square lattice of parameter a. The excess Gibbs function, ΔG, created in the system by the formation of this nucleus is

$$\Delta G = 4\iota a \sqrt{\nu}. \ \ \sigma + 2\nu \ \sigma_e - \nu a^2 \iota \ \Delta f$$

where σ is the Gibbs function per unit area of prism face, σ_e similarly that for the basal surface and $\Delta f \ \ \Delta h \Delta T / T^o_m$ in the Gibbs function/unit volume of crystal; Δh is the enthalpy of fusion/unit volume at the equilibrium melting temperature T^o_m and $\Delta T = T^o_m - T_C$ where T_C is the crystallization temperature. Then the dimensions of the critical nucleus are given by

$$\frac{\partial \Delta G}{\partial \nu} = \frac{\partial \Delta G}{\partial \iota} = 0 \tag{1}$$

leading to $\iota^* = \dfrac{4\sigma_e}{\Delta f} \simeq \dfrac{\sigma_e \ T^o_m}{\Delta h . \Delta T}$ and $\Delta G^* = \dfrac{32\sigma^2 \sigma_e}{(\Delta f)^2}$ (2)

The critical thickness of the primary nucleus increases as $(\Delta T)^{-1}$. It is also of the correct order of magnitude (other quantities can be derived independently). At first it was suggested that the observed lamellar thicknesses were those of the primary nuclei but it was very soon shown that a change in crystallization temperature (from solution) produced a sharp change in crystal thickness, i.e. the process controlling lamellar thickness occurs at the growing surface[8,1]. The usual model for this has been that of secondary nucleation[8]. This has considered the laying down of a chain-folding molecule, stem by stem, to form a strip on smooth growth faces like those produced on polyethylene and many other polymers, grown from solution. Provided the strip laid down does not protrude beyond its substrate, then the excess Gibbs' function due to laying down the first stem (which is assumed not to contain a fold)

is $\Delta G_1 = - a^2 \nu \iota \Delta f + 2a\iota \sigma$ (3)

and $\Delta G_2 = a^2 v (2\sigma_e - \iota \Delta f)$ for all subsequent stems in the same strip. If we are to form a stable crystal, eventually there must be a decrease in Gibbs' function. This can only be achieved if ΔG_2 is negative accordingly $2\sigma_e - \iota \Delta f < 0$

and $\ell = 2\sigma_e/\Delta f + \delta \ell$ (4)

The results of various theories differ only in their values of $\delta \iota$, which is generally small in relation to the preceding term. It may be computed by considering the steady state advance of a long strip as chain segments are added in sequence. For one thickness, the steady state flux $S(\iota)$ depends strongly upon the activation energy to lay down the first strip, i.e. the rate depends upon $\exp \Delta G_1/kT$. We may use $S(\iota)$ as a weighting factor for ι and so compute the average value and variance of the secondary nuclear thickness. This is taken to be close to the observed crystal thickness firstly because the length of the macromolecular chain precludes major readjustment of the conformation laid down in nucleation and, secondly, because it has been calculated that the thickness of successive strips is stable agains fluctuation.

Values of lamellar thickness from equation (4) are typically some tens of nm, varying approximately as ΔT^{-1} This sensitivity to crystallization temperature allows the crystallization history to be traced in part by inspection of the distribution of lamellar thickness. This can be attempted from X-ray diffraction etc. but the information is often most usefully and readily obtained from the melting endotherm. The thickness of polymer lamellae depresses the melting point typically by 10-20K, according to the equation[1].

$T_m = T^{\circ}_m (1 - 2\sigma_e/\Delta h.\iota)$ (5)

Thin lamellae are metastable and not in equilibrium. They form because they grow the fastest but can become more stable by thickening. Such behaviour is well known, occurring in part by local melting and recrystallization, and causes related modification of properties.

Calorimetry is also a most valuable guide to the interpretation of properties because not all lamellae in the sample are equivalent. Polydispersity (distribution of molecular length) is very important in practice, leading to segregation of different molecular species into particular locations. It is possible for these to crystallize at different temperatures (because of different kinetics) and be immediately identifiable in the melting endotherm. In general a combination of thermal analysis and modern methods of electron microscopy provides a good capability of analysing the contribution and nature of lamellar structure and hence an interpretation of physical properties.

The kinetics of growth are among the most studied aspects of polymeric crystallization. There is particular interest in the rate of advance of the melt/crystal interface which is compared with observations of the linearly increasing diameter of spherulites (Fig. 2). Following observations of discontinuous changes in the growth rate of good fractions of polyethylene at a precise supercooling, a body of theory has been developed to account for three regimes of behaviour, known as I, II and III[9,10]. Regime I is when strip completion rates are fast compared to initiation, i.e. nuclei are well separated on a growth face, leading to planar growth fronts. Conversely, many growth centres close together make a rough surface in regime II[9]. Regime III intervenes when the separation of centres (niches) approaches molecular widths[10]. (It is for these unfavourable conditions

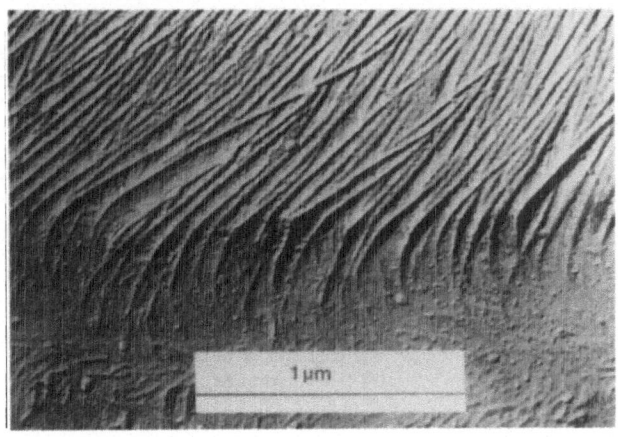

Fig. 2. Detail of the edge of a spherulite of polyethylene growing from the
melt (Courtesy of Dr A.S. Vaughan).

that the 0.65 probability of adjacent re-entry chainfolding has been
found[7]). The experimental distinction between the three regimes rests
upon the different exponents κ_g in equations of the form

$$G = G_o \exp \frac{U^*}{R(T-T_\infty)} \exp \frac{\kappa \Delta G^*}{kT} \qquad (6)$$

which all regimes satisfy.

Here the first exponential relates to transport of molecules to the growth
face. In the second exponential.

$$\Delta G^* = \frac{4a\sigma\sigma_e}{\Delta f} = \frac{4a\sigma\sigma_e}{\Delta h \Delta T} T_m^o \qquad (7)$$

is the Gibbs' function of the classical secondary nucleus; for regimes I and
II, $\kappa = 1$, but $\kappa = \frac{1}{2}$ for regime II[9,10,1]. The difference is because the
rate of completion of a strip is proportional to the nucleation rate in
regimes I and II but only to the square root in regime II.

Measurements claiming to show the three regimes and their transitions
continue to appear[11]. Of course detailed agreement with theory would
provide strong support for the relevant theories. At the same time these
are receiving a strong challenge from a new approach which produces a
similar formalism from different premises. This has arisen from observation
of crystal habits and the appreciation that, at higher crystallization
temperatures, solution grown polyethylene crystals develop rounded and,
therefore, rough growth faces although this is well known e.g. for
polyamides. It has been pointed out[12] that, under conditions in which $\delta\ell$
in equation (4) is small, i.e. $2a\sigma/kT \gg 1$ to discriminate strongly against
higher ℓ values, niches in a two dimensional crystal would be highly
favoured for growth and would rapidly grow out to smooth surfaces.

As polymer crystals do not generally behave in this way and very recent findings show that individual crystals grown at high temperature from the melt are rounded[13] (Fig. 3) then either they cannot be regarded as two dimensional layers or revision of nucleation theories is necessary. It has been proposed, following computer simulations of growth, that the dependencies of ℓ on ΔT^{-1} equation (4), and G on $1/T\Delta T$, equations (6) & (7), would follow from a so called 'roughness pinning' model[12]. In this all stem lengths are explored by the system but the crystal can only advance if the profile normal to the edge is filled out. A further stimulus has come from the availability of monodisperse paraffins. These have allowed precise growth data on very well defined systems, confirming that chainfolding sets in for chain lengths as low as C_{150} and the preference for conformations with ends excluded from lamellae[14]. They have also revealed minima in growth and nucleation rates as functions of temperature[15]. Evidently the crystal growth of polymers has entered an exciting phase of new results and new interpretations.

The lamellar nature of crystalline polymers provides both the basis of explanation of conventional properties and ways to produce exceptional properties from unusual morphologies. As one example, environmental stress cracking can be cited. Linear polyethylene is vulnerable to accelerated cracking in the presence of organic solvents. This is favoured both by the relatively sharp fold surfaces and the propensity of short molecules to segregate together. Much improvement can be gained by the incorporation of branches into the chain. Ethyl and longer branches are mostly excluded from lamellae making fold surfaces more diffuse and able to blunt cracks while the basis of segregation is changed to high branch content[1].

A second example is the production of high modulus and high strength fibres. These require high orientation and absence, or at least bridging with covalent bonds, of weak chainfolded interfaces. This can be achieved in various ways. Linear polyethylene drawn to high values >~20 has sufficient crystal bridges to produce an axial Young's modulus of c.70 GPa[15]. Polyethylene of sufficiently high molecular mass made into gels with suitable entanglement density can also be drawn to very high levels[17] (>10C) achieving both high modulus and high strength (7 GPa)[18]. Alternatively lyotropic liquid crystal polymers have been aligned to produce fibres with little or no folding and exceptional mechanical properties from

Fig. 3.   Individual lamella of polyethylene grown from the melt at 130°C
          after extraction from a quenched matrix. (Courtesy of Mr M.M.
          Shahin).

aromatic polyamides[19]. Such new materials may take polymers into the
field of structural materials, especially if creep can be inhibited.

The first major use of polyethylene was as an electrical insulator.
Nowadays there is great interest in the electroactive properties of polymers. Thermotropic liquid crystal polymers can align under electric and
magnetic as well as stress fields raising many possible applications[21].
Solid state polymerization can, in certain circumstances, generate macroscopic single polymer crystals from suitable monomers among which the
polydiacetylenes have been most studied[20].

Our complex society demands ever more sophisticated materials.
Polymers are well placed to respond to the need for e.g. high performance
composites and electroactive organic properties[21]. Satisfactory performance will only result, however, when the internal organization can be ade-
quately controlled. To that end studies of the crystal growth of polymers
have many practical as well as intellectual challenges ahead.

REFERENCES

1. D. C. Bassett, Principles of Polymer Morphology, Cambridge University
    Press (1981).
2. O. Gerngross, K. Herrman, and W.Z. Abitz, Physik.Chem.Bio, 371 (1930).
3. A. Keller, Phil.Mag, 2:1171 (1957).
4. E. W. Fischer, Z.Naturforsch, 12a:753 (1957).
5. P. H. Till, J.Polymer.Sci, 17:447 (1957).
6. Faraday Disc.Chem.Soc, 68 (1979).
7. J. D. Hoffman, C. M. Gruttman, and E. A. Di Marzio, Disc.Faraday Soc,
    68:210 (1979).
8. J. I. Lauritzen and J.D. Hoffman, J.Res.Natl.Bur.Stand, 64a:73 (1960).
9. J. D. Hoffman, L. J. Frolen, G. S. Ross, and J. I. Lautitzen, J.Res.
    Natl.Bur.Stand, 79a:671 (1975).
10. J. D. Hoffman, Polymer, 24:3 (1983).
11. P. J. Phillips and Y. H. Kao, Polymer, 27:1969 (1986).
12. D. M. Sadler, Nature, 326:174 (1987).
13. D. C. Bassett, R. H. Olley, I. A. M. Al Rahell, Polymer 29:1539 (1988).
14. G. Ungar and A. Keller, Polymer, 27:1985 (1986).
15. G. Ungar and A. Keller, Polymer, 28:1899 (1988).
16. I. M. Ward, in: Integration of Fundamental Science and Technology, L.A.
    Kleintjens and P.J. Lemstra (eds), Elsevier, 634 (1985).
17. P. Smith, P. J. Lemstra, B. Kalb, and A. J. Pennings, Polym.Bull.I, 733
    (1979).
18. A. V. Savitskil, I. A. Gorshkova, I. L. Frovola, G. N. Shmikk, and A.
    F. Loffe, Polym.Bull., 12:195 (1984).
19. W. B. Black, in: Flow Induced Crystallization in Polymer Systems, R.L.
    Miller (ed), Gordon and Breach, 245 (1979).
20. D. Bloor, in: Developments in Crystalline Polymers 1, D.C. Bassett
    (ed), Applied Science (1982) p.151.
21. H. J. Coles, in: Developments in Crystalline Polymers 2, D.C. Bassett
    (ed), Applied Science (1988) p.297.

GROWTH OF SHAPED CRYSTALS

R. S. Feigelson

Center for Materials Research
Stanford University
Stanford, USA

I. INTRODUCTION

Shaped crystal growth refers to the growth of single crystals with a predetermined cross sectional configuration designed for a specific application. The quest to develop such techniques has been driven by the need to minimize costs associated with device fabrication such as cutting or polishing, the loss of expensive materials during these fabrication processes, and the damage created by machining processes.

Perhaps the first intentionally shaped crystals were the metal filaments grown by Von Gomperz [1] in Germany in 1922 through a hole in a mica plate positioned on the melt surface. In 1938, Stepanov [2] in the USSR, began studies on the growth of shaped crystals using mechanical means to control the forces of surface tension. In his approach, the melt column was crystallized outside of the container by inserting wettable or nonwettable dies of different shapes into the melt.

In the United States, work on growing sapphire filaments and single crystal superalloy jet engine turbine blades began in the late 1960's [3,4]. The former led several years later to the development of the Edge-Defined Film Growth Process (EFG) [5] for the production of sapphire tubes, filaments, ribbons and other shapes for special purposes. The EFG process (which is similar to the Stepanov method) and the unidirectional casting method for single crystal turbine blades were the first shaped crystal growth methods to find practical commercial applications.

Aside from these two applications, shaped crystal growth technology would probably have remained a laboratory curiosity were it not for the energy crisis in the mid-seventies and the emergence of an urgent need to produce high quality silicon for solar cell applications. The cost of producing silicon single crystals had to be reduced dramatically to make these devices competitive with other energy sources. Increased growth rates compared with conventional methods and reduction in machining costs and material losses were necessary. Development of production methods for the preparation of high quality single crystal silicon ribbon at very high growth velocities was therefore a highly desirable approach to minimize fabrication costs. As a result, existing methods such as the Stepanov, EFG, and Dendritic Web Processes were studied for this application, and new processes such as the Ribbon to Ribbon, Ribbon Against Drop, and Silicon on Ceramic methods were developed.

This lecture will cover the various techniques used for growing shaped crystals including the Stepanov, EFG, single crystal casting, laser-heated pedestal fiber growth, capillary shaping, and VLS methods, as well as some special techniques for growing Si ribbon. Also discussed will be some of the underlying physical principles governing these crystal growth processes.

A comprehensive survey and bibliography on the science and technology of shaped crystal growth was published a number of years ago as a special issue of the Journal of Crystal Growth [6] and is an excellent starting point for those interested in pursuing this subject further.

II. THE STEPANOV METHOD

The first serious, detailed work on shaped crystals, began with Stepanov and his co-workers [7-15]. They simultaneously studied the development of the technology for preparing shaped single crystals and the physical phenomena and theory involved. The Stepanov method involves the growth of crystals with a cross sectional configuration determined by a die-shaped melt column which is usually above the melt level in the crucible. Several versions of Stepanov's method are illustrated diagrammatically in Figure 1. Versions A-E can be used to grow cylinders, tubes, and ribbons while F was used to obtain thin films. Shaped semiconductor crystals of Ge [10,15,16], InSb [17], GaAs [18], and Si [19] were grown by a variety of these techniques. The effectiveness of the method and the quality of the crystals produced depends on a number of factors including the crystal structure of the material to be grown, the shape required, and the wettability and reactivity of the melt with the shaper material. Both wetted and nonwetted shapers have been used and the degree of wettability determines the shape of the meniscus, and the cross sectional configuration of the growth interface as seen in Figure 1 where θ is the wetting angle. When the melt does not react with the shaper, then crystals with near perfect shape and with few inclusions can be grown. For the semiconductor crystals mentioned above, graphite and glassy carbon shapers are commonly used.

In common with other crystal growth methods, it is important to know the relationship between the chemical and physical properties of the materials to be grown and the processing parameters so that high quality crystals can be produced economically. Extensive measurements and/or calculations have been made on the temperature distributions in these systems, thermoelastic stress formation in crystals having different configurations, capillary effects, and impurity incorporation. The shape of the melt column is determined by its capillary properties which are influenced by surface tension, density, melt viscosity, and impurities ($\alpha = 2G/pg$), and the interaction between the melt and shaper material, i.e., the wetting angle, chemical reactivity, etc.

The basic conditions required to produce melt columns of specific shapes were found by solving Laplace's capillary equations [14,20-22]. From these calculations, the column configurations for various diameters and melt pressures could be determined as well as the maximum height of the melt column, the stability and shape of the growth interface and the conditions required for maintaining constant shape of both the column and crystal. Many of these theoretical calculations were experimentally verified, thereby demonstrating the effectiveness of the mathematical

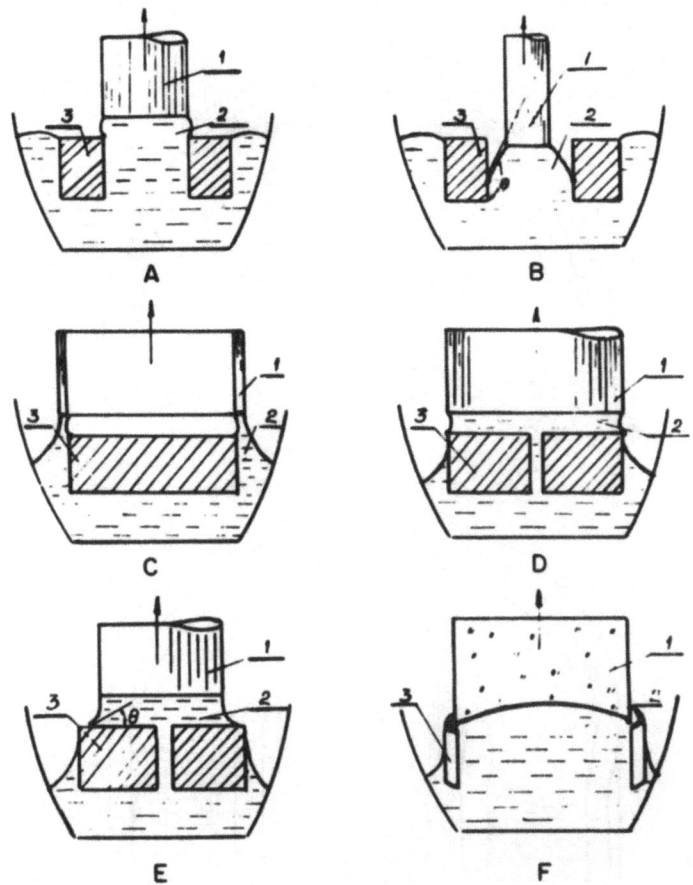

Fig. 1.   Diagrams for various versions of Stepanov's method: (1) hard
          crystals; (2) melt; (3) shaper [2].

approach and giving credence to its ability to predict the way changes in
processing conditions affect crystal size, shape, and quality. The final
crystal shape was found to depend on the properties of the melt column and
the temperature gradients in the system.

     The melt column shape in Stepanov growth can be changed in a number
of ways besides changing the configuration of the shaper, including
passing electric current through the interface [23], applying ultrasonic
vibrations, and by using electromagnetic fields [24]. The crystal shape
may also be changed by changing the wettability of the melt or its
viscosity with the use of melt additives.

     As with other crystal growth methods, growth rate anisotropy gives
rise to faceted crystals which can affect the shape of the final crystal.
Facet and growth ridge formation is a function of the growth orientation,
supercooling, and temperature gradients in the liquid and solid. When
growing Ge crystals by the Stepanov method, it was found [25] important to
carefully align the melt column, temperature field, and growth orientation
to minimize the formation of lattice defects such as dislocations and
twins. Dislocations, caused by plastic deformation during growth can
often be caused by thermoelastic stresses. This happens when the maximum
stress in certain regions of the crystal exceeds the critical threshold
for heterogeneous dislocation formation at a given temperature. These

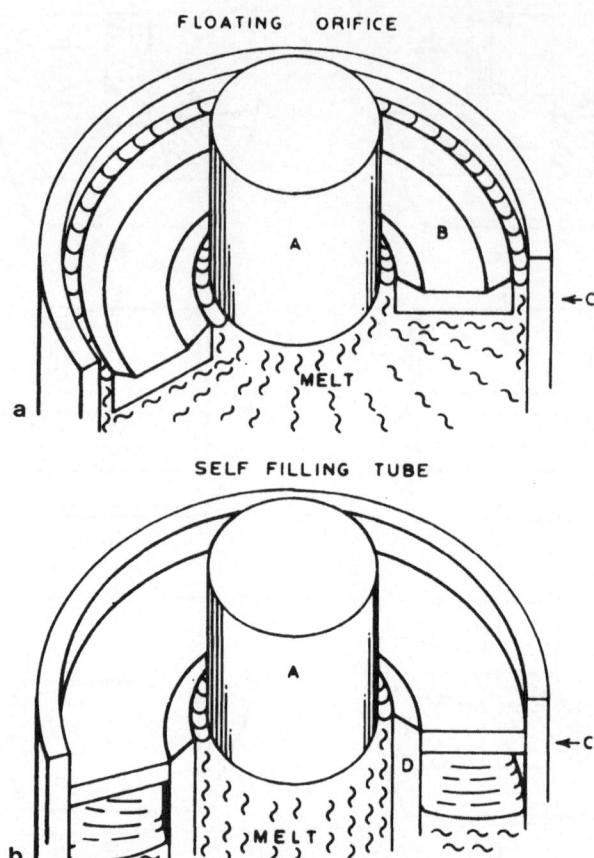

FLOATING ORIFICE

SELF FILLING TUBE

Fig. 2. (a) Floating orifice technique and (b) self-filling tube: A is
the filament, B is the floating orifice, C is the crucible, D is
the capillary tube [3].

stresses have been found to be an order of magnitude greater in thin
ribbons compared to a circular cross section. In a circular rod $\sigma_t \alpha R$,
whereas in a ribbon $\sigma_t \alpha b^2/t$ where R is the radius of the rod, b is the
width, and t the thickness of the ribbon [26].

III. EFG GROWTH

Details about the invention and development of the EFG process for
sapphire growth are described by LaBelle [3]. Initially, a method for
growing sapphire fibers was sought for high strength fiber-matrix
composite applications. A Floating Orifice Technique [27] was tried first
as shown in Figure 2a. This method had some serious drawbacks for
production applications since as the melt is consumed the level drops as
does the orifice. This causes a change in temperature at the
solidification interface and mechanical instability due to the orifice
submerging.

To address the above problems and the desire to be able to have a
continuous process, a modification was investigated called the Self-
Filling Tube Technique [28] as illustrated in Figure 2b. In this version

278

the interface position was fixed and the melt is supplied to the interface region by capillary action, eliminating changes in temperature and mechanical instabilities as the melt level drops. A key to the success of this method was that the melt had to wet the capillary tube material and be nonreactive. Molybdenum was found to meet this criteria and sapphire crystals with 5 ppm Mo were grown using this technology. The refined method had a number of advantages including: (1) an ability to grow a number of different shapes such as ribbons, tubes, and fibers; (2) a macroscopically stable interface; (3) a controllable heat flow and temperature gradient in the growing crystal; (4) continuous melt recharging where thermal perturbations at the growth interface are damped out due to its remote location from the bulk of the melt; (5) many crystals can be grown simultaneously (as many as 54 at one time have been reported [3]); (6) dendritic growth can be achieved [29] by supercooling the capillary region thereby maximizing growth rate; and (7) in ribbon growth the linear pull rate is not dependent on ribbon width, only ribbon thickness since the latent heat is removed per unit area. In spite of these positive features, some dimensional instability in the growing crystal was encountered because of a lateral movement of the crystal with respect to the melt due to convective currents in the inert atmosphere used. Although guiding mechanisms added were partially successful at reducing the effect, it did not completely resolve the problem and the crystal diameter could exceed the size of the orifice. Diameter variations were also caused by temperature fluctuations at the interface.

A significant observation was made by LaBelle during the course of this work that led to the EFG method [3]. It was found that the crystal could grow several times larger than the diameter of the orifice if the angle between the die surface and its normal was 75° or greater. If it was 30° or less, it could not exceed the orifice diameter. One could limit the crystal size therefore by either making this angle very small (a small cone angle which is mechanically unstable) or making it 90° and letting the melt flow out across the top surface until it reaches the edge where it stops. This latter mode, EFG, allows diameter or shape control simply by controlling the outer dimension and shape of the die.

The EFG process as developed by LaBelle and co-workers [3,30] is similar to Stepanov's method, particularly version D in Figure 1. The major difference lies in the fact that the outside of the shaper the die determines the shape of the crystal in EFG, while the melt column shape was the central focus in Stepanov's work as shown in Figure 3. A significant advance in shaped crystal growth was made when LaBelle recognized [31] that when the die shaper was wetted by a melt it provided an effective means of isolating the growth interface from the bulk melt affording a better way of controlling the cross sectional shape of the

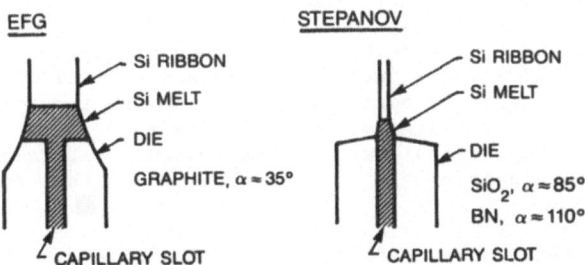

Fig. 3.   Schematic representation of silicon ribbon growth in the EFG and Stepanov processes, α is the contact angle of the die material with the molten silicon [31].

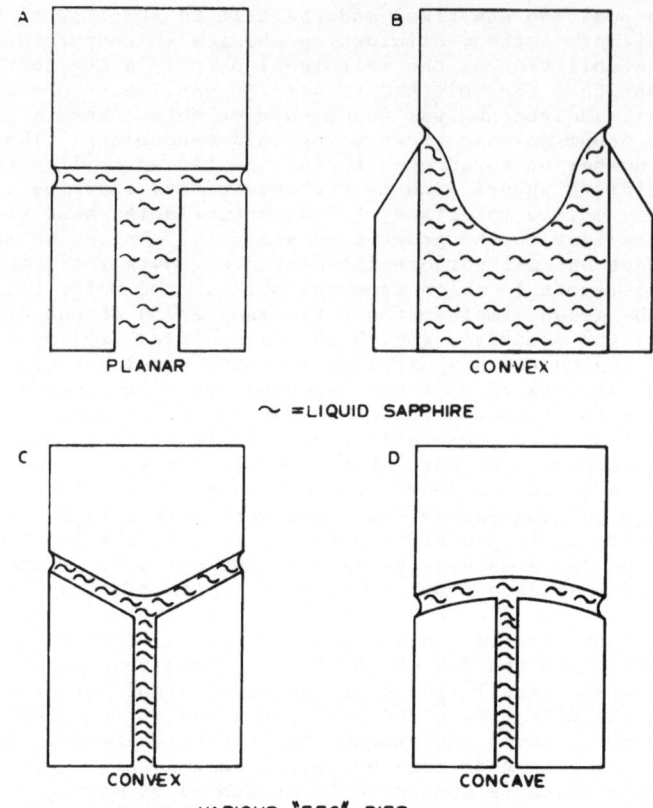

A

B

PLANAR

CONVEX

~ =LIQUID SAPPHIRE

C

D

CONVEX

CONCAVE

VARIOUS "EFG" DIES

Fig. 4.   EFG dies used to vary interface shape by die design [3].

crystal. In a relatively short period of time after this discovery, the
EFG method became a commercially viable technology. In order to control
the crystal shape and develop an efficient, cost effective manufacturing
process, a fundamental understanding of the phenomena involved was needed
and the theoreticians were brought in very early in the development of
this method. The basic EFG method is illustrated schematically in Figure
4a for a solid of any desired cross section. The shape of the interface
can be simply varied as shown in Figure 4b-d. In Figure 5a, an
illustration of the first EFG die used for tube growth is shown with
internal holes directing the melt to the die surface. Figure 5b shows a
die design for a seven bore tube and the melt column shape.

The EFG process therefore has the virtues of the self-filling tube
method plus other advantages including complete control of cross sectional
shape as shown in Figure 5, optimized heat flow and impurity distribution
by control of capillary design as shown in Figure 6. Control of interface
shape as shown in Figures 3a-d and a distribution coefficient of unity can
be achieved through appropriate capillary design so that impurities

Fig. 5.   a) The first EFG tube die used. The die is 3.0 mm outside
          diameter and 1.5 mm inside diameter.
          b) Die design for seven bore tubing growth comparing the melt
          column shape to crystal shape.

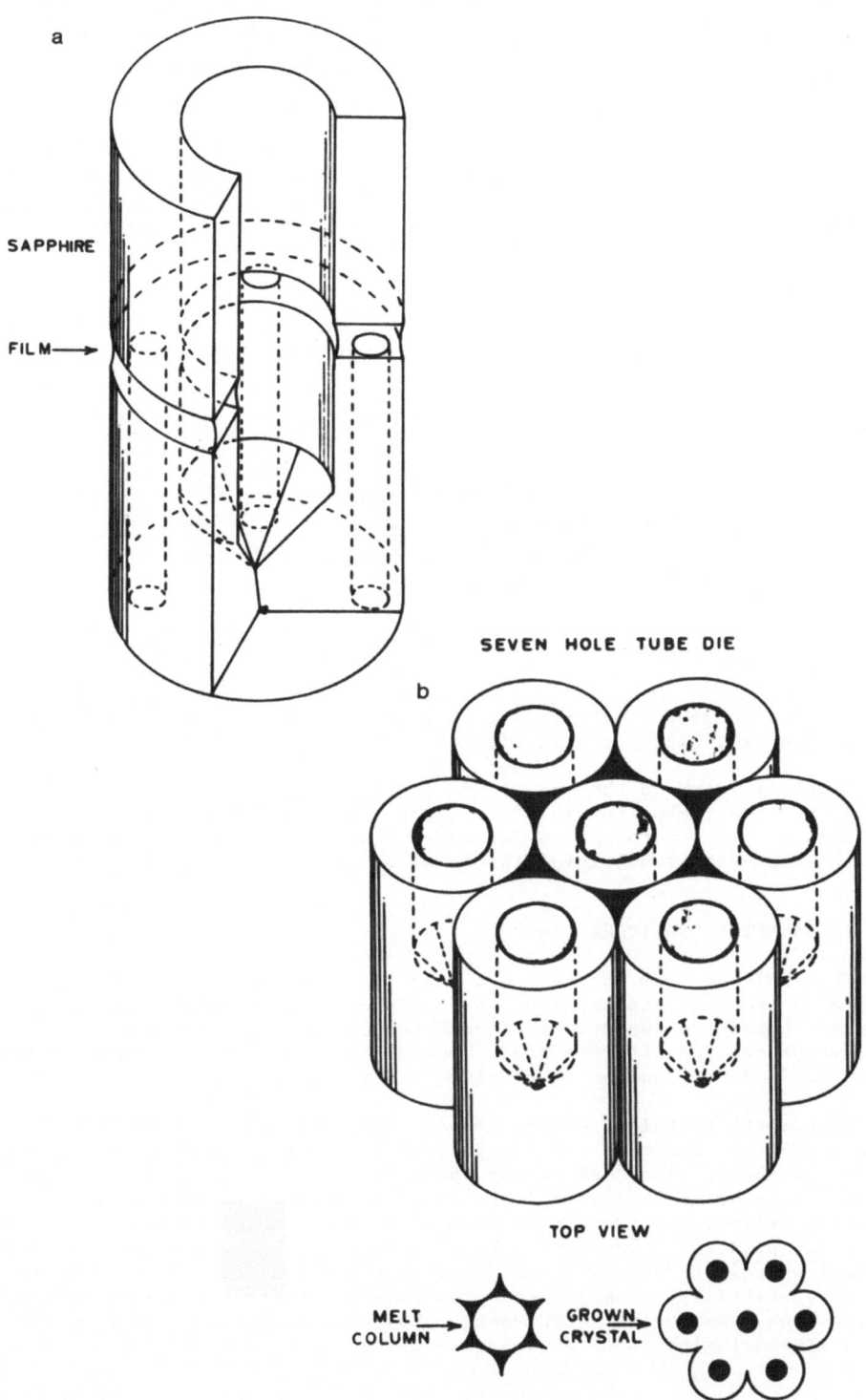

a

SAPPHIRE

FILM→

SEVEN HOLE TUBE DIE

b

TOP VIEW

MELT
COLUMN →

GROWN
CRYSTAL →

Fig. 5. See Legend on previous page [3].

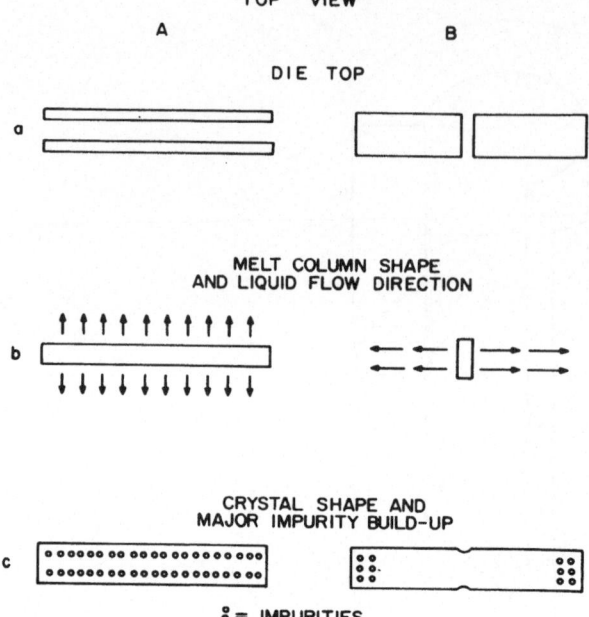

Fig. 6.   The influence of die design on liquid flow and impurity distribution [3].

rejected at the growth interface cannot diffuse back to the melt as shown by Swartz et al. [32] for eutectic growth.

Extensive work on the use of EFG for the growth of low cost single crystal silicon ribbon for solar cells has been on-going, principally by Mobil Tyco, for more than a decade and has concentrated on the very high speed growth of multiple ribbons of solar quality material.

IV.   FUNDAMENTAL CONSIDERATIONS

The scale-up from laboratory set-up to commercial process usually requires a thorough knowledge of the fundamental phenomena involved in the process - hence the theoretical foundation needs to be developed to solve problems associated with the scale-up and perturbations introduced outside the controlled environment of the laboratory.

Like most crystal growth processes, shaped growth methods have a large number of dependent and independent parameters which effect growth stability, shape, size, and growth rate. Both the material and system properties determine the ease a specific product can be manufactured. As mentioned before, the transition from a research scale process to commercial production requires a basic understanding of the physical and chemical principles involved including mass and heat transport effects, capillary stability, chemical reactivity, meniscus height and stability, supercooling, thermoelastic stresses, etc. Although there are many shaped growth methodologies, many of the underlying principles are common to most of them.

The melt growth of shaped crystals usually occurs by a meniscus-controlled process. Figure 7 shows a cross section, schematic view of the

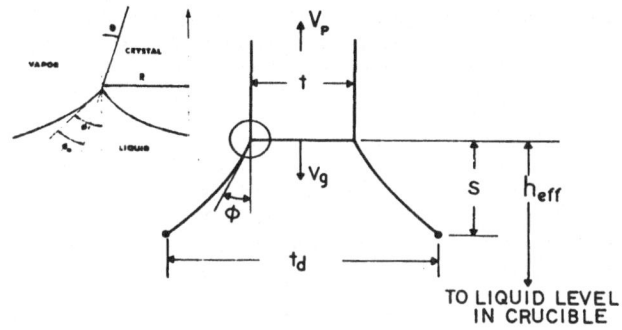

Fig. 7. Schematic cross-section of crystal growth from a die shaper of dimension t_d showing geometrical variables which determine the crystal dimension t. [33]

EFG or Stepanov processes. In this type of process, the angle ϕ between the crystal and liquid-free surfaces at the crystal-melt-vapor juncture, as shown in Figure 7, influence the crystal size and shape. Surek et al. [33] developed a general theory based on linear perturbation analysis for the time evolution and stability of the crystal shape in meniscus-controlled growth. It was found that a crystal cannot grow with a uniform cross section unless the meniscus angle $\phi = \phi_0$ where ϕ_0 is a characteristic and anisotropic property of the material. Table 1 shows some typical values of ϕ_0 for a number of common materials. The crystal size t in Figure 7 will increase when $\phi > \phi_0$ and decrease when $\phi < \phi_0$. While the basic physical processes of determining crystal shape are the same for the Czochralski, float-zone, Stepanov, and EFG methods, which are all meniscus-controlled processes, the degree of shape control and the range of cross sectional shapes which can be grown stably does vary. These differences relate to heat flow and the exact shape of the meniscus.

Surek et al. [33] used the following set of equations to describe the time dependent crystal shape in a meniscus-controlled process:

$$dt/d\tau = 2\ V_g\ \tan(\phi - \phi_0) \tag{1}$$

$$\phi = \phi(t,\ td,\ s,\ h_{eff}) \tag{2}$$

$$ds/d\tau = V_p - V_g \tag{3}$$

$$V_g = V_g(t,\ S,\ \phi,\ V_p,\ H) \tag{4}$$

Table 1. Measured Values for ϕ_0

Material	Orientation	ϕ_0 (deg)	Ref.
Si	<111>	11	[34]
Ge	<111>	13	[34]
Al$_2$O$_3$	[0001]	12	[35,36]
YAG	<100>	8	[35]
LiNbO$_3$	[0001]	4	[35]

where Vg is growth velocity in the pull direction, Vp is the velocity of
pulling the crystal, t is crystal dimension (ribbon thickness and rod
diameter), td is the edge-defining die dimension (point of attachment of
melt to die), s is the meniscus height (height of growth interface above
die top), h_{eff} hydrostatic pressure (effective height of growth interface
above the melt level in crucible), H a heat transfer coefficient in the
system, and τ a time variable. Anisotropy was neglected in their
analysis. They found that because the lower end of the meniscus is
constrained by the die shaper in the EFG and Stepanov methods, the
condition where $d\phi/dt < 0$ leads to shape stability for a wide range of
growth conditions. Whether to use a wetted or nonwetted die depends, they
found, on other process parameters such as h_{eff}. Greater shape stability
was predicted for large values of meniscus height and high growth rates
when wetted dies with $h_{eff} > 0$ were used. Tatarchenko et al. [37],
studying the crystallization stability (capillary thermal) in the Stepanov
method, found it easy to grow shaped crystals of constant cross section
when the relation between crystal and shaper sizes was $R/r_0 > 1/2$ and that
self-stabilization can take place when forced cooling of a local region of
the crystal is used, and at increased growth rates.

In the growth of ribbon-shaped crystals the geometry of the crystal
is largely controlled by the meniscus length across the width of the
ribbon and the position of the ribbon edges [38]. It is therefore
necessary to control this parameter if ribbon of uniform thickness and
width are to be produced.

V.   SILICON RIBBON GROWTH

The need for low cost, large grain or single crystal solar cell
quality silicon in a form requiring little fabrication led to a dramatic
increase in effort on shaped crystal growth development. Not only were
the Stepanov and EFG methods studied for this application and high
efficiency cells produced at growth velocities up to 7.5 cm/min, but a
large number of other potentially viable methods were invented and
studied. The Capillary Action Shaping Technique (CAST) [39] is similar to
the EFG process while another one is based on an inverted version of the
Stepanov process [40]. High efficiency cells have been made by many of
these methods but the cost factor will be the determining factor in which
technique if any will be satisfactory. A brief discussion of the
Dendritic Web, Silicon on Ceramic, Ribbon Against Drop, Ribbon to Ribbon,
and Horizontal Ribbon Growth methods is given below.

A.   Dendrite Web Process

The silicon Dendritic Web ribbon growth process [41-43] relies on
crystallographic and surface tension forces rather than shaping dies to
control crystal shape. Ribbons 2-4 cm wide and 0.1-0.2 mm thick have
obtained AMI conversion efficiencies of 15.5%. Figures 8a and 8b show a
schematic depiction of this process. When a dendritic seed is placed in
a melt and supercooled by a few degrees, the seed first spreads
laterally to form a button. When the seed is raised, two secondary
dendrites propagate from each end of the button and they, together with
the button, form a support from which the film forms and solidifies.
The temperature difference across the melt surface fixes the ultimate
web width and the web grows from a meniscus height of 6-7 mm as shown in
Figure 8. Growth direction is near the [211] because of
crystallographic and heat flow factors. Flatter melt profiles permit
larger web widths but larger thermal stresses cause twisting or curling.
The silicon web effectively segregates metal impurities to the melt so

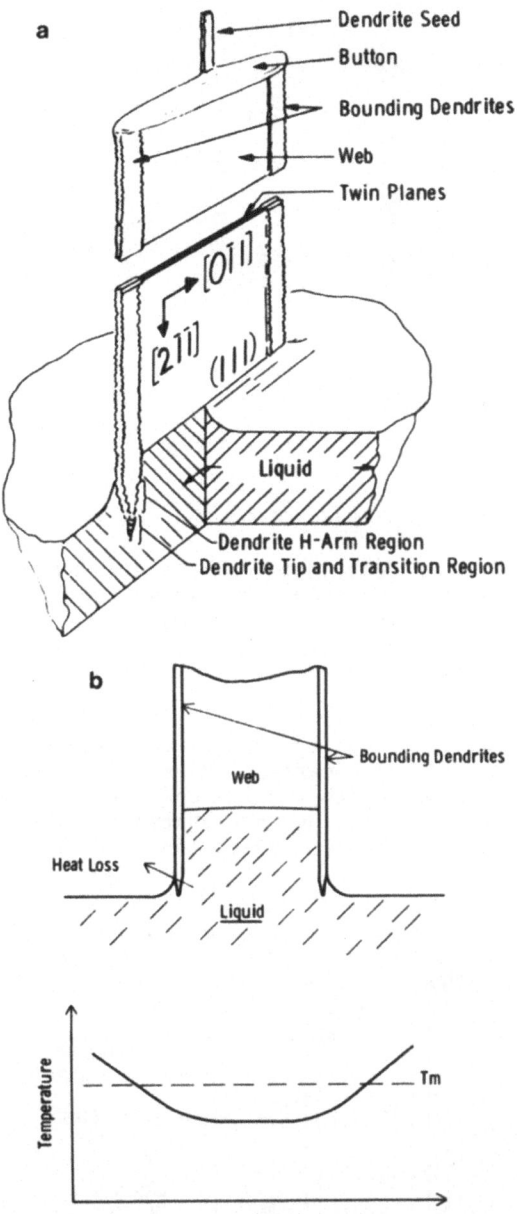

Fig. 8. a) Schematic depiction of dendritic web growth. Seed button, bounding dendrites and crystallographic orientation of the ribbon are illustrated.
b) Illustration of the way the lateral temperature distribution in melt and lateral heat loss from the ribbon influence the dendritic web widening process [41].

that cheaper starting material can be used. The growth velocity is controlled by the latent heat of fusion and can reach 5 cm/min. One disadvantage of the web process is that the dendritic edges must be removed.

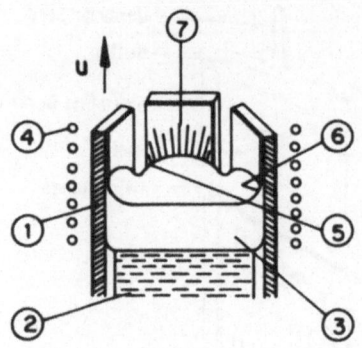

Fig. 9.   Schematic sketch of the RAD pulling process for a quasi-
cylindrical configuration: (1) ribbon substrate, (2) Si pedestal,
(3) molten zone, (4) working coil, (5) freezing isotherm, (6) Si
meniscus, (7) Si layer and grain boundaries [44].

B.   Ribbon Against Drop Method and Silicon on Ceramic Processes

The Ribbon Against Drop (RAD) Method was described by Belouet [44-
45]. This process involves deposition of thin silicon sheets on a ribbon-
like carbon substrate as illustrated in Figure 9. The ribbons are not
really single crystal but have highly oriented grains elongated along the
growth direction with the [$\bar{1}$12] being close to the growth axis.

The Silicon on Ceramic Process (SOC) [46] is like the RAD method
except a ceramic substrate is used to support the growing silicon ribbon
as illustrated in Figure 10. The ceramic, usually mullite, is coated with
a carbon layer in the regions where the silicon layer is wanted to
facilitate the molten silicon wetting the substrate. This coated
substrate is lowered into the molten silicon and then pulled at a constant
velocity of up to 9 cm/min.

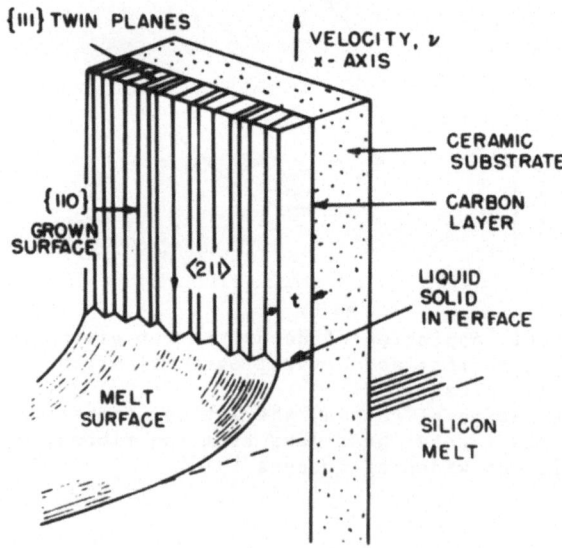

Fig. 10.   Sectional schematic of SOC crystal growth by dip-coating from
the melt [46].

286

The type of substrate material suitable for use with this process is limited by the severe thermal and chemical conditions involved.

C. Ribbon to Ribbon Growth

The Ribbon to Ribbon (RTR) Growth Method [47] is basically a Laser-Heated Float-Zone technique. A polycrystalline silicon ribbon is first grown on a temporary substrate by chemical vapor deposition and then separated from it during cooling. A narrow molten zone is then established across the polycrystalline ribbon as shown in Figure 11 by a pair of scanned laser beams. Large grained (several mm wide) silicon ribbon can be crystallized at growth rates as high as 8.5 cm/min. At higher growth rates (to 13 cm/min) dendritic structures are encountered. Temperature profiles are optimized by using auxiliary heaters and heat shields.

D. Horizontal Ribbon Growth

The Horizontal Ribbon Growth (HRG) Method [48-49] for silicon single crystal growth involves pulling a ribbon horizontally from a free melt surface which extends beyond the edge of the crucible as shown in Figures 12a and 12b. The ribbon can be grown continuously from a thin seed by cooling the upper surface of the crystal and hence forming a crystal growth front which has a wedge shape. This wedge has a small ratio of length to thickness (L/t). The process does not require a die, only the flat melt surface and control of the freezing isotherm. Since the latent heat of solidification is dissipated to the atmosphere from the growth interface, extremely fast growth rates 10-40 cm/min for single crystal and 85 cm/min for polycrystalline ribbon can be obtained. The ribbon thickness was in the range 0.4-2 mm and typically 10-30 mm wide. Typical dislocation densities were 10^5 cm^{-2} and conversion efficiencies of 9-10% AMI were obtained.

All the above have some positive and negative advantages. Remember that for the solar cell applications, economics is the principal

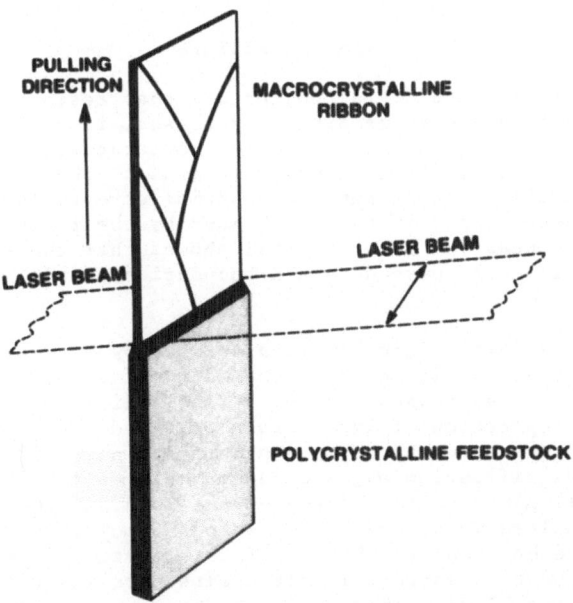

Fig. 11. RTR crystal growth method [47].

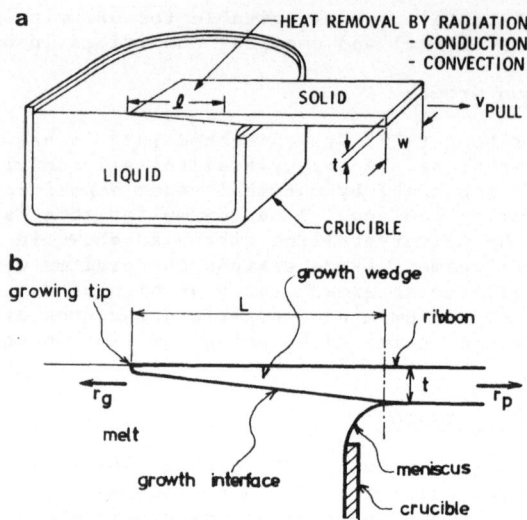

Fig. 12.　a) Cross-sectional sketch of horizontal ribbon pulling from a melt (w = half-width).
b) Schematic of the growth wedge used to define variables in heat flow analysis [48].

driving force and sacrifices in perfection, both crystallographic and chemical, are permissible as long as efficiency is high and cost low. Therefore, research workers are pushing all the above technologies to their limits. Extensive studies have been made on the relationship between the growth conditions and defect structures, and theoretical relationships have been derived to help guide in the selection of processing parameters which allow higher perfection at faster growth rates.

VI.　UNIDIRECTIONAL CASTING OF SINGLE CRYSTALS

In 1960 Ver Snyder and Guard [50] showed that resistance to creep, the gradual elongation due to mechanical stresses, in certain nickel-based alloys which have high strength at elevated temperature ($\sim 1000°C$) could be dramatically increased if these alloys were made so that all the grain boundaries are parallel to the applied uniaxial stress, such as the centrifugal stresses caused in turbine blades by their high rates of rotation. In 1967 Kear and Piearcey [51] showed that these alloys would be even stronger if there were no grain boundaries at all, i.e., a single crystal.

Turbine blades were originally manufactured by a standard casting method as illustrated in Figure 13a. The melt solidified first at the mold walls and proceeded inward leaving a fine-grained polycrystalline structure. The preparation of large grain oriented nickel-based superalloy blades were made by using a scheme shown in Figure 13b in which unidirectional solidification was induced by using a temperature gradient created by a chill plate at the bottom of the mold and the heat source on top. After the molten metal was poured into the mold, it was slowly lowered out of the hot zone so that the first crystals form on the chill plate and grow upward as large columnar grains parallel to the centrifugal forces acting on the blade. This grain structure prevents the crystals from being pulled apart by these forces. By placing a constriction

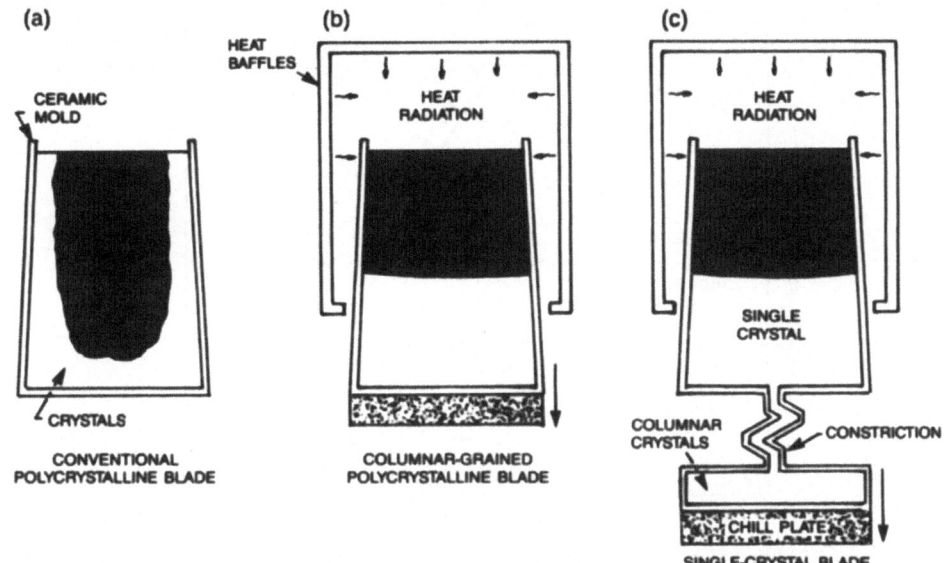

Fig. 13. Techniques for casting turbine blades a) conventional method;
b) unidirectional solidification; c) unidirectional
solidification with grain selection [50].

between the chill plate and mold it was possible, as shown in Figure 13c,
to select only one appropriately oriented grain to proceed into the mold,
thereby creating a single crystal turbine blade.

This method was also used to prepare oriented eutectic superalloys
such as nickel-aluminium-molybdenum where molybdenum solidifies in the
form of aligned filaments in a nickel-aluminium matrix.

Like most crystal growth processes the Single Crystal Casting Method
is controlled by two key parameters, the thermal gradient G at the melt-
solid interface and the growth rate R. The relationship between G, R, and
the structure is shown diagrammatically in Figure 14. The type of ceramic
mold and core material used is also critical because of the high
temperatures involved.

VII. SINGLE CRYSTAL FIBERS

There has been a continuing interest in the development of solid-
state materials for the generation, transmission, detection, and
conversion of optical signals over a broad range of wavelengths and power
levels. Single crystals have played a major role in most of these
applications either in the form of bulk crystals (three-dimensional) or
epitaxial thin films (two-dimensional). Interest in the growth of small
fiber crystals, which might be viewed as being nearly one-dimensional, is
much more recent, and has been stimulated, in large part, by their
potential for use in a variety of fiber optic applications such as laser
sources, electro-optic modulators, switches, couplers, isolators,
transmission lines, remote sensors, etc. Signal processing elements made
from single crystal fibers, when used in conjunction with low loss glass
fiber transmission lines, may lead to important new devices. Some
potential applications for single crystal fibers were discussed by Goodman
[52].

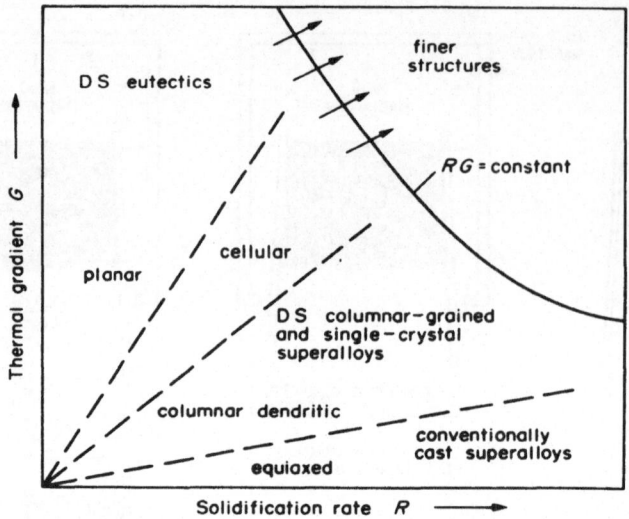

Fig. 14.   Growth rate-temperature gradient effects on grain shape and the
refining of microstructures by high cooling rates [50].

For optical applications, it is important that fibers be produced in
specific lengths, diameters and orientations. While there is a wide
variety of techniques which can be used to produce single crystal fibers,
not all allow control of these parameters sufficiently to be useful. For
this reason, vapor and solution growth methods are generally undesirable.
Fortunately several melt growth methods can satisfy these requirements
relatively easily.

Melt growth techniques which have been successfully used to produce
single crystal fibers include: (1) the EFG method [5] mentioned
previously, (2) pulling through a die [1], (3) Float-Zone (pedestal)
Growth [53-55], (4) Capillary Drawing [56] (Figure 15), and (5)
Pressurized Capillary-Fed Growth [57] (Figure 16). The choice of method
will ultimately depend on the physical and chemical properties of the
material to be grown. Clearly some methods or their adaptations are more
versatile than others.

Perhaps the most versatile of the melt growth techniques used for
fiber growth is the Float-Zone Method. When the fiber diameter is smaller
than the source rod diameter from which it grows, as shown in Figure 17,
it is also known as pedestal growth. Of all the methods mentioned above,
it alone does not require crucibles or furnace components which can lead
to contamination and confinement stress problems. In addition, crystals
of congruently and incongruently melting crystals can be grown, and the
composition of the crystal can be controlled by controlling the
composition of the starting material. With the small crystal size and
focused heat source, steep temperature gradients, and hence rapid growth
rates typically in the order of mm/min, can be achieved.

There are several types of heat sources which can be used to produce
molten zones suitable for use with the Pedestal Growth Method. For fiber
growth they must be able to produce zones having dimensions comparable to
that of the fiber and source rod diameters. Resistance, induction,
electron beam, focused lamp, and laser heating are all possible. It is
difficult, however, to produce small zones with steep gradients using

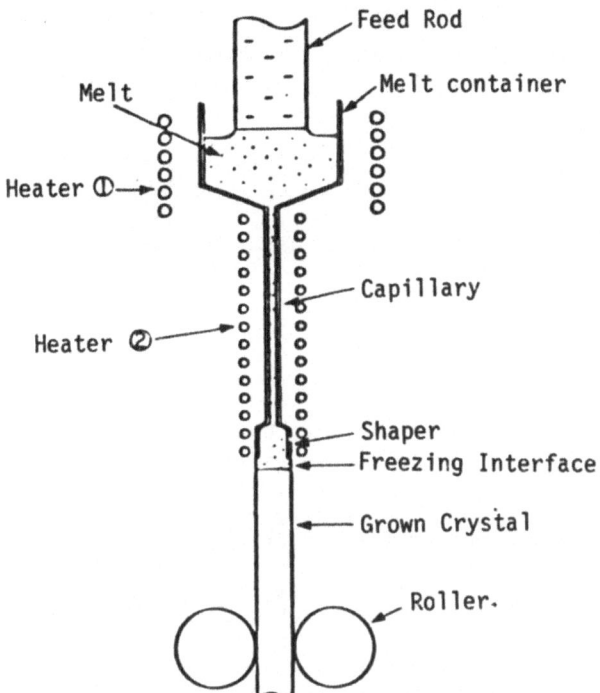

Fig. 15.   Capillary drawing method [56].

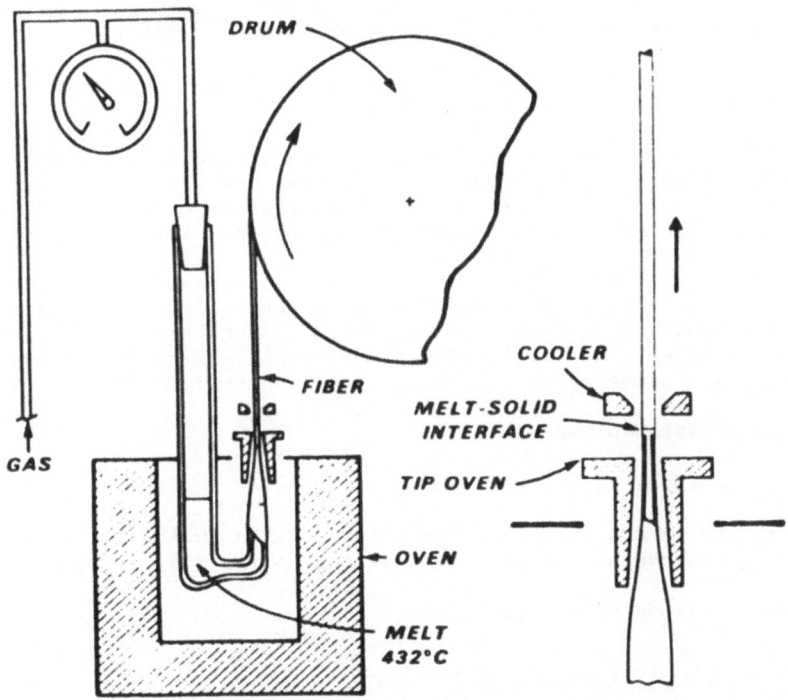

Fig. 16.   Pressurized Capillary-Fed Growth [57].

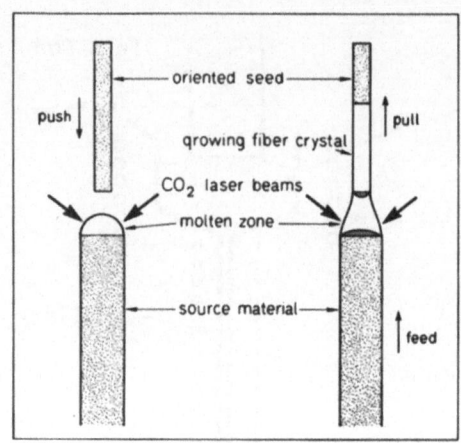

Fig. 17.   Schematic drawing of laser heated pedestal growth (Float zone)
method.

resistance heating, induction heating requires either a conducting sample
or a susceptor and electron beam techniques require a vacuum chamber.

Laser heating was first used for crystal growth by Eickhoff and Gurs
[58] for the Float-Zone growth of ruby, and soon afterwards by Gasson and
Cockayne [59] for other oxides. It is an ideal heat source because it can
be tightly focused directly onto the sample with a beam size comparable
with fiber dimensions (which may vary from a few microns to several mm),
can be used in ambient, inert, reactive, or vacuum atmospheres and is
available in power levels which can readily melt any known material whose
dimensions are of the order of the beam size. Growth can be achieved by
either moving the laser beam or the source rod and fiber.

The Laser-Heated Pedestal Growth (LHPG) Method, while not the only
approach for single crystal fiber growth, is one of the most versatile and
perhaps simplest of crystal growth methods imaginable. With it, single
crystal fibers of over 50 different materials have been grown to date
including oxides, halides, borides, carbides, metals, and semiconductors
as given in Table 2. A more detailed description of the history of fiber
growth and the LHPG method can be found in Refs. [55,60].

The typical shape of the molten zone in pedestal growth is
illustrated in Figure 18. The dimensional stability of a growing fiber,
similar to most other shaped growth methods, requires that the zone length
and volume be kept constant. The steady-state growth of a fiber crystal
with constant cross section in a meniscus controlled process requires:

(a) conservation of mass,

$$r_f = R_s (v_f/V_s)^2,$$
(5)

where r_f and v_f are radius of the fiber and its pull rate respectively and
R_s and V_s the radius and push rate of the source rod;

(b) conservation of energy,

$$Q_s = Q_f + Q_m = A\rho_s \Delta H_f \frac{dx}{dt} + AK_\ell \left(\frac{dT}{dx}\right)$$

$$= AK_s \left(\frac{\delta T}{\delta x}\right)_s = \text{const.,}$$
(6)

292

Table 2. Representative List of Crystals Grown by the LHPG Method

Material	Melting Temp °C	Orientation	Diam (μm)
Oxides:			
Nd:YAG	~1940	[111],[100]	6-600
Al_2O_3	2045	a,c	55-600
$Ti^{+3}:Al_2O_3$	2045	c	200-800
$LiNbO_3$	1260	a,c	10-800
$Nd:LiNbO_3$	1260	c	800
Li_2GeO_3	1170	a,c	100-600
$Gd_2(MoO_4)_3$	1157	[110]	200-600
$CaSc_2O_4$	2200	a,b,c	100-600
$Nd:CaSc_2O_4$	2200	c	600
$SrSc_2O_4$	2200	-----	600
YIG	1555	[110]	100-600
$Eu:Y_2O_3$	2410	c	500-800
Nb_2O_5	1495	------	700-1700
$Bi_2Sr_2CaCu_2O_8$	900	[110]	50-1000
$BaTiO_3$	1618	c[hexagonal and cubic phases]	300-800
Fluorides:			
BaF_2	1280	[110]	200-600
CaF_2	1360	[111]	600
High melting metals and semiconductors:			
LaB_6	2715	------	200
Nb	2468	------	200
B_9C	~2400	------	200

where Q_s is the heat flux in the crystal away from the growth interface, Q_m is the heat flux from the melt toward the interface, Q_f is the latent heat of crystallization, A is the area of interface, ρ_s is the density of solid, ΔH_f is the latent heat, K_ℓ and K_s are the thermal conductivity of the liquid and solid, respectively, and $(dT/dx)_s$ and $(dT/dx)_\ell$ are the temperature gradient in the solid and liquid respectively; and

(c) shape stability,

$$\phi = \phi_o,$$

where ϕ, which is related to surface tension of the melt and its ability to wet the crystal, is the angle between the meniscus and the growth axis and ϕ_o is a material constant independent of fiber growth rate, diameter and zone length but not crystallographic orientation. The optimum steady-state interfacial wetting angle can be expressed by the relationship

$$\cos\phi_o = \sigma_{sg}^2 + \sigma_{\ell g}^2 - \sigma_{s\ell}^2/2\sigma_{sg}\sigma_{\ell g}, \tag{7}$$

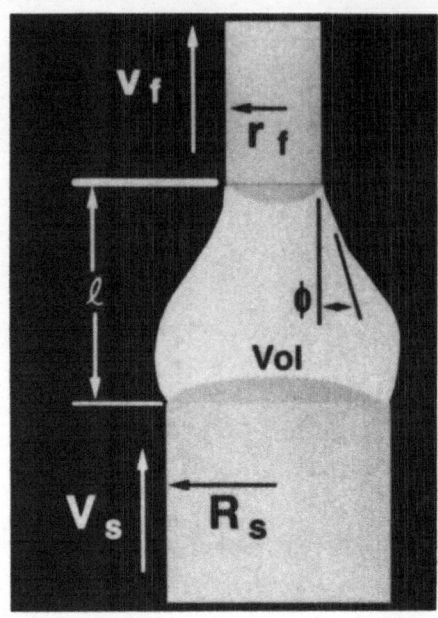

Fig. 18. Shape and stability parameters for pedestal growth where r_f and
R_s are fiber and source rod radius, v_f and V_s the fiber pull
rate and source rod push rate, ℓ the zone length and ϕ the
meniscus angle.

where σ is the interfacial free energy between solid-solid, solid-gas and
liquid-gas interfaces and anisotropy is neglected.

Measured values for ϕ_o for a few representative materials were given
in Table 1.

Deviations from steady state, i.e., perturbations in zone length or
volume lead to diameter variations such that

$$dr/dt = v \tan(\phi - \phi_o) \neq 0. \tag{8}$$

Considerable work on the shape and stability of both horizontal and
vertical molten zones have been reported [61-68]. A detailed analysis of
the stability of a vertical floating zone of cylindrical symmetry under
the influence of gravity has been given by Heywang [61], who found that
for small diameter rods with $r_f = R_s$ (vol $= \pi r^2 1$), the maximum stable zone
length (l_{max}) increases linearly with diameter. For large diameter rods,
$l_{max} = 2.84 \, (\gamma/\rho g)^{1/2}$, where γ and ρ are surface tension and density of
the liquid respectively. In the case of zero gravity, Rayleigh [62] and
Pfann and Hagelbarger [64] calculated that $l_{max} = 2\pi r$. Coriell et al.
[66] showed that in the limiting case for small diameter rods the
stability of a zone under gravity can be characterized by the
dimensionless Bond number, $B = \rho g d^2/4\gamma$. For typical fibers $B \ll 1$ (3×10^{-2}
for a 1.5 mm Al_2O_3 fiber) and gravity can be neglected.

Surek and Coriell [67] studied the shape stability of various float-
zone configurations for silicon crystal growth. Using linear perturbation
analysis they studied the effect of small deviations from steady-state on
the stability of the floating zone under the influence of gravity, i.e.,
whether these perturbations decay with time and the system returns to
steady-state conditions or are amplified. They showed that for a stable

system, these perturbations lead to an exponentially decaying oscillation about steady-state. A set of homogeneous and inhomogeneous equations of the form

$$\Delta(r, \phi, vol) = K_1 e^{-\alpha t} \sin(\beta t + \theta_1), \tag{9}$$

$$\Delta(r, \phi, vol) = \frac{K_2}{(\omega^2 - \alpha^2 - \beta^2)^2 + 4\alpha^2\omega^2} \times \sin(\omega t + \theta_2), \tag{10}$$

respectively were derived from a series of coupled differential equations for Δr, $\Delta\phi$, and Δ volume where Δ represents a small deviation from steady state, α and β the decay constant and period of oscillation respectively and ω the imposed fluctuation. If $\omega \gg \alpha$ and $\omega \gg \beta$, the externally induced perturbations are much faster than the natural frequencies of the system and the perturbation will not significantly affect the fiber dimensions. When $\omega \approx \alpha$ or β, the effect can be large.

Fejer [35] made an in-depth study of the stability and dynamics of the miniature molten zone used in laser-heated pedestal fiber growth. Neglecting the gravitational contribution to the total energy, a reasonable approximation for small diameter zones, he found that for a given molten zone length the decay coefficient tends to decrease with the diameter reduction ratio r_f/R_s, eventually becoming negative (indicating unstable growth) at a critical value of r_f/R_s that depends on ϕ_0. Materials with higher values will tolerate greater reduction ratios as shown in Figure 19.

For the growth of oxides and fluorides by the LHPG method, zone stability is usually achieved at reduction ratios in the range 1/2 to 1/3. Short fibers of very small diameter ($\sim$5 to 10 μm) have also been grown at a ratio of 1/10 [55].

A material that melts congruently, has a low vapor pressure at its melting point, and evaporates congruently, will generally be easy to grow in single crystal form by the LHPG method. The LHPG method is also an

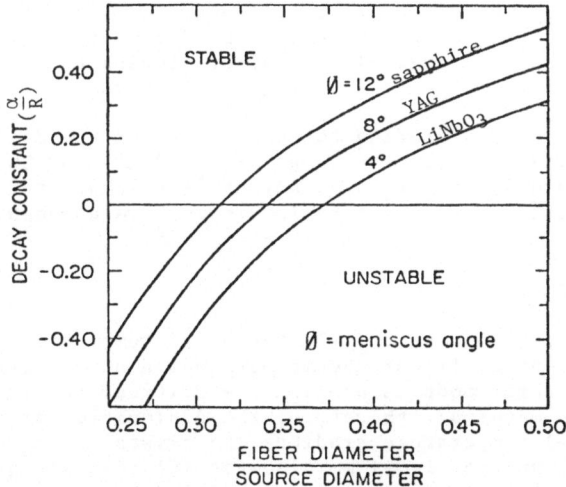

Fig. 19. Plot of decay constant as a function of fiber diameter/source diameter ratio for three values of ϕ showing the growth stability versus diameter reduction ratio [35].

excellent technique for producing single crystals of incongruently melting compounds. If we start with a source rod of such a compound, the melt that is formed will produce a crystalline phase that is initially different from that of both the melt and the source rod. The melt composition will change as growth proceeds, and the liquidus temperature will drop until it reaches the peritectic decomposition temperature. At this point the compound desired will crystallize and steady-state growth is achieved because the source rod and crystal will now have the same composition [55]. Since the melt becomes a solution, growth rates are reduced from an average of 1 mm/min for a pure melt to less than 0.1 mm/min. Temperature stability is a critical factor if high quality crystals are to be produced.

If a melt evaporates incongruently, i.e., losing one or more components preferentially, crystals will be more difficult to grow by the LHPG method due to the large surface-to-volume ratio of the melt. If the rate of evaporation is moderate, high quality fiber crystals can be grown if an excess of the more volatile component is incorporated into the source material. The melt composition can be kept constant if it is replenished with the vaporizing component at a rate equal to that lost from the melt surface.

There are some materials that are not suitable for growth by the LHPG method due to thermodynamic or kinetic considerations. Many important IR materials melt at too low a temperature or exhibit significant vapor or dissociation pressures at their melting temperatures. In these cases other single-crystal fiber techniques that avoid the formation of a vapor-melt interface, such as growth in a capillary tube or die, might be more appropriate.

Single crystal "whiskers" with diameters in the micron range were of great interest about 30 years ago because of their great structural perfection and resulting near theoretical strengths. Nabarro and Jackson [69] defined whisker growth as "any fibrous growth of a solid". They distinguished between "proper whiskers, which grow spontaneously or as a result of applied stress from the bulk solid, and those whiskers which grow from the vapor, solution or melt". A number of authors have shown that dislocation densities in Czochralski grown crystals decreased substantially with crystal diameter for Al [70], Cu [71], Sb [72] and Ge [73]. It is also well-known that reducing the diameter (necking-down) of a Czochralski grown silicon crystal just after seeding and before widening out to final diameter leads to dislocation-free material.

Inoue and Komatsu [74] studied the dependence of the dislocation density on crystal diameter (in the range of 0.5-7 mm) for Czochralski grown KCl and Sr-doped KCl crystals. They believed that their results (Figure 20) could best be explained by Tsivinsky's relationship [75]

$$N = \frac{\alpha}{b} \nabla T_n - \frac{2\tau_{cr}}{Gb} \frac{1}{D} , \tag{11}$$

where N is the dislocation density, ∇T_n the axial temperature gradient, α and b the coefficient of linear thermal expansion and Burgers vector respectively, G the shear modulus and τ_{cr} the critical resolved shear stress. For a given material, therefore, the dislocation density is a function of the axial temperature gradient and crystal diameter. Sobakar and Tsivinsky [74] found for 1.5-14mm diameter KCl crystals grown with a ∇T_n of 150-500°C/cm that the dislocation densities ($6 - 8 \times 10^5$ cm^2) remained unchanged. Inoue and Komatsu, however, found reduced dislocation densities when the diameters were less than about 2 mm, with axial gradients of 60 and 20°C/cm (Figure 20). As the temperature gradients

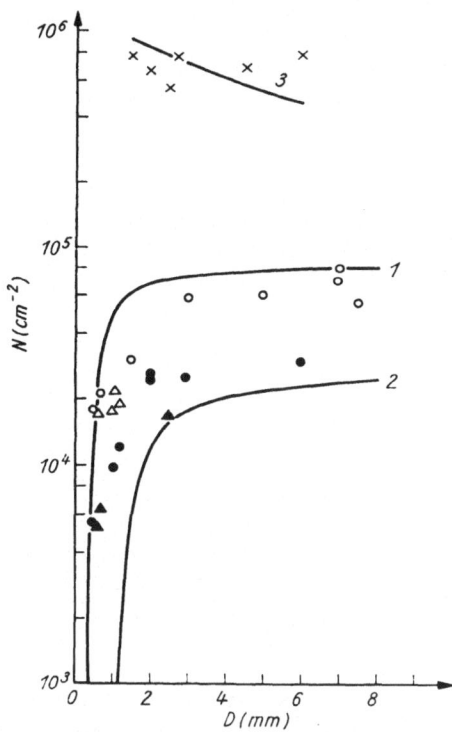

Fig. 20. Dependence of dislocation density on diameter of KCl crystals
from [22]. Experimental values for pure KCl (O,$\triangle$) and Sr-doped
KCl ($\bullet$,$\blacktriangle$) (Merck and Johnson-Matthey, respectively).
Experimental data (X) by Sobakar and Tsivinsky. Curves 1 and 2
calculated by Eq. (6) with ∇T_n= 60 and 20°C/cm, respectively
[74].

increase only very much smaller diameter fibers will have low dislocation
densities. For a ∇T_n of 1000°C/cm, such as was estimated for the LHPG
growth of LiNbO$_3$ fibers [77], dislocation densities in fibers with
properties similar to KCl would not drop substantially according to Eq.
(11) until their diameters reached dimensions of about 20 μm.

Another reason why the density of line defects is low in fiber
crystals is because defects propagating at even a small divergence angle
from the fiber axis will eventually grow out of the crystal. Only defects
propagating along the growth axis or generated during growth will remain
in the crystal. Such a case was found in the growth of Nb$_2$O$_5$ fiber
crystals [55] where lamellar twins grew out of the fiber after about a cm
of growth.

Kim [78] studied the formation of microdefects (swirls) in 4-5 mm
diameter Si crystals grown by the pedestal technique. While these crystal
diameters were still rather large compared with fiber dimensions, they
found no defects at the crystal rim although a uniform density of
microdefects (3 x 10^6 cm^{-3}) was found in the center of the crystals. They
believed this was due to out-diffusion of the point defects responsible
for these microdefects. This suggests that at much smaller crystal
diameters < 1 mm, Si crystals might be produced without microdefects.

VIII.   GROWTH OF ORGANIC NONLINEAR OPTIC CRYSTALS IN CAPILLARIES

The growth of fiber crystals in capillary tubes requires that the material be not only stable on melting, but also be non-reactive with the capillary tube material, and have a thermal expansion or contraction which will not lead to significant stress build-up or shrinkage.

Several attempts have been made to grow optical fibers in capillary tubes which when appropriately chosen also serve to act as a cladding to enhance its light-guiding properties. In 1974 Stevenson and Dyott [79] reported on the preparation of an optical fiber waveguide with a single crystal core of meta-nitroaniline. They first prepared a hollow glass capillary of a lanthanide flint type glass by drawing down a machined rectangular cross section preform by standard glass forming methods. The cross section of the preform was monitored and bore diameters of < 20 μm were easily achievable in lengths up to 450 mm. These capillaries were then filled with molten meta-nitroaniline at temperatures above 112°C by either feeding downwards under gravity or upwards by capillary action. Nucleation of the melt was followed by unidirectional solidification by slow cooling. While the mNa crystals were single in lengths to 200 mm, voids lying between the crystal and glass wall were found to be the major physical defects in the fiber. These voids are probably formed due to the large difference in volume (12.8%) between the liquid and solid. The core was always along the <001> crystallographic axis. These fibers may be useful for a range of nonlinear optical applications including SHG, parametric oscillation, and optical linear electro-optic modulation.

Cesium iodide crystal fiber arrays have also been successfully produced by this method for special multiplex light-guiding applications [80]. Hollow glass fibers are first bundled together to form rods of dimensions typically 5 cm in diameter and 1 mm in length. The fiber bundles are then radially compressed so that a close-packed hexagonal array is formed. The resulting honeycomb structure is cut into slabs of varying thickness and placed into CsI melts where it is drawn up into the $\sim$ 10 μm diameter capillaries (10^6/in.^{2}) and unidirectionally solidified.

IX.   VLS GROWTH

In 1964 Wagner and Ellis [81] discovered a new growth mechanism known as the Vapor-Liquid-Solid (VLS) Mechanism. This process, an excellent review of which is given in [82], can be illustrated by considering a gold particle placed on a (111) silicon substrate as shown in Figure 21. When heated, the gold alloys with the substrate as required by the Au-Si binary phase diagram (eutectic at 370°C) forming a droplet of solution of Si in Au. If this sample is subjected to a H_2 + $SiCl_4$ atmosphere, reduction of $SiCl_4$ occurs at the surface of the liquid drop. The droplet becomes supersaturated with silicon and the excess silicon is deposited on the (111) plane forming a faceted interface with the liquid droplet rising over the original substrate surface leading to the formation of a

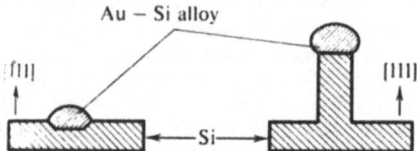

Fig. 21.   Schematic diagram of growth of whiskers by the vapor-liquid-solid mechanism [82].

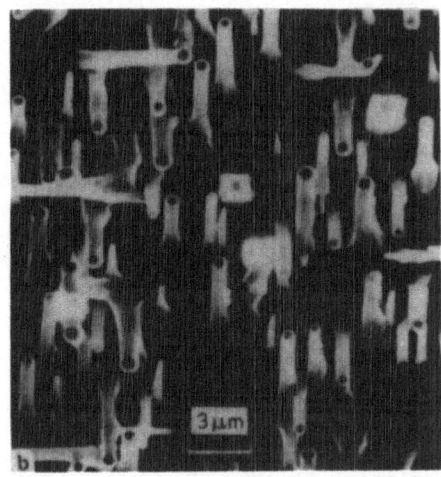

Fig. 22. GaAs whiskers on the (001) substrate of GaAs, plan view (the angle of incidence of the beam in the scanning electron microscope is 0°). The whiskers grow principally in four inclined <111> directions, with only some of them growing along the normal to the (001) face [83].

prismatic Si column as shown in Figure 21. VLS growth depends on the structure and properties of the V-L and L-S interfaces. An example of GaAs whiskers on a (001) GaAs substrate is shown in Figure 22 [83]. The whiskers or fibers produced by VLS growth have, to date, been mostly interesting for scientific studies (the thermodynamic and kinetic factors involved in crystallization) although their use in device applications may some day be possible.

Four stages comprise the VLS process: (1) mass transport in the gas phase, (2) condensation and chemical reaction at the V-L interface, (3) diffusion in the liquid phase, and (4) the incorporation of atoms into the crystal lattice, the last stage, being the slowest, controls the overall rate of the process. Another important factor affecting growth is that the condensation coefficient for the liquid surface be much higher than the solid surface.

The growth rate of Si whiskers depends on their diameter and reduces to 0 at some critical values depending on supersaturation. A dramatic reduction in growth rate occurs below $\sim$ 1 μm diameter indicating that the vapor pressure and solubility of the crystallizing substance in the liquid increases as whisker diameter and saturation decreases. This can be expressed as

$$\frac{\Delta\mu}{kt} = \frac{\overline{\Delta\mu}}{kt} - \frac{4\Omega\alpha}{kt}\frac{1}{d} \tag{12}$$

where $\Delta\mu$ is the chemical potential difference between Si in the vapor and the whisker, $\overline{\Delta\mu}$ the same at a plane interface ($d \rightarrow \infty$), α the specific free energy of whisker surface and Ω the atomic volume of Si.

Experimental data suggest that most or all true whisker growth from the vapor phase is by the VLS mechanism, in other words, a liquid phase must be present. Other factors such as screw dislocations, microtwins, poisoning impurities and mechanical stresses, etc., play a secondary role.

The fact that the diameter of a whisker grown by VLS depends on the liquid drop diameter permits the controlled growth of whiskers. Arrays of oriented whiskers have been grown. In addition to Si, whiskers of Ge, GaAs, GaP, SiC, B, and CdSe, etc., have been grown. Wagner and Ellis [81] established requirements for the liquid forming impurities which must satisfy VLS growth including: (1) the distribution coefficient of impurity must be much less than unity, (2) the equilibrium vapor pressure over liquid alloy should be small, (3) the impurity must be chemically inert, and (4) the contact angle of liquid on the solid tip must be large (for stable growth between 95-120°).

X. REFERENCES

1. E. von Gomperz, Z. Physik, 8:184 (1922).
2. P. I. Antonov and S. P. Nikandrov, J. Crystal Growth, 50:3 (1980).
3. H. E. LaBelle, Jr., J. Crystal Growth, 50:8 (1980).
4. F. L. Ver Snyder and M. E. Shank, Mater. Sci. Eng., 6:213 (1970).
5. H. E. LaBelle, Jr. and A. I. Mlavsky, Mater. Res. Bull., 6:581 (1971).
6. Shaped Crystal Growth, J. Crystal Growth, 50:297 (1980).
7. A. V. Stepanov, Zh. Tech. Fiz., 29:382 (1959).
8. A. L. Shach-Budagov and A. V. Stepanov, Zh. Tech. Fiz., 29:394 (1959).
9. A. V. Stepanov, "Future of Metal Treatment", Leningrad (1963).
10. S. V. Tsivinskii and A. V. Stepanov, Fiz. Tverd. Tela, 7:194 (1965).
11. B. M. Golzman, Opt.-Mech. Prom., 11:45 (1958).
12. S. V. Tsivinskii, Yu. I. Koptev and A. V. Stepanov, Soviet Phys.-Solid State, 8:449 (1966).
13. Yu. I. Koptev and A. V. Stepanov, Fiz. Tverd. Tela, 9:3007 (1967).
14. V. A. Tatarchenko, A. I. Saet and A. V. Stepanov, in: "Proc. Conf. on Manufacturing Semiconductor Crystals by Stepanov's Method and their Application to Instrument-Making Industry", A. F. Ioffe Physico-Technical Inst., Acad. Sic. USSR, p. 83, Leningrad (1968).
15. P. I. Antonov and A. V. Stepanov, Izv. Akad. Nauk SSSR, Ser. Neorg. Mater., 2 (1966).
16. J. Boatman and R. Goundry, Electrochem. Tech., 5:98 (1967).
17. Yu. G. Nosov, P. I. Antonov and A. V. Stepanov, Izv. Akad. Nauk SSSR, Ser. Fiz., 33:2008 (1969).
18. P. C. Goundry, in: "Proc. Intern. Symp. on GaAs" (1966).
19. L. M. Zatulovskii, P. M. Chaikin and others, Izv. Akad. Nauk SSSR, Ser. Fiz., 33:1998 (1969).
20. S. V. Tsivinskii, Inzh.-Fiz. Zh., 5:59 (1962).
21. P. I. Antonov, V. A. Tatarchenko, A. I. Saet and A. V. Stepanov, Ref. [10], p. 42.
22. V. A. Tatarchenko and S. K. Brantov, Izv. Akad. Nauk SSSR, Ser. Fiz., 40:1468 (1976).
23. L. P. Egorov, L. M. Zatulovskii and P. M. Chaikin, IV Akad. Nauk SSSR, Ser. Fiz., 35:466 (1976).
24. P. P. Artychevskii, D. Ya. Kravetskii and L. M. Zarulovskii, Izv. Akad. Nauk SSSR, Ser. Fiz., 35:469 (1971).
25. P. I. Antonov, N. S. Grygorjev and A. V. Stepanov, Izv. Akad. Nauk SSSR, Ser. Fiz., 37:2328 (1973).
26. P. I. Antonov and V. M. Krymov, in: "Proc. 5th Conf. in Processes in Growth and Synthesis of Semiconductors and Films", Novosibirsk (1978).
27. H. E. LaBelle, Jr., US Patent No. 3,527,574 (1970).
28. H. E. LaBelle, Jr., US Patent No. 3,471,266 (1969).
29. H. E. LaBelle, Jr. and A. I. Mlavsky, US Patent No. 3,650,703 (1972).
30. H. E. LaBelle, Jr., US Patent No. 3,591,348 (1971).
31. H. E. LaBelle, Jr., Mater. Res. Bull., 6:581 (1971).

32. J. C. Swartz, B. Siegel, A. D. Morrison and H. Lingertat, J. Electron. Mater., 3:309 (1974).
33. T. Surek, S. R. Coriell and B. Chalmers, J. Crystal Growth, 50:21 (1980).
34. T. Surek and B. Chalmers, J. Crystal Growth, 29:1 (1975).
35. M. M. Fejer, PhD Thesis, Standford University (1986).
36. A. B. Dreeben, K. M. Kim and A. Schujko, J. Crystal Growth, 50:126 (1980).
37. V. A. Tatarchenko and E. A. Brener, J. Crystal Growth, 50:33 (1980).
38. E. M. Sachs, J. Crystal Growth, 50:102 (1980).
39. T. F. Ciszek, G. H. Schwettke and K. H. Yang, J. Crystal Growth, 50:160 (1980).
40. J. M. Kim, S. Berkman, H. E. Temple and G. W. Cullen, J. Crystal Growth, 50:212 (1980).
41. R. G. Seidensticker and R. H. Hopkins, J. Crystal Growth, 50:221 (1980).
42. D. L. Barrett, D. R. Hamilton and E. H. Meyers, J. Electrochem. Sol., 118:952 (1971).
43. R. G. Seidensticker, J. Crystal Growth, 39:6 (1977).
44. C. Belouet, J. Crystal Growth, 50:279 (1980).
45. C. Belouet, J. J. Brissot, R. Martres and Ngo-Tich Phuoc, in: "Proc. Photovaltaic Solar Energy Conf.", pp 164-175, Luxembourg (1977).
46. J. Zook, B. G. Koepke, B. L. Grung and M. H. Leipold, J. Crystal Growth, 50:260 (1980).
47. A. Baghdadi and R. W. Gurtler, J. Crystal Growth, 50:236 (1980).
48. C. E. Bleil, J. Crystal Growth, 5:99 (1969).
49. B. Kudo, J. Crystal Growth, 50:247 (1980).
50. F. L. Ver Snyder and R. W. Guard, Trans. Am. Soc. Metals, 52:485 (1960).
51. B. H. Kear and B. J. Piearcey, Trans. AIME, 238:1209 (1967).
52. C. H. L. Goodman, Solid State Electron Devices, 2:129 (1978).
53. J. S. Haggerty, Production of Fibers by a Floating Zone Fiber Drawing Technique, Final Report NASA-CR-120948, May (1972).
54. C. A. Burns and J. Stone, Appl. Phys. Letters, 26:318 (1975).
55. R. S. Feigelson, Growth of Fiber Crystals, in: "Crystal Growth of Electronic Materials", E. Kaldis, ed., p. 127, North-Holland, Amsterdam (1985).
56. Y. Mimura, Y. Okamura, Y. Komazawa and C. Ota, Japan. J. Appl. Phys., 19:L269 (1980).
57. T. J. Bridges, J. S. Hasiak and A. R. Strand, Opt. Letters., 5:1985 (1980).
58. K. Eickhoff and K. Gurs, J. Crystal Growth, 6:21 (1969).
59. D. B. Gasson and B. Cockayne, J. Mater. Sci., 5:100 (1970).
60. R. S. Feigelson, W. L. Kway and R. K. Route, Opt. Eng., 24:1102 (1985).
61. W. Heywang, Z. Naturforsch., 11a:238 (1956).
62. Lord Rayleigh, Phi. Mag., 34:145 (1892).
63. R. E. Green, J. Appl. Phys., 35:1297 (1964).
64. W. G. Pfann and D. W. Hagelbarger, J. Appl. Phys., 27:12 (1965).
65. S. R. Coriell and M. R. Cordes, J. Crystal Growth, 42:466 (1977).
66. S. R. Coriell, S. C. Hardy and M. R. Cordes, J. Colloid Interface Sci., 60:126 (1977).
67. T. Surek and S. R. Coriell, J. Crystal Growth, 37:253 (1977).
68. K. M. Kim, A. B. Dreeben and A. Schuiko, J. Appl. Phys., 50:4472 (1979).
69. F. R. N. Nabarro and P. J. Jackson, in: "Growth and Perfection of Crystals", R. H. Doremus, B. W. Roberts and D. Turnbull, eds., Wiley, New York (1958).
70. S. Howe and C. Elbaum, Phil. Mag., 6:1227 (1961).
71. A. G. Basariya, B. K. Kapanadze and V. V. Sanadze, Soviet Phys.- Cryst., 18:411 (1973).

72. S. V. Tsivinsky, G. A. Sobakar and B. N. Aleksandrov, Kristall Tech., 8:621 (1973).
73. F. D. Rossi, RCA Rev., 19:349 (1958).
74. T. Inoue and H. Komatsu, Kristall Tech., 14:1511 (1979).
75. S. V. Tsivinsky, Fiz. Metallov Metalloved., 25:1013 (1968).
76. G. A. Sobakar and S. V. Tsivinsky, Izv. Akad. Nauk SSSR., Ser. Fiz., 36:580 (1972).
77. Y.-S. Luh, R. S. Feigelson, M. M. Feher and R. L. Byer, J. Crystal Growth, 78:135 (1986).
78. K. M. Kim, J. Appl. Phys., 50:1135 (1979).
79. J. L. Stevenson and R. B. Dyott, Electron. Letters, 10:449 (1974).
80. M. Tripp, private communication (1986).
81. R. S. Wagner and W. C. Ellis, Appl. Phys. Lett., 4:89 (1964).
82. A. A. Chernov, "Modern Crystallography III Crystal Growth", Springer Verlag, Berlin (1984).
83. T. Muranoi and M. Furukoshi, J. Electrochem. Soc., 127:2295 (1980).

TECHNOLOGIES BASED ON ORGANOMETALLIC VAPOR PHASE EPITAXY

G. B. Stringfellow

University of Utah
Salt Lake City
Utah 84112, USA

1.   INTRODUCTION

The organometallic vapor phase epitaxy (OMVPE) technique originated
from early work of Manasevit [1]. This technique, by which compound
semiconductor epitaxial layers are grown using organometallic group III
and organometallic or hydride group V molecules to transport the elements
to the heated substrate, is also sometimes called metalorganic chemical
vapor deposition or MOCVD. The class of compounds containing the group
III or group V elements combined with methyl, ethyl, butyl, and other
simple radicals is termed the organometallic compounds in common
terminology. Since we are solely interested in the epitaxial growth of
single crystalline layers for semiconductor device applications, the term
OMVPE seems more appropriate. One sometimes also encounters the terms
MOVPE and OMCVD in the literature. All four acronyms refer to exactly the
same growth process.

Device quality semiconductors were not obtained by OMVPE until the
homoepitaxial growth of GaAs with $77°K$ mobilities of 120,000 cm^2/Vs was
demonstrated in 1975 by Seki et al. [2]. This led to serious work in
several laboratories to develop both GaAs for FETs and AlGaAs for
injection lasers and solar cells. The growth of high quality AlGaAs
proved difficult, and it was not until 1979 that Dupuis and Dapkus [3]
reported the first low-threshold current density injection lasers
operating cw at room temperature. Shortly thereafter, Saxena et al. [4],
in 1980, reported the fabrication of GaAs/AlGaAs solar cells with
efficiencies of 23% at a concentration of 369 suns, comparable to the
results obtained in liquid phase epitaxial (LPE) material. These early
successes led to a flood of interest and activity in this area. Duchemin
et al. [5] and Hirtz et al. [6] first showed that OMVPE could also be
successfully employed for In containing materials, such as GaInAs and
GaInAsP for lasers and FETs. Early efforts were troubled by adduct
formation between triethylindium (TEIn) and PH_3. This resulted in the
development of low pressure (approximately 76 Torr) OMVPE or LPOMVPE
[5,6]. Other approaches to solving these problems led to the development
of adduct sources by Moss and Evans [7] which allow the growth of In
containing III/V alloys in atmospheric pressure (APOMVPE) reactors.
Today, the preferred solution is simply to use trimethylindium (TMIn), an
approach pioneered by Hsu et al. [8,9], Kuo et al. [10], and Sacilotti et

al. [11]. No problems with adduct formation are experienced in the growth
of InP, GaInAs, and GaInP by APOMVPE.

In more recent years even lower pressure OMVPE techniques have been
developed. Kamon et al. [12] found that OMVPE performed at pressures as
low as 1 to 10 Torr allowed selective growth, i.e., on a SiN_x or SiO_2
masked GaAs substrate, GaAs and AlGaAs grow only where the GaAs is
exposed. This is expected to be useful for the growth of planar
structures containing lasers, detectors, and FETs for future optoelectronic integrated circuits (OEICs). At the extreme lower limit, Tsang
[13] converted a stainless steel molecular beam epitaxy (MBE) machine
operating at ultra-low pressures to OMVPE by adding alkyl group III and
group V sources with cracking furnaces to produce atomic species injected
into the vacuum in a beam directed toward the substrate. This process is
called chemical beam epitaxy (CBE) by Tsang. When only the group V
sources are replaced by gases the technique is termed gas source MBE by
Panish and co-workers [14]. When only the group III sources are replaced
by alkyls, the pioneers K. Takahashi and co-workers prefer the name MOMBE
[15]. In a real sense, CBE belongs to the OMVPE family to be discussed in
this manuscript.

OMVPE has assumed an increasingly dominant position among compound
semiconductor epitaxial growth techniques during the last decade. The
rate of development of the technique is surprisingly high. A rough
estimate shows the doubling time is approximately 2.5 years, as obtained
from an approximate count of the number of reactors in use versus calendar
year. Today, OMVPE is used for the commercial production of a wide
variety of semiconductor devices from double heterostructure GaAs/AlGaAs
lasers and quantum well lasers, to solar cells and two-dimensional
electron gas (2DEG) transistors. This rapid acceptance of OMVPE is based
on the ability of the technique to meet the demands for semiconductor
devices: high purity, abrupt interfaces, versatility, and the capability
for large-scale, uniform, reproducible, and economical epitaxial growth
for research and production operations. In the remainder of this paper,
these attributes will be analyzed on the basis of the fundamental aspects
of OMVPE.

II.   FUNDAMENTAL ASPECTS OF OMVPE

Hydrodynamics and Mass Transport

The hydrodynamic aspects of the OMVPE process are complex. A
thorough understanding of the gas velocities and mass transport rates in
various parts of the flowing gas stream can only be reliably predicted
using extremely complex numerical modelling techniques. The Navier-Stokes
equation (momentum conservation), mass continuity equation, and energy
transport equation must be solved simultaneously [16]. Recently, a few
efforts to model real OMVPE systems in this manner have been successful
[17]. Unfortunately, such calculations are still in the realm of the
dedicated hydrodynamicist. They are not within the easy grasp of the
typical crystal grower and the complex mathematics involved often obscures
any physical insight or simple interpretation of experimental
observations. The efforts to develop simpler models have largely centered
on the concept of the boundary layer, a quasi-static layer of gas adjacent
to the growth interface through which the source molecules diffuse to
supply nutrient for the epitaxial process. This boundary layer does not
exist in any real sense [16], but can serve as a zeroeth order model to
facilitate the interpretation of experimental data [18]. More recently,
an intermediate approach has proven useful. Van de Ven et al. [19] have
successfully solved the complex hydrodynamic problem using simplifying

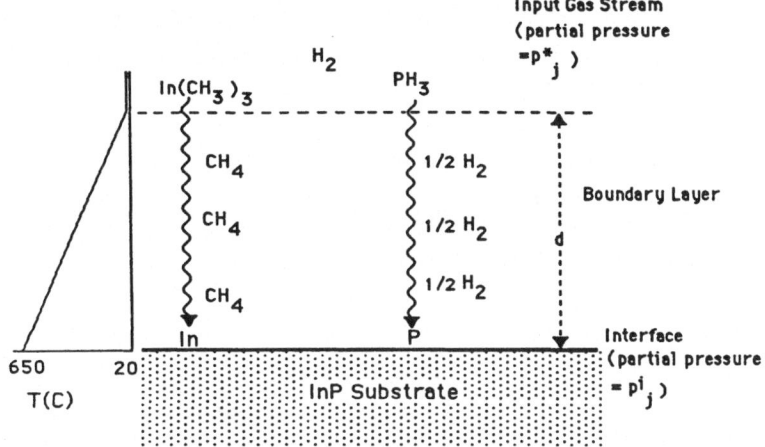

Fig. 1. Schematic diagram of OMVPE growth of InP using TMIn and PH₃.

assumptions appropriate to the conditions existing in most OMVPE systems. This has resulted in rather simple, analytical solutions giving the growth rate as a function of position on the susceptor in terms of the flow parameters, diffusion coefficients, and substrate temperature. The simple calculations give results which agree well with experimental observations. Many of the results are similar to those expected using the simple boundary layer model. For the sake of illustration, the boundary layer model is sufficient to describe qualitatively many phenomena occurring in an OMVPE reactor. In fact, it is only a simple representation of the mass transport step in the growth process in which the mass transport coefficient is represented as a diffusion coefficient, D, divided by a characteristic length, d.

Figure 1 is a schematic illustration of an OMVPE reactor showing the boundary layer and the overall chemical reactions occurring during the OMVPE growth of InP. A characteristic feature of OMVPE is the high degree of supersaturation of the input vapor phase, i.e., for the growth process shown in Figure 1, the overall growth reaction may be written,

$$In(CH_3)_3(v) + PH_3(v) = InP(s) + 3CH_4,$$ (1)

and the inequality below is obeyed,

$$p^{*v}_{TMIn}p^{*v}_{PH_3}a^s_{InP} >> K_{InP}.$$ (2)

p^{*v}_j represents the input partial pressure of component j, a^s the activity in the solid, and K the equilibrium constant. At the vapor/solid interface, the input molecules are decomposed, and equilibrium is approximately established [20]. Thus Eq. (2) would be replaced by the mass action expression, with the partial pressures at the interface denoted p^i_j,

$$p^i_{In}(p^i_{P_4})^{1/4}/a^s_{InP} = K_{InP}.$$ (3)

The driving force for any crystal growth process is the super-saturation in the nutrient phase. For OMVPE, the driving force to restore equilibrium, i.e., the supersaturation, is conveniently written in terms of the chemical potential of the input vapor phase above that of the equilibrium gas at the interface, $\Delta\mu^*$. Since the chemical potential, μ, is expressed,

$$\mu_j = \mu^o_j + RT\ln(p_j/p^o_j), \tag{4}$$

where the superscript o indicates the standard state, the supersaturation for the OMVPE growth of InP can be written

$$\Delta\mu^* = RT\ln(p^o_{In}p^o_{P_4}{}^{1/4}/p^i_{In}p^i_{P_4}{}^{1/4}). \tag{5}$$

The maximum rate at which InP solid can be produced is simply the amount which would allow the entire entering gas stream to establish equilibrium with the growing solid, and is thus fundamentally limited by thermodynamics and the total amount of gas flowing through the OMVPE reactor. Ordinarily, the growth rate is considerably less than that calculated from thermodynamics. Kinetics, both surface reaction rates and diffusion through the boundary layer, are not fast enough to allow equilibrium to be established throughout the system at all times. This situation is illustrated by Figure 2a, where $\Delta\mu$ is plotted versus reaction coordinate. This allows the schematic representation of the overall, thermodynamic driving force for the action, $\Delta\mu^*$, being broken into two terms, one driving the diffusion process ($\Delta\mu_D$) and the other driving the surface reactions ($\Delta\mu_S$). As mentioned above, even in cases with a large supersaturation in the input vapor phase, i.e., $\Delta\mu^* \gg 0$, near equilibrium conditions may exist at the growing solid surface. This simply requires that the interface kinetics be much more rapid than the diffusion kinetics, i.e., the two processes proceed at the same rate with $\Delta\mu_S \ll \Delta\mu_D$. This situation, termed diffusion limited growth, is shown schematically in Figure 2b. Using ordinary growth conditions, with temperatures between approximately 550 and 800°C, this is the normal situation for OMVPE growth of InP [21], and for other III/V semiconductors as well.

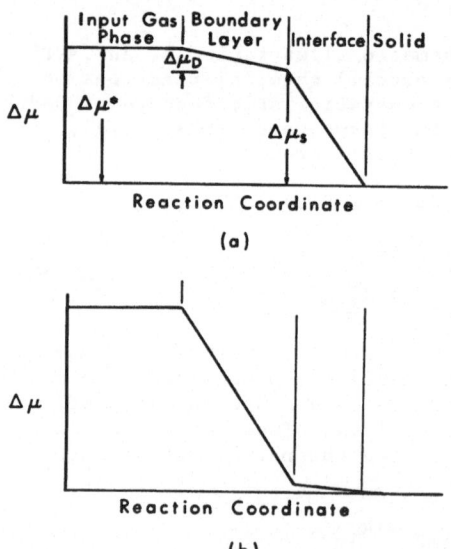

Fig. 2. Schematic diagram of chemical potential versus reaction coordinate showing the drop in chemical potential required for each step in the growth sequence. The cases illustrated are the general case (a) and the case of rapid surface kinetics (b). (After Stringfellow, Ref. [20].)

The overall reaction occurring during OMVPE growth of InP is given in Eq. (1). However, the growth process is considerably more complex. The removal of CH_3 from the TMIn and H from the PH_3 are shown schematically in Figure 1. However, knowledge of the actual mechanisms, including whether the reactions are homogeneous (occurring in the vapor phase) or heterogeneous (occurring at the solid/vapor interface) is extremely limited. The actual reactions occurring have only recently been discovered [22-25].

The key experiment has been the study of InP growth from TMIn and PH_3 in a D_2 ambient, where a mass spectrometer is used to trace the D to determine the reaction paths [22-24]. Similar deuterium tracer experiments have been used to determine the GaAs growth reactions using trimethylgallium (TMGa) and AsH_3 [25].

The technique is perhaps most easily illustrated by first considering the pyrolysis of PH_3 in D_2 with no TMIn present, by simply passing the gas mixture down a heated SiO_2 tube and examining the reaction products using the mass spectrometer. If PH_3 were to partially decompose homogeneously in the vapor phase, H atoms would be produced which would react with the D_2 producing HD molecules. From Figure 3, it is seen that no HD is detected even when the PH_3 is totally pyrolyzed. Above $650^{\circ}C$, exchange reactions occur which produce HD as seen in Figure 3. In this temperature range HD is observed even in H_2/D_2 mixtures. The fact that only H_2 is produced during PH_3 clearly indicates that pyrolysis occurs by adsorption of PH_3 on the SiO_2 walls, with subsequent release of H atoms. These remain adsorbed to the SiO_2, reacting with each other to create H_2 which is rapidly released into the vapor. This is the first experiment which directly and clearly traces the pyrolysis of a group V hydride. For InP-coated walls the reaction is also purely heterogeneous and occurs at much lower temperatures. In addition, we observe the pyrolysis to be identical for H_2, D_2, and N_2 carrier gasses, which provides additional evidence of the heterogeneous nature of the pyrolysis.

The results of a similar experiment conducted for TMIn pyrolysis in D_2 are much more complex to interpret [24]. The products, with partial pressures plotted versus pyrolysis temperature in Figure 4, are principally CH_3D and C_2H_6. The rate of pyrolysis is found to be dependent on the carrier gas, being faster in H_2 than in D_2 and slowest in He.

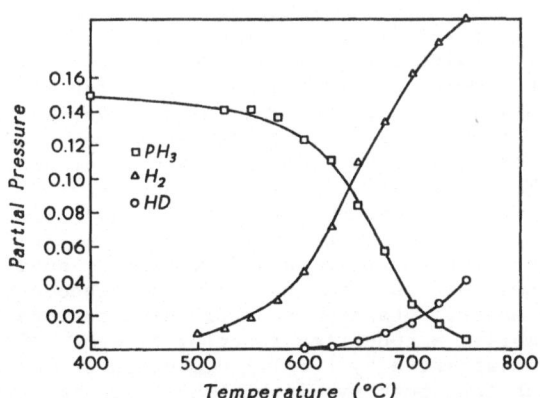

Fig. 3. Species observed in mass spectrometer versus temperature after passing PH_3 in a D_2 ambient through an OMVPE reactor.

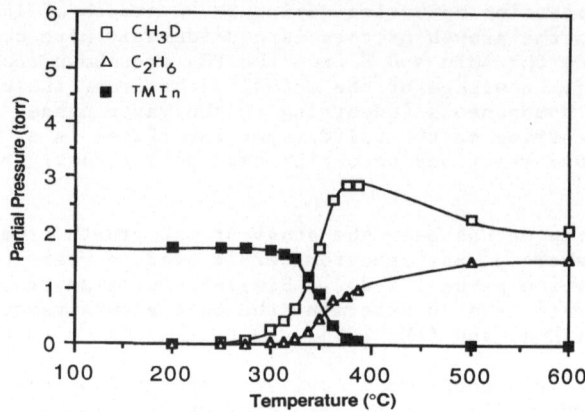

Fig. 4. Decomposition products observed in the exit of an OMVPE reactor where only TMIn and D_2 are present, versus temperature.

Significantly, the activation energy is also dependent on the carrier gas species. In addition, the pyrolysis rate depends very little on surface area. These results indicate that the pyrolysis occurs by a homogeneous process [24]. Since the carrier has a large effect, the reaction cannot occur by simple homogeneous fission, as previously suggested [26]. Comparison of the experimental data from these and other experiments with the results of kinetic reaction modelling suggests the pyrolysis of TMIn occurs by a chain reaction involving radical attack of the TMIn [24]. The initiation step involves the homogeneous fission of TMIn. The CH_3 radicals react with D_2 to form CH_3D and D radicals. The D radicals play a key role in the chain reaction by attacking TMIn to form the hypervalent species $DIn(CH_3)_3$,

$$D + In(CH_3)_3 \rightarrow DIn(CH_3)_3 \rightarrow InCH_3 + CH_3 + CH_3D. \tag{R1}$$

The termination reactions simply involve the recombination of 2 CH_3 or 2 D radicals or the reaction of a CH_3 plus a D radical.

For the pyrolysis of TMIn and PH_3 together under typical growth conditions with a V/III ratio (PH_3 to TMIn ratio in the input gas stream) of 30 with InP coating the SiO_2 reactor walls, only CH_4 and H_2 are observed [22,23]. Clearly, the PH reacts with TMIn before the first CH_3 radical is produced. Apparently an adduct is formed for a very short time, since other data indicate that no stable adduct is formed [27]. Before the PH_3 and TMIn molecules can be separated, the elimination reaction occurs releasing CH_4. It is also instructive that the PH pyrolysis temperature dramatically decreases, from 700 to 425°C as the $TMIn/PH_3$ ratio increases from 0 to 1.0. The TMIn is clearly assisting the PH pyrolysis. The TMIn pyrolysis temperature decreases less dramatically from 375 to 300°C as PH_3 is added to the system. The entire TMIn pyrolysis reaction may occur homogeneously, or the last stage of the pyrolysis involving the non-volatile CH_3InPH may occur heterogeneously on the InP surface. Since both homogeneous and heterogeneous reactions result in CH_4, these experiments cannot distinguish between the two possibilities. Similarly, a concerted reaction between TMGa and AsH_3 during GaAs growth is suggested by the 1:1 depletion of Ga and As from the gas phase and the fact that both pyrolyze at the same temperature when present simultaneously. This adduct mechanism has not been postulated previously; however, Nishizawa et al. [28] observed a new, unidentified, molecule or fragment in IR adsorption spectra during OMVPE growth of GaAs from TMGa and AsH_3 which they thought might be an adduct. The earliest mechanistic studies of Schlyer and Ring [29], also demonstrated that the heterogeneous reaction of TMGa and PH_3 occurs on the GaP surface at low temperatures (< 270°C) by exactly such a mechanism.

III. ANALYSIS OF ADVANTAGES AND DISADVANTAGES OF OMVPE

In this section, characteristics of the process; purity, the ability to grow abrupt interfaces, etc., discussed in Section I, will be analyzed in terms of the fundamental aspects of OMVPE. In this manner, insight into the inherent strengths and weaknesses of the process can be compared rather than simply considering only the present state-of-art, which literally changes daily.

Purity

The purity of OMVPE grown III/V semiconductors continues to improve. At the present time purity is limited mainly by the hydride sources. Contamination by the large amounts of C present has been addressed [30-34]. It is significant only for AlGaAs, where levels of 10^{17} cm^{-3} are always observed [30]. For GaAs, growth at low temperatures of approximately 600°C yields very low C levels [31]. Interestingly, the growth of carbon-free InP [34] and GaInAs [32,33] requires higher, rather than lower, growth temperatures. This significant difference is not understood. Oxygen contamination was a major early problem for the OMVPE growth of AlGaAs [30]. The strong Al-O bond results in the incorporation of essentially all oxygen reaching the growing interface [35]. This gives oxygen concentrations of $> 10^{20}$ cm^{-3} in the solid for oxygen levels in the gas stream as low as 1 part per million. Today, oxygen contamination of AlGaAs is controlled using molecular sieves and liquid metal bubblers [36] to purify the hydrides, which are notorious sources of H_2O [30,31]. An important reactor consideration is to provide for loading the reactor from an oxygen-free environment (glove box or vacuum interlock) to avoid H_2O adsorption on the graphite and SiO_2 components of the growth reactor. These precautions and the use of high growth temperatures [37] have resulted in the growth of high quality AlGaAs [30,37], AlGaInAs [38], and AlGaInP [39,40] for use in laser devices, with wavelengths moving into the visible spectrum, and 2DEG transistors.

The other major problem with purity is related to impurities in the organometallic sources themselves. We expect liquid sources to be superior in purity to gaseous sources, since liquids are easily purified by distillation techniques and the areas of the cylinder walls, which are known to adsorb water, are relatively small. This resulted in early GaAs grown using $AsCl_3$, a liquid, having some of the highest purities ever reported as summarized in an excellent review by Cardwell [41]. The organometallic group III source purity has improved dramatically as more effort has been expended by the suppliers to improve their synthesis and purification techniques. Recently, Alpha/Ventron developed a new synthesis route for TMIn which immediately resulted in dramatically improved InP purity in several laboratories [42,43]. Similar efforts on TMIn purification in England have also yielded high purity InP [44]. As a result, the InP grown by OMVPE, is now the purest ever produced by any technique, as evidenced by the standard measure of the total ionized impurity concentration, the 77°K electron mobility. The value of 264,000 cm^2/Vs for OMVPE InP [44] is a record. The electron mobilities of several semiconductors are compared in Table 1. In addition to the bulk mobilities, the mobilities of 2DEGs measured for the GaAs/AlGaAs and GaInAs/InP systems are also listed. For comparison, the highest mobilities recorded for materials grown by other techniques are also

Table 1. Mobilities Measured in III/V Semiconductors Grown by Various Techniques

All Mobility Units are cm^2/Vs
Temperatures are indicated (°K)

Material	OMVPE	MBE	ClVPE*	HVPE**	LPE***
GaAs(77°K)	210,000[73]	163,000[74]	200,000[75]	200,000[75]	200,000[75]
InP(77°K)	264,000[44]	55,050[77]	129,520[78]	71,000[79]	94,000[80]
	105,000[76] (CBE)				
GaInAs(77°K)	75,000[81]	53,800[82]	79,000[83]	50,000[84]	70,000[85]
GaAs/AlGaAs 2DEG(4°K)	450,000[86]	>10^6[70]	---	---	---
GaInAs/InP 2DEG(4°K)	92,000[87] 130,000[88] (CBE)	---	106,000[89]	---	---
GaInAs/AlInAs 2DEG(4°K)		92,000[90]			

*   ClVPE = Chloride Vapor Phase Epitaxy
**   HVPE = Hydride Vapor Phase Epitaxy
*** LPE   = Liquid Phase Epitaxy

listed. As mentioned previously, the weak-link is currently the hydride sources. This may be improved as these gaseous sources are replaced by liquid organometallic group V sources such as tertbutylphosphine (TBP) [45-47], tertbutylarsine (TBAs) [48], and diethylarsine (DEAs) [49]. These sources will be discussed in more detail in the Safety Section below.

Versatility

The versatility of OMVPE for the growth of III/V semiconductors is well-known. Since OMVPE can be used for the growth of essentially all III/V alloys containing Al, Ga and In combined with P, As and Sb, it claims the honor of being the most versatile of all epitaxial techniques. Other aspects of the versatility of OMVPE are the ability to grow selectively, i.e., only on openings in SiO_2 or SiN_x masks [12], and the ability to grow thermodynamically metastable alloys. The reader is referred to previous publications [50-52] where this aspect of OMVPE has been thoroughly discussed. A final important aspect of versatility is the ability to grow at low temperatures, an area increasingly important for the growth of very fine structures and low band gap alloys. Growth of GaAs below 525°C presents problems using AsH_3, since the pyrolysis rate limits the growth process [25,53], yielding poor morphologies. This situation is expected to improve as the group V organometallic sources are developed. For example, TBAs is found to pyrolyze at temperatures fully 250°C lower than for AsH_3[48]. Other, organometallic sources will undoubtedly be developed with even more rapid low temperature pyrolysis rates.

IV. INTERFACE ABRUPTNESS

A revolution in semiconductor device design has occurred during the last decade based on the development of man-made superlattice structures which give properties much superior to homogeneous, or random, alloys. Compositional superlattices of alternating layers of GaAs and AlGaAs, so-called multiple quantum-well structures, yield injection laser devices with shorter wavelengths, lower threshold current densities, and higher quantum efficiencies. The 2DEG, developed at the interface of the high band gap, heavily n-type AlGaAs and the very pure, low band gap GaAs, has a much higher electron mobility than for the normal n-type GaAs used in FETs. This results in transistors with both higher speed and trans-conductance. Thus, a requirement for modern epitaxial techniques is the ability to produce extremely abrupt interfaces. Since MBE has demonstrated the ability to grow truly atomically abrupt interfaces with no transition region, this has been the goal of OMVPE. Progress was slow and steady until within the last 1-2 years several laboratories have demonstrated atomically abrupt interfaces. In 1985 workers at Sony in Japan [54], using APOMVPE, demonstrated truly atomically abrupt interfaces between GaAs and AlAs. In 1986, Tsang, using CBE [55], showed that atomically abrupt interfaces could be obtained even in the more difficult GaInAs/InP system where both the group III and the group V elements must be rapidly switched simultaneously to avoid the occurrence of GaInAsP at the interface. Late in 1986, Miller et al. [56] showed that APOMVPE could also produce atomically abrupt interfaces in this system. Similar results have been reported very recently by Wang et al. [57].

Several requirements must be met to achieve such abrupt interfaces in APOMVPE. Naturally, the first requirement is that the gas be switched in times of less than 1 second, since the growth rate is typically a few Å per second. Second, no eddy currents can exist in the reactor. Recirculating gas patterns act as memory cells. Dopants such as H_2Se [58] and DETe [59] can adsorb on the surface. This also results in undesirable memory effects during switching of dopants during superlattice growth. Third, pressure transients occur during switching between vent and run lines operating at even slightly different pressures. The pressures in both lines must be nearly identical. Automatic control of the pressure differential is used in some reactors [56]. Finally, the growing surface must be nearly atomically flat during growth. Of course, the growing surface cannot truly be atomically flat; however, the growth must occur by 2-dimensional nucleation [18,51,60] where the growing islands on the surface are only 1 atomic layer in height.

V. SAFETY

A major obstacle to the commercialization of the OMVPE process has been a concern for safety. The use of AsH_3 and PH_3 is extremely dangerous. Not only are the hydrides of P, and especially As, highly toxic, but the gasses are contained in high pressure cylinders [51]. A mistake could result in the entire contents of the cylinders being dumped into the laboratory atmosphere. In the past, the issues have been effectively handled by commonsense safety precautions such as placing the cylinders in vented cabinets, the use of breathing apparatus when the cabinet is opened, toxic gas monitoring systems, and special valves which limit the maximum flow rates so the cylinders cannot be accidentally vented in a short time. However, in today's emotional climate where "absolute" safety is demanded, such measures are not satisfactory. Absolute safety requires infinitely expensive safeguards. Today it is not unusual for the specially-equipped room and the safety equipment to cost far more than the OMVPE reactor itself. In several locations in the US presently available safety measures are not acceptable. Permission to use the As and P hydrides has been denied no matter how stringent the safety measures. Clearly another approach is needed. The most promising approach involves the substitution of group V organometallic sources for the hydrides. These organometallic liquids have modest vapor pressures which are high enough to allow their use as effective sources at normal carrier gas flow rates, but low enough to limit the rate at which the molecules can enter the atmosphere in case of an accident [51]. In addition, preliminary data indicates them to be much less toxic. For example, data for TBP indicates its toxicity limit (LC_{50}) to be above 1100 ppm, as compared to 11 ppm for PH [61].

A significant effort is being devoted to the replacement of AsH_3 and PH_3 with more suitable group V sources. Bhat [62] has reported that the simplest solution, the use of a solid As source, is not suitable for growth of device quality GaAs and AlGaAs by OMVPE. The material has morphological problems and is contaminated by carbon. The carbon presumably comes from the CH_3 radicals adsorbed to the GaAs surface from TMGa pyrolysis. As discussed above, H from the hydride is responsible for their removal as CH_4. Since the elemental As provides no atomic H, C incorporation is dramatically increased. Similar problems are apparently encountered with the use of trimethylarsenic (TMAs). Several workers have reported the successful growth of As containing III/V semiconductors using TMAs [10,63-66]. However, increased impurity incorporation is reported [65,66], in part due to the increased C incorporation. The first attempts to use alkylphosphines in OMVPE employed trimethylphosphine and triethyl-phosphine [67]. Early reports claimed that phosphide growth had been achieved, but later publications cast doubt on these assertions. In particular, it was shown that GaInAs grown from trimethylindium-triethylphosphine and arsine contained no phosphorus [7] and that if the phosphorus trichloride was removed from a trimethylindium-trimethylphosphine growth system, liquid indium droplets were formed [67].

A recent report of the use of diethylarsine (DEAs) by Bhat et al. [68] is more encouraging. In this molecule, the single H atom attached to the As, along with the 2 ethyl radicals, is expected to provide the H necessary to oxidize the CH_3 to CH_4, which escapes from the surface into the vapor before the C can be incorporated into the solid. Similar to the case of elemental As, morphological problems are observed, especially for growth above $600°C$. However, the DEAs apparently pyrolyzes more rapidly than AsH_3 allowing the use of very low V/III ratios (approximately unity) in the input vapor phase for the growth of good morphology layers of GaAs at low temperatures. The purity is quite acceptable for a newly synthesized organometallic source. Carrier concentrations as low as 5×10^{14} cm^{-3} have been reported with $77°K$ mobilities as high as 67,000 cm^2/Vs. Under some growth conditions, the level of C contamination still appears to be higher than for GaAs grown under equivalent conditions using AsH_3. This may indicate the need for 2 H atoms bonded to the As with a single organic radical. Recent use of isobutylphosphine (IBP) and TBP for the growth of high quality InP [45-47] and TBAs for GaAs growth [48] indicates that no additional carbon incorporation can be attributed to these organometallic group V sources. These molecules have 2 hydrogens and 1 butyl radical bonded to the group V atom. For InP growth using TBP and GaAs growth using TBAs, lower V/III ratios were possible than with PH_3 and AsH_3. This, plus direct pyrolysis studies, indicates that the tertiarybutyl-V molecules decompose at lower temperatures than the hydrides.

VI. APPLICATIONS

The OMVPE technique is already in production for several GaAs/AlGaAs devices including: solar cells, injection lasers for compact disk players, quantum-well lasers, and 2DEG FETs. The latter device results are particularly interesting since high performance requires an extremely abrupt interface and high GaAs purity. Workers at Sony, where the technique is currently in production, have reported values of $4°K$ mobility of $> 400,000$ cm^2/Vs [69], somewhat lower than the highest reported for similar structures in MBE material of over 10^6 cm^2/Vs [70]. A more meaningful comparison is provided by the transconductance. However, in terms of a more practical parameter, the transconductance of 2DEG FETs, OMVPE grown structures are similar to those grown by MBE [71]. The technique also appears promising for devices in other materials systems such as lasers and detectors in the GaInAsP/InP system. The best AlGaInP/GaInP lasers are reported using OMVPE-grown material [72].

OMVPE will become the dominant production technique for most devices. In the near future this will involve mostly discrete devices: lasers, light emitting diodes, detectors, solar cells, heterojunction field effect transistors, and heterojunction bipolar transistors. In the more distant future, these devices will be integrated to produce digital and analog integrated electronic circuits and, eventually, even optoelectronic integrated circuits (OEICS) where lasers, detectors, optical modulators, and electronic switching devices will be integrated together on the same chip.

VII. SUMMARY

The OMVPE technique has been described in terms of our current fundamental understanding of the process. On this basis, the fundamental advantages and problems with OMVPE have been discussed. To summarize, thermodynamics controls the overall growth process, determining the maximum growth rate, and, in many cases, the solid composition produced from a given gas phase composition at a specified temperature. The actual growth rate is always lower than the thermodynamically determined upper limit. Under normal growth conditions, the growth rate is controlled by the mass transport and gas flow dynamics of the OMVPE reactor, which are difficult to understand using simple models. The most fruitful approach is to use approximate solutions to the coupled mass, and momentum, and energy transport equations, which appears to give reasonably accurate solutions, and gives physical insight into the important aspects of this problem. The chemical reactions by which the III/V semiconductor is produced from organometallic group III sources and either hydride or organometallic group V sources are complex. The group III organometallics apparently interact with the group V hydrides before pyrolysis. The final stages of pyrolysis appear to occur at the solid-vapor interface.

The current status of OMVPE may be summarized by stating that extremely high purity materials, sometimes the purest ever reported, are grown by OMVPE. Interface abruptnesses approaching 1 atomic monolayer, necessary for some superlattice device structures, have been reported for both the GaAs/AlGaAs and GaInAs/InP systems. The range of materials which can be produced is impressive. It extends to essentially all III/V compounds and alloys, including thermodynamically metastable alloys. For low pressure operation, the deposition process is selective, i.e., layers can be grown in local regions using insulator masking techniques. The technique appears promising for large scale production with many wafers produced per run, although the development of such reactors is still at an early stage of development. Safety issues are currently limiting the development of the technique, but with the discovery of effective new organometallic group V sources, the safety problems will be significantly decreased. These new group V organometallic sources also offer the promise of allowing OMVPE growth at substantially lower temperatures than the current lower limit of approximately 525°C.

REFERENCES

1. H. M. Manasevit, Appl. Phys. Lett., 12:156 (1968).
2. Y. Seki, K. Tanno, K. Iida and E. Ichici, J. Electrochem. Soc., 122:1108 (1975).
3. R. D. Dupuis and P. D. Dapkus, Appl. Phys. Lett., 32:406 (1978).
4. R. R. Saxena, V. Aebi, C. B. Cooper, J. J. Ludowise, H. A. Van der Plas, B. R. Cairns, T. J. Maloney, P. G. Border and P. E. Gregory, J. Appl. Phys., 51:4501 (1980).
5. J. P. Duchemin, J. P. Hirtz, M. Razeghi, M. Bonnet and S. D. Hersee, J. Crystal Growth, 55:64 (1981).
6. J. P. Hirtz, J. P. Duchemin, B. de Cremoux, T. Pearsall and M. Bonnet, Electron. Lett., 16:275 (1980).
7. R. H. Moss and J. S. Evans, J. Crystal Growth, 55:129 (1981).
8. C. C. Hsu, R. M. Cohen and G. B. Stringfellow, J. Crystal Growth, 63:8 (1983).
9. C. C. Hsu, R. M. Cohen and G. B. Stringfellow, J. Crystal Growth, 62:648 (1983).
10. C. P. Kuo, R. M. Cohen and G. B. Stringfellow, J. Crystal Growth, 64:461 (1983).
11. M. Sacilotti, A. Mircea and R. Azoulay, J. Crystal Growth, 63:111 (1983).
12. K. Kamon, S. Takagishi and H. Mori, Japan. J. Appl. Phys., 25:L10 (1986).
13. W. T. Tsang, J. Electron. Mater., 15:235 (1986).
14. M. B. Panish, H. Temkin and S. Sumski, J. Vac. Sci. Technol., B3:657 (1985).
15. K. Takahashi, Inst. Phys. Conf. Ser. No. 79, p. 73, Adam Hilger Ltd., Bristol and Boston (1986).
16. F. Rosenberger, "Fundamentals of Crystal Growth", Springer-Verlag, Berlin (1979).
17. H. Moffat and K. F. Jensen, J. Crystal Growth, 77:108 (1986).
18. G. B. Stringfellow, in: "Crystal Growth", 2nd Edition, B. R. Pamplin, ed., p. 181, Pergamon Press, Oxford (1980).
19. J. van de Ven, G. M. J. Rutten, M. J. Raaijmakers and L. J. Giling, J. Crystal Growth, 76:352 (1986).
20. G. B. Stringfellow, J. Crystal Growth, 70:133 (1984).
21. C. C. Hsu, R. M. Cohen and G. B. Stringfellow, J. Crystal Growth, 63:8 (1983).
22. N. Buchan, C. A. Larsen and G. B. Stringfellow, Appl. Phys. Lett., 51:1024 (1987).
23. N. Buchan, C. A. Larsen and G. B. Stringfellow, J. Crystal Growth, 92:605 (1988).
24. N. Buchan, C. A. Larsen and G. B. Stringfellow, J. Crystal Growth, 92:591 (1988).
25. C. A. Larsen, N. Buchan and G. B. Stringfellow, Appl. Phys. Lett., 52:480 (1988).
26. M. G. Jacko and S. J. W. Price, Can. J. Chem., 41:1560 (1963).
27. C. A. Larsen and G. B. Stringfellow, J. Crystal Growth, 75:247 (1986).
28. J. Nishizawa and T. Kurabayashi, J. Electrochem. Soc., 130:413 (1983).
29. D. J. Schl... and M. A. Ring, J. Electrochem. Soc., 124:569 (1977).
30. G. B. Stringfellow, J. Crystal Growth, 55:64 (1981).

31. P. D. Dapkus, H. M. Manasevit and K. L. Hess, J. Crystal Growth, 55:10 (1981).
32. C. P. Kuo, J. S. Yuan, R. M. Cohen, J. Dunn and G. B. Stringfellow, Appl. Phys. Lett., 44:550 (1984).
33. K. L. Fry, C. P. Kuo, R. M. Cohen and G. B. Stringfellow, Appl. Phys. Lett., 46:955 (1985).
34. K. L. Fry, C. P. Kuo, C. A. Larsen, R. M. Cohen and G. B. Stringfellow, J. Electron. Mater., 15:79 (1986).
35. D. W. Kisker, J. N. Miller and G. B. Stringfellow, Appl. Phys. Lett., 40:614 (1982).
36. J. R. Scheely and J. M. Woodall, Appl. Phys. Lett., 41:88 (1982).
37. M. J. Tsai, M. M. Tashima and R. L. Moon, J. Electron. Mater., 13:437 (1984).
38. M. D. Scott, A. G. Norman and R. R. Bradley, J. Crystal Growth, 68:319 (1984).
39. I. Hino and T. Suzuki, J. Crystal Growth, 68:483 (1984).
40. M. Ikeda, Y. Mori, M. Takiguchi, K. Kaneko and N. Watanabe, Appl. Phys. Lett., 45:661 (1984).
41. M. J. Cardwell, J. Crystal Growth, 70:97 (1984).
42. C. H. Chen, M. Kitamura, R. M. Cohen and G. B. Stringfellow, Appl. Phys. Lett., 49:963 (1986).
43. L. D. Zhu, K. T. Chan and J. M. Ballantyne, Appl. Phys. Lett., 47:47 (1985).
44. E. J. Thrush, C. G. Cureton, J. M. Trigg, J. P. Stagg and B. R. Butler, Chemtronics, 2:62 (1987).
45. C. A. Larsen, C. H. Chen, M. Kitamura, G. B. Stringfellow, D. W. Brown and A. J. Robertson, Appl. Phys. Lett., 48:1531 (1986).
46. C. H. Chen, C. A. Larsen, G. B. Stringfellow, D. W. Brown and A. J. Robertson, J. Crystal Growth, 77:11 (1986).
47. C. H. Chen, C. A. Larsen and G. B. Stringfellow, GaAs and Related Compounds, 1986, Vol. 83, p. 75, Inst. of Phys., London (1987).
48. C. H. Chen, C. A. Larsen and G. B. Stringfellow, Appl. Phys. Lett., 50:218 (1987).
49. R. Bhat, M. A. Koza and B. J. Skromme, Appl. Phys. Lett., 50:1194 (1987).
50. G. B. Stringfellow, J. Crystal Growth, 65:454 (1983).
51. G. B. Stringfellow, "Organometallic Vapor Phase Epitaxy: Theory and Practice", Academic Press, Boston (1989).
52. M. J. Cherng, H. R. Jen, C. A. Larsen and G. B. Stringfellow, J. Crystal Growth, 77:408 (1986).
53. H. Krautle, H. Roehle, A. Escobosa and H. Beneking, J. Electron. Mater., 12:215 (1983).
54. A. Ishibashi, Y. Mori, M. Itabashi and N. Watanabe, J. Appl. Phys., 58:2691 (1985).
55. W. T. Tsang, J. Electron. Mater., 15:235 (1986).
56. B. I. Miller, E. F. Schubert, U. Koren, A. Ourmazd, A. H. Dayem and R. J. Capik, Appl. Phys. Lett., 49:1384 (1986).
57. T. Y. Wang, K. L. Fry, A. Persson, E. H. Reihlen and G. B. Stringfellow, Appl. Phys. Lett., 63:2674 (1988).
58. C. R. Lewis, M. J. Ludowise and W. T. Dietze, J. Electron. Mater., 13:447 (1984).
59. J. N. Miller, D. Houng, G. Hom, D. Kisker and G. B. Stringfellow, Electron. Mater. Conf., Fort Collins, Colo., Paper C-6.
60. G. B. Stringfellow, in: "Crystal Growth of Electronic Materials", E. Kaldis, ed., Chapter 18, Elsevier Science Publishers, Amsterdam (1985).
61. CRC Handbook of Safety, N. V. Steere, ed., The Chemical Rubber Co., Cleveland, Ohio (1967).
62. R. Bhat, J. Electron. Mater., 14:433 (1985).
63. C. B. Cooper, M. H. Ludowise, V. Aebi and R. L. Moon, Electron. Lett., 16:20 (1980).

64. M. J. Cherng, R. M. Cohen and G. B. Stringfellow, J. Electron. Mater., 13:799 (1984).
65. J. Baumann, private communication.
66. C. Blaauw, C. Miner, B. Emmerstorfer, A. J. Spring Thorpe and M. Gallant, Can. J. Phys., 63:664 (1985).
67. K. W. Benz, H. Renz, J. Weidlein and M. H. Pilkuhn, J. Electron. Mater., 10:185 (1981).
68. R. Bhat, M. A. Koza and B. J. Skromme, Appl. Phys. Lett., 50:1194 (1987).
69. Y. Mori, F. Nakamura and N. Watanabe, J. Appl. Phys., 60:334 (1986).
70. S. Hiyamizu, J. Saito, K. Nanbu and T. Ishikawa, Jpn. J. Appl. Phys., 22:L609 (1983).
71. H. Takakuka, K. Tanaka, Y. Mork, M. Arai, Y. Kato and S. Watanabe, Trans. Electron. Dev., ED-33:595 (1986).
72. K. Kobayashi, S. Kawata, A. Gomyo, I. Hino and T. Suzuki, Electron. Lett., 21:931 (1985).
73. S. K. Shastry, S. Zemon, D. G. Kenneston and G. Lambert, Appl. Phys. Lett., 52:150 (1988).
74. E. C. Larkins, E. S. Hellman, D. G. Schlom, J. S. Harris, M. H. Kim and G. E. Stillman, Appl. Phys. Lett., 49:391 (1986).
75. G. E. Stillman, L. W. Cook, T. J. Roth, T. S. Low and B. J. Skromme, "GaInAsP Alloy Semiconductors", T. P. Pearsall, ed., Chapter 6, John Wiley and Sons Ltd., New York (1982).
76. Y. Kawaguchi, H. Asahi and H. Nagai, Inst. Phys. Conf. Ser. No. 79, 79 (1985).
77. J. S. Roberts, P. A. Claxton, J. P. R. David and J. H. Marsh, Electron. Lett., 22:506 (1986).
78. L. L. Taylor and D. A. Anderson, J. Crystal Growth, 64:55 (1983).
79. T. J. Roth, B. J. Skromme, T. S. Low, G. E. Stillman and L. M. Zinkiewicz, in: "Semiconductor Growth Technology", E. Krikorian, ed., Proceedings of SPIE, 323:36 (1982).
80. L. F. Eastman, in: "Proceedings of the NATO Sponsored InP Workshop", cited by J. L. Benchimol, M. Quillec and S. Slempkes, J. Crystal Growth, 64:96 (1983).
81. J. P. Andre, E. P. Menu, M. Ermann, M. H. Meynadier and T. Ngo, J. Electron. Mater., 15:71 (1986).
82. T. Mizutani and K. Hirose, Japan. J. Appl. Phys., 24:L119 (1985).
83. H. M. Cox, S. G. Hummel and V. G. Keramidas, J. Crystal Growth, 79:900 (1986).
84. H. Jurgensen, D. Schmitz, M. Heyen and P. Balk, Inst. Phys. Conf. Ser. No. 74, 199 (1984).
85. J. D. Oliver and L. F. Eastman, J. Electron. Mater., 9:693 (1980).
86. N. Kobayashi and T. Fukui, Electron. Lett., 20:887 (1984).
87. L. D. Zhu, P. E. Sulewski, K. T. Chan, K. Muro, J. M. Balantyne and A. J. Sievers, J. Appl. Phys., 58:3145 (1985).
88. W. T. Tsang, A. M. Chang and J. A. Ditzenberger, Appl. Phys. Lett., 49:960 (1986).
89. M. Takikawa, J. Komeno and M. Ozeki, Appl. Phys. Lett., 42:280 (1983).
90. A. Kastalsky, R. Dingle, K. Y. Cheng and A. Y. Cho, Appl. Phys. Lett., 41:274 (1982).

CRYSTAL GROWTH IN TECHNOLOGY

ELEMENTARY SEMICONDUCTORS: SILICON

L. J. Giling

University of Nijmegen
Experimental Solid State Physics III
Toernooiveld, 6525 Ed Nijmegen, The Netherlands

The 20th century and probably also the 21st century with good reason may be called the silicon age. The onset to the silicon development was slow however. It had begun already in the beginning of the 19th century with the discovery of the elemental form of silicon. It then not only took a good century to define what silicon really was (some forms in reality were silicides), but also to determine its physical and chemical properties. But from there on, events developed progressively, however.

The major demand for silicon came from the steel industry. In the early part of this century steel manufacturers developed the electric arc process for the reduction of SiO_2 (mainly in the form of quarzite or sand) with carbon according to the reaction:

$$SiO_2 + 2C \rightarrow Si + 2CO.$$

In this way metallurgical silicon (98 to 99%) was and still is created and used as additive for the steel fabrication. Nowadays its production amounts to several hundred thousand tons per annum.

Also in the beginning of this century the first reports were published on the semiconductor behavior of silicon and germanium and their detector properties described. These publications started what now is being recognized as one of the biggest impacts in the technological development of the 20th century viz the creation of the solid state electronic industry. First microwave (radar) detectors were fabricated, directly followed by electrical diodes. Then the invention was made of the transistor effect in 1949 by Bardeen and Brattain [1], which, together with the theoretical work of Shockley [2] gave insight into the role of electrons and holes in the solid. From then on it became clear that in order to have control over the electrical properties, the fabrication of useful devices totally depended on the quality and purity of the materials used, and that both silicon and germanium had to be purified to below the ppb level. This had never been done before with solid material.

The first breakthrough in this respect came by the discovery of Pfann [3] that germanium could be purified by zone melting (Figure 1). By moving a molten zone from one end of a rod to the other, impurities which normally prefer to remain in the liquid, are swept to the other end. Because of its lower melting temperature germanium was preferably used and

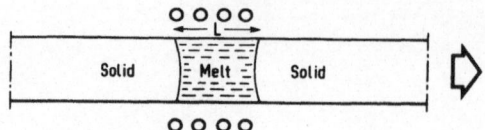

Fig. 1. The principle of zone melting [14]. A molten zone of length L is moved through the material to be purified. The impurities preferably stay in the liquid, so that the solidified fraction will become more pure. By repeating the process several times the zone refined material eventually becomes highly purified.

remained the sole master of the scene up to 1960, in particularly because it could easily be doped p and n-type, using an alloying technique (Figure 2). In this way the germainium transistors lead the way into the area of electronic devices.

The need for devices which could work at higher temperature, however, more and more prompted research on higher bandgap materials such as silicon. Two difficulties had to be overcome: (i) the purity of the material, since even zone-melting had not really been able to remove boron from this material, and (ii) the strong tendency of silicon to oxidize, which made the alloying technique for p/n junction formation impossible.

The second breakthrough came in 1954 by the development of an alternative purification method called the Siemens process [4]. Here metallurigcal silicon first is transformed into trichlorosilane by reaction with hydrogen chloride in a fluid bed reactor at about $300°C$

$$Si + 3HCl \rightleftharpoons SiHCl_3 + H_2.$$

The $SiHCl_3$ (b.p. $31.8°C$) is easily purified by distillation, removing in this way also the boron impurities, ending up with semiconductor grade $SiHCl$.

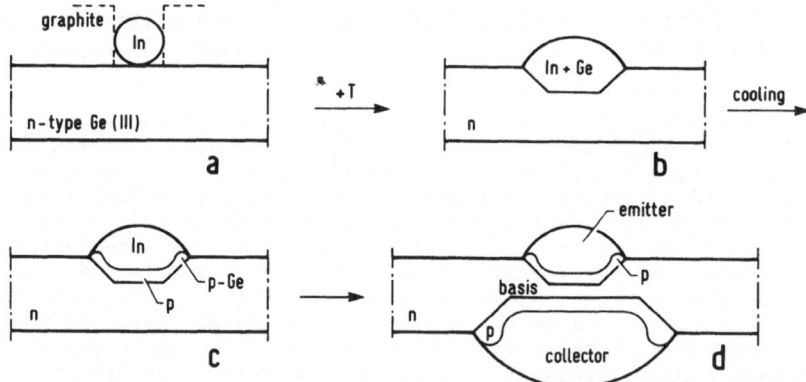

Fig. 2. The alloying technique for making p–n junctions [14]. (a) An indium droplet is placed on the surface of an n-type (111) germanium wafer. (b) Upon heating the indium dissolves some germanium forming an indium-germanium alloy inside a flat-bottomed etch pit. (c) During cooling a thin layer of germanium, heavily p-type doped with the indium, is deposited on the n-type substrate, forming a p-n junction. (d) A p–n–p transistor is made by alloying from both sides.

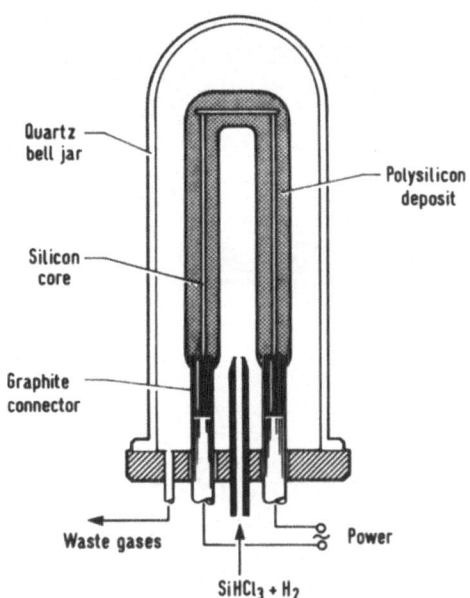

Fig. 3. Deposition of pure polycrystalline silicon on a hot wire by
decomposition of SiHCl$_3$.

The reaction scheme given above is of course highly over-simplified, since the gas phase contains all kinds of silicon-hydrogen-chloride species, but dominant amongst them at 300°C is the trichlorosilane.

The next step, and the first application of chemical vapor deposition (CVD), is the preparation of pure silicon from SiHCl$_3$ formed by reduction at temperatures above 1000°C in hydrogen. At these temperatures the above given equilibrium shifts to the left and silicon is deposited again. In the Siemens process the mixture of SiHCl$_3$ and H$_2$ is transported to a reactor (Figure 3) where a silicon wire is heated by an electric current. The chemical reaction takes place at high temperatures leading to silicon deposition in the form of polysilicon on the wire. To keep the temperature constant the current must be increased with increasing diameter.

In this way rods of over 10 cm diameter have been obtained, with lengths up to 1.5 m. These rods can directly be used in a float zone process to yield rods of single crystalline material of resistivities > 100 cm. Todays production of this very pure silicon stock material is estimated 5000 tonnes per annum [5]. This, of course, is far less as compared to the fabrication of metallurgical grade silicon, but its value added is much higher.

The third breakthrough came with the introduction of the so-called planar technology on silicon. The early disadvantage of the easy formation of silicon dioxide was turned into an advantage. It appeared that not only the SiO$_2$ layer chemically was very stable, but also that the density of electrically active surface states at the Si-SiO$_2$ interface was low so that it did not significantly influence the electrical performance of the device.

At high temperatures silicon is oxidized with oxygen or water vapor giving a well adhering 1 μm thick SiO$_2$ film. In this layer selectively windows are etched using photolithographic techniques followed by HF

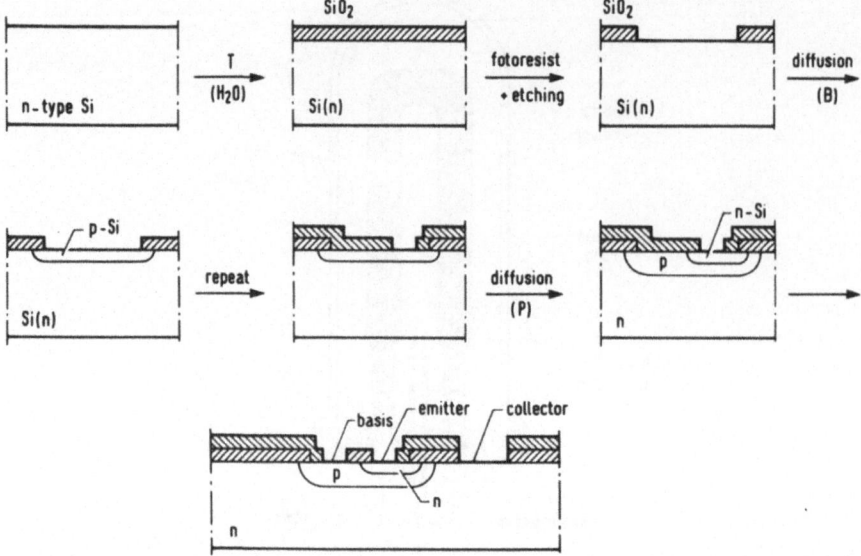

Fig. 4.   Planar technology [14].   Oxidation of silicon gives a thin, well
          adherent SiO_2 layer.   Small windows can be etched into the oxide
          layer using photolithographic techniques followed by HF etching.
          p-type diffusion can be performed with boron, n-type diffusion
          with phosphorus compounds, e.g., from the gas phase.   The process
          can be repeated by using different masks to select the diffusion
          area.

etching.   The silicon itself is not etched by HF.   By gas phase diffusion,
p or n-type doping can be achieved inside these windows, the SiO_2-layer
effectively blocks all dopants.   Then by regrowth of a thin SiO_2-layer and
repeating the process one can obtain the desired geometries (Figure 4).
Nowadays very small devices can be made using high precision photo-
lithography.   On one piece of silicon several electrical devices can be
combined giving the integrated circuit or I.C. (Figure 5).   The packing
density of the active components has become so high that the earlier main
frame computers have been reduced to table top personal computers.

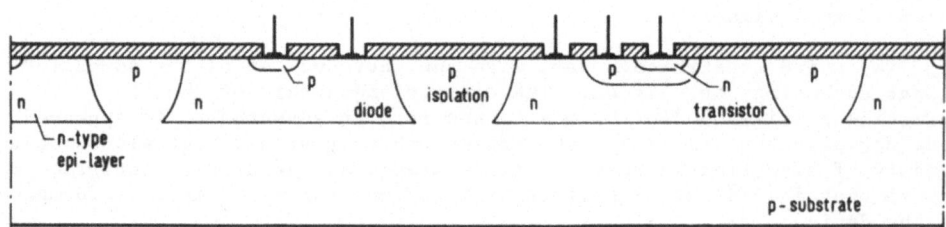

Fig. 5.   Integrated circuit [14].   Many diode and transistor structures
          made in an n-type layer.   Isolation between these devices is
          achieved by the diffusion of p-type dopant through the n-type
          layer.

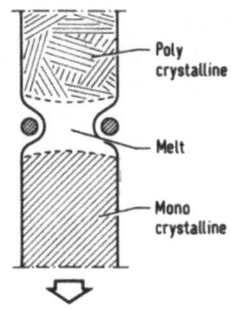

Fig. 6. The needle-eye technique during FZ-pulling.

CRYSTAL GROWTH

Float Zoning

The high melting point of silicon (1412°C) offers a main problem for crystal growth, because at these temperatures it significantly reacts with all container materials. Also quartz slowly dissolves in the silicon melt mainly by the formation of volatile SiO

$$SiO_2 + Si \rightarrow 2SiO.$$

By turning the horizontal float zone system vertically, Theurrer [6] could get rid of the crucible problem. The molten zone is now stabilized by the surface tension force. Mind that the growth direction normally is not up but down. Experimentally it appears that this molten zone is stable up to a length of 30 mm. By applying in addition electromagnetic forces from the R.F. heating coil, the molten zone can be made considerably longer. This is especially important for the growth of large crystal diameters which are grown with the so-called needle-eye technique (Figure 6). The smaller dimension of the molten zone is necessary in connection with the centrifugal forces due to the rotation of feed stock and counter rotation of the drawn single crystal. So a large diameter feed rod melts at the narrow coil, whereas the crystal grows thick again after passage of the needle-eye. The flat R.F. coil can be opened at the end of the growth so that the crystal can be removed from the system.

The actual geometry of the floating zone strongly depends on the growth rate and the dimensions of the R.F. coil. The smaller the diameter of the coil the better is the coupling of the field with the melt. With a proper combination of rod diameter and coil dimension, rotation speed and growth rate, the shape of the growing interface of the melting rod is curved as indicated in Figure 7. If the R.F. coil diameter is too large,

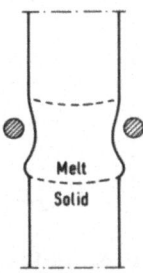

Fig. 7. Concave melt-solid interface at the growth front.

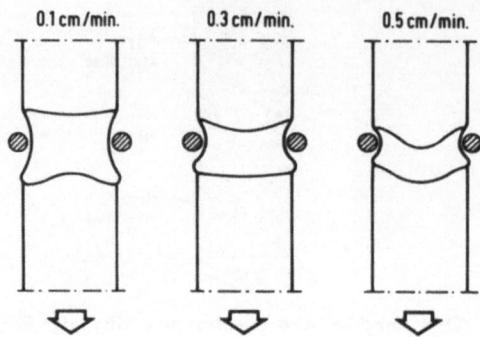

Fig. 8. Influence of growth rate on melt-solid interface.

the coupling of the field is more concentrated at the rim and both interfaces become sharply convex until either the rod does not melt any longer in the middle, or the connection is broken between the feed top and the crystalline rod at the bottom [7]. The influence of the growth rate is indicated in Figure 8. Here the amount of material to be melted and later on to be crystallized plays a role via the heat of melting (at the feed-end) and the heat of crystallization at the growth interface. In equilibrium the heat balance is zero, but under growth conditions the limited heat conductivity of the melt and limited convective flow prevent a perfect exchange. This means that at higher growth rates the heat developed at the growing interface cannot get away and will locally, i.e., in the middle, melt a small portion of the crystal, giving the desired concave interface shape. At the feed end the interface in the middle will not receive enough heat to melt, so at high growth rates it becomes more convex, but also the whole interface is lowered more and more to the plane of the R.F. coil in order to have more profit of the heat introduced by the electromagnetic field.

The shape of both interfaces of course also strongly depends on the convective flow pattern. Four factors play a role in this respect (see Figure 9):

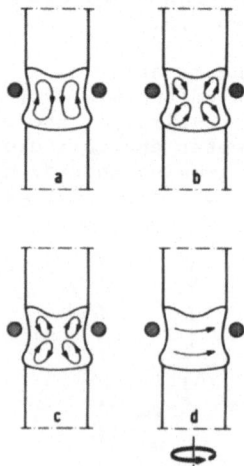

Fig. 9. Effect of convective flow on melt-solid interface as caused by:
a) free convection; b) Marangoni flow; c) R.F. field; and
d) rotations.

1. Free convection. The temperature difference creates density
    variations which move in the gravity field. Near the coil the
    temperature is highest, the melting and growth interface have a lower
    temperature.
2. The Marangoni flow, i.e., the surface tension driven flow. This
    effect is quite severe and is always present whenever a melt has a
    temperature gradient on its surface. The reason is that the surface
    tension is a strong function of temperature. At lower temperatures
    the surface tension is much higher than at higher temperatures. This
    creates a flow from the low surface tension area (high temperature)
    to the high surface tension side (low T).
3. The electrodynamic driven flow. The R.F. field drives a flow in the
    coil plane radially inwards to the melt.
4. The centrifugal flow due to the rotations. The faster the flow, the
    more the flow cells are confined to the rim of the melt, but is also
    creating more mixing and removing of the heat of crystallization,
    i.e., the middle part of each interface becomes cooler and melts
    slower or grows faster, respectively, until instead of a concave, a
    bulged convex center part is developed. The undercooling can become
    so large that even faceting can take place at the growth interface
    (i.e., growth of the low index (111) plane). This is undesirable, as
    here steps move fast over the surface and trap all kinds of
    impurities. The result is that in this part of the crystal the
    dopant and impurity distribution differs from the rest of the crystal
    (higher dopant concentration in faceted area).

The zone refining action, i.e., the redistribution of impurities
along the length of the rod can be mathematically treated as follows:

When a melt is in equilibrium with a solid, the concentration of
impurities in melt and solid usually are different. This is entirely
determined by the positions of the solidus and liquidus in the T-x diagram
of the binary silicon-impurity (Figure 10). For the small concentrations
we are talking about, these curves can be approximated by straight lines.
The equilibrium distribution between solid (s) and liquid (l) defines the
segregation coefficient k_o:

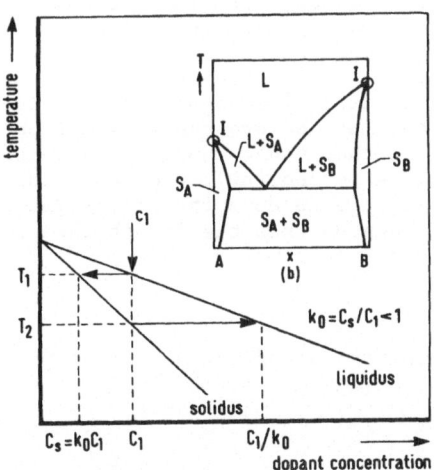

Fig. 10. Part of the solidus-liquidus of the binary silicon-impurity.
The insert schematically gives the complete T-x diagram.

$$k_o = \frac{c_s}{c_1}.$$

Assume now that the rod originally contains a uniformly distributed impurity with concentration c_o, that the length of the molten zone is ΔL, and that the whole length of the rod is L. In addition the cross section is assumed to be constant and unit over the length of the rod.

The number of impurities N in the molten zone is given by

$$N = c_1 \cdot \Delta L.$$

When the zone is moved over a length dx the number of impurities is enlarged at the melting end by $c_o dx$ and diminished at the solidifying end with $c_s dx = k_o c_1 dx$, so that the change in the number of impurities is given by

$$dN(x) = c_o dx - k_o c_1 dx$$

$$dN(x) = (c_o - \frac{k_o N(x)}{\Delta L}) dx.$$

Integrating:

$$\int_{N_o}^{N} \frac{dN}{c_o - (k_o N/\Delta L)} = \int_0^x dx$$

gives as final result

$$\frac{c_o - c_s(x)}{c_o(1 - k_o)} = e^{-\frac{k_o x}{\Delta L}}.$$

For repeated passes the mathematics become somewhat cumbersome but the limiting value in the solid (c_∞) is determined by

$$c_\infty = A e^{BL/\Delta L}$$

where A and B are complicated constants containing k_o, ΔL, L and c_o.

It is important to note here that the above given analysis is only valid if full equilibrium segregation is obtained. This will certainly not be true for actual growth conditions. All will depend on the diffusion velocities of the impurities in melt and solid as compared to the growth rate. A small diffusion coefficient in the liquid means that the impurities will accumulate before the growth front, giving rise to a much higher concentration than under equilibrium conditions (i.e., the onset for constitutional supercooling), whereas a too small diffusion coefficient in the solid, breaks down the solid-liquid exchange and trapping occurs. This happens when the growth rate (G) is larger than the diffusion over one atomic layer:

$$G > \frac{D}{x} \qquad x \cong 2\text{\AA}$$

In those cases one introduces an effective segregation coefficient k_{eff} defined by

$$k_{eff} = \frac{c_s}{c_1}.$$

Float zoning nowadays usually is no longer used as a purification method. The starting CVD poly-silicon material is already so pure that the poly-silicon only has to be transformed into a single crystal and one passage of a float zone is enough. One starts with a single crystalline

seed crystal which is tipped into the molten zone of the feed-rod. By a
necking procedure (fast pulling), which reduces the diameter of the seed
crystal to about 3mm, and pulling this out for a few centimeters, all
dislocations still present in the seed crystal, move to the surface of the
neck and a dislocation free seed remains. Most likely the dislocations
are removed by a climb mechanism involving silicon interstitials, which,
due to the rapid cooling of the neck, are in a highly supersaturated
state.

Czochralski Pulling

Although the highest purity single crystalline silicon is obtained
with the FZ-technique, most of the silicon material on the market is grown
by the Czochralski (CZ) technique. The main reason for this is that the
melt is confined in a container, so that there are no stability problems
anymore as compared with the FZ technique (Figure 11).

The main disadvantage is the presence of the same crucible as this
slowly dissolves and introduces electrically active impurities in the
melt. This means that the resistivity of CZ-silicon is limited to values
below 100 Ωcm, what, for most applications, is good enough.

Originally CZ-Si was contaminated with carbon. The source for this
carbon was the carbon support of the quartz crucible, which emitted at
the high growth temperature all kinds of hydrocarbons when hydrogen was
used as a purge gas. By using argon instead of H_2 this problem was
solved. Now only oxygen remains as the main impurity in CZ-Si. It even
precipitates in the form of SiO_2 at lower temperatures forming small
clusters in the crystal.

The heat balance (Figure 12) for the CZ-configuration is given by [8]

$$L \frac{dm}{dt} + \kappa_1 \frac{dT}{dz_1} A = \kappa_s \frac{dT}{dz_s} A$$

where L = latent heat of fusion; (dm/dt) = amount of material which
crystallizes per unit time; κ_1, κ_s = heat conductivities in liquid (1) and
solid (s); (dT/dz_1),(dT/dz_s) = temperature gradients near growth interface
in liquid and solid; and A = area of the crystal. When ρ is the density of

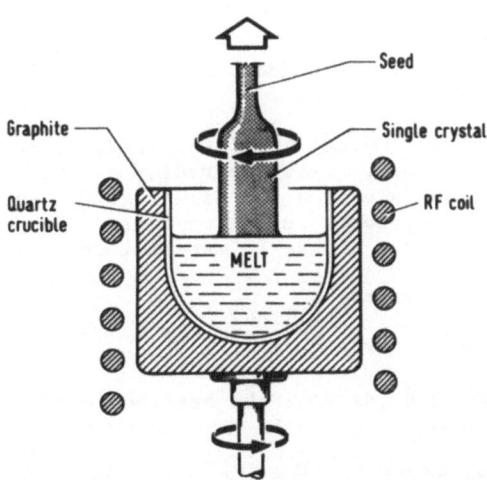

Fig. 11. Czochralski pulling (CZ technique).

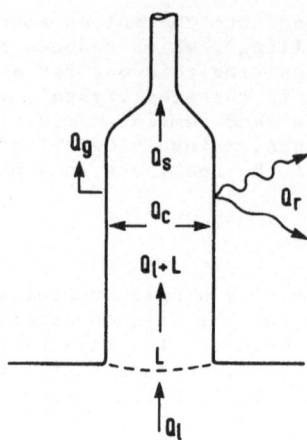

Fig. 12. Heat balance during CZ-growth: Q_l = heat abducted from liquid, L = latent heat of fusion, Q_c = heat conduction in solid, Q_r = heat-loss by radiation, Q_g = heat-loss by conduction to gas, and Q_s = heat conducted to seed.

the silicon and v the growth velocity then (dm/dt) = v.A.ρ, so that the heat balance is written as

$$A(v \cdot \rho \cdot L + \kappa_1 \cdot \frac{dT}{dz_1}) = \kappa_s \cdot \frac{dT}{dz_s} \cdot A.$$

The heat loss from the crystal is determined by radiation and conduction to the argon gas and for a small amount to the seed end. The first two losses dominate and are proportional to the crystal surface area, so proportional to the radius r if a cylinder is grown. This means that

$$\kappa_s \cdot \frac{dT}{dz_s} \cdot A = C \cdot r$$

where C is a constant. The heat balance now becomes:

$$v \cdot \rho \cdot L \pi r^2 + \kappa_1 \cdot \frac{dT}{dz_1} \cdot \pi r^2 = C \cdot r$$

$$v \cdot \rho \cdot L + \kappa_1 \cdot \frac{dT}{dz_1} = \frac{C'}{r} .$$

For normal experimental conditions the first term dominates, which means that when the growth rate is enlarged, the diameter of the crystal will reduce and vice versa.

A special case exists during the first stage of crystal growth when the conical top end is grown. Here the radius of the crystal becomes larger, i.e., its top surface area increases, giving an easier way to get rid of the heat which is present in the crystal. When the top area is approximated by πr^2 then the change in material is given by

$$\frac{dm}{dt} = \frac{d}{dt} \text{ (volume x density)}$$

$$= \frac{d}{dt} \cdot (\pi r^2 \cdot z \cdot \rho) = 2\pi r \rho z \frac{dr}{dt} + \pi r^2 \rho \frac{dz}{dt} .$$

For a constant growth rate (dz/dt), the heat balance equation in this special case becomes [8]

$$L\{2\pi r z \rho \frac{dr}{dt} + \pi r^2 \rho \frac{dz}{dt}\} + \kappa_1 \frac{dT}{dz_1} \pi r^2 = (R + C)\pi r^2 + Q_o$$

where R is the heat loss by radiation, C by conduction and Q_0 is the amount of heat flow into the seed end. This equation has the form

$$r \frac{dr}{dt} - C_1 r^2 = C_2$$

where C_1 and C_2 are constants. The solution of this equation is given by [8]

$$r^2 = C_3 \exp(2C_1 t) - \frac{C_2}{C_1}$$

where C_3 again is a constant.

This solution tells us that the surface area can grow exponentially with time in the first stages of the pulling. This exponential growth can be avoided by a careful adjustment of the pulling rate in this period.

CVD

Chemical vapor deposition is a technique by which a solid material can be deposited on a support. The solid material is formed from gaseous reactants by a chemical reaction, which explains the terms "chemical" and "vapor". The gaseous reactants only react with each other at high temperatures, the driving force being the change in free energy of the chemical reaction. The reaction conditions are such that at these conditions the gas phase is supersaturated with the material to be deposited. The actual growth rate, however, is limited by one of the kinetic barriers in the growth process.

Although a lot of devices are made on a silicon wafer by in-diffusion, ion implantation, proton bombardment and oxidation, there are still a number of applications where CVD is useful, as, for instance, in high performance diodes. Here a high ohmic thin epitaxial layer is needed on a low ohmic substrate. The I-V characteristic of this epitaxial diode combines the advantages of a low series resistance in the forward mode because of the low resistivity of the substrate and the high breakthrough voltage in the reverse mode (Figure 13).

The CVD technique is exclusively useful for the fabrication of doping superlattices and strained layer superlattices made of silicon and germanium.

In Figure 14 a schematic view is given of the reaction paths leading to the epitaxial growth. The following steps can be discerned:

1. Transport of the gas mixture through the cell to the hot zone.
2. Diffusion through the gas phase to the hot crystal surface, coupled with chemical reactions as the gas is heated up. Formation of a chemical equilibrium mixture close to the surface.
3. Adsorption on the crystal surface.
4. Surface diffusion to step sites.
5. Incorporation in the lattice.
6. Desorption of reaction products.
7. Transport of the products away from the crystal surface.

For a proper growth the following conditions are essential:

1. The transport should be laminar, otherwise not only diffusion of reactants will determine the growth rate but also convective flow. This will give rise to growth non-uniformity [9,10].

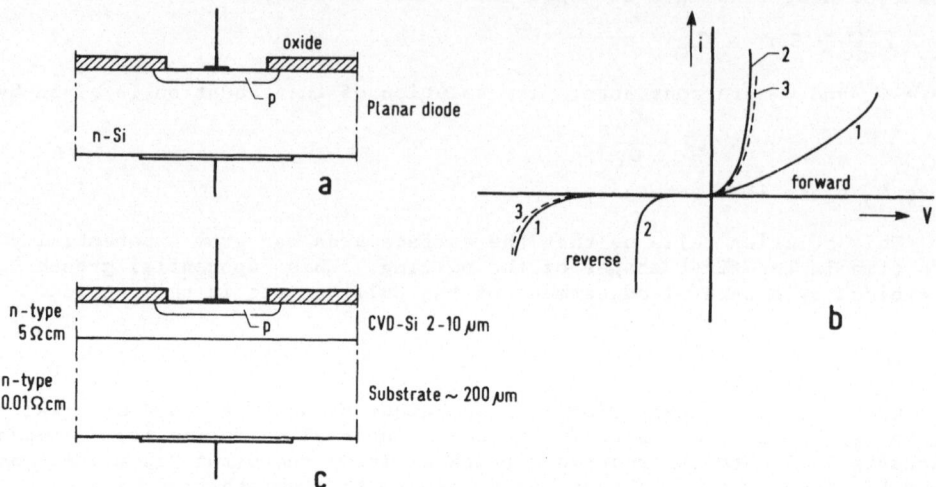

Fig. 13. Forward and reverse currents for diodes made in different
substrate materials: curve 1 - low doped (high ohmic) n-type
substrate showing a small forward current, but high breakthrough
voltage (diode structure a); curve 2 - high doped (low ohmic) n-
type substrate with a high forward current, but low reverse
voltage (also diode structure a); curve 3 - (dashed) combined
advantages, diode structure c.

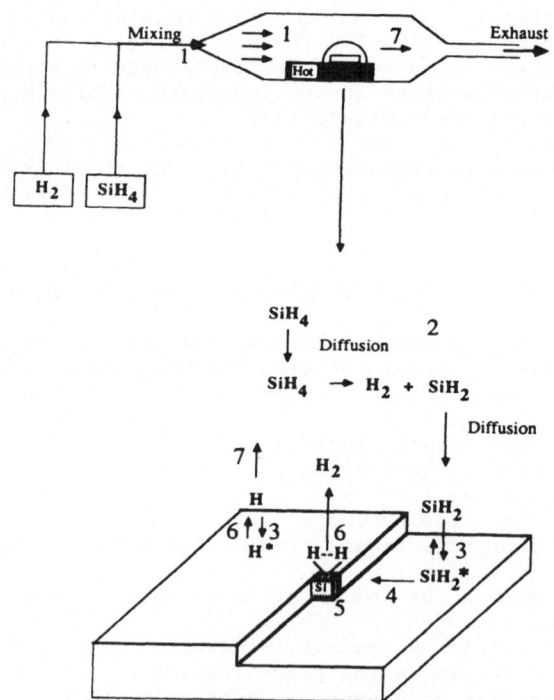

Fig. 14. Schematic view of the growth of silicon from silane by CVD [13].

2. Crystal growth should take place by a step growth mechanism, i.e., by surface diffusion of the adsorbed growth species over the low index plane to the steps on the surface. Only in this case defect free growth and a controllable doping can be attained.
3. The coverage of the crystal surface by impurities should be low.

The transport problem will not be treated here, as this has been done extensively, quite recently in other publications [10,11]. It is, however, essential not only to have laminar flow as indicated above, but also to avoid memory cells in the form of return flows, as these vortexes will act as a persistant memory cell for reactants. This is detrimental for sharp dopant or compositional transitions.

The second point is less evident. Step growth should be compared with rough growth, where atoms are directly incorporated from the vapor into the lattice. This will always happen whenever the arriving atoms form two or more bonds with the surface atoms. Each site then is a kink site and no re-adjustment is possible. Low index planes, such as (111) or the reconstructed (001), only have one bond available. Here the silicon growth species only can form one bond and due to the much lower adsorption energy, they can diffuse over the surface to the step where they will form a second bond and become incorporated. During the diffusion path each atom can adjust itself to the atomic arrangement of the lattice, i.e., a perfect match is prepared during the trip. Also the amount of impurities

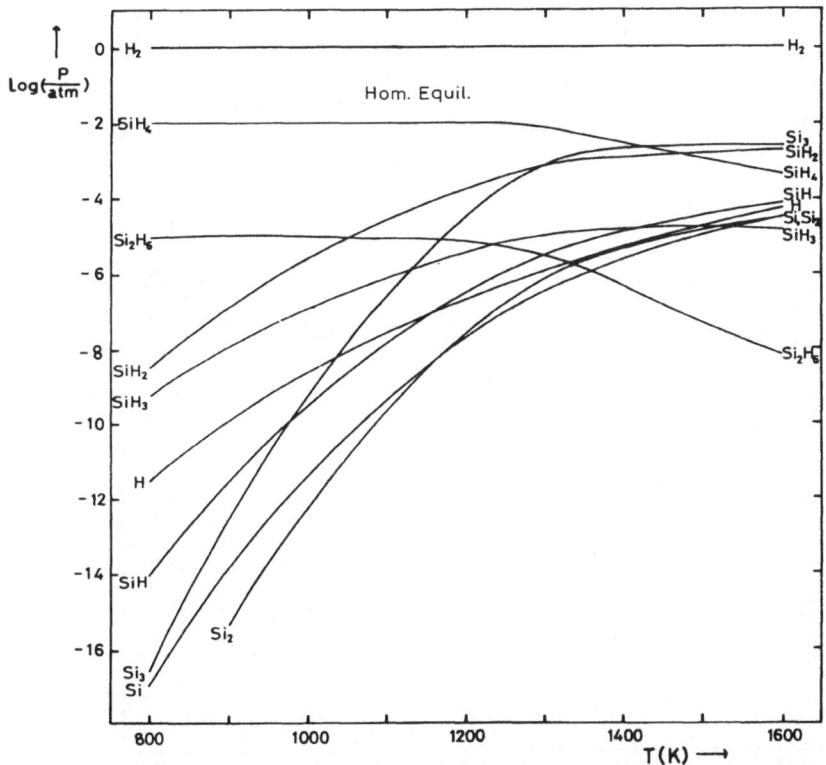

Fig. 15. Equilibrium composition of 1% SiH_4 in H_2 as a function of temperature at a total pressure of 1 bar. The interaction with solid silicon is not allowed (homogeneous equilibrium approach) [12].

on such a surface, which contains one unsaturated bond per surface site, is much lower. This considerably contributes to defect free growth, as the change that the epitaxial feeling gets lost by a monolayer of impurities which covers the surface is minimzied. Similarly the adsorption-desorption process for dopant atoms is only possible whenever one adsorption bond is made. For two bonds the adsorption energy is so large, that desorption becomes very unlikely, and dopants are trapped.

Figure 15 gives equilibrium calculations of the gas phase composition for the growth of silicon from silane. It is seen that at low temperatures quite a large portion of SiH_4 is undecomposed, at higher temperatures the silicon radical Si_3 and SiH_2 dominate. Note the low partial pressure of H at all temperatures. The adsorption calculations (Figure 16) which are more appropriate for the crystal growth process, show a completely different picture. Here adsorbed H dominates at all temperatures, whereas the main growth species at all temperatures is the SiH_2 radical. In Figure 16 the equilibrium situation is depicted; for growth conditions, however, all coverages of the silicon species increase considerably (Figure 17). Now the coverage of SiH_2 and Si approach a few per cent of the crystal surface, what indeed can lead to a growth rate of a few microns per minute [12]. The calculated hydrogen coverage in both cases is quite high, only at elevated temperatures and low pressures it is reduced considerably. In practice the growth results do not really point to a high surface coverage of H, so the calculated amounts may be too high.

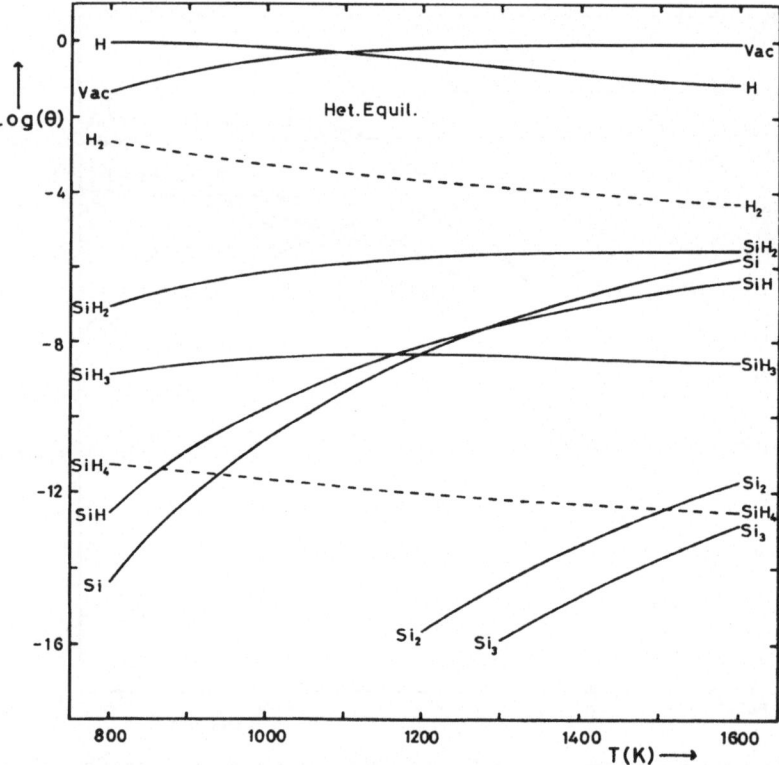

Fig. 16. Equilibrium surface composition of silicon (111) as a function of temperature in the Si-H-system at 1 bar total pressure. The dotted lines refer to physisorption [12].

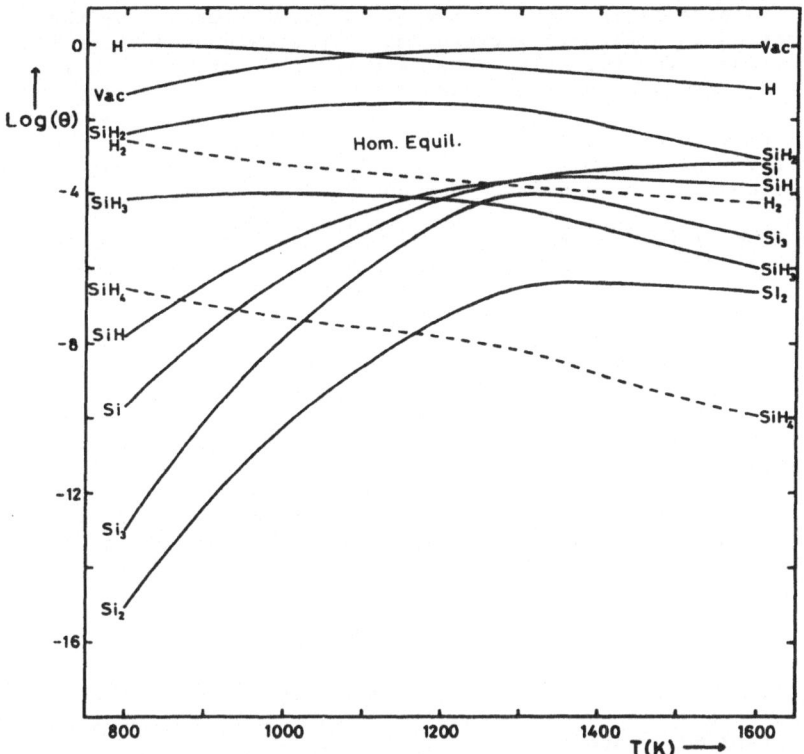

Fig. 17. Surface coverage of silicon during the crystal growth process. The coverage is calculated using the partial pressures as calculated for the gas phase in the homogeneous approach of Figure 15.

The (001) surface is more complicated as here a "chemical" reconstruction has taken place (Figure 18). Adsorption on such a surface - with the reconstruction fully intact - resembles growth on a (111) surface, but when the partial pressures increase also the reconstruction can be broken and the adsorption process becomes more intricate. The

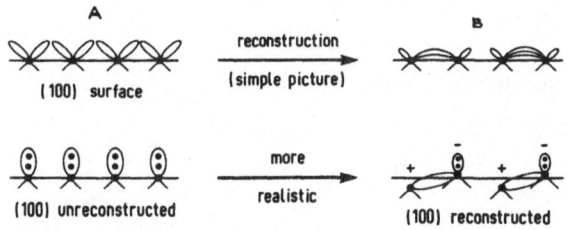

Fig. 18. Dimer formation of a Si (001) surface. When the bonds are broken (A) the crystal is atomically rough and the crystal growth process does not proceed via steps. In (B) the (001) surface resembles a (111) surface, and steps are formed because the step free energy is non-zero.

outcome of the process is that for the usual growth conditions the surface is reconstructed in a (2 x 1) pattern and step growth occurs for low supersaturations. For slightly higher supersaturations the 2 x 1 reconstructions at some local positions is broken up giving rise to additional nucleation in between the steps. Such 2-dimensional surface-nuclei give rise to an extra step creation mechanism and thus to a less controllable way of growth [13]. So on a (001) surface in general lower supersaturations will be required than on (111) surfaces.

It will be clear that there is a competition between adsorbed impurities and adsorbed growth species. Reducing the H_2 pressure helps considerably in this respect. In Figure 19 the influence of H_2-pressure on the H-coverage is given. The H-coverage reduces considerably at lower H_2-pressures. When the same amount of growth species is introduced, the relative contribution of growth species on the surface will be much higher. In other words, the diffusion path of SiH_2 to the step is less hindered by adsorbed H atoms at lower H_2-pressures, and a more perfect growth is to be expected. This indeed has been observed: the growth of single crystalline silicon could be extended to much lower growth temperatures (< 1000°C). At these low temperatures oxygen becomes a seriously adsorbed species. Very low concentrations of O_2 or H_2O in the gas phase (< 1 ppb) are required to avoid full oxidation of the silicon layer at these temperatures.

CONCLUSION

The growth of silicon from the melt has become very sophisticated, large diameter (6") rods can be grown routinely, and doping during growth no longer offers major problems.

The growth of silicon by CVD on the other hand is still open for improvements. Lower growth temperatures are preferred requiring alternative growth techniques. High vacuum growth, plasma and laser assisted growth are serious possibilities here although their mechanisms are still not fully understood. One thing is clear however, the presence of oxygen in the gas phase, even in very little amounts, will prevent growth in a carrier gas being performed at low temperatures (< 600°C). Here the only way out is the cleanness of the high vacuum chamber.

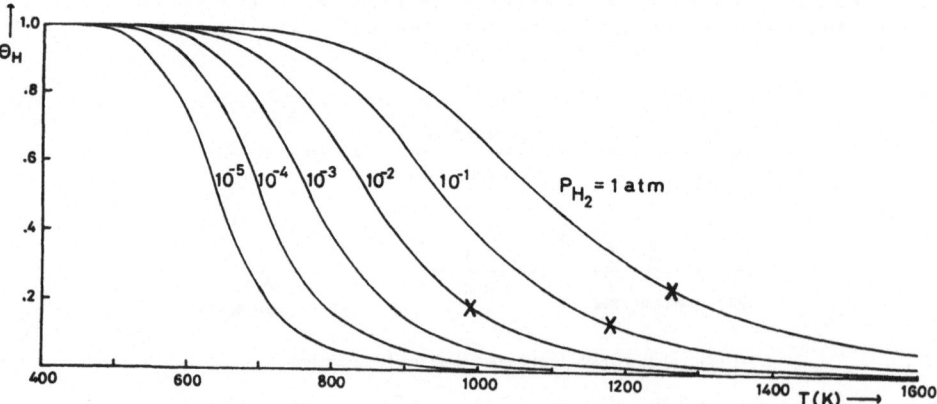

Fig. 19. Influence of H_2-pressure on the coverage with monoatomic H.

REFERENCES

1. J. Bardeen and W. H. Brattain, Phys. Rev., 75:1208 (1949).
2. W. Shockley, Bell System Techn. J., 28:435 (1949).
3. W. J. Pfann, "Zone Melting", Wiley, New York (1958); 2nd Edition (1966).
4. F. Bischoff, DBP 1.102.117 - 1.140.549 filed 18 May (1954).
5. K. C. Barraclough, 6th International Summer School on Crystal Growth Edinburg 1986, in: "Advanced Crystal Growth", P. H. Dryburgh, B. Cockayne and K. C. Barraclough, eds., Chapter 19, Prentice Hall Int., Hampstead (1987).
6. H. C. Theuerer, USP 3060123 filed Dec. 17 (1952).
7. W. Dietze, W. Keller and A. Mühlbauer, in: "Crystals, Growth, Properties and Applications", Part 5, Chapter I, J. Grabmaier, ed., Springer Verlag, Berlin (1981).
8. W. R. Runyan, "Silicon Semiconductor Technology", p. 40, Texas Instruments Electronics Series, McGraw-Hill Book Co., New York (1965).
9. L. J. Giling and J. van de Ven, 6th International Summer School on Crystal Growth, Edinburg 1986, in: "Advanced Crystal Growth", P. H. Dryburgh, B. Cockayne and K. C. Barraclough, eds., Chapter 12, Prentice Hall Int., Hampstead (1987).
10. J. van de Ven, G. M. J. Rutten, M. J. Raaymakers and L. J. Giling, J. Crystal Growth, 76:352 (1986).
11. L. J. Giling, in: "Crystal Growth of Electronic Materials", E. Kaldis, ed., Chapter 6, Elsevier Science Publishers B.V. (1985).
12. L. J. Giling, H. H. C. de Moor, W. P. J. H. Jacobs and A. A. Saaman, J. Crystal Growth, 78:303 (1986).
13. H. H. C. de Moor, "Thermodynamics and Kinetics of CVD Processes of Si and GaAs", Thesis Catholic University Nijmegen, The Netherlands, Jan 28 (1987).
14. J. Bloem and L. J. Giling, in: "Current Topics in Materials Science", E. Kaldis, ed., Vol. I, Chapter 4, North Holland Publ. Co. (1978).

TECHNOLOGY OF CRYSTAL GROWTH:

BINARY SEMICONDUCTORS: II-VI COMPOUNDS

Carlo Paorici

Physics Department
University of Parma
43100 Parma, Italy

1.   INTRODUCTION

This contribution will give a short account of the bulk technology of the II-VI semiconductor compounds. For the epitaxial technology the reader is referred to [1].

The II-VI compounds of interest for applications are limited to Zn and Cd chalcogenides, ZnO and $Hg_xCd_{1-x}Te$. With the exception of this latter compound (*), all of them are characterized by a large value of the forbidden gap, ranging from 1.4 eV of CdTe to 3.66 eV of ZnS. Such large values, which include the visible spectrum, are due to the large degree of ionicity in the chemical bonds, especially if comparison is made with elementary (Si, Ge) and III-V semiconductors. When considering that these gaps are always direct, which means a high probability of optical transitions, the interesting luminescent and photoconductive properties of these materials can be easily understood.

The physical properties of these compounds are relatively well-known. Reviews can be found in [1-3]. For ZnO and CdTe, see also [4] and [5] respectively.

Lattice structure. These compounds can have either the hexagonal (wurtzite) structure (as in CdS, CdSe, ZnSe and ZnO) or the cubic (zincblende) structure (as in CdTe and ZnTe). In ZnS both structures are observed, zincblende being stable at lower temperatures. Under high pressure, the NaCl-type structure has often been observed. In addition to the hexagonal and/or cubic closed-packed modifications, the II-VI compounds often crystallize with longer repeat periods, giving rise to polytype structures. Finally, metastable wurtzite structures have been observed in the first stages of growth of those compounds whose stable structure is zincblende.

Electrical properties. With the exception of CdTe, which can exhibit both n- and p-type conduction, all wide-gap II-VI semiconductors can be made either n-type (ZnS, ZnSe, CdS, CdSe) or p-type (ZnTe). This fact has

_____
(*) This narrow gap alloy, whose interesting applications as I.R. photodetectors are currently being studied, will not be discussed here. For an introduction, see [2,3].

up to now prevented the fabrication of p-n junctions and related applications.

The main applications of II-VI semiconductors are in: 1) solar-cell absorbers (CdTe) and windows (CdS,$Zn_xCd_{1-x}S$); 2) photovoltaic detectors (heterojunctions based on, for example, p-InP/n-CdS; p-$CuInSe_2$/n-CdS; p-GaAs/n-ZnSe); 3) LEDs (ZnS, ZnSe, ZnTe, ZnS_xSe_{1-x}); 4) LASER-window materials (ZnSe, CdTe); 5) electroluminescent displays (ZnS, ZnSe, ZnS_xSe_{1-x}); 6) photoconductive detectors (CdS, CdSe, CdS_xSe_{1-x}, $Zn_xCd_{1-x}S$); 7) non-linear optical devices, based on ZnO, ZnS (optical waveguides and modulators); on ZnS, CdS, CdTe (bulk modulators and switches; and 8) nuclear detectors (CdTe).

2. PHASE RELATIONSHIPS

The liquidus T-X diagrams are more or less known for all Zn and Cd binary chalcogenides, but the P-T projections are only known for the Cd:Te and Zn:Te systems. In Figure 1, the T-X and P-T projections are reported for the Cd:Te system. With respect to III-V compounds, which exhibit a rounded liquidus curve close to the melting point, here (due to higher ionicity) a pointed maximum is observed. The three-phase boundaries (loops) are given in terms of Cd (above) and Te (below) pressure. P_{Cd}^0 and $P_{Te_2}^0$ are the vapor pressures of pure Cd and Te respectively. The existence region of solid CdTe is also reported, as calculated [6] on the assumption of a Frenkel disorder in the Cd sublattice.

For a better comprehension, points A_1, A_2, A_3 and A_4 are reproduced in the three diagrams of Figure 1.

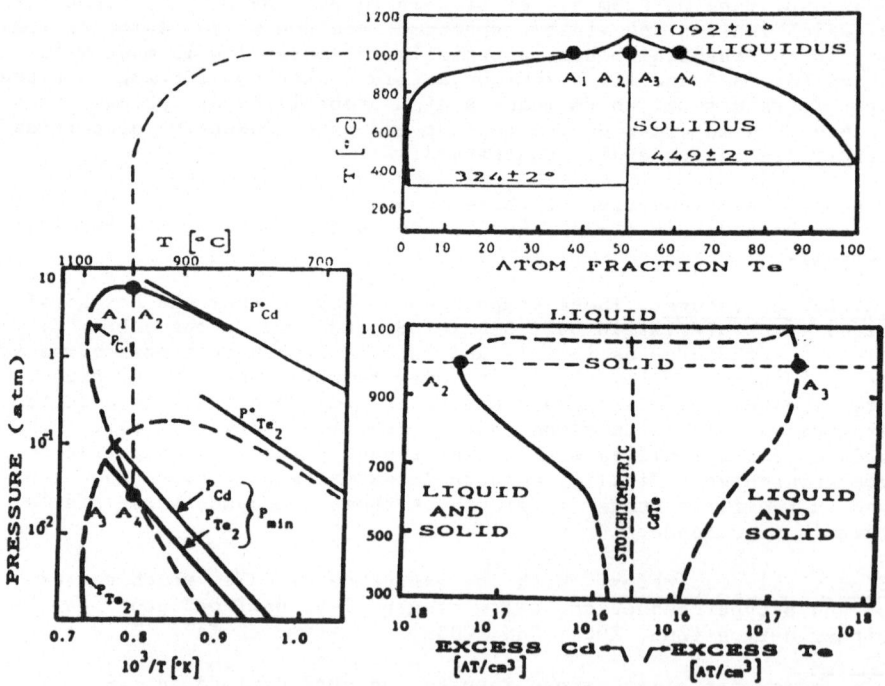

Fig. 1.   T-X projection (above), T-P projection (below, left) and existence region (below, right) of the Cd:Te system (after [6]).

In the T-range of interest for growth processes, Zn and Cd binary chalcogenides dissociate congruently according to the reaction:

$$AX(s) = A(g) + 1/2 \, X_2(g) \tag{1}$$

(A = Cd, Zn; X = S, Se, Te). Due to the very narrow existence region, the dissociation constant K_{AX} can be assumed to depend only on temperature, hence

$$P_A P_{X_2}^{1/2} = K_{AX}(T). \tag{2}$$

Further, because of the constraint

$$P_A + P_{X_2} = P \tag{3}$$

(P being the overall pressure), from (2) and (3) one gets

$$P = P_A + K_{AX}^2/P_A^2 = P_{X_2} + K_{AX}/P_{X_2}^{1/2}. \tag{4}$$

It is easy to verify that P goes through a minimum when $2P_A = P_{X_2}$.

Because of (1), a solid phase AX(s) is stable provided that either $P_A < P^0_A$ or $P_{X_2} < P^0_{X_2}$.

The Cd:Te P-T and P-X projections seem to be quite representative of the general phase relationships in II-VI binary chalcogenides, provided the maximum pressure and maximum temperature values are shifted to higher values when passing from Cd to Zn and from Te to S. However, unlike CdTe, the existence regions of the other II-VI compounds lie either on the Cd (Zn)-rich side, as in ZnS, ZnSe, CdS and CdSe or on the chalcogen-rich side, as in ZnTe. This fact has been related to the type of conductivity. By assuming a dominant Schottky-Wagner disorder (i.e., neutral and ionized vacancies in both sublattices), it can be shown [6-7,1] that (with the exception of CdTe) only one type of free carriers (electrons for ZnS, ZnSe, CdS, CdSe; holes for ZnTe) can control the electroneutrality condition in part of the existence region. For the remaining part of the existence region the electroneutrality condition is approximately given by a balance between ionized vacancies of opposite sign (defect compensation and semi-insulating behavior).

The electrically active intrinsic defects are also responsible for the automatic compensation of added donor or acceptor impurities (self-compensation).

3.   GROWTH TECHNOLOGY

Both a) melt and solution (from excess-component) growth and b) vapor growth are possible. Further, because of the solubility of II-VI compounds in molten metals, inorganic salts and water, "flux" and "hydrothermal" growth can be considered. Since these latter techniques are of minor importance, they will not be discussed here (on the subject see [1]).

3.1 Melt Growth

The melting points of the II-VI compounds are (in °C): ZnS(1830), ZnSe(1520), ZnTe(1295), CdS(1475), CdSe(1239), CdTe(1092). With the exception of CdTe and ZnTe, melt growth has to be carried out under high pressure to prevent vapor dissociation. Further, as no satisfactory encapsulant has so far been discovered, pulling from the melt is

337

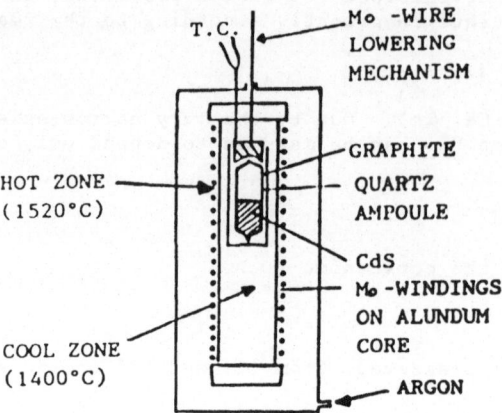

Fig. 2. Growth of CdS by the "soft ampoule" method, also applicable to
ZnSe (after [8]).

impractical, and recourse has to be made to Bridgman-type techniques. An
example is given in Figure 2, in which the quartz ampoule, softened at
high temperature, is counterpressured by graphite powder.

Growth methods under pressure have been reviewed by Fisher [6]. The
pressures and temperatures involved, which can reach 200 atm and 2000°C,
require the use of ad-hoc autoclaves. By these methods large (up to 10 cm
both in length and diameter) binary and ternary (mixed) crystals have been
grown.

Drawbacks of the method are equipment costs and reactions between
melt and silica or graphite crucibles.

Melt growth also includes multipass vertical zone melting,
successfully employed for purifying and growing CdTe [5].

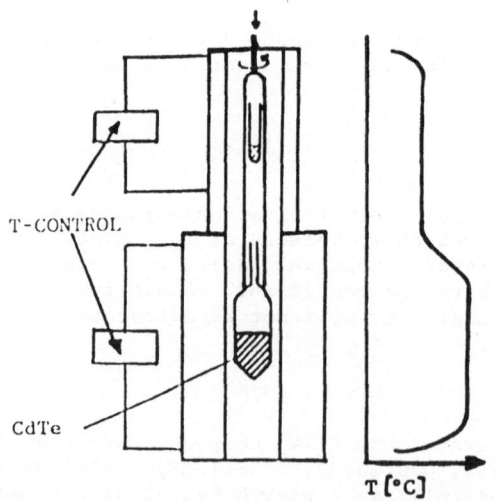

Fig. 3. Modified Bridgman for growing CdTe (after Kyle [5]).

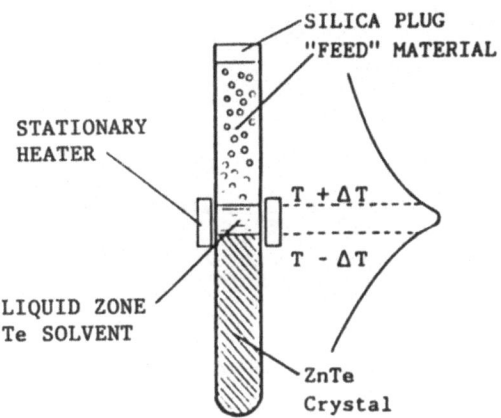

Fig. 4. Travelling solvent method (TSM) for growing ZnTe (after Triboulet and Didier [5]).

In the case of CdTe and ZnTe, no (inert-gas) pressure is necessary, however vapor dissociation and mass transport to cold points of the closed container have to be prevented. This can be done by making use of excess-component reservoirs held at temperatures for which, in the growth zone, either $P_A < P_A^0$ or $P_{X_2} < P_{X_2}^0$ (Figure 3).

3.2 Growth from Excess-Component Solutions

A drawback of melt growth is the tendency of II-VI compounds to undergo, upon cooling, a transition from wurtzite to zincblende structure. This favors twinning and stacking faults particularly in Zn chalcogenides. This undesired phenomenon can be avoided by growing the crystals at lower temperature, which can be done with either vapor-phase or excess-component-solution growth techniques.

CdTe ingots containing large single crystal grains have been grown (growth rate: 0.3 - 2 cm/day) from Te-rich solutions by TSM (Figure 4). Growth from Cd-rich solutions are impractical because of too small a solubility of Te in molten Cd.

An advantage of TSM is the reduction of contamination.

3.3 Vapor Growth

3.3.1 Open-tube (OT) techniques. The OT growth systems are very simple. Various modifications have been reported. In the Frerichs modification [9], Cd or Zn are separately vaporized and carried into the growth region by a flow of $H_2 + H_2S$. Here crystallization follows by direct synthesis according to reactions such as (Figure 5):

$$Cd(g) + H_2S(g) = CdS(s) + H_2(g).$$

Sublimation methods (Figure 6) make use of inert gases (Ar, N_2) to carry the vapors of a polycrystalline source into the growth region. Sublimation is sometimes coupled with chemical transport when transport agents (HCl, H_2, etc.) are added to the carrier gas. In sublimation methods, charge preheating is often required to adjust the stoichiometry.

The crystal size increases with ΔT and the gas flow of either carrier gas or CVT agents or both.

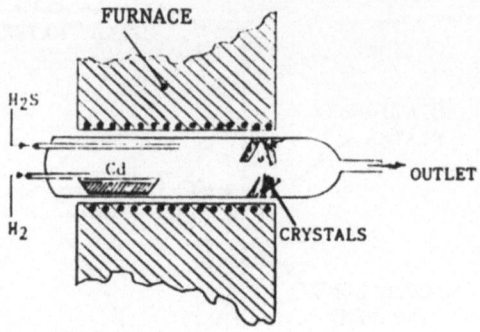

Fig. 5. Open-tube technique for direct synthesis of II-VI chalcogenide
         crystals (after Frerichs).

     Reported growth parameters are:

a)   source and growth temperatures (°C): ZnS(1200-1350/900-1300);
     ZnSe(1200-1275/800-1175); ZnTe(1050-1100/800-1050); CdS(1100-
     1200/920-1150); CdSe(900-1070/750-1050); CdTe(850-1050/700-900). The
     lower temperatures are used when CVT agents are added to the carrier
     gas;
b)   flow rates (1/hr): 1-50 (most used values: 5-20);
c)   growth times (hr): 1-5; maximal size: plate area: 2 x 1 cm^2;
d)   crystal habit: prisms, plates, needles, whiskers, hollow and skeletal
     forms, dendrites.

     3.3.2 Closed-tube (CT) techniques. 1) Sublimation techniques:
various modifications are reported [1-3,5,7]. A typical vertical-ampoule
version is shown in Figure 7. The vaporized mass is transported from
source to growth region by diffusion and convection.

     Reported growth parameters are:

a)   charge temperature (°C): ZnS(1350-1360); ZnSe(1200-1280); ZnTe(1050-
     1150); CdS(1050=1150); CdSe(900-1000); CdTe(800-950);

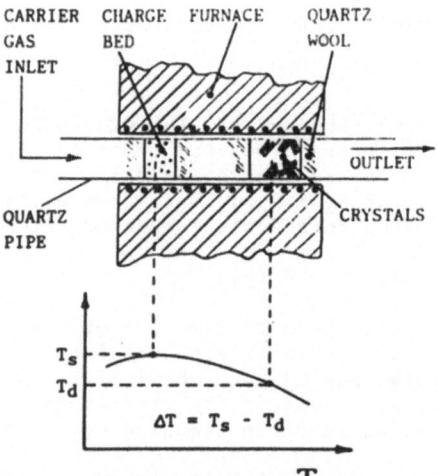

Fig. 6. Open-tube technique for growing II-VI chalcogenides (after Reed
         and Lafleur [1]).

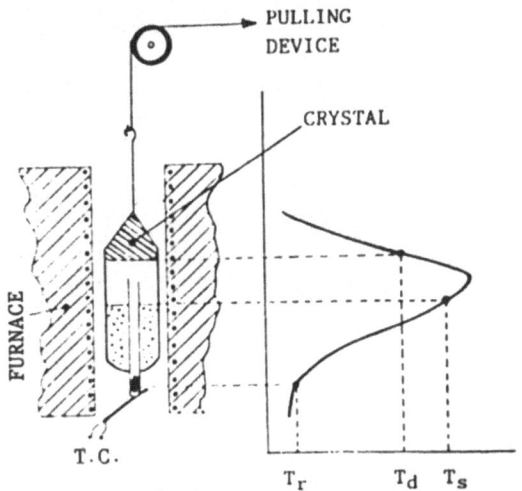

Fig. 7. Closed-tube sublimation; vertical-ampoule version with excess-component reservoir.

b) apparent undercooling: $(T_s - T_d(°C))$: 10-20;
c) temperature gradient at the growing interface: 1-5°C/cm;
d) pulling rate: 0.2-0.5 mm/hr; this quantity has not to exceed the maximum growth rate;
e) post-growth (maximum) cooling rate: 50°C/hr;
f) growth rate: 50-150 hr.

Single crystals with average size of 5 x 5 x 3 up to 20 x 20 x 10 mm^3 are commonly grown, at growth rates of 10^{-8} - 10^{-7} mol/cm^2.s.

The total pressure in the ampoule is generally less than 0.1 atm.

Stoichiometry adjustments can be made with the use of excess components kept in reservoirs at suitable temperatures (T_r), such as to favor a growth process close to P_{min} conditions. The selection of the P_A (P_{X_2}) pressure and T_r is performed by means of rel. (2) and (4), for $P_A = P_{X_2}$.

Stoichiometry adjustments can also be obtained by drilling a small orifice in the ampoule, close to the source region, and placing the ampoule in vacuum. The excess component will outdiffuse, favoring the obtainment of the P_{min} conditions.

To favor monocrystalline growth, the growth end of the ampoule is supplied either with a narrow neck (Figure 8A), or with a capillary tube directly joined to the tip (Figure 8B). When the capillary is initially open (Figure 8C) and the ampoule is placed in vacuum, seeding forms by self-sealing.

The most favorable shape of the growing interface is convex to the vapor. Such a shape is obtained by a slow pulling of the ampoule. However, since the growth temperature is generally much smaller than the melting point, the growing interface is often faceted.

For a theoretical discussion of the growth phenomenology, the reader is referred to Ref. [10-11].

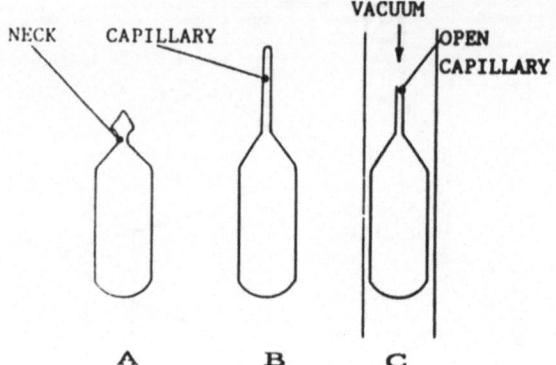

Fig. 8. Various types of ampoule for sublimation growth (see text).

2) CT-sublimation for growing ternary solid solutions: with the exception of ZnS-ZnTe, ZnS-CdSe, ZnS-CdTe, ZnSe-CdTe and CdS-CdTe, all other ternary II-VI compounds give rise to complete solid solutions.

The preparation of large single crystals of these solutions is difficult. Basically two methods are reported. The former, suitable for solutions of binary constituents having not too different vapor pressures (e.g., CdS-CdSe, ZnSe-ZnTe) is based on separating the sources and passing them separately through a maximum temperature region (Figure 9). Graded solutions can thus be grown.

The latter method, suitable for solutions of binary constituents having markedly different vapor pressures (e.g., ZnS-CdS; ZnSe-CdSe), makes use of variable charge temperatures (Figure 10). The two binary constituents are placed in the same source region, and the source temperature is raised while keeping the growth temperature constant. For example, in the case of ZnSe-CdSe, the source temperature is increased from 1140°C up to 1310°C, thus obtaining a single crystal graded solution rich in Zn in its end portion.

3) Chemical-vapor-transport (CVT) techniques: these techniques proved suitable for growing small samples (less than 1 cm^3) of sufficiently good quality for solid-state physics experiments. The possibility of obtaining

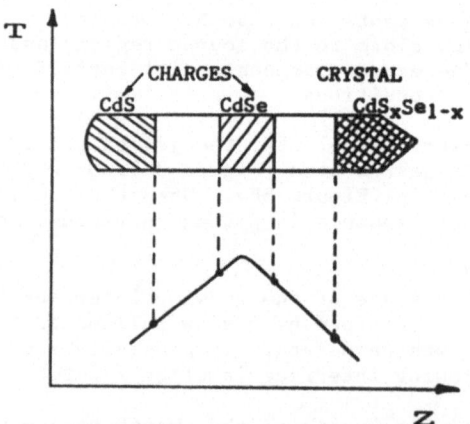

Fig. 9. Growth of graded ternary alloys by the "separated-source" method.

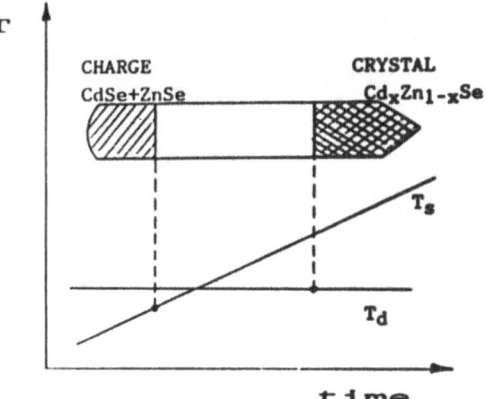

Fig. 10.   Growth of graded ternary alloys by the source-temperature
           increasing method.

"large" crystals in a reproducible way is still to be proved, though large
samples have occasionally been grown.

The advantages of CVT techniques are in the possibility of growing
ternary solutions of any desired composition, and in the ease of doping
control.   Drawbacks are the irreproducibility and the incorporation of
traces of transport agent.

Basically (Figure 11), it consists in reacting, at T_s, a polycrystalline charge AX (A = Cd, Zn; X = S, Se, Te, O) with a small amount
of "transport agent", typically I_2, though Br_2, HCl, HBr, H_2, NH_4Cl,
$H_2 + I_2$, etc. can be used as well.

The charge is vaporized according to reactions of the type:

$$AX(s) + I_2(g) = AI_2(g) + 1/2X_2(g).$$

The vapor phase composition will depend on the equilibrium constant,
K_p, and since K_p depends in its turn on temperature according to the
relationship:

$$\frac{d\ln K_p}{dT} = \frac{\Delta H}{RT^2}$$

(ΔH being the enthalpy change associated with the CVT reaction; for these
reactions ΔH is generally positive), when the vapors are transported to
the colder end of the ampoule, at T_d, the CVT reaction is reversed, and
some AX solidifies.   Under suitable thermal and compositional conditions,
the solidification can be controlled to yield a number of small single
crystals.

Of course, this description is oversimplified.   In general, many
chemical reactions contribute to a CVT process.   Further, the
crystallization and the growing interface stability depend on both
interface kinetics and mass transport through the vapor phase.   Mass
transport is in itself a complicated phenomenon, in which diffusion and
free convection overlap.   A complete theoretical understanding of CT-CVD
processes is to date still lacking, in spite of much research work.
Reviews on both theoretical and experimental work can be found in [11].
The author's articles [12-13] can also be consulted.

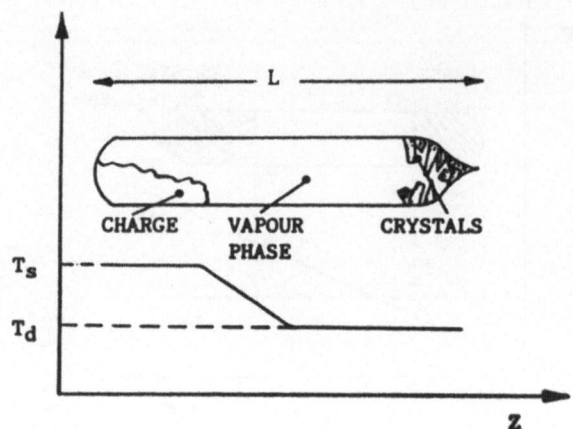

Fig. 11. Closed-tube CVT arrangement; the vapor phase contains the gaseous species due to the heterogeneous reaction between charge and transport agent.

The relevant growth parameters reported for these CT-CVT techniques are as follows:

a) charge temperature (°C): ZnS(900-1000); ZnSe(800-850); ZnTe(700-800); CdS(800-900); CdSe(750-900); CdTe(700-800);
b) apparent undercooling: ($\Delta T = T_s - T_d$, °C);
c) growth time (hr): 150-350;
d) ampoule geometry (mm): $\emptyset$:10-30; L:80-200;
e) dT_ddz (°C/cm): 1-20;
f) transport agents (mg/cm^3 of ampoule): I_2(1-5); Br_2(3, for growing ZnO); NH_4Cl (0.2-0.8);
g) overall pressure (atm): 0.5-2;
h) average mass transport (mol/cm^2.s): 10^{-8} - 10^{-7}.

4. CONCLUSION

After a summary of structural and electric properties and main applications of the wide-gap binary and ternary II-VI semiconductor compounds, the most important aspects of their bulk technology have been reported and discussed.

However, no mention can be found here of either epitaxial technology or bulk technology of narrow-gap II-VI compounds, for which the reported references should be consulted.

5. REFERENCES

1. H. Hartman, R. Mach and B. Selle, Wide-gap II-VI compounds as electronic materials, in: "Current Topics in Materials Science", E. Kaldis, ed., Vol. 9, North-Holland, Amsterdam (1982).
2. B. Ray, "II-VI Compounds", Pergamon, Oxford (1969).
3. M. Aven and J. S. Prener, eds., "Physics and Chemistry of II-VI Compounds", North-Holland, Amsterdam (1967).
4. W. Hirschwald, Zinc Oxide, in: "Current Topics in Materials Science", E. Kaldis, ed., Vol. 7, North-Holland, Amsterdam (1981).

5. A. Zanio, Cadmium Telluride, in: "Semiconductors and Semimetals", R. K. Willardson and A. C. Beer, eds., Vol. 13, Academic Press (1978).
6. D. De Nobel, quoted in [5].
7. F. A. Kröger, "The Chemistry of Imperfect Crystals", North-Holland, Amsterdam (1964).
8. A. G. Fischer, Methods of growing crystals under pressure, in: "Crystal Growth", B. R. Pamplin, ed., Chap. 13, Pergamon (1975).
9. R. Frerichs, Phys. Rev., 72:594 (1947).
10. E. Schönherr, in: "Crystals, Growth, Properties and Application ", Vol. 2, Springer, Berlin (1980).
11. M. M. Faktor and J. J. Garrett, "Growth of Crystals from the Vapour", Chapman and Hall, London (1974).
12. C. Paorici, V. Pessina and L. Zecchina, Crystal Res. and Tech., 21:1149 (1986).
13. C. Paorici and C. Pelosi, Rev. Physique Appliquee, 12:155 (1977).

LIST OF SYMBOLS

P	overall pressure
P_A, P_{X_2}	actual partial pressures of gaseous species A, X_2
$P_A^0, P_{X_2}^0$	equilibrium partial pressures of gaseous species A, X_2
T	temperature
R	gas content
T_s	source temperature
T_d	deposition (or crystallization) temperature
ΔT	difference between source- and deposition temperature ($T_s - T_d$)
T_r	reservoir temperature
K_p	equilibrium constant
ΔH	enthalpy change

CRYSTAL GROWTH IN TECHNOLOGY

BINARY SEMICONDUCTORS: III-V COMPOUNDS

L. Zanotti

Istituto Maspec C.N.R.
Parma, Italy

INTRODUCTION

In recent years, electronic technologies have been confronted with the need of highly integrated, high-speed devices. The biggest part has been made of silicon. But there are limits to the physical capabilities of silicon in some critical applications, and new materials to substitute silicon are being sought for. Studies have shown that III-V compounds with special intrinsic properties (such as high electron mobility, direct and, in some cases, wide energy gap, high resistance to elevated temperature and radiation, the possibility of being prepared directly with a high resistivity), give superior performances in some specific applications in both electronic devices (FET), microwave devices, integrated circuits, Gunn diodes and optical devices (LEDs, LDs, IR detectors, solar cells) (Table 1).

III-V compounds are mainly used as subtrates on which the active regions of the semiconductor devices are created by ion implantation or by epitaxial processes. However, most applications necessitate high quality substrate material with specific, uniform and stable electrical properties, while at the same time the crystal defect content must be kept low. Typical specifications for GaAs wafers presently marketed are given in Table 2. Specifications actually depend on the device to be fabricated, with, for instance, lower defect densities required for FETs, lasers and APDs than for LEDs and solar cells. Moreover, as most electronic process technologies employ equipment for handling 3-inch and larger wafers, there is a growing demand of large wafers, so that modern processing equipment may be introduced. In this contribution problems of crystal preparation are discussed, followed by consideration of further development.

CRYSTAL PREPARATION

III-V compounds generally have high dissociation pressure at their melting points, therefore special methods are necessary to grow single crystals from the melt. Although several methods of growth have been tried, no other, more practical method has been found in industrial scale with practical growth rates except the "Horizontal Bridgman" (HB) method and the Czochralski (CZ) method.

Table 1. III-V Compound Semiconductor Materials together with Fields of
Applications

Material	Functions, Characteristics	Applications	Notes
GaAs	High carrier mobility, direct gap transition, high intrinsic resistivity	High speed high frequency devices: - FETS - Gunn diodes - ICs - CCds - Varactors - Schottky barrier diodes optoelectronic devices: - laser diodes - LEDs - visible light and UV detectors - solar cells Hall effect devices ultrasonic transducers substrates	FETs, Hall devices, Gunn diodes, laser diodes, varactors have been commercialized formation of oxide layers difficult grown by VPE, LPE, MBE
GaP	Bandgap corresponding to visible wavelengths	LEDs (red, orange, green) display elements infrared ray modulation elements, high temp. semiconductor elements	LEDs have been commercialized high defect densities present problems
InP	High mobility, high saturation, high resistivity, lattice aligned with InGaAsP	High speed, high frequency devices: - Gunn diodes - FETs substrates	high defect densities grown by VPE, LPE
GaN	Large bandgap	Blue light LEDs	p-type crystals unobtainable
GaSb		Tunnel diodes, substrates, light detectors	grown by VPE, LPE
InN		Yellow light, LEDs	grown by VPE
InAs	High carrier mobility	Hall effect devices/Gunn diodes/infrared laser diodes	
InSb	High carrier mobility	Hall effect devices/ infrared ray detectors/ filters magnetoresistive elements	all devices have been commercialized
ALN	Large bandgap high acoustic velocity	UV LEDs SAW devices	p-type crystals unobtainable

Table 2. Typical Specifications for Substrate GaAs Products

Category	Specifications	Notes
Dimensions	50+ - 0.3 mm dia.	
	76+ - 0.6 mm dia.	
Shape	cylindrical	when LEC-grown
Resistivity	greater than 10^7 ohm-cm	undoped
Defect density	less than 5-20 x 10^3/cm^2	HB grown
	less than 1 x 10^5/cm^2	LEC grown
Dopant	none	for ICs
	none, Cr, 0.5-5 ppm	for FETs, ICs
	Si,Zn,Cd,S,Se,Te	for LEDs
Planarity	less than 30 microns	3-inch wafers, both sides polished

Bridgman (Boat) and Related Methods

Up to 1979, GaAs substrate crystals were exclusively produced by HB or the boat growth method, giving a crystal shaped like a cylinder cut in half lengthwise, and aligned in the (111) direction, meaning that to obtain wafers oriented in the (100) direction the crystal had to be cut diagonally, yielding specimens of an irregular shape, Figure 1a. Although the HB method yielded crystals with low defect densities, they were limited in size and in quality by silicon contamination from the container, which is normally quartz.

Its advantages, on the other hand, are: a) the stoichiometry of the melt is easiy controlled by adjusting the As partial pressure, b) the shape of the crystal is entirely determined by the boat, so that expensive devices to regulate the crystal size are unnecessary, and c) the defect density may be kept low by keeping the temperature gradient low, (i.e., 2 to 10°C/cm). Specifically there are three different HB methods for GaAs boat growth. In Figure 2, the two-temperature (2T-HB), the three-temperature (3T-HB) horizontal Bridgman methods and the "Gradient Freezing" (GF) method are compared; the vertical axis represents temperature and in the 2T-HB and 3T-HB methods the furnace moves to the left, cooling the boats, from the right, while in the GF method, the temperature as a whole is slowly lowered. Tf is the melting point of GaAs, Ga, Gb and Gc are temperature gradients in the growth zones and DB indicates a diffusion barrier necessary to preserve the partial pressure of the gas in the high temperature region.

The conditions for growth by 3T-HB, possibly the most widely used HB method, of low defect density (EPD less than 2000/cm^2) single crystal GaAs in the (111) direction with a cross-section area of 5 cm^2 or more are as follows [1]:

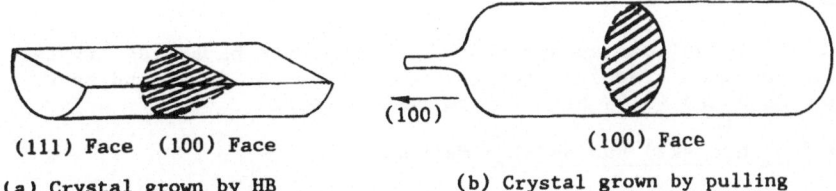

(111) Face (100) Face   (100) Face

(a) Crystal grown by HB   (b) Crystal grown by pulling

Fig. 1. Orientations of crystals grown by HB and by pulling.

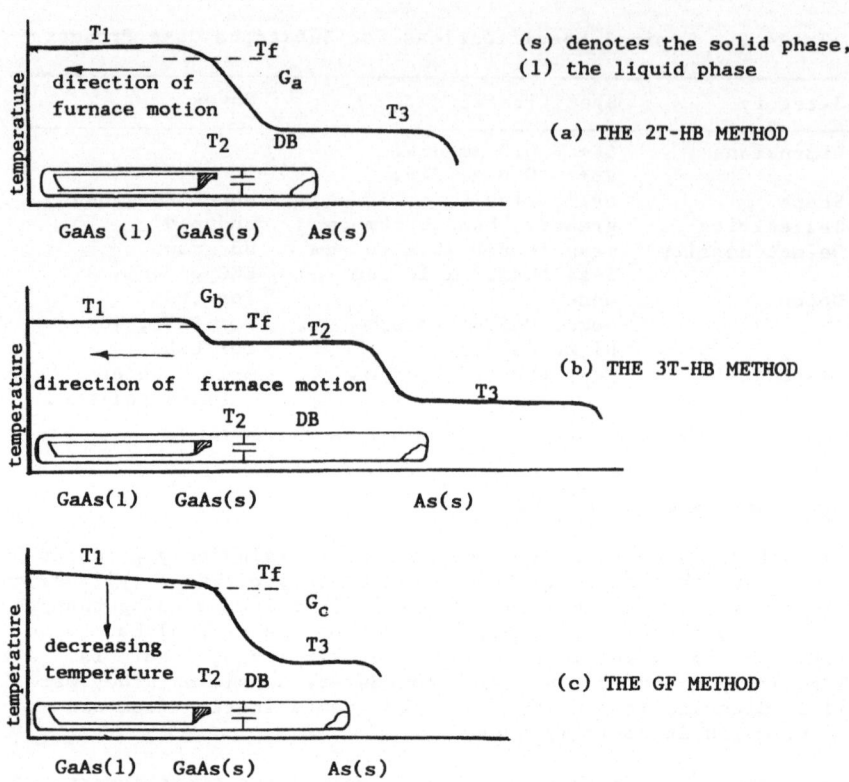

Fig. 2. The 2T-HB, 3T-HB and GF methods of GaAs crystal growth.

1) Temperature of the second zone should be between 1100 and 1200°C.
2) The temperature gradient at the crystal-melt interface should be between 2 and 10°C/cm.
3) The growth rate should be between 2 and 10 mm/hr.
4) The cooling rate after crystal growth should be less than 50°C/hr.

Crucial for all the Bridgman methods is the rigorous and proper control of the melt stoichiometry, i.e., regulation of the As partial pressure in the growth ampoule.

Among the progress in the HB technique it is worth mentioning advantages obtained by using boats of rectangular cross-section to produce crystal ingots from which nearly-rectangular shapes are sliced, thus reducing material waste [2]. In addition to this, multi-zone furnaces assisted by microprocessor systems were seen to improve the thermal conditions so that planar freezing interfaces were realized and consequently the introduction of dislocations in the crystal was minimized.

Recently developed apparatus for arsenide and phosphide crystal growth by vertical freezing techniques (vertical Bridgman and heat-exchanger methods) [3,4] has also to be mentioned.

Pulling (Czochralski) and Related Methods

By pulling or the Czochralski (CZ) method crystals grown in the (100) direction can be directly cut into circular (100) wafers, as shown in

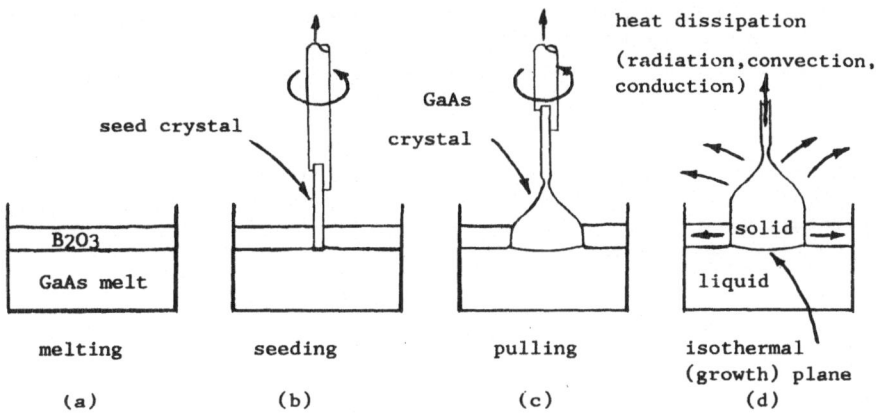

Fig. 3. Principle of LEC growth method.

Figure 1b. A major advantage of using circular crystals is that many of the processes used for silicon wafers can be directly applied to GaAs. On the other hand, while in most respects the pulling methods are superior to the boat method, the Czochralski method yielded crystals with more lattice defects, mainly because of the sharp temperature gradient at the solid-liquid interface during pulling. As shown in Figure 3, the special feature of the Liquid Encapsulated Czochralski (LEC) method is the use of a liquid encapsulant, which floats on the melt beneath the inert gas maintained at more than the dissociation pressure of the melt to avoid the dissociation of the growing single crystal and the melt. The required properties of the encapsulant are: to be non reactive with respect to the crucible and the material to be grown, to be stable at high temperatures, to have a lower density than the melt, to be capable of wetting the grown crystal, to be transparent. As incapsulant, boron oxide is usually used. However, its viscosity is very high at temperatures below $1000°C$, therefore other materials such as alkali oxides, alkali florides or an eutectic mixture of these materials are used for pulling single crystals at low melting points (Table 3).

As shown in Figure 3 pulling is started by bringing a seed crystal in contact with the melt through the boron oxide layer and after "necking" in which the crystal is kept thin to remove as many seed defects as possible, the crystal diameter is increased by regulating the temperature.

Table 3. Examples of Compound Semiconductors Grown by the LEC Method

Crystals	Encapsulant	Pressure (gas) (atm)	Dia. Max. Size (mm)	Length Max. Size (cm)
GaAs	B_2O_3,$BaCl_2$ $CaCl_2$,$BaCl$+KCl	1-100 (N_2,Ar,He)	100	40
GaP	B_2O_3	40-55 (He,N_2)	62	20
InP	B_2O_3	25-35 (He,N_2)	80	15
InAs	B_2O_3	1-10 (N_2,Ar,He)	20-50	10
GaSb	KCl,$NaCl$	1(N_2,He)	20-50	10
$GaAs_{(1-x)}Px_x$	B_2O_3		20-35	5
PbTe	B_2O_3	1(N_2)	25	10

Table 4. A Comparison of LEC and HB Methods Used Commercially for GaAs Crystal Growth

	LEC Methods HP LEC*	MP LEC*	HB Methods 3T-HB
Gas pressure	15-70 atm (N_2)	3-10 atm (N_2)	1 atm (As_4)
Temperature gradient at the liquid-solid interface (°C/cm)	100-300	50-100	2-10
Synthesis	direct	injection method	2-zone method
Max. residual Si conc. (cm^3)	$< 10^{16}$	$< 10^{16}$	10^{16}
Residual B conc. (cm^{-3})	10^{16}	10^{16}	
Average dislocation density (cm^{-2}):			
– undoped	$2-5 \times 10^4$	$1-3 \times 10^4$	less than 10^4
– Si doped	$0.1-5 \times 10^3$	$0.1-1 \times 10^3$	$0-2 \times 10^3$
– Cr-O doped	$2-5 \times 10^4$	$1-3 \times 10^4$	$0.1-1 \times 10^4$
Wafer shape	circular	circular	D-shaped
Grown in (100) directions	possible	possible	difficult
Diameter (inches)	3-4	2-4	2 (trimmed)
Growth speed (mm/hr)	3-15	3-15	3-5

* HP = high pressure, MP = medium pressure

The LEC method is used in the growth of crystals other than GaAs; the more important examples are given in Table 3, along with growth conditions and crystal size. In Table 4 the LEC and HB methods are compared.

In recent years, the necessity of using automated process control in pulling GaAs and other crystals has come to be widely recognized. The trend is to provide pullers with a loading cell which constantly weighs the crystal during growth (Figure 4). The signal issued from the loading cell is amplified and transmitted to a microcomputer linked to the crystal pulling mechanism. The computer compares the measured crystal mass with the programmed mass and adjusts the furnace power level, thereby maintaining a constant relationship between the length and the weight of the crystal. The computer system can also drive the pulling and rotation speed of the seed and the crucible shaft and control each step of the thermal cycle.

PURIFICATION OF RAW MATERIALS AND SYNTHESIS PROCESSES

One important aspect of semiconductor materials is impurity. Degree of purity has a direct bearing on controlling the carrier and stability of the materials, and is a decisive factor in quality, dependability and durability of the device. Silicon raw materials at a purity as high as ten nines are readily available (i.e., concentration of impurities is less than 1×10^{10} at/cm^3) but in III-V compounds, raw materials currently used have a purity of 6 nines, or 7 nines at the most. Companies are still studying now to obtain a higher purity of the elemental components by electrochemical, zone refining, vacuum backing, distillation and metal-organic processes. The synthesis process and container contamination are also crucial for the obtainment of high purity material. The use of pyrolitic boron nitride (pBN) crucibles combined with in-situ synthesis from the elements either by injection methods (medium-pressure pullers) or

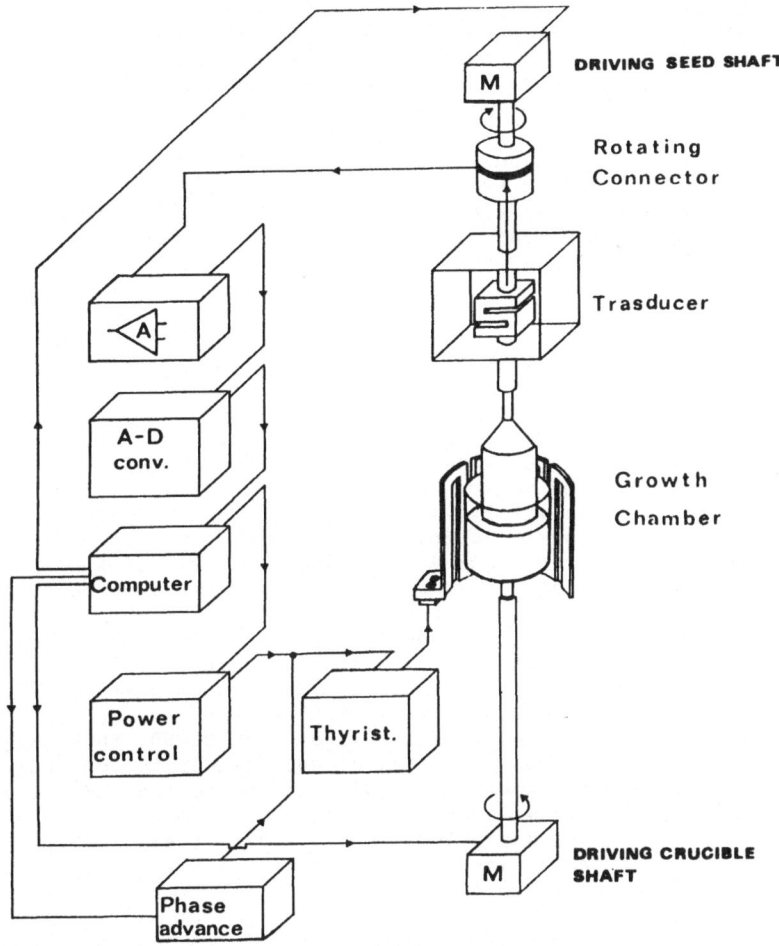

DRIVING SEED SHAFT

Rotating
Connector

Trasducer

Growth
Chamber

Computer

Power
control

Thyrist.

DRIVING CRUCIBLE
SHAFT

Phase
advance

A-D
conv.

Fig. 4. Schematic of automated puller for III-V compound LEC growth.

by direct synthesis (high-pressure pullers) [5], and subsequent pulling-from-the-melt growth in a single operation made possible a consistent decrease in the contamination level.

PREPARATION OF SEMI-INSULATING GaAs (SI GaAs)

The high mobility expected of GaAs devices is due not only to high electron mobilities but also to a highly resistive GaAs substrate (1×10^7 - 10^8 ohm/cm) which means lower capacitance. At present SI GaAs is mainly used as substrate for devices such as FET and high speed integrated circuits (IC), while future demands are also foreseen for optoelectronic integrated circuits. In these cases a high resistivity substrate makes device isolation relatively simple. Besides high resistivity and high carrier mobility (1×10^5 - 1×10^6 cm^2/V.sec) substrates for ICs are required with low impurity concentrations, low defect density and in particular, so as to utilize the advantages of ion implantation process, they must be stable against heat treatments, such as the annealing (typically at 800°C) required after ion implantation to activate the implanted impurities. The residual impurity content and the presence of

353

deep levels induced by native structural defects, have a determining role in the compensation mechanisms which give rise to semi-insulating properties. In the earlier work on GaAs crystal growth, silicon, which acted as a shallow donor impurity, could not be kept out of the melt. Thus to obtain a sufficiently resistive substrate material the crystal had to be doped with chromium, an acceptor impurity, thus balancing the effect of the silicon. The silicon impurities were present mainly due to the use of quartz crucibles in the synthesis/growth process. Thereupon quartz had to be replaced by pBN crucibles, as a result the silicon impurity level has been lowered by an order of magnitude, making unnecessary the chromium doping. In pBN-LEC crystals carbon (a shallow acceptor) is the main electrically active impurity ($1-5 \times 10^{15}/cm^3$) and the resulting electrical properties are dependent on the presence of the main donor deep level, labeled EL2, whose origin formerly was correlated with native antisite defect As(Ga) and, later with a complex defect (As precipitate clusters around As(Ga) antisites) [6]. An EL2 concentration above $1 \times 10^{16}/cm^3$ can be obtained if the donor shallow impurity content is maintained down to $1 \times 10^{16}/cm^3$. Consequently a natural compensation (dopant unnecessary) of shallow residual acceptors by EL2 deep donors is possible when the GaAs crystal is grown in high purity conditions, (very low shallow-donor impurity content) while still maintaining a shallow acceptor as dominant shallow impurity. Today it is known that the EL2 concentration in the growing crystal is strictly correlated with melt composition. It was shown that high resistivity crystals can be obtained from melts in which the As atom fraction was higher than 0.480 and that high resistivity material, having both high Hall mobility values (6000 $cm^2/V.sec$) and good thermal stability (16 hours at 860°C) can be achieved only in material grown from melts in which the As atom fraction was higher than 0.505 [7,8]. On the other hand, low resistivity, p-type material is obtained from melts having a Ga atom fraction higher than 0.480, thus suggesting that under these conditions a shallow acceptor impurity (carbon) or an acceptor native defect predominates [9]. It can be concluded from this that an entirely semi-insulating ingot can be grown from melts whose As atom fraction is higher than 0.500 whereas a high resistivity, n-type/low resistivity, p-type transition may be expected to result from crystals grown from Ga-rich melts.

DISLOCATION REDUCTION PROCESSES

In order to fabricate ICs which operate properly, it is important to reduce the scattering in FET characteristics such as threshold voltages on individual wafers [10]. To do so it is absolutely necessary that wafers with low defects are employed. The presence of dislocations also favor and accelerate the degradation of LEDs and LDs by diminishing the average diffusion length of minority carriers [11]. These requirements served to accelerate R&D on growth techniques to lower defect densities in GaAs ingots, and induced deeper investigations to understand the phenomena which give rise to dislocation formation.

At present the main causes of dislocation formation are thought to be related with thermal stresses wich give rise to glide dislocations [12], and with the coalescence of point defects (grow-in dislocations) [13]. In particular thermal stress effects are crucial in LEC process, where due to the need to dissipate the latent heat during growth, thermal-gradient induced stress cannot be reduced below the critical yield-stress limit (above which the thermal stress is relieved by plastic deformation). Table 5 summarizes the ways for practical dislocation density reduction.

354

Table 5. Means to Reduce the Dislocation Density

Reduction of thermal stress	- Reduction of dT/dz (shields, afterheaters, low pressure) - Reduction of dT/dr (bottom heater)
Reduction of point defects	- Stoichiometry control - As atmosphere - Si, In doping
Increase of mechanical stress	- "Solution hardening" (S, Se, Te doping)
Decrease of fluctuations of growth conditions	- Automated crystal growth - Magnetically stabilized growth
Optimization of growth parameters	- Dislocation-free seed - Necking - Slow cooling

APPLICATION OF MAGNETIC FIELDS

One of the major problems with the LEC method is the high temperature gradient at the melt-crystal interface, which induces convection currents in the melt and consequently causes fluctuations in dopant/impurity segregation and slight changes in Ga/As stoichiometry, which lead to the formation of growth striations.

A quantitative evaluation of local dopant concentrations, down to $1 \times 10^{16}/cm^3$ has been recently achieved by SEM observation in the EBIC mode [14] offering interesting experimental evidence of inhomogeneity induced by melt temperature fluctuations.

One measure found to be effective for removing convection in the liquid during the LEC growth process is the application of magnetic fields around the growth environment.

Similarly to the case of silicon growth, the magnetic field is important for suppressing the temperature fluctuations at the growth interface, thereby eliminating irregularities such as grown striations (melt temperature fluctuations were reduced from about 15°C to a range of 0.1°C by the application of 1.200-1.300 Gauss magnetic field [15], by using a superconductive electromagnet, essentially eliminating the repeated local melting and recrystallization that led to high defect density).

CONCLUSIONS

One trend that has become rather apparent in the semiconductor industry is the trend away from relying on a single material, e.g., silicon, and towards diversification to different materials developed for different applications. Though the expected sales of III-V compound devices are forecast to remain within a few percent of the total Si-based devices, GaAs and related compounds will certainly play a major role in certain applications where high speeds and/or low power consumption, as well as high resistance to radiation and other damage, are important.

There are still many problems in III-V compounds substrate manufacture remaining to be solved, i.e., increase of crystal sizes, reduction of lattice defects, improvement in chemical homogeneity and decrease of impurity background. Steps in this direction are the use of large low pressure LEC or "HOT WALL" pullers provided by multiple heaters which allows a reduction of the temperature gradient at the melt-crystal interface. Melt convection-induced instability phenomena can be faced by employing magnetic fields in vertical and horizontal versions, while impurity contamination can be strongly reduced by eliminating all silica and graphite parts used in today's standard growth environments.

In the longer term, mixed crystals such as InGaAs, InAsP and InGaAsP, the lattices of which match GaAs may become of crucial importance. Greater electron mobilities may be realized and there are more degrees of freedom which may be utilized for device optimization, whether for lasers/LEDs, microwave or future optoelectronic devices. Even so, however, unless growth of mixed crystal ingots and other new processes like: electroepitaxy, crystal growth in space, III-V compound deposition on Si substrates, see rapid progress, GaAs, InP and GaP wafer substrates will continue to dominate.

REFERENCES

1. S. Akai, Y. Nishida and K. Fujita, US Patent 4:158,851 (1979).
2. T. Toyoshima, J. Nakagowa, S. Mizuniwa and S. Okubo, Hitachi Cable Rev., 1:45 (1982).
3. W. A. Gault, E. M. Momberg and J. E. Clemans, J. Crystal Growth, 74:491 (1986).
4. C. P. Khattack, J. Lagowski, J. H. Wohlgemuth, S. Mil'htein, V. E. White and F. Schmid, GaAs and Related Compounds, 1985, M. Fujmoto, ed., Inst. Phys. Conf. Ser. No. 79, p. 31, Bristol (1986).
5. R. L. Lane, Seimicon. International, October (1984).
6. H. C. Gatos and J. Lagowski, Mat. Res. Soc. Symposium on "Microscopic Indentification of Electronic Defects in Semiconductors", S. Francisco, CA (1985).
7. R. N. Thomas, H. M. Hobgood, G. W. Eldrige, D. L. Barrett, T. T. Braggins, L. B. Ta and S. K. Wang, Semiconductors and Semimetals, in: Semi-insulating GaAs", R. K. Willandson and A. C. Beer, eds., Vol. 20, p. 1, Academic Press, Orlando (1984).
8. C. G. Kirkpatrick, R. T. Chen, D. E. Homes, P. M. Asbeck, K. R. Elliott, R. D. Fairman and J. R. Oliver, Ibidem p. 159.
9. D. E. Holmes, K. R. Elliott, R. T. Chen and C. G. Kirkpatrick, "Semi-insulating III-V Materials" Evian 1982, S. Makram-Ebeid and B. Tuck, eds., p. 19, Shiva Publ. (1982).
10. S. Miyazawa, "Semi-insulating III-V Materials", Hakone 1986, H. Kukimoto and S. Miyazawa, eds., p. 3, North-Holland Publ. Co., Amsterdam (1986).
11. L. Hollan, J. P. Hallais and J. C. Brice, "Current Topics in Materials Science", E. Kaldis, ed., Vol. 5, pp. 28-44, North-Holland Publ. Co., Amsterdam (1980).
12. A. S. Jordan, J. Crystal Growth, 49:631 (1980).
13. L. J. Giling, J. L. Wehyer, A. Montree, R. Fornari and L. Zanotti, J. Crystal Growth, 79:271 (1986).
14. C. Frigeri, 5th Oxford Conf. on Microscopy of Semiconducting Materials, Oxford (1987).
15. K. Terashima and T. Fukuda, J. Crystal Growth, 63:423 (1983).

Abbreviations

APD   Avalanche photodiode
CZ   Czochralski or pulling method
EBIC Electron beam induced current
EPD   Etch pit density
FET   Field effect transistor
GF   Gradient freezing method
HB   Horizontal Bridgman or boat method
LD   Laser diode
LEC   Liquid encapsulated Czochralski method
LED   Light emitting diode
pBN   Pyrolitic boron nitride
SEM   Scanning Electron Microscope

AN INTRODUCTION TO MOLECULAR BEAM EPITAXY

M. Ilegems

Institut de micro-et opto-électronique, Ecole Polytechnique
Federale/Swiss Federal Institute of Technology CH-1015
Lausanne, Switzerland

1 INTRODUCTION

The molecular beam epitaxial technique (MBE) has been developed over
the past 20 years in response to the need for a crystal growth method
capable of depositing very thin films of semiconductor materials with near-
perfect surface morphology and near-perfect control of layer thicknesses,
layer compositions, and doping levels needed for the fabrication of advanced
semiconductor devices. While the initial studies were carried out mainly on
III-V compounds and their ternary and quaternary alloys, the technique has
since found widespread application also for the growth of II-VI and IV-VI
semiconducting compounds and alloys, as well as for the deposition of
silicon and germanium films. Other applications involving the deposition of
epitaxial metal films, dielectrics and superconductors have emerged recently
and are being actively pursued in many laboratories.

World wide activity in the area of MBE has grown tremendously over the
past decade following the pioneering work of J.R. Arthur and A.Y. Cho at
Bell Laboratories during the 1968-1973 time period. The number of published
papers, for example, has grown from less than 10 per year in the period
1970-1972 to well over a thousand per year in 1984 and thereafter. A number
of review articles[1-3] and books[4,5] are listed at the end of this chapter
for further reference.

In these lecture notes we will consider mainly the growth of III-V
compounds and alloys, taking the GaAs/AlGaAs and GaInPAs/InP systems as
generic examples.   Excellent reviews of Silicon epitaxy by MBE can be found
in the articles by Bean[6] and Shiraki[7].

2. PRINCIPLES OF CRYSTAL GROWTH BY MBE

2.1 Definitions

Molecular beam epitaxy can be defined as an epitaxial growth process
which involves the reaction of one or more thermal beams of atoms or
molecules with a crystalline surface under high vacuum conditions. Depending
of the nature of the sources used to produce the thermal beams one distinguishes between:

(a) <u>solid source (conventional) MBE</u> based on the use of solid sources (for example, gallium or silicon) heated by thermal radiation or by electron beam impact, and

(b) <u>gas source MBE</u> where the evaporation materials are introduced in gaseous form and decomposed into their elements either in a hot zone crucible (for example, decomposition of AsH_3 into arsenic and hydrogen) or directly at the surface of the substrate (for example, decomposition of trimethylgallium into gallium and methyl radicals on the heated substrate surface)

Figure 1 shows schematically the principal components of a solid source and gas source MBE system for the growth of AlGaAs and GaInPAs compound alloys. In the solid source system[1-5] (fig. 1a), the substrate, which is heated in high vacuum to temperatures in the range from 550 to 725°C, is exposed to molecular beams of Ga, Al, and As (mainly in the form of As_4) which are evaporated out of Knudsen cell type heated crucibles. Layer composition and growth rate are controlled by the temperatures of the Al and Ga cells, while the As cell temperature must be kept sufficiently high to maintain an As overpressure over the growing substrate. Doping levels in the layer are controlled by separate n-and p-type evaporation cells. Shutters in front of the cells are used to establish or shut off the evaporant beams instantly, thereby making it possible to achieve extremely abrupt heterostructure interfaces and doping transitions.

In the Ga-In-P-As gas source system[8] (fig. 1b), the group III elements are introduced at low pressures (typically 10 to 100 mbar) in the growth chamber in the form of metalorganic gaseous compounds such as triethylgallium (TEG) and trimethylindium (TMIn), while the group V elements are introduced as hydrides, in this case arsine and phosphine, at intermediate to high pressures (100 to 1000 mbar). The hydrides are decomposed in a hot zone crucible held at temperatures above 900°C before reaching the substrate, while the group III elements reach the substrate in metalorganic compound form and decompose catalytically on the substrate surface. In this system, growth rates are controlled by the group III element beam fluxes, while the layer compositions are determined by the Ga to In and P to As ratios at the given growth temperature. Again, the group V equivalent beam pressure over the substrate should be sufficiently large to prevent layer decomposition, and the doping levels are controlled by separate thermally heated solid cells. Layer transitions are realized in this system either by gas source switching or by means of shutters.

Apart from the full gas source system illustrated in figure 1b, hybrid systems[9] using elemental group III and hydride group V sources can be used. Experiments are also underway with the objective to replace the solid source doping cells by gas cells, which would then allow full control of the growth sequence using gas source switching only.

During the epitaxial deposition process, the background pressure in the growth chamber extends over the range from 10^{-7} mbar in solid source MBE to 10^{-5} mbar for gas source MBE due to the presence of the hydrogen and hydrocarbon decomposition products. At these pressures, the mean free paths of the molecules range from approximately 50 m down to 50 cm, and are thus much larger or at most comparable to the dimensions of the growth chamber. In this pressure range, the behavior of the particles in the growth chamber is entirely determined by surface collisions, and particle to particle interactions in free space can be neglected. This feature constitutes the main distinction with the alternative epitaxial growth techniques for these materials such as atmospheric pressure or low-pressure (1 to 100 mbar) chemical vapor deposition (CVD) where mean free paths range from 50 μm down to less than 0.1 μm (Fig. 2).

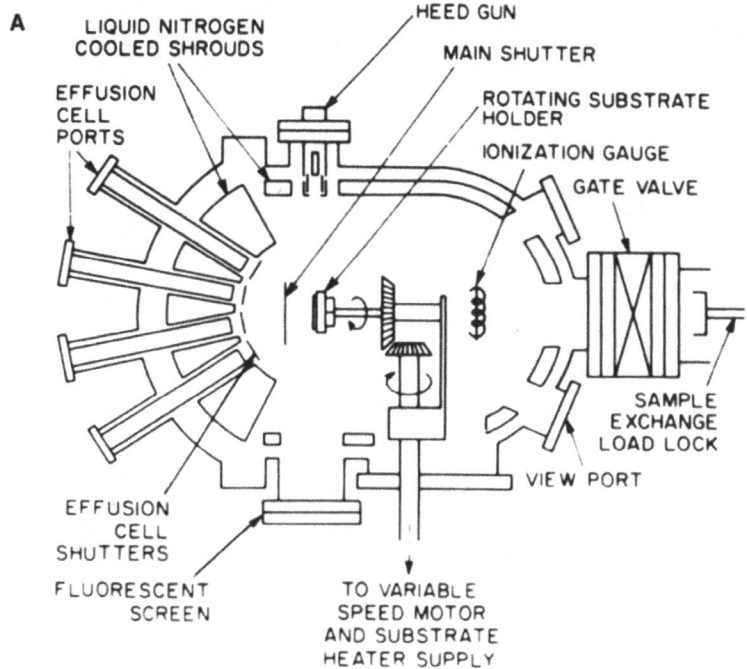

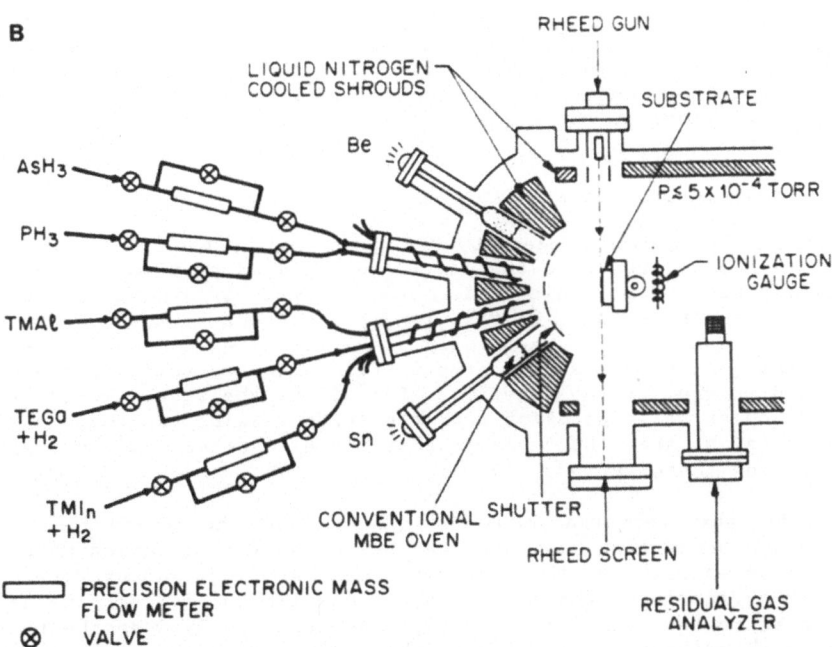

Fig. 1.(a) schematic of a solid source MBE growth chamber showing the thermally heated effusion cells and the rotatable heated substrate holder in the growth position. (b) schematic of a gas source MBE system for growth of GaInPAs quaternary alloys showing the gas system for introduction and flow control for the group III and group V elements. After Cho[10] and Tsang[8].

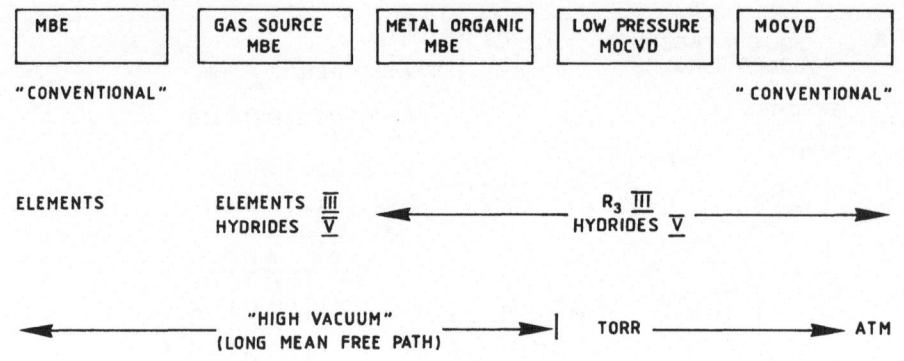

Fig. 2. Typical pressures during growth for conventional MBE, gas source
MBE or CBE, low pressure MO-CVD and atmospheric pressure MO-CVD.
After Panish[9].

There are significant differences between various gas source MBE techniques depending on whether the elements constituting the film arrive at the
substrate surface in elemental form (for example, as Ga, As, As$_2$, As$_4$, etc.
molecules) or as constituent elements of a more complex molecule (for
example, in the form of trimethylgallium, trimethylaluminum, etc.). In the
first case, the surface reactions are identical to those found in conventional solid-source MBE, whereas in the second case, the surface reaction
kinetics and surface chemistry may be significantly different. For this
reason, the term chemical beam epitaxy (CBE) has been proposed for the MBE
technique based on the simultaneous use of hydride and metalorganic sources.
In the discussion which follows we will concentrate mainly on the conventional MBE technique, and we refer to references 8-9 and 11-15 for more
detailed descriptions of the gas source and CBE approaches.

2.2 Description of the Growth Process

The epitaxial growth process by MBE involves a series of events taking
place at the heated substrate surface[10,16]: (1) adsorption of the
constituent atoms and molecules; (2) surface migration and dissociation of
the adsorbed molecules; and (3) fixation of the atoms to the substrate in
crystallographically and energetically preferred sites. These processes
result in the growth of a single crystal film with a crystallographic structure related to that of the substrate. For all of the steps listed, reverse
reactions such as re-evaporation from the surface exist and compete with the
growth process, so that the overall growth rate is determined by a precise
balance between these various events.

The different steps that come into play during the growth are
controlled by the surface chemistry and by the surface reaction kinetics,
rather than by equilibrium thermodynamic considerations. However, even in a
high vacuum environment, the thermodynamic equilibrium forces give an
indication as to the behavior to be expected, especially when the residence
lifetime of the molecules arriving at the surface is sufficiently long to
allow equilibrium with the substrate surface to be reached[16].

The interaction of the impinging flux with the surface of the growing
film is caracterized by its sticking coefficient, S, (Fig. 3), which is
defined as the ratio between the "sticking density", F_s (number of atoms per
cm^2 per second which stick to the growth surface) and the incident flux
density F_i (number of atoms per cm2 per second ariving at the surface), i.e.

362

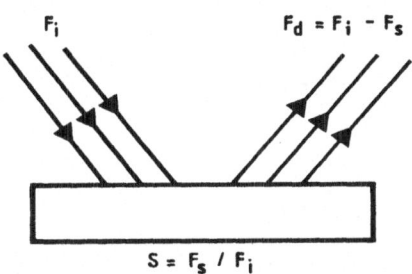

Fig. 3. Definition of the sticking coefficient $S = F_s/F_i$ where F_i = incident flux, F_d = desorbed flux, F_s = number of atoms incorporated in the growing film.

$$S = F_s / F_i \qquad (1)$$

The sticking coeficient depends critically on the nature of the inter-action of the impinging beam with the film surface and of the substrate temperature. Elements whose vapor pressure at the growth temperature is lower than the incident beam flux equivalent pressure generally have sticking coefficients close to unity, while the inverse is true for higher vapor pressure materials. For this evaluation, the incident beam flux equivalent pressures are calculated using the standard Knudsen-Langmuir relation[10]

$$P_{equi} = (2.\pi.M.R.T_s)^{1/2}.J/N_a \qquad (2)$$

where P_{equi} is the beam equivalent pressure, M the molecular weight, R the gas constant, N_a Avogadro's number, T_s the temperature of the substrate surface and J the incident flux. If the pressure is expressed in Pa, the molecular weight in atomic mass units, and the flux in molecules.cm^{-2}.s^{-1}, then expression (2) becomes

$$P_{equi} = (M.T_s)^{1/2}.(J/2.635.10^{20}) \qquad (3)$$

The incident flux arriving at the substrate can be calculated from the vapor pressure data of the elements contained in the effusion cell under the assumption that the vapor in the effusion cell is at equilibrium with the condensed solid or liquid phase and that ideal Knudsen effusion conditions apply. For a cell of orifice area A_c, situated at a distance ℓ from the substrate, and at a temperature T_c, the flux of molecules or atoms striking the substrate per cm^2 per second is given by:

$$J = 8.387.10^{19}.p_c.A_c/[\ell^2.(M.T_c)^{1/2}] \qquad (4)$$

where p_c is the cell pressure in Pa and M is the atomic mass number of the source material. In the above expression, it is assumed that the substrate is normal to the direction of the incident beam.

Under typical GaAs growth conditions, M = 70, A = 5 cm^2, ℓ = 15 cm, and T_c = 1203 K for the Ga cell. The corresponding equilibrium pressure of Ga at this temperature is 0.125 Pa and the calculated arrival rate of Ga atoms on the substrate equals J_{Ga} = 8.10^{14} atoms/cm^{2}s. The growth rate may then be calculated knowing the incident Ga flux, provided that a sufficient quantity of As molecules is supplied simultaneously to permit the growth of a single crystal layer, by the expression

$$R = \alpha_{GaAs} \cdot S_{Ga} \cdot J_{Ga} \qquad (5)$$

where $\alpha_{GaAs} = 0.45 \cdot 10^{-15}$ nm.cm^2 for growth on the GaAs (001) plane and J_{Ga} is expressed in atoms/cm^{2}s.

Thus for the example given, assuming unity Ga sticking coefficient:

$$R = 0.36 \text{ nm/s} = 1.3 \text{ } \mu\text{m/hour}.$$

Under ideal Knudsen effusion conditions, the angular distribution of the molecular effusion obeys a cosine law. Therefore, to achieve optimum thickness and composition uniformity, the substrate should be continuously rotated during growth and should be slightly offset with respect to the main axis of the effusion cells.

The substrate temperature, T_S, for optimum growth conditions must be high enough to ensure that the atoms sticking to the surface maintain a sufficient surface mobility to allow them to settle in their equilibrium positions, and should not be too high to cause excessive re-evaporation of the impinging species. When the substrate temperature is too low, polycrystalline or amorphous growth results, the actual temperature for the loss of epitaxy depending critically on the cleanliness of the growing surface and on the density of background impurities in the gas phase.

The use of a sufficiently high growth temperature will also be effective in displacing the equilibrium conditions for the Ga oxidation reaction

$$4.GaAs + Ga_2O_3 = 3.Ga_2O + 2.As_2 \qquad (6)$$

sufficiently to the right to avoid incorporation of oxygen which is known to form deep levels in GaAs and to have a deleterious effect on the electrical and optical properties of the layers.

In the case of GaAs, the minimum temperature to achieve oxygen desorption lies around 530°C as determined from the observation of reflection high energy electron diffraction spectra (RHEED). At temperatures below the oxygen desorption temperature, T_{OX}, the oxygen containing background species will stick to the GaAs growth surface and be incorporated in the epitaxial layers. In the case of the growth of AlGaAs, the corresponding desorption temperatures are much higher due to the stability of the Al_2O_3 oxide as compared to the Ga_2O_3 oxide.

To achieve good layer quality, it is therefore essential to maintain an ultra-clean vacuum environnement. This requires thorough baking and outgasing of the system and sources after every exposure of the growth chamber to air, high temperature outgasing of the mounting blocks and substrates before introduction, loading and unloading of the wafers via separate loading chambers, and extensive cryopaneling around the growth zone to ensure very high pumping speeds.

Despite these precautions, the ultimate layer qualities are generally not achieved until after about 10 μm of layer deposition, which is effective in fully coating and passivating all surfaces exposed to the beams. For this reason, the system should be designed so as to assure reliable operation and should be equipped with large capacity effusion cells to minimize down-time for source replenishment.

For normal growth rates of around 1 to 2 μm/hour, the optimum substrate temperature T_S for the realization of highest purity and highest mobility GaAs layers lies in the range 600-640°C . When the growth temperature is substantially lowered below these values increased incorporation of deep level traps results which may be Ga-vacancy or oxygen related. Somewhat higher temperatures, in the range 640-700°C are required to obtain AlGaAs material with the best luminescent properties and highest radiative efficiencies. Since the luminescence behavior is dictated by the minority carrier properties, this fact also points to a continuous decrease in the incorporation of deep level defects, and to an improvement of crystallographic quality, as the growth temperature is increased.

The last growth parameter to be considered is the As to Ga flux ratio. Generally, under the conditions stated above, the optimum layer quality is achieved at the lowest As flux intensity which is just sufficient to maintain As-stabilized growth conditions (i.e. $J_{As4}/J_{Ga} = 1$ to 2 for a growth temperature of 600°C). This optimum growth region on (001) oriented surfaces is characterized by the appearance of a c(2 x 8) electron diffraction pattern (see section 3.2 below). Higher As fluxes produce a c(4 x 4) structure and lead to an increased incorporation of deep levels, while substantially lower As fluxes produce a c(8 x 2) structure corresponding to Ga-rich conditions, and may lead to a build-up of excess gallium on the surface with Ga-droplet formation and unacceptable surface morphology degradation.

2.3 Film stoichiometry

For a simplified view of the growth process it can be assumed that at relatively low growth temperatures all the incident group III atoms stick on the growth surface and only enough group V elements adhere as necessary in order to give stoichiometric growth. The excess group V species supplied are desorbed without interfering with the growth process, so that the control of stoichiometry is essentially automatic, at least in the case of compounds or alloys containing only one group V element. The above model holds provided the growth temperature is above the congruent evaporation temperature of the film which insures that the group V elements are preferentially desorbed; in practice therefore, it is sufficient to provide an overpressure of the group V species during deposition.

The simple interpretation given above is substantiated by the results of modulated quadrupole mass spectrometry measurements[2,17] whereby one can directly observe the nature and intensity of the fluxes desorbing from the substrate. Considering the interaction of Ga and As with GaAs, it is found that Ga impinging on the surface obeys a first order desorption process. If F_i is the incident Ga beam flux density, the rate of change of the gallium surface population, n_{Ga}, is then given by the equation

$$dn_{Ga}/dt = -n_{Ga}/\tau + F_i \tag{7}$$

where τ designates the surface resident lifetime and $\Gamma = n_{Ga}/\tau$ the surface desorption rate.

The flux desorbing from the substrate upon exposing the substrate to the gallium beam is then given by

$$\Gamma(t) = F_i.[1 - \exp(-t/\tau)] \tag{8}$$

When the gallium flux is turned off the desorbing flux decays to zero intensity with an exponential time constant τ. Figure 4 shows the shape of the desorbed flux in reponse to an incident rectangular pulse of Ga at two different temperatures, yielding values for the Ga surface resident lifetime of 7 and 10 seconds at 885 and 904 K, respectively. Under similar

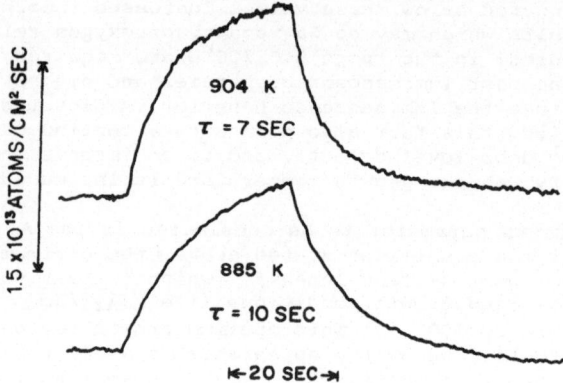

Fig. 4. Desorbed pulse shape from an incident rectangular pulse of Ga on a
GaAs substrate. After Arthur[17].

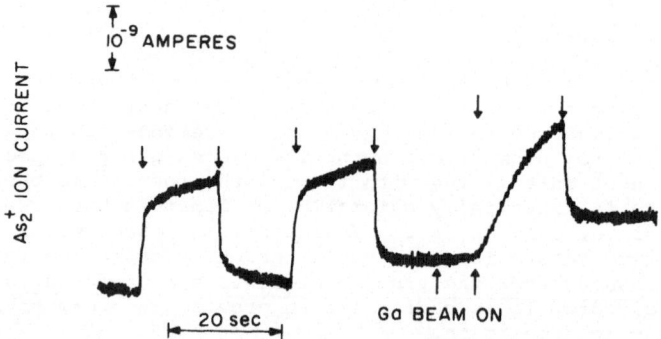

Fig. 5. Desorbed pulse shapes from incident rectangular pulses of As_2 on
GaAs; prior to the third As_2 pulse approximately 0.5 monolyer of Ga
was deposited as indicated. After Arthur[17].

conditions, using As_2 or As_4 incident beam fluxes, one would find that
essentially all incident As molecules re-evaporate instantly either as As_2
or As_4, which means that the corresponding As surface lifetimes are
extremely short. In the presence of a steady state Ga adatom population,
however, the As_2 and As_4 molecules will bind to the surface and will be
incorporated in the growing layer. The mechanism is illustrated in figure 5,
where one sees that in the absence of a Ga flux the As flux is nearly
completely desorbed, while the desorption is surpressed during simultaneous
deposition of Ga.

A model for the growth of GaAs proposed by Foxon and Joyce[2] is
illustrated schematically in figure 6. When GaAs is grown from Ga and
dimeric As (Fig. 6a) the reaction is one of dissociative chemisorption of
As_2 molecules on single Ga atoms. Since the process considered is first
order, the sticking coefficient of As_2 is proportional to the Ga flux. As
stated before, excess As is re-evaporated leading to the growth of stoichiometric GaAs. For GaAs grown from Ga and the tetrameric As_4 the process is
more complex (Fig. 6b). Pairs of As_4 molecules react on adjacent Ga sites.
Even when excess Ga is present there is a desorbed As_4 flux and the maximum
sticking coefficient for As_4 is around 0.5. For very low As/Ga flux ratios,

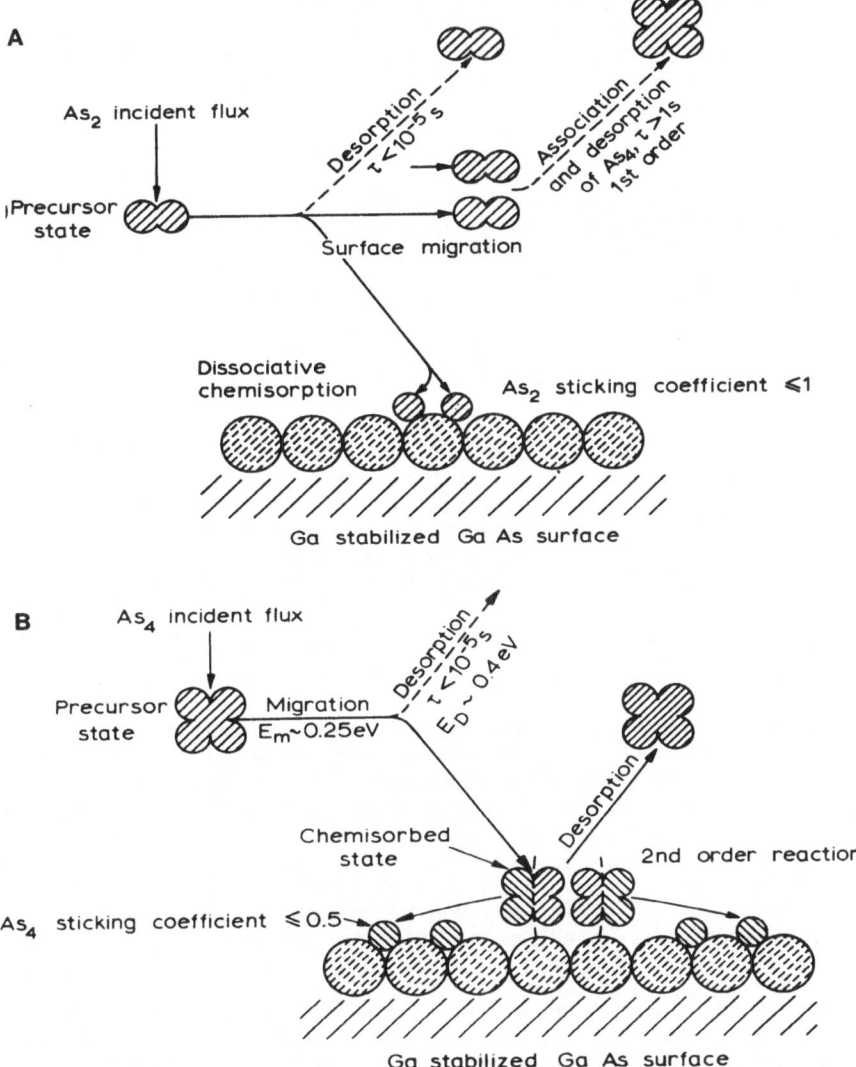

Fig. 6. (a) Model for the growth of GaAs from Ga and As_2. (b) Model for the growth of GaAs from Ga and As_4. After Foxon and Joyce[2].

when the As_4 surface population is small compared to the number of Ga sites, the growth rate limiting step is the encounter and reaction probability between the As_4 molecules. In practice however, $J_{As4} > J_{Ga}$ and there is a high probability that arriving As_4 molecules will find adjacent sites occupied by other As_4 molecules and the desorption rate becomes proportional to the number of As_4 molecules being supplied. Growth thus proceeds by adsorption and desorption of As_4 via a bimolecular interaction resulting in one As atom sticking for each Ga atom. As the growth temperature is increased, As_2 is lost by desorption, resulting in an increased Ga surface adatom population. Analysis of the surface using reflection high energy electron diffraction shows that at low temperatures and high As/Ga flux ratios an "As-stabilized" (2 x 4) or c(2 x 8) surface structure is observed, while at high temperatures and low As/Ga ratios a (4 x 2) or c(8 x 2) "Ga-stabilized" surface structure develops.

2.4 Alloy growth

The growth of ternary or quaternary alloys sharing one common group V element, such as $Al_xGa_{1-x}As$ or $Al_xGa_yIn_{1-x-y}As$, can be treated in a manner similar to that of the growth of the binary compounds, provided the growth temperature is sufficiently low so that the sticking coefficients of the different group III elements remain close to unity. The thermal stability of the less stable of the binary compounds in the alloy puts a limit as to the maximum growth temperature that can be used. It is clear also that for precise layer composition control, it is necessary to take into account the differences in the desorption rates of the different group 3 elements involved, since the highest vapor pressure component will desorb preferentially.

Up to now there has been comparatively little work published concerning the growth mechanisms in systems with more than one group V element. A priori, one would expect the relative incorporation ratios of the volatile group V elements to be dictated by the thermodynamic equilibrium reactions for the formation of the ternary or quaternary alloys, which would imply that the controlled growth of alloys where there exist large differences in the free energies of formation of the end compounds would be very difficult, if not impossible. The limited experimental evidence available so far suggest however that this is not the case, and that surface kinetics, rather than equilibrium thermodynamics, dominate the growth reactions.

In order to achieve epitaxial growth under optimal conditions, the lattice constant of the alloy to be deposited should nearly exactly match that of the substrate, the maximum allowable relative deviation $\sigma = [a_{layer} - a_{substrate}]/a_{substrate}$ being of the order of 5.10^{-4} to 1.10^{-3}. Apart from the AlAs-GaAs system where the lattice constants of the end compounds differ by less than 2.10^{-4}, this lattice matching condition imposes stringent controls on the layer composition. In the case of the GaInAs and GaInPAs systems, which can be lattice matched to either InP or GaAs substrates, the composition range over which perfect epitaxy can be obtained is illustrated in figure 7.

The main difficulty encountered in the growth of these systems thus lies with the precise control of the beam intensities emanating from the different effusion cells. Despite the excellent results that have been achieved in some laboratoires with the conventional MBE technique so far, interest for the growth of ternary and quaternary alloys with 2 or more group V elements is shifting to the newly developed gas source MBE or CBE techniques where the beam intensities should be more readily controllable than it is the case for solid source evaporation.

2.5 Silicon epitaxy

While the epitaxial growth of elemental semiconductor materials such as silicon and germanium is conceptually simpler than that of the compound semiconductors and alloys where stoichiometry control and composition control play a large role, the development of silicon molecular beam epitaxy initially evolved rather slowly because of severe difficulties associated with substrate preparation and with the control of dopant incorporation, as well as because of the existence of very advanced competing vapor phase epitaxy techniques.

368

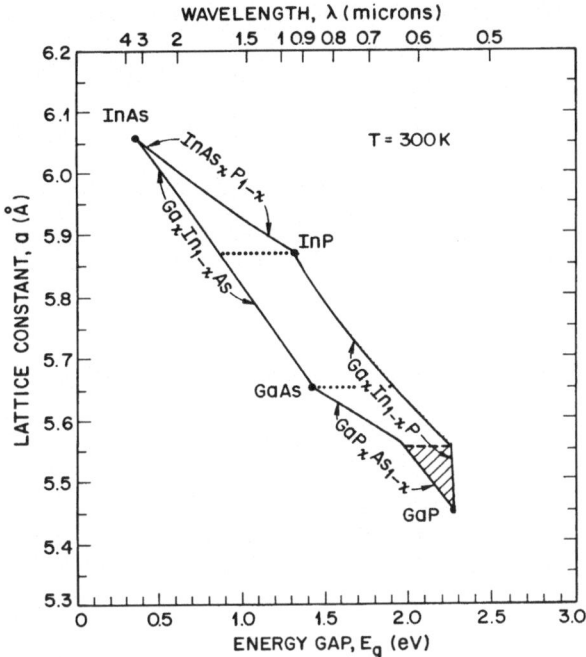

Fig. 7. Lattice constant versus bandgap in the GaInPAs system with alloy composition as a running parameter.

During the last 5 years or so however, important progress has been achieved and the silicon MBE technique has emerged as a very powerful method for the fabrication of new high speed and high frequency devices. A variety of epitaxial metal and insulator films have been succesfully grown on silicon, and the emergence of Si/Ge heteroepitaxy has added a viable and previously inexistent semiconductor heterostructure capability to the Si materials system. At the same time, advances in the design and scaling of the growth apparatus to allow the treatment of large diameter (100 mm) Si wafers with automated batch-type loading and unloading sequences and fast throughput, have raised the expectation that Si MBE may become a cost-competitive commercial VLSI processing technique[18].

The growth of device quality silicon MBE films requires ultra high vacuum conditions of 10^{-9} to 10^{-10} mbar with particularly low oxide and hydrocarbon partial pressures, as well as particular care in the preparation of the starting surface. Usually, the substrate is first chemically cleaned and placed through a load-lock aperture in a separate substrate preparation chamber where an appropriate treatment is carried out to remove the few monolayers of residual oxygen and contaminants that remain on the surface. In early systems, this cleaning step was achieved by simply heating the substrate briefly to approximately 1250°C, at which point the oxide decomposed and evaporated and residual carbon diffused inward away from the surface. Because of problems associated with this high temperature thermal etching step, which causes impurity diffusion and generates crystal defects such as slip line dislocations and substrate warpage, alternative approaches where wafers are sputter cleaned by low energy inert gas bombardement and re-ordered by a 5 to 10 minutes anneal at 850 to 900°C have been developped. Another technique used involves repetive chemical cleaning and oxidation of the silicon surface before growth, ending with an oxidation step which leaves a very thin oxide film free of carbon which may then be thermally desorbed in the growth chamber at about 850°C.

Following anneal, the sample is cooled to the deposition temperature which lies in the range from 550 to 750°C for homoepitaxy. Typical deposition rates lie around 1 to 2 μm/h, with the Si flux being generated by electron-beam evaporation of an elemental silicon charge, rather than by conventional thermal evaporation. Maximum growth rates are limited by the characteristics of the deposition source and by the occurence of splashing of the molten Si charge when a critical e-beam current density is exceeded. By using e-beam sweeping techniques, coupled with appropriate substrate rotation during deposition, thickness uniformities better than ±5% can be achieved over 100 mm diameter wafers[19].

Dopants are introduced during growth by opening shutters in front of standard resistively heated effusion cells as in conventional III-V compound MBE, or by using low-energy ion beam implantation during growth. With effusion cells, doping is controlled either by changing the heating power to the cells or by switching between cells. With ion doping, levels can be electronically programmed by modulating the raster scanning speed. The main silicon flux and the dopant fluxes are measured continuously during growth and provide the feedback signals needed to program and stabilize the beam fluxes.

The commonly used evaporated dopants are gallium and aluminum for p-type and antimony for n-type; amongst the other more conventional silicon dopants, boron has too low a vapor pressure to be thermally evaporated and phosphorous and arsenic have too high a pressure resulting in unacceptably-high background doping contamination. While the use of these dopants has allowed the fabrication of device-quality materials exhibiting sharp doping profile transitions, problems remain in reaching high doping concentrations and in controlling surface segregation phenomena. In addition, because the sticking coefficients for these elements tend to be low and depend exponentially on substrate temperature[7], extremely tight control of both evaporation cell and substrate temperature is required which is difficult to achieve. For this reason, ion doping is used almost exclusively in newer systems. In this case, the ion energy is adjusted to implant the dopant a few atomic layers deep to avoid re-evaporation and to limit radiation-induced damage and the need for subsequent annealing steps.

Besides the advantages of Si MBE for the growth of complex homoepi-taxial Si on Si devices, potential applications have also been demonstrated in hetereroepitaxial growth such as for instance, silicon on saphire, $(Ca,Ba,Sr)F_2$ on Si, and $CoSi_2$ or $NiSi_2$ on Si. These metal disilicide layers have the metallic characteristics of high conductivity and Schottky barrier formation, and are stable in contact with silicon up to high processing temperatures, which opens up possibilities for improved circuit metallization and the formation of buried interconnect lines and active gates. Most appealing for Si MBE is its potential application in building Si-based multilayered quantum-well and superlattice devices based on the use of Ge-Si alloys or on the use of modulation doping, where the doping density and type is varied periodically to yield n-p doping superlattices or controlled quantum wells.

3. SURFACE STUDIES

3.1 Overview

A number of analytical tools are available to study the surfaces of the layer in-situ before and during growth, such as Auger electron spectroscopy

(AES), secondary ion mass spectrometry (SIMS), low energy electron
diffraction (LEED) and reflection high energy electron diffraction (RHEED).
The RHEED technique especially has proven to be extremely useful since it
provides information on the atomic structure of the surface and since it can
be used continuously during deposition to provide a real-time image of the
growth process.

3.2 RHEED pattern formation

In RHEED, an electron beam with an energy in the range 5 - 50 keV is
incident at a glancing angle of 1° to 2° to the crystal surface. Under these
conditions, the component of electron momentum normal to the surface is
sufficiently small to ensure that the penetration depth is only of the order
of a few atomic layers.

The conditions for constructive interference of the elastically
scattered electrons from the surface is given by the Laue diffraction
condition

$$k - k' = g \qquad (9)$$

where k and k' are the wavevectors of the incident and scattered
electron beams, respectively, and g designates any vector of the reciprocal
lattice. In three dimensions, this condition gives rise to the well-known
Laue X-ray diffraction patterns which consist of a series of spots
corresponding to reflections from different crystal planes.

In two dimensions, the relationship between incident and diffracted
wavevectors expressed by equation (9) may be be visualised using the Ewald
construction in the reciprocal lattice. In the case where the interaction of
the electron beam is essentially with a two-dimensional atomic surface, the
reciprocal lattice is composed of rods in reciprocal space in a direction
normal to the real surface. Fig. 8 shows the Ewald sphere and the reciprocal
lattice rods in the case of a simple square atomic surface arrangement. The
reciprocal lattice rods are of finite thickness due to lattice imperfections
and thermal vibrations; the Ewald sphere is also of finite thickness
representing the energy spread of the incident electron beam.

The de Broglie wavelength of an electron, λ, is related to the electron
energy, E, by the relation

$$\lambda^2 = h^2/(2.m.E) \qquad (10)$$

where m is the electron mass and h the Planck constant. In laboratory units

$$\lambda = 1.2265 / (E)^{1/2} \qquad (11)$$

where λ is expressed in nm and E in eV. For example, for an electron
accelerating voltage of 15 kV the corresponding electron wavelength is 0.01
nm.

The radius of the Ewald sphere is equal to the electron momentum k
given by

$$k = p / \hbar = 2\pi / \lambda \qquad (12)$$

For the example given, one finds k = 627 nm^{-1}. This radius is very much
larger than the spacing of the reciprocal lattice rods which is given by
$2\pi/a$ where a is the lattice constant of the surface mesh. For an unrecon-
structed GaAs (001) surface with a lattice constant a = 0.565 nm, the
distance between adjacent rods will be 11 nm^{-1}. As a result, the inter

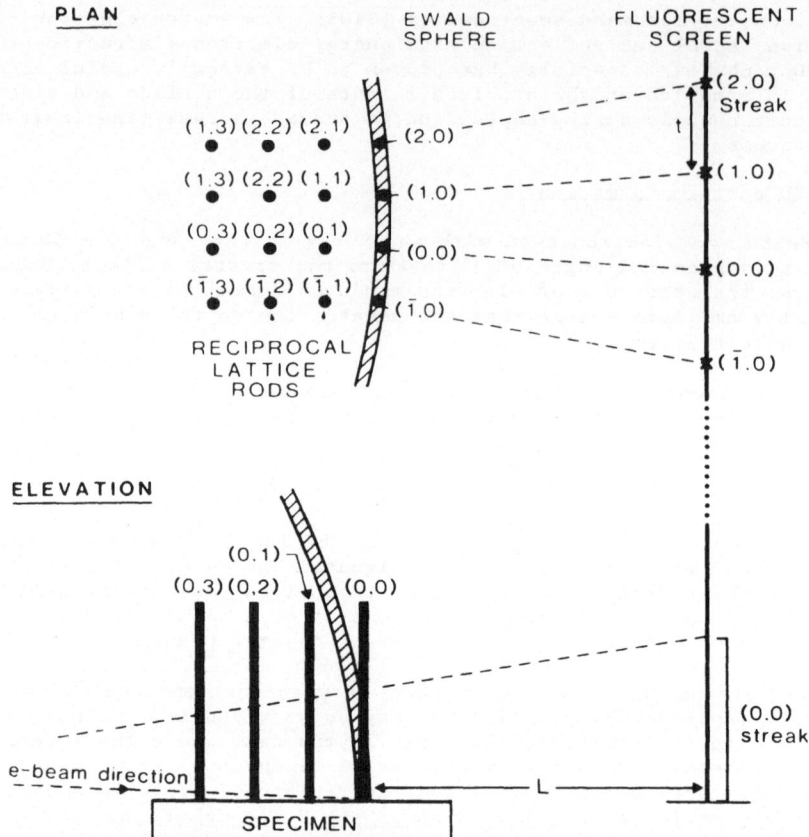

Fig. 8. Schematic representation showing the intersection of the Ewald
        sphere with the lattice rods in reciprocal space and the resulting
        streaked diffraction pattern on the RHEED fluorescent screen [ref.
        20].

section of the Ewald sphere and the rods occurs for some distance along
their lenght, resulting in a streaked, rather than a spotty diffraction
pattern. Examples of diffraction patterns obtained on the GaAs (001) surface
are shown in Fig. 9.

     If the distance beween the crystal surface and the screen is L and the
separation of the streaks is t then the periodicity of the crystal surface
is given by

     $a = \lambda.L / t$

     In general, the free semiconductor surface reconstructs to a more
stable surface atomic arrangement which exhibits a periodicity which is
different from that of the bulk crystal[20,21]. The detailed image of the
reconstructed surface can be inferred from the observation of diffraction
patterns taken at different azimuths. In the case of GaAs in the [100] and
[111] directions, the crystal is formed with alternative layers of Ga and As
atoms, and stable surface structures rich in either Ga or As atoms are
observed during MBE growth. On the (001) surface, the so-called Ga-

stabilized surface has a centered c(8x2) or (4x2) structure and the As-
stabilized surface has a c(2x8) or (2x4) structure. These results relating
surface structure and surface composition have been confirmed by mass
spectrometric and Auger studies.

3.3 Surface ordering

The (001) and (111) surfaces of GaAs and related compounds exhibit a
variety of possible surface reconstructions depending on growth conditions
or post-growth treatment. A reconstructured surface is characterized by the
fact that the surface atoms re-arrange themselves to produce a periodicity
different from that of the underlying crystal. By convention, a surface
reconstruction designated by GaAs (001)-(m x n) means that a GaAs crystal is

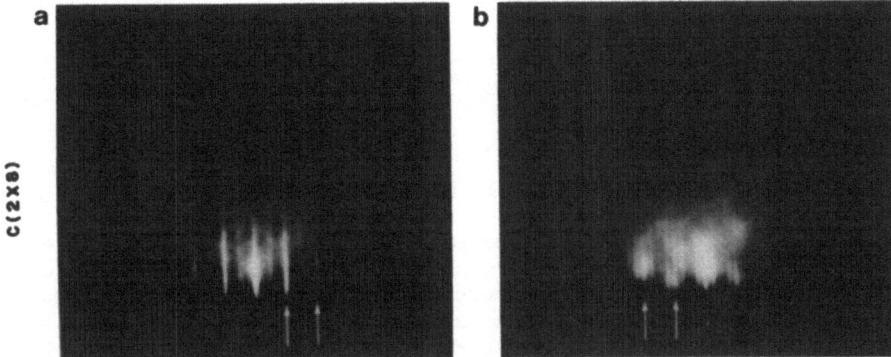

Fig. 9. RHEED patterns of a (001) GaAs surface grown with an As-stabilized
        c(2x8) or (2x4) surface reconstruction taken (a) along the [1$\bar{1}$0]
        azimuth (half order diffraction lines) and (b) along the [110]
        azimuth (quarter order diffraction lines). After Cho[10].

orientated with the [001] direction normal to the surface and has a surface
structure whose unit mesh is m x n times larger than the underlying bulk
strucure. Such surface meshes may be centered, in which case the notation
would be (001) - c(m x n).

In initial studies[1], 2 main reconstructions were identified on the
(001) surface, which were interpreted as GaAs (001)-c(2 x 8)As for the As-
stabilized surface and (001)-c(8 x 2)Ga for the Ga-stabilized surface. These
patterns were obtained during growth and could be reversibly interchanged by
varying the As to Ga flux ratio with constant substrate temperature or by
varying the substrate temperature with constant flux ratio (Fig. 10). The
transition from a c(2 x 8) to a c(8 x 2) structure is quite complex and may
involve many intermediate structures.

Some controversy remains concerning the detailed interpretation of the
RHEED diagrams; in particular, the distinction between c(2 x 8)As and (2 x
4)As structures and between c(8 x 2)Ga and (4 x 2)Ga structures on the (001)
surface is not always clear. It has been argued that during growth the real
surface may consist of domains corresponding to either a c(2 x 8)As or a (2
x 4)As reconstruction which coexist on the surface[21].

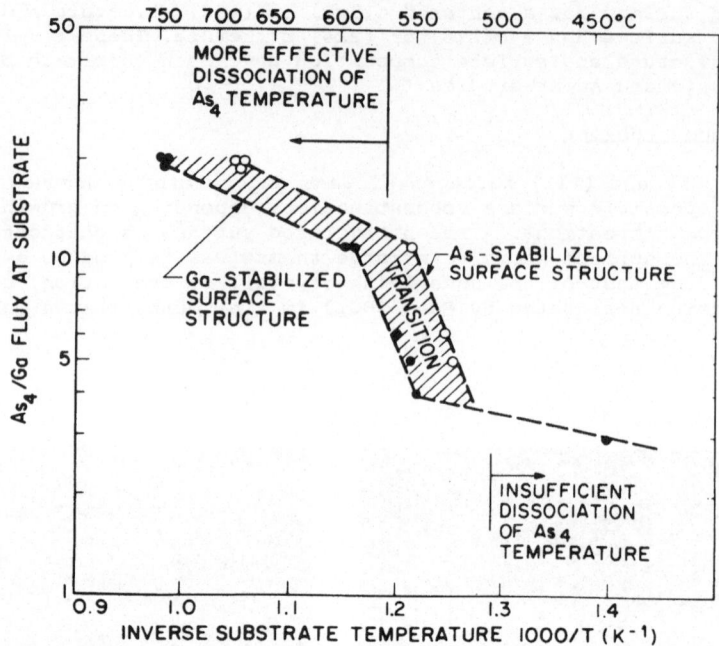

Fig. 10. As₄/Ga molecular beam flux ratio as a function of substrate temperature showing when the transition between As-stabilized and Ga-stabilized structures occurs on the (001) GaAs surface. Beam fluxes were measured by an ion gauge at the substrate position with a Ga flux corresponding to a growth rate of approximately 1 µm per hour. After Cho[10].

Studies on well oriented crystal surfaces have established that for both Ga- and As-stabilized structures the twofold periodicity in the surface pattern occurs in the [110] direction parallel to the plane containing the dangling bonds on the surface. This suggests that this periodicity may be due to dimerization of adjacent As or Ga atoms produced by a bond-pairing mechanism.

3.4 RHEED pattern intensity oscillations

Much interest has recently been focused on the oscillations in the intensity of the RHEED patterns during growth[23]. On monitoring a growing (001)-(2 x 4)As surface it was found that the amplitude of the oscillations was greater for an incident beam in the [110] azimuth compared to the [1̄10] azimuth. In all cases the effect is most pronounced in the specularly reflected beam. Figure 11 shows how the oscillations commence immediatly after growth, i.e. upon opening the Ga shutter, and how the initial intensity is recovered once growth is stopped.

Although the oscillations are damped, they may be followed for many cycles and correspond exactly to monolayer growth (one layer of Ga + As

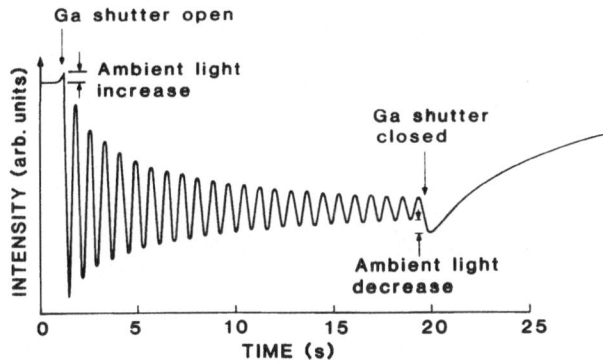

Fig. 11. Intensity oscillations of the specular beam in the RHEED pattern
from a GaAs (001)-(2 x 4)As reconstructed surface obtained in the
[110] azimuth. The period exactly corresponds to the growth of a
single Ga + As layer [ref. 23].

atoms). These oscillations may thus be used to accurately follow the layer
growth by MBE and to calibrate growth rates and layer thickness with
monolayer precision.

The oscillations in the intensity of the RHEED patterns are related to
changes in surface roughness during growth. The model proposed is shown in
figure 12. The equilibrium surface existing before growth is smooth, giving
a high reflectivity of the specularly reflected beam. At the beginning of
growth, nucleation islands will form at random positions on the surface,
leading to a decrease in reflectivity. The islands will grow and ultimately
produce another smooth surface. The minimum in reflectivity is thought to
correspond to a 50% coverage of the surface by the growing islands. The
observed damping in the reflectivity oscillations is attributed to the
nucleation of new islands before the underlying layer is complete, and
oscillations will ultimately cease when the islands become distributed over
several uncomplete atomic layers. Upon stopping growth, the surface atoms
rearrange to complete the atomic layers, and the reflectivity returns to its
high initial value. It is believed that most of the steps developing on the
growing surface lie along the [1$\bar{1}$0] direction, thereby explaining why the
oscillations are most easily observed along the [110] direction.

4. DOPANT INCORPORATION

4.1 Unintentional impurities[24]

Undoped GaAs layers grown in present-day MBE systems with solid sources
typically show p-type conductivity with a free hole concentration around
1.10^{14} cm^{-3}. The residual impurity responsible for this behavior is believed
to be carbon which forms a shallow acceptor with a binding energy around 26
meV (Table 1). The presence of carbon is presumably due to a a reaction of
CO or CO_2 present in the background gases of the system with As or Ga on the
growing surface. Other background impurities, such as Mn, Si, or various
other unidentified shallow donors, generally depend on the previous history
of the system or can be traced to contamination due to certain components
heated to high temperatures during growth, and can be reduced or eliminated
below detectable levels by careful system design and growth procedures.

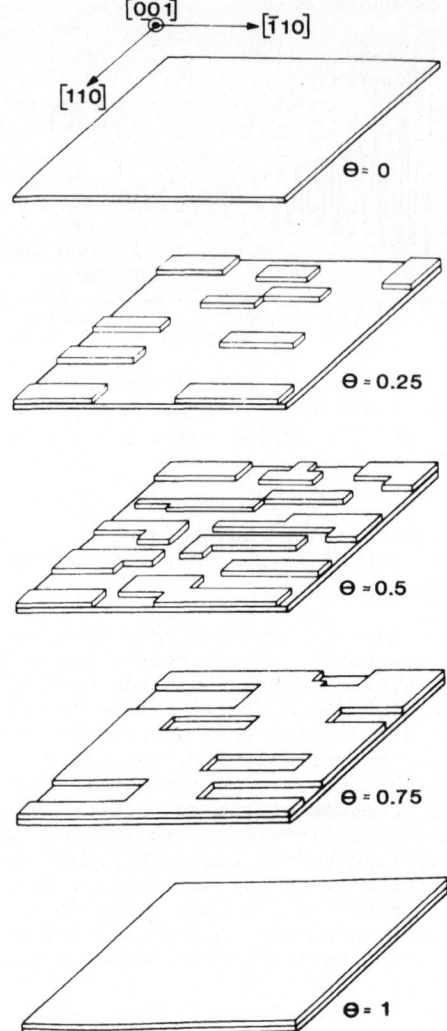

Fig. 12. Evolution of the surface coverage during GaAs monolayer growth. θ represents the fractional monolayer surface coverage. Secondary nucleation of new islands is possible before completion of the first layer, leading to the damping of the intensity oscillations. After Neave et al.[23].

4.2 Shallow acceptors[24]

Beryllium is generally used as a p-type dopant in GaAs and AlGaAs and produces a shallow level approximately 28 meV above the valence band. The sticking coefficient and electrical activity (i.e the ratio of ionized to neutral Be atoms in the layer) are equal to unity. up to doping concentrations in the mid 10^{19} cm^{-3} range. Higher doping concentrations of Be have been achieved[25] by lowering the growth temperature below 500°C.

For materials grown at lower temperatures such as InP and GaInAs, magnesium can be used as an alternative to beryllium, with the advantage of being less toxic while yielding layers of comparable electrical and optical

Table 1. Binding energies for dopant atoms occupying substitutional sites in GaAs

donors:	on Ga-sites:	Si, Ge, Sn	5.71 meV
	on As-sites:	S, Se, Te	5.71
acceptors:	on As-sites:	C	26.0 meV
		Si	34.5
		Sn	171.0
	on Ga-sites:	Be	28.0
		Mg	28.4
		Zn	30.7
		Cd	34.7

quality. With increasing growth temperature, the sticking coefficient of Mg decreases rapidly to fall below 10^{-3} at substrate temperatures above 600°C.

Manganese can be used as a p-type dopant at concentrations up to around $1 \cdot 10^{18}$ cm^{-3}; above these concentrations the surface morphology of the layers is strongly degraded presumably because of a Mn-As complex formation. This reason, together with the fact that the Mn-acceptor level is relatively deep, precludes the use of this dopant for device applications. The group II elements Zn and Cd, which are widely used in liquid and vapor phase epitaxy, have vapor pressures which are many orders of magnitude above that of gallium at the substrate temperature. Attempts to dope with these materials by normal thermal beam evaporation during MBE growth have been unsuccessful.

4.3 Shallow donors[24]

Silicon is the most widely used n-type dopant in GaAs and related materials. It predominantly enters the lattice as a substitutional donor on Ga-sites, and has the desirable features of having a close to unity incoporation coefficient and very low diffusivity. The maximumum doping level that can be achieved lies around $5 \cdot 10^{18}$ cm^{-3} at normal growth temperatures and is determined by the solubility limit of Si in GaAs.

A certain degree of self-compensation occurs with Si doping because of Si incorporation as an acceptor on Ga sites. Under normal growth conditions, the degree of self-compensation is around or below 10%[26]. Silicon acceptor incoporation is favored by growing at high substrate temperatures and low As$_4$/Ga flux ratios.

Because of the low self-compensation, excellent mobilities can be achieved in Si-doped layers. Figure 13 shows as an example experimental 77 K mobilities measured in Si- and Sn-doped MBE layers, together with semi-empirical curves drawn for different values of the compensation ratio $(N_D + N_A)/n$ where N_D and N_A represent the donor and acceptor impurity concentrations and n the free electron concentration.

The best results reported for non-intentionally doped MBE layers at the time of this writing[27] are in excess of 160'000 cm^2/Vs at 77 K with a peak value of 216'000 cm^2/Vs at 46 K, very close to the best mobilities achieved in GaAs material grown by other techniques.

Germanium as another possible n-type dopant shows a much larger degree of self-compensation than silicon, and will change its dopant behavior from n-type to p-type when changing from As-stabilized to Ga-stabilized growth conditions. Because of this amphoteric behavior, both the substrate

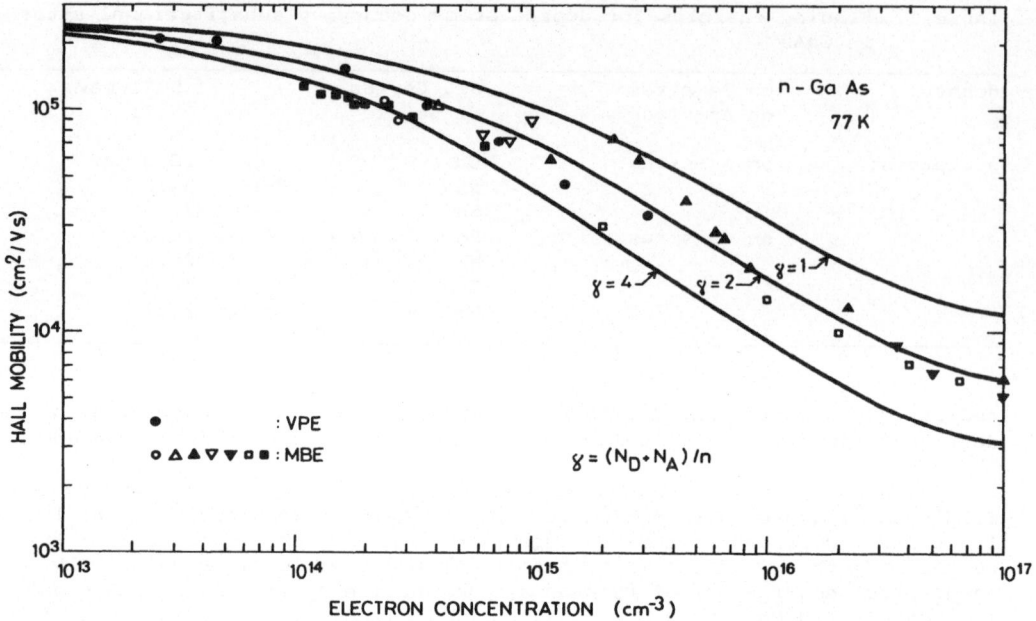

Fig. 13. Empirical curves relating the 77 K mobility in GaAs to the free
electron concentration as a function of the compensation ratio.
The experimental data are for Si and Sn doped MBE layers [ref.
24].

temperature and the As_4 to Ga flux ratio must be closely controlled to
achieve and maintain a desired electron concentration.

Tin is non-amphoteric in GaAs and is widely used as a donor because of
its relatively low evaporation temperature. Sn-doped GaAs layers exhibit
excellent luminescent properties due presumably to the low degree of compensation. The incorporation mechanism of Sn is, however, fairly complex and
doping proceeds via a Sn-accumulation layer which "floats" on the substrate
surface during growth[28]. The steady-state Sn-surface concentration depends
on substrate temperature, surface As coverage and the incident Sn flux. As a
result, abrupt changes in doping level can not be achieved by simply
changing the incident Sn-flux and some other methods such as stopping
growth, varying the substrate temperature, or predepositing Sn atoms, must
be used. The control of Sn-doping profiles is thus much more difficult than
is the case for Si-doping.

The group VI species S, Se, and Te can not be incorporated using
elemental doping sources due to the high vapor pressure of these elements at
normal growth temperatures. Exchange doping with S, Se, and Te has however
been demonstrated using PbS, PbSe and PbTe as dopant sources[28]. These
chalcogenide compounds are evaporated and deposited as molecules on the
substrate surface; doping then proceeds by incorporation of the group VI
elements on As-sites followed by subsequent re-evaporation of Pb.

4.4 Transient doping profiles

Because of their low diffusivity at normal growth temperatures and
nearly complete absence of surface accumulation effects, Be and Si are
ideally suited to achieve abrupt, near-atomic-plane doping transitions as

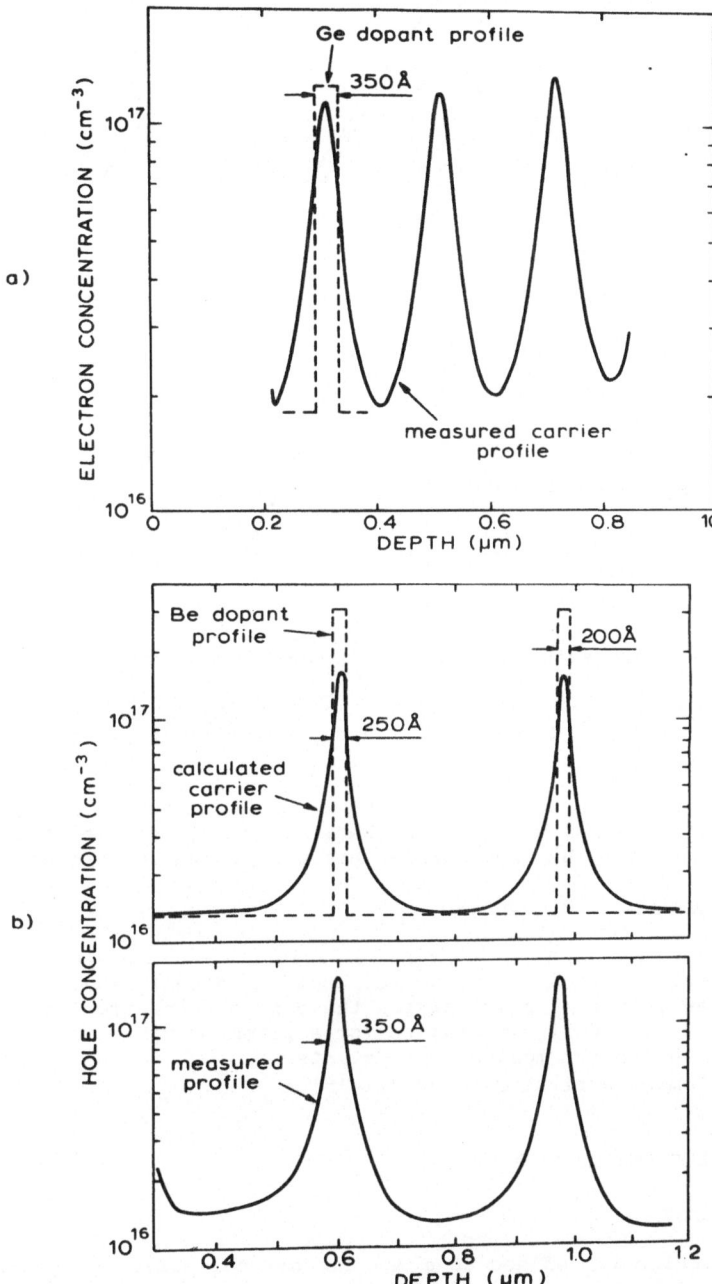

Fig.14. Measured profiles of periodically doped GaAs layers: (a) Ge-doped.
(b) Be-doped [ref. 29].

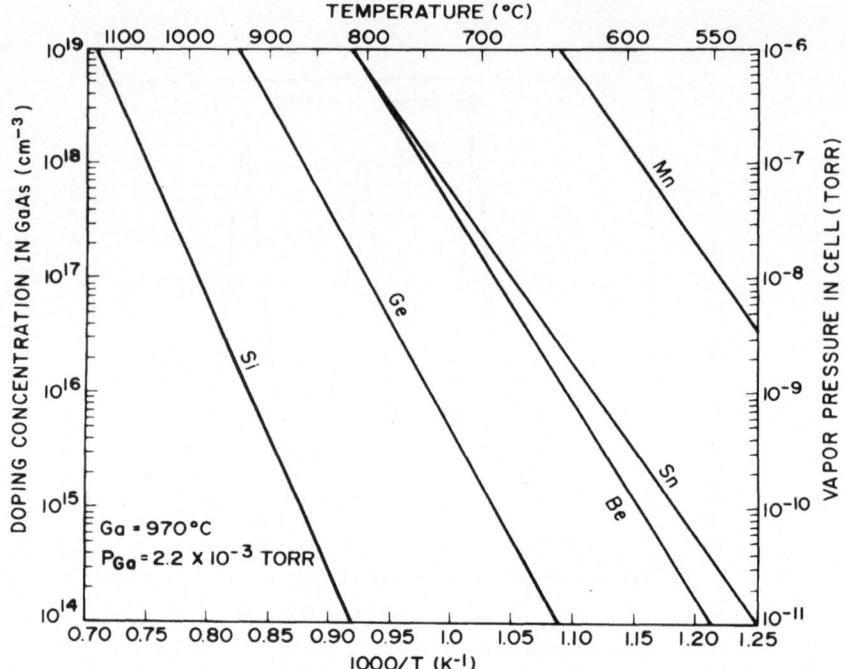

Fig. 15. Doping concentration in GaAs as a function of dopant effusion cell
temperature assuming unity sticking coefficients [ref. 20].

needed in superlattice, quantum well, and selectively doped heterojunction
structures.

The behavior achieved using Ge and Be doping pulses is illustrated as
an example in figure 14, which shows that very narrow carrier concentration
spikes, broadened only by free carrier diffusion, can be obtained.

4.5 Dopant concentration control

For dopants with unity sticking coefficient where the dopant concentration is directly proportional to the incident dopant flux, it is possible
to establish an universal chart giving the doping concentration in GaAs as a
function of effusion cell temperature for a given growth rate. The result is
shown in Fig. 15 for the most common dopants, Si, Ge, Be, Sn and Mn, and for
a group III element arrival rate of 3.10^{15} $cm^{-2}s^{-1}$.

5. SUPERLATTICES AND QUANTUM WELLS

5.1 Introduction

A superlattice is defined as a one dimensional periodic structure
consisting of alternating ultrathin layers whose period is less than the
electron mean free path.   When the above condition is fulfilled, and with
the assumption of ideal interfaces, the entire electron system enters into a
quantum regime and exhibits a number of novel and very interesting features.

A detailed study of superlattice and quantum well properties[30-32] clearly goes beyond the scope of these lectures, and the discussion in this section will be limited to aspects wich are relevant to MBE crystal growth.

The initial studies of Esaki and Tsu dealt with two types of super-lattices: doping superlattices and compositional superlattices, as illustrated in Fig. 16. In either case, a superlattice potential is introduced by a periodic variation of impurities or composition during epitaxial growth. The introduction of this superlattice potential perturbs the band structure of the host material, with the degree of perturbation being dependent on the amplitude and periodicity of the doping or compo-sition variation. Since the superlattice period, l, is usually much greater than the original lattice constant, the Brillouin zone is divided in a series of minizones, giving rise to narrow allowed subbands, separated by forbidden regions in the conduction and/or valence band of the host material. This results in a modified energy-wave vector relationship for the conduction electrons as shown schematically in figure 17.   In an extreme case where the potential wells are sufficiently apart from each other, the allowed bands become discrete states, and the electrons become completely two-dimensional.

5.2   Compositional superlattice

Figure 18 illustrates the subband formation in a superlattice with equal well and barrier widths as a function of well width, calculated for the case of a periodic square-well potential with a 0.4 V barrier height and

(a) DOPING SUPERLATTICE

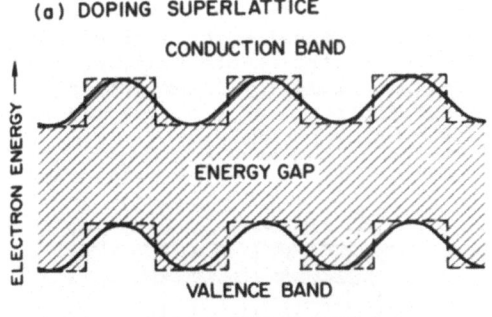

(b) COMPOSITIONAL SUPERLATTICE

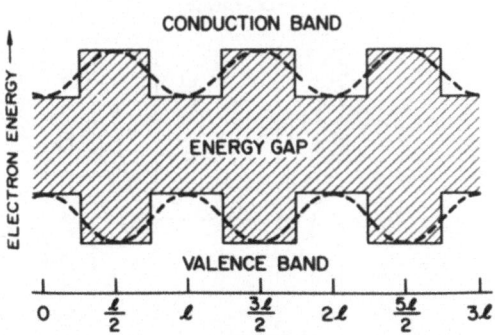

Fig. 16. Spatial variation of the conduction and valence band edges in two types of superlattices : (a) a doping superlattice of alternating n-type and p-type layers, (b) a compositional superlattice of alternating crystal composition [ref. 30].

a 0.1 m_0 electron mass. This figure clearly shows the existence of bands whose energy width decreases with increasing well width.

Superlattices of the type indicated in figure 16 form the basis for the study of a large family of novel physical and device concepts in semiconductors, and are finding increasing applications in advanced electronic and optoelectronic devices.

Compositional superlattices have been formed in many III-V semiconducting systems such as GaAs-AlGaAs, InAs-GaSb, and InP-GaInPAs, for example. From early on, it was realized that a clean and atomically smooth interface between layers was an essential requirement to achieve abrupt potential steps with little undesirable localized states. The continued refinement of the advanced epitaxial deposition techniques has played a crucial role in making the realization of near ideal superlattice structures possible.

The GaAs/AlGaAs system has been most widely studied so far because of the facility with which lattice-matched heterojunctions can be fabricated. A transmission electron microscope cross section of a GaAs-AlAs superlattice structure with approximately 11 nm period is shown in Fig. 19, and illustrates the high degree of perfection and the near-atomically flat nature of the interface between the successive layers that can be achieved.

5.3 Doping superlattices [34]

A doping superlattice is formed by a periodic sequence of n- and p-doped layers, interleaved with undoped zones of the same semiconductor material (n-i-p-i crystals). These doping superlattices exhibit some of the main features of the compositional superlattice, but differ in one important aspect: their electronic properties are tunable, and can be varied by the application of an electric field, by illumination, or by carrier injection. A large number of novel applications are expected to derive from the tunability of the electronic properties.

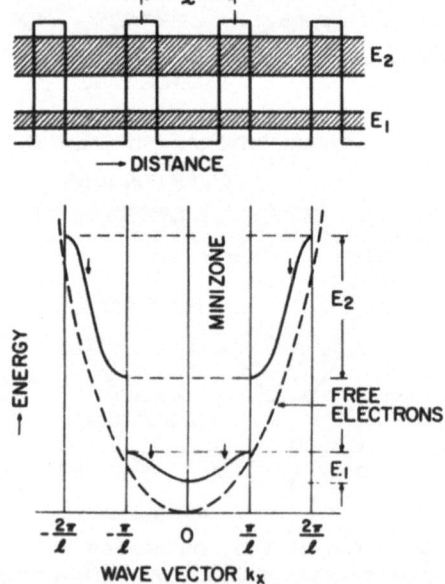

Fig. 17. Potential profile of a superlattice and its energy-wavevector relationship in the minizones [ref. 30].

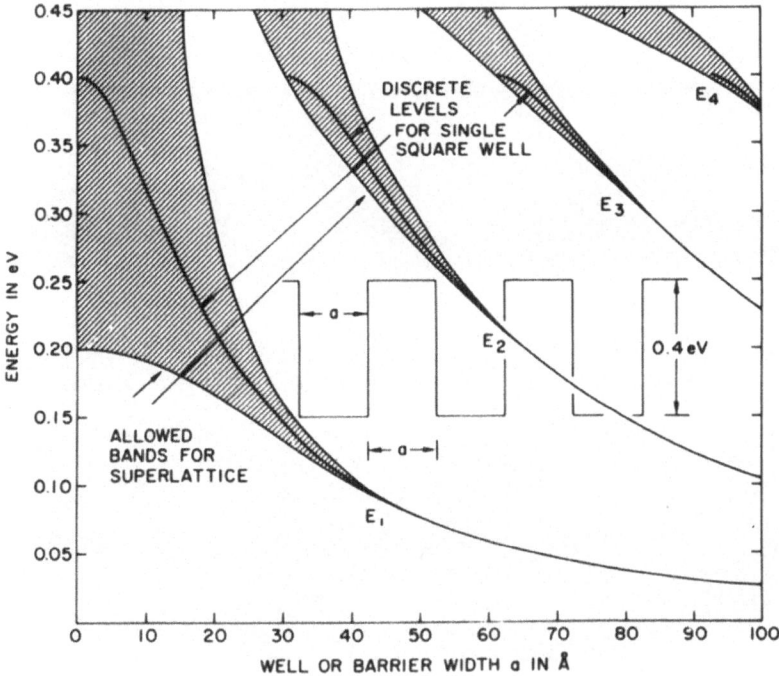

Fig. 18. Allowed energy bands, E_1, E_2, E_3, and E_4 calculated as a function
of well or barrier width, a. Discrete energy levels for a square
well are also shown [ref. 30].

5.4   Modulation doping

In superlattices it is possible to spatially separate the free carriers
from their parent impurity atoms by locating the doping impurities in the
regions of the potential barriers. As a result, the ionized impurity
scattering is reduced, so that high mobilities and high carrier
concentrations can be simultaneously achieved.

This concept was first succesfully implemented by Dingle[35] in modu-
lation-doped GaAs/AlGaAs superlattices as illustrated schematically in Fig.
20-a. Modulation doping was achieved by synchronizing the Si (n-dopant) and
Al fluxes during MBE growth, so that the dopant is introduced only in the
AlGaAs layer and is absent from the lower-bandgap GaAs layer. The resulting
structures showed a much enhanced mobility as compared to conventional
structures with comparable carrier concentration, as well as the
suppression of mobility drop-off at low temperatures due to the absence of
ionized impurity scattering.

The modulation doping concept was subsequentially applied to single-
well heterostructures of the type illustrated in Fig. 20-b, and applied
towards the development of a new high-speed field effect transistor
structure[36,37] called HEMT (high electron mobility transitor) or TEGFET
(two-dimensional electron gas effect transistor), whose performances are
significantly improved compared to traditiononal MESFET devices made of the
same parent material. Structures of this type have also been used to demonstrate the quantized Hall effect and have led to the precise determination
of the fine structure constant using the two-dimensional electron gas in a
GaAs/AlGaAs single quantum well[38].

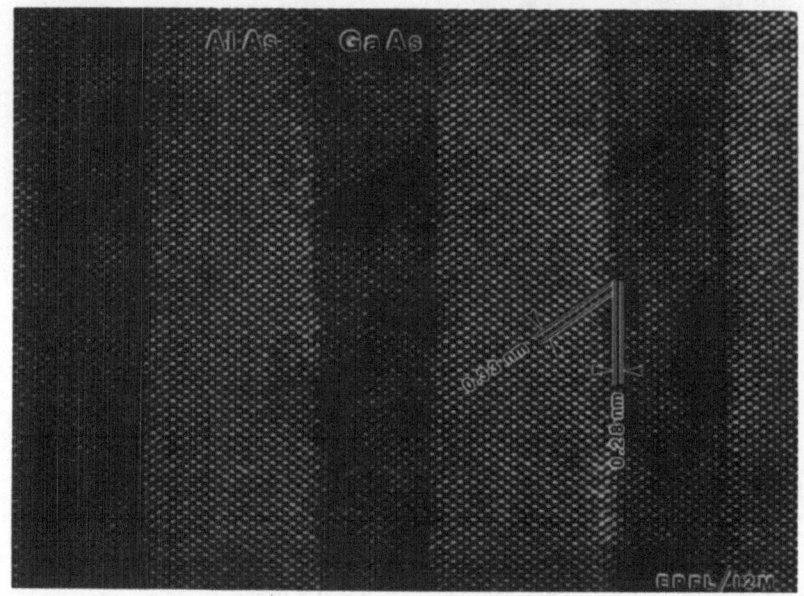

Fig. 19. Transmission electron micrograph of a GaAs/AlAs superlattice; the superlattice period is approximately 11 nm [ref. 33].

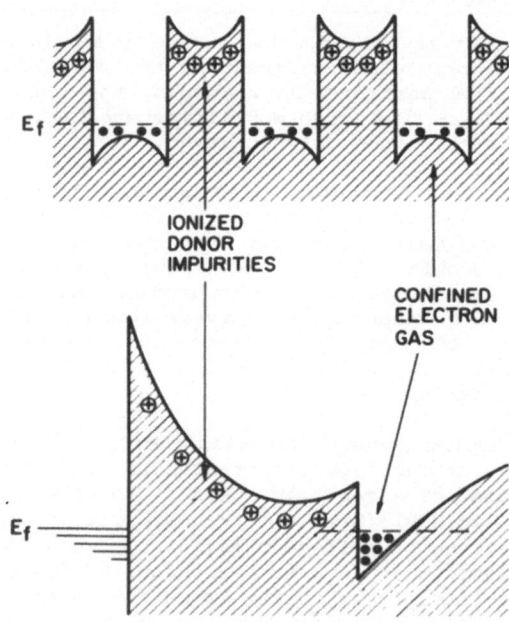

Fig. 20. Modulation doping (a) in a superlattice, (b) in a single hetero-junction with an attached Schottky surface barrier [ref. 30].

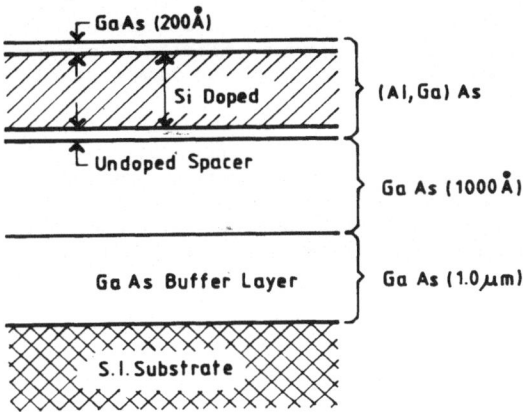

Fig. 21. Schematic diagram of the layer deposition sequence for a single well GaAs/AlGaAs structure [ref. 37].

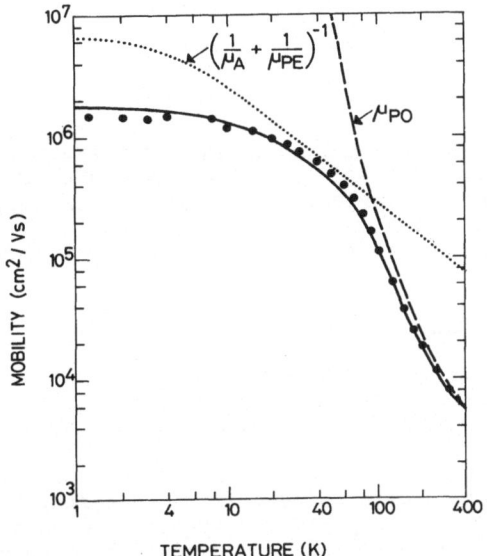

Fig. 22. Maximum mobilities attained in modulation doped GaAs layers as a function of temperature. The theoretical mobility limit as determined by piezoelectric scattering is estimated at around 6.10^6 cm^2/Vs [ref. 40].

Figure 21 shows a cross section of a high mobility single interface 2D-electron gas structure. The growth starts with the deposition of an about 1 μm thick undoped buffer layer to provide a barrier against the propagation of defects or impurities from the substrate. A very thin (4 to 20 nm) undoped AlGaAs layer containing approximately 30% AlAs is then deposited without interruption, followed by a n-type layer of the same composition doped with a suitable donor such as Si to the level required to provide the desired sheet electron concentration. The role of the undoped AlGaAs separation or spacer layer is to reduce the Coulombic interaction between the free electrons in the potential well and the ionized impurities in the n-type AlGaAs, thus resulting in improved electron mobilities. In the case of transistor devices, a final n-type GaAs cap layer is added to facilitate the formation of ohmic contacts.

The highest mobilities obtained in these systems[39] lie around 5.10^6 cm^2/Vs at temperatures below 1 K, more than a factor of 20 higher than the best mobilities that can be achieved in bulk GaAs at any temperature (Fig. 22). These results are critically dependent on the purity of the undoped GaAs and of the AlGaAs spacer layer, as well on the structural quality and interfacial properties of the interfaces. These mobilities thus provide an extremely valuable tool to qualify the materials and to guide towards the optimization of the growth, preparation and processing procedures.

The superlattice and modulation doping concepts are finding increasing applications in the field of high frequency devices based on the HEMT or TEGFET principle as described above, as well as in areas such as quantum-well diode lasers, superlattice avalanche photodiodes, and multiple resonant tunneling devices and oscillators. These results have opened up a new area of interdisciplinary investigations in the fields of materials science and device physics, and have provided new challenges towards the achievement of ever greater control and perfection of the epitaxial techniques.

6. SPECIAL GROWTH TECHNIQUES: STRAINED LAYERS, GaAs ON Si

The MBE growth technique offers a number of interesting possibilities which can be exploited for the fabrication of novel electronic and optoelec-tronic devices. These include in-situ growth of two-dimensional patterns by SiO_2 masking[1] or shadow masking[41] of the substrate, epitaxial overgrowth on top of etched or partially processed structures, pseudomorphic growth of strained layers[42] of materials with different lattice constants, and heteroepitaxy exhibiting low defect densities on a wide variety of single crystal substrates (saphire, fluoride salts).

6.1 Strained layer epitaxy

Two modes of epitaxial growth are possible when the epitaxial layer and the substrate share the same crystal structure but differ in lattice constant. In the first mode, both the substrate and the layer retain their bulk structure. In this situation, the normal fourfold bonding of the atoms can not be maintained along the interface and a certain number of atoms are left with one or several dangling bonds. If we represent the tetragonally bonded covalent group IV or group III-V crystals by a square array of atoms in two-dimensions as shown in Fig. 23, then rows of improperly bonded atoms appear at the interface and form misfit dislocation lines which may thread into the epitaxial layer. These dislocations may act as trapping sites for electrons or holes, thereby severely degrading the electrical properties of both the interface and the epitaxial material.

In a second mode, the epitaxial layer distorts elastically to match the lattice constant of the substrate, resulting in pseudomorphic growth. When the overgrowth has a larger lattice constant than the substrate, the epitaxial layer compresses along the interfacial or growth plane so that all atoms retain their fourfold bonding. To compensate for this compression, the epitaxial layer planes spread slightly farther apart perpendicular to the interface. Strained layer epitaxy lowers the energy of the interfacial atoms at the expense of the elastic strain energy stored in the epitaxial layer.

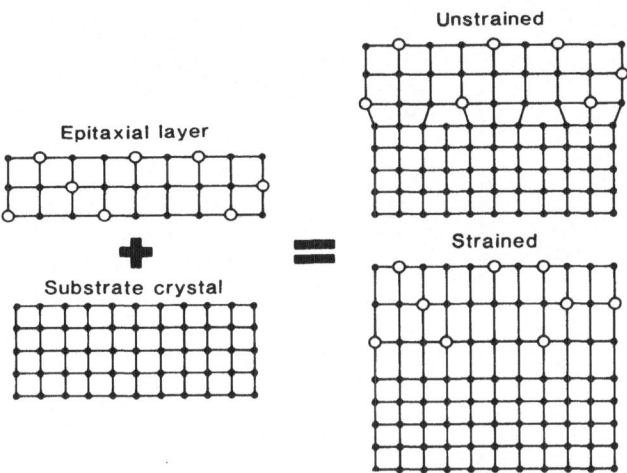

Fig. 23. Two possible modes of lattice mismatched epitaxy. Top right: in conventional epitaxy the layers are undeformed and lattice mismatch produces dislocations at the interface. Bottom right: in strained layer epitaxy the alloy layer deforms to match the atomic spacings of the substrate. After Bean[43].

As the thickness of the strained epitaxial layer increases, the cumulative strain energy builds up until at some point the interface shears, creating a large density of misfit dislocations, and the epitaxial layer reverts to it unstrained state.

The maximum thicknesses up to which strained layer epitaxy can be sustained under equilibrium conditions can be theoretically predicted [44,45]. For the heterostructure to relax from a strained to an unstrained state, bonds along the entire heterostructure interface must be broken. This represents a high-energy step and constitutes a substantial barrier to reaching equilibrium. In practice, grown-in defects and dislocations will provide a means for more gradual relaxation. Dislocations migrate and multiply under the influence of strain, and gradually convert regions of the heterostructure from a strained into an unstrained state. Therefore, to maintain the metastable strained state over the largest possible layer thickness, the epitaxial layer should be grown virtually without dislocations and at a low growth temperature where vibration-induced disorder and interdiffusion are reduced.

Two of the most studied strained layer systems are those formed by alternative epitaxial deposition of Ge_xSi_{1-x} and Si layers unto a silicon substrate[43] and by epitaxial deposition of thin layers of $Ga_xIn_{1-x}As$ on GaAs and InP substrates.

The lattice constants of Si and Ge differ by 4.2% and follow Vegard's law in the alloy. Under these conditions, the maximum critical thickness up to which epitaxy of Ge_xSi_{1-x} on Si can be maintained increases with decreasing Ge content in the alloy from about 1 µm for an alloy with x = 0.1 to less than 10 nm for an alloy with x = 0.5 (Fig. 24). The absence of dislocations can be verified by cross-sectional transmission electron microscopy in which crystals are imaged edgewise to reveal the interfaces between the layers. The absence of dislocations gives clear evidence for strained layer growth, which may be further confirmed by means of x-ray

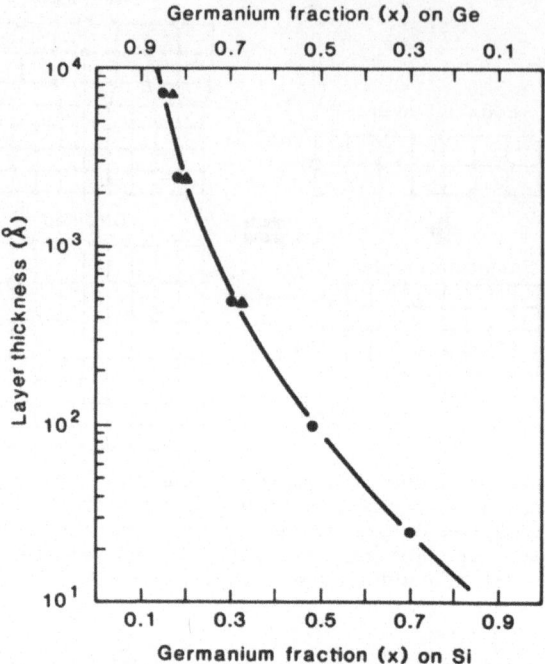

Fig. 24. Experimentally determined critical thicknesses up to which defect-
free Ge$_x$Si$_{1-x}$ epitaxial layers can be maintained on Si (bottom
scale) and Ge (top scale) substrates [ref. 43].

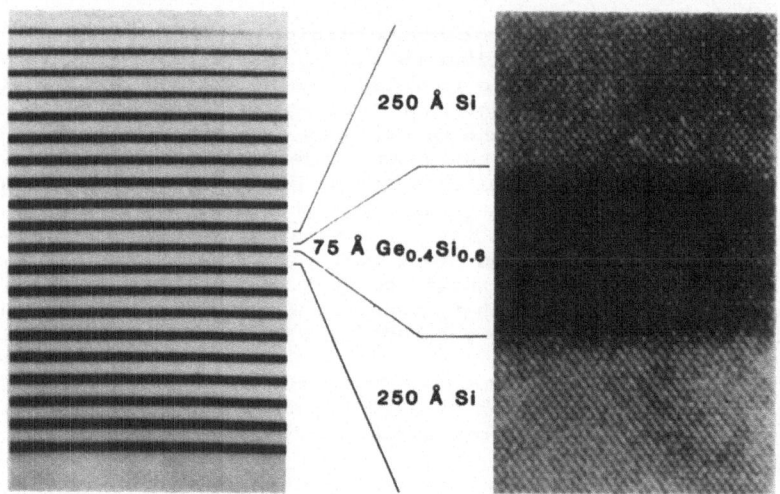

250 Å Si

75 Å Ge$_{0.4}$Si$_{0.6}$

250 Å Si

Fig. 25. Cross-sectional micrograph of a 20 period Ge$_x$Si$_{1-x}$/Si strained
layer superlattice. The enlarged image at right resolves the
individual atomic rows [ref. 43].

reflectance or ion backscattering techniques. The important result is that
the maximum critical thicknesses for strained-layer growth are much larger
than predicted by equilibrium theory, and are in a range that is sufficient
to be useful in electronic devices.

Many heterostructure devices require repeated regularly alternating strained layers to form a strained layer superlattice. In such a superlattice, made up for example of alternating Ge_xSi_{1-x} and Si layers, the Si layers practically retain their undeformed bulk structure provided they are sufficiently thicker (3 to 5 times) than the GeSi strained alloy. The strain in adjacent alloy layers is therefore decoupled, and defect-free superlattices can be designed using the same critical thickness criteria as shown in Fig. 24 for single strained layers.

Figure 25 shows a cross sectional TEM micrograph of a 20-period strained-layer GeSi/Si superlattice. The left micrograph is a conventional low-magnification view of the entire array. The right micrograph is made in phase contrast-mode where individual atomic rows and channels are imaged as dark and light dots. Both micrographs show perfect crystalline order, with abrupt planar interfaces between the layers.

In addition to its interest for understanding growth processes, the technique of strained-layer epitaxy provides a means to tailor the properties of the material such as the bandgap, its bandstructure and its optical and electronic properties within a given range to the specific application desired. Many interesting applications such as photodetectors[46] or high mobility microwave devices[47] may be expected to be developped with these materials in the near future.

6.3 GaAs on Si heteroepitaxy

The epitaxial growth of GaAs on Si is generating much interest because of the many advantages inherent to Si substrates such as larger size, lower cost, better thermal conductivity, and better mechanical strength. The promise of integrating specific GaAs functions in a compatible Si IC process sequence also appears as very attractive for the development of future mixed GaAs/Si optoelectronic integrated circuits.

Very fast progress has been achieved in recent years in the growth of device quality material with good electrical and optical properties and excellent surface morphology[48-52]. In addition, a number of test devices and circuits including bipolar and field effect transistors, junction lasers and detectors, have already been demonstrated. Despite these results, several long term aspects such as reliability and degradation under stress, which are especially important for minority carrier devices operating at high current levels, still need to be further investigated before the potential of this approach can be fully established.

When a polar compound semiconductor material such as GaAs or InP is grown on a nonpolar substrate such as Si or Ge, several new problems arise which are absent in homoepitaxial growth. They include the problem of the formation of antiphase domains on the compound side of the interface, the problem of misfit dislocation generation at the interface due to the large differences in lattice constants, and the deformations induced upon cooling because of the differences in thermal expansion coefficients of the two materials.

Antiphase disorder[50]

Antiphase disorder is characterized by a change in the sublattice occupation. A crystal such as GaAs consists of two interpenetrating face-centered cubic sublattices, which differ from each other in the spatial orientation of the four tetrahedral bonds that connect each atom to its four nearest neighbours situated on the other sublattice. As shown for example in

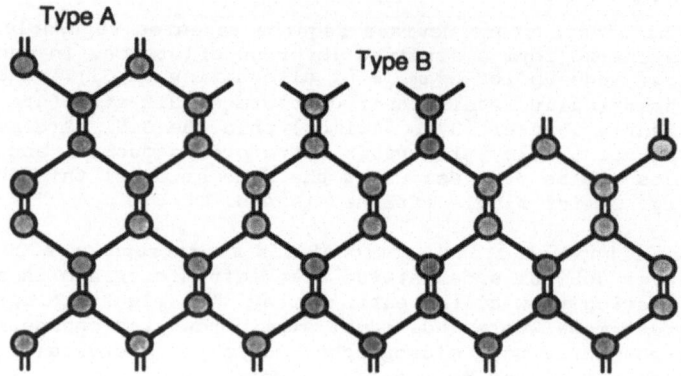

Fig. 26. Two kinds of bond configuration at [011]-oriented single height atomic steps on a (100) Si or Ge surface. For a type-A step, the dangling bonds run parallel to the step, while for the type-B step they run perpendicular. After Kroemer[50].

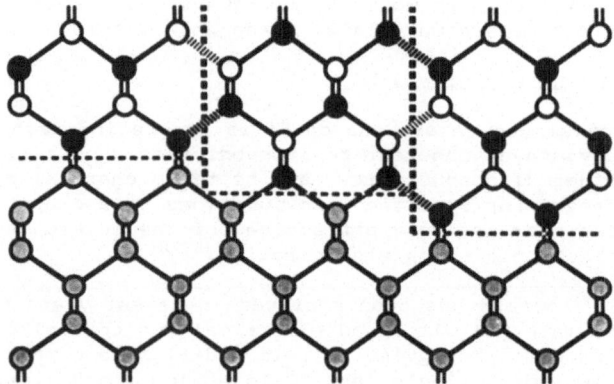

Fig. 27. Mechanism of APB formation during polar on non-polar growth due to the presence of single height steps on the substrate [ref. 50].

figure 26, for the atoms with the bond orientation indicated by A the dangling bonds lie in the plane perpendicular to the plane of the figure, while for the atoms labeled B the dangling bonds extend outwards left and right in the plane of the figure.

The presence of antiphase disorder manifests itself in the occurence of domains where the Ga and As atomic positions are inverted. The interface between domains with opposite sublattice allocation then forms a two-dimensional structural defect which is called an antiphase boundary (APB). Since Ga-Ga bonds acts as acceptors and As-As bonds act as donors, such an APB constitutes a highly compensated doping sheet, which is bound to have a deleterious effect on the electrical properties of the layers.

In a simple model of GaAs growth on Si or Ge by MBE, one would expect the first atomic layer bound to the surface to be an As layer due to the stronger chemical affinity of As for Si as compared to As to Ga. An antiphase boundary is thus expected to be generated at every single-height step on the substrate surface. Conversely, the occurence of antiphase disorder during growth on the (100) surface can be avoided if the silicon surface contains only double height atomic steps. For reasons that are not

390

yet quite understood, experimental evidence indicates that perfectly double stepped (100) surfaces with dangling bonds parallel to the step edge (type A in Fig. 26) can can be obtained by vacuum annealing following careful chemical etching of the silicon substrate. This indicates that the formation of type A double-height steps is energetically favorable compared to type B or single height steps due to the occurence of surface reconstruction.

Misfit dislocations and thermal expansion

Misfit dislocations form during growth because of the lattice mismatch between the epitaxial film and the substrate. In addition to these grown-in dislocations, the existence of a large difference in thermal expansion coefficients (approximately 59% larger for GaAs than for Si) causes tensile stress to develop in the epitaxial layer during cool-down. This tensile stress partially relaxes during cooling by the formation of additional misfit dislocations at the interface. Above a given critical layer thickness, however, these relaxation mechanisms are insufficient and macroscopic cracks develop in the epitaxial layer.

Several techniques have been developed to reduce the density of misfit dislocations in the epitaxial overgrowth. They include the use of tilted substrates, where the angle of tilt is adjusted so as to favor the formation of misfit dislocations lying in the plane of the interface, and the use of the strain field of a pseudomorphic superlattice to bend the threading dislocations that escape from the interface and confine them in the plane of the superlattice layer.

The edge dislocations generated during growth are predominantly of 2 types: The so-called type I dislocation has its Burgers vector in the plane of the nominal substrate orientation, while the second type II dislocation has its Burgers vector inclined at 45° with respect to the substrate orientation. The type I dislocation is preferred because it is more effective in accomodating the misfit and because the dislocation lines run along the substrate-epi interface and do not propagate in the epitaxial overgrowth so that only the interfacial region is degraded.

The role of steps in tilted substrates is thus to act as nucleation centers which serve to preferentially induce the type I dislocation. Since one in-plane dislocation is required every 25 atomic rows to accomodate the 4.2% misfit between GaAs and Si, a substrate tilt of approximately 2.3° which provides a minimum of one step every 25 planes should be sufficient. In practice, larger tilt angles, usually 3 to 4°, are used. As the steps on the tilted orientations of the (100) plane tend to run preferentially along the [011] direction, dislocations will be principally aligned along this direction when the substrate is tilted from (100) towards the (011) plane (Fig. 28-a). Alternatively, a tilt towards the [001] direction can be used, in which case steps would run along the [011] and [01$\bar{1}$] directions (Fig. 28-b), thereby further increasing the density of nucleation sites.

A typical growth procedure runs as follows. Substrates oriented 3 to 4° off (100) are chemically etched to leave a final oxide surface layer, which is then thermally desorbed immediatly prior to growth. Growth starts by the predeposition of a nucleating As or Ga pre-layer, followed by the growth at low temperature (400°C) of a thin 100 nm buffer layer. The buffer layer is partially relaxed during deposition, and further relaxes during heat-up and thermal anneal at the final growth temperature of around 600°C. Cross-sectional micrographs show that under these conditions a majority of the misfit dislocations terminate at the surface of the first buffer layer and do not propagate into the final overgrowth. Optimum results reported

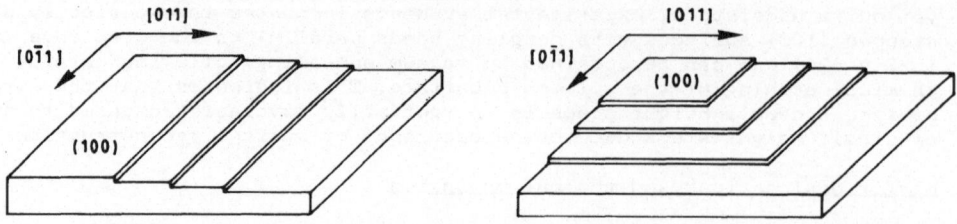

Fig. 28. Schematic diagrams showing the effect of tilting a (100) substrate surface (a) towards [011] or (b) towards [001]. In the latter case, since steps run preferentially along [011], a two-dimensional staircase is created with ledges in both the [011] and [01$\bar{1}$] directions. After Fisher et al.[49].

indicate that dislocation densities of the order of 10^5 to 10^6 cm^{-2} in the top epitaxial layer can be obtained. Additional reductions in dislocation density can be achieved by using a strained layer superlattice layer before starting the final epitaxial deposition. Although the mechanisms involved are not entirely clear, it appears that the strain field associated with the superlattice reflects or terminates many of the threading dislocations, and thus may lead to a further reduction in the density of dislocations in the higher lying layers.

7. CONCLUSIONS

We will close this chapter by giving a summary of the principal features of the MBE growth technique in order to illustrate its principal advantages and limitations.

As strong points of this technique, we can note the following features:

(a) the growth rate is very low, in general around 1 μm per hour or approximately 1 monolayer per second. This allows changes in composition or doping to be made to within atomic dimensions, as well as extremely precise control of layer thicknesses.

(b) in comparison with other growth techniques the growth temperature is low (from 550 to 650°C for GaAs), so that interdiffusion effects are minimized.

(c) a superior surface morphology can be achieved, with extended atomically flat regions. This allows the fabrication of structures and superlattices with near perfect interfaces.

d) an excellent thickness uniformity (typically below 1% variation over a 5 cm diameter wafer area) can be achieved by rotating the substrate during growth.

(e) a precise control and good reproducibility of growth rates, doping levels and compositional or doping profiles can be achieved by using large capacity and well regulated effusion cells.

(f) abrupt compositional or doping variations can be realized by shuttering the effusion cells; except for certain dopants which exhibit surface accumulation effects, there is no equilibrium layer present which would exerce a retarding influence.

(g) sequential deposition of different materials is possible, for example metal films or dielectric films on semiconductors, without removing the substrate from vacuum and thus maintaining the chemical integrity of the various interfaces.

(h) geometrical control of layer growth in 2 dimensions is possible through the use of masks.

(i) regrowth over non-planar surfaces is possible without altering the surface structure. This is an important feature, for example, in the case of regrowth over periodic grating corrugations used in distributed feedback lasers.

(j) a wide range of surface analytical probes can be installed in the growth system so that the chemical and structural properties of the films can be monitored before and during growth. For example, layer-by-layer monitoring of the deposition sequence is possible by following the variations of the RHEED intensities in real time during deposition.

(k) a high degree of process automation is possible.

A number of difficult problems remain however which need to be overcome before the MBE technique can fully qualify as a production tool. These include the elimination of various surface defects created during growth, the origin of which is still a subject of some controversy, and the achievement of the precise composition control needed for growth of lattice-matched heterojunctions in systems such as GaInAs/InP or GaInPAs/InP

At present, it appears that solutions to these problems are at hand by the introduction of gas source MBE systems where a precise control of the group III and group V fluxes is possible by controlling the gas flows into the system. The replacement of solid Ga sources by organometallic Ga gaseous compounds also has been shown to eliminate those defects related to Ga spitting from the effusion cells. Excellent results have already been demonstrated in the laboratory for the growth of ternary and quaternary phosphorous-based materials using gas-source or chemical beam epitaxy, which indicate that the gas source MBE techniques will also play a major role in the development of optoelectronic devices in this materials family.

Parallel to the technological advances realized in MBE, the other vapor phase epitaxy methods, in particular the low-pressure and atmospheric pressure metalorganic chemical vapor deposition (MO-CVD) method[53], also underwent a spectacular development in the past few years. In several aspects, such as layer purity, luminescent properties, and interface abruptness, these techniques have equalled the results achieved with MBE. The understanding and control of the chemical reactions in the gas phase, and the achievement of very uniform deposition over large area wafers, are however more difficult with the vapor phase techniques, and further work will be needed before a definite evaluation of the respective advantages and disadvantages of each approach can be established.

In comparison with these other techniques, we believe that MBE is capable of greater control over the growth of thin layers and variations in alloy composition, and will excel as well in the ability to precisely tailor doping profiles. The thinnest superlattice structures will ultimately probably be best produced by MBE rather than by vapor phase deposition. Similarly, for high speed logic circuits, where device performance and uniformity of device characteristics over the surface of the wafer are the determinig factors, MBE also appears as a stronger candidate. These inherent features of the MBE technique thus make it well suited to the research environment and to the demonstration of new concepts involving thin heterojunction layers and abrupt interfaces. Against these advantages one should consider the higher costs in terms of capital investment, the lower throughput with present design systems, and the relative complexity of the MBE apparatus and growth procedures. For these reasons, the alternative MO-CVD vapor phase deposition techniques will continue to develop side by-side with MBE in the future, with each technique probably finding its own specific application niche for industrial device and circuit fabrication.

REFERENCES

1. A. Y. Cho and J.R. Arthur, Molecular beam epitaxy, in "Progress in Solid State Chemistry," volume 10, E.H.J. McCaldin, editor (Pergamon Press, New York 1975), pp. 157-191

2. C. T. Foxon and B.A. Joyce, Fundamental aspects of molecular beam epitaxy, in "Current Topics in Materials Science," E. Kaldis, editor (North Holland, Amsterdam, 1981), vol. 7, pp. 1-68

3. K. Ploog, Molecular beam epitaxy of 3-5 compounds, in "Crystals: Growth," Properties and Applications (Springer-Verlag, 1980) pp. 75-162

4. L. L. Chang and K. Ploog, editors, "Molecular Beam Epitaxy and Heterostructures," Nato Advanced Science Institutes Series, vol. E 87 (Nyhoff, Dordrecht, 1985)

5. E. H.C. Parker, editor, "The Technology and Physics of Molecular Beam Epitaxy" (Plenum Press, New York, 1985)

6. J. C. Bean, J. Crystal Growth 81, 411-420 (1987)

7. Y. Shiraki, "Silicon molecular beam deposition," ref. 5, pp. 345-386

8. W. T. Tsang, J. Appl. Phys. 60, 4182-4185 (1986)

9. M. B. Panish, Progress in Crystal Growth and Characterization 12, 1-28 (1986)

10. A. Y. Cho, "Growth and properties of 3-5 semiconductors by molecular beam epitaxy," ref. 4, pp. 191-226

11. N. Pütz, H. Heinecke, M. Weyers, M. Heyen, H. Lüth and P. Balk, J. Crystal Growth 74, 292 (1986)

12. D. Huet and M. Lambert, J. Electron. Mat. 15, 37-40 (1986)

13. W. T. Tsang, J. Electron. Mat. 15, 235-245 (1986)

14. M. B. Panish, J. Crystal Growth 81, 249-260 (1987)

15. W. T. Tsang, J. Crystal Growth 81, 261-269 (1987)

16. R. Heckingbottom, "The application of thermodynamics to MBE," ref. 4, pp. 71-104

17. J. R. Arthur, J. Appl. Phys. 39, 4032 (1968)

18. J. C. Bean and P. Butcher, "Proc. 1st Int. Symp. on Si MBE" (The Elec-trochem. Soc., Pennington, New Jersey, 1985), pp. 429-437

19. M. Tabe, J. Vac. Sci. Technol. B3, 975-980 (1985)

20. A. Y. Cho, "Introduction to MBE," ref. 5, pp. 1-13

21. R. Ludeke, R.M. King, and E.H.C. Parker, MBE Surface and Interface Studies, ref. 5, pp. 555-628

22. J. R. Arthur, Surface Science A 43, 449 (1974)

23. J. H. Neave, B.A. Joyce, P.J. Dobson, and N. Norton, Appl. Phys. A 31, 1 (1983)

24. M. Ilegems, "Properties of 3-5 layers," ref. 5, pp. 83-142

25. J. L. Lievin and F. Alexandre, Electron. Lett. 21, 413 (1985)

26. R. Nottenburg, H.J. Bühlmann, M. Frei, and M. Ilegems, Appl. Phys. Lett. 44, 71 (1984)

27. E. C. Larkins, E.S. Hellman, D.G. Schlom, J.S. Harris, M.H. Kim, and G.E. Stillman, J. Crystal Growth 81, 344-348 (1987)

28. C. E.C. Wood, "Dopant incorporation, Characteristics, and Behavior," ref. 5, pp. 61-82

29. J. J. Harris, "3-5 Microwave devices," ref. 5, pp 425-465

30. L. Esaki, "Compositional Superlattices," ref. 5, pp. 143-184

31. L. Esaki, "Semiconductor superlattices and Quantum wells through development of MBE," ref. 4, pp. 1-36

32. K. Ploog, G.H. Döhler, "Compositional and doping superlattices in III-V semiconductors," Adv. in Physics 32, 285-359 (1983)

33. P. A. Buffat, P. Stadelman, J.D. Ganière, D. Martin and F.K. Reinhart, Microsc.Semicond. Mat. Conf., Inst. Phys. Conf. Ser. 87, 207-212 (1987)

34. G. H. Döhler, "Doping Superlattices," ref. 5, pp. 233-274

35. R. Dingle, in "Advances in Solid State Physics/Festkörperprobleme," H.J. Queisser, ed., (Pergamon/Vieweg, Braunschweig, 1975), vol. 15, pp. 21-48

36. D. Delagebeaudeuf and N.T. Linh, IEEE Trans. Electron Dev. 29, 955 (1982)

37. H. Morkoc, "Modulation doped AlGaAs/GaAs field effect transistors," ref. 4, pp. 625-676

38. D. C. Tsui, H.L. Störmer, and A.C. Gossard, Phys. Rev. Lett. 48, 1562 (1982)

39. J. H. English, A.C. Gossard, H.L. Störmer, and K.W. Baldwin, Appl. Phys. Lett. 50, 1826 (1987).

40. K. Lee, M.S. Shur, T.J. Drummond, and H. Morkoc, J. Appl. Phys. 54, 6432 (1983)

41. W. T. Tsang and M. Ilegems, Appl. Phys. Lett. 31, 301 (1977)

42. G. C. Osbourn, J. Vac. Sci. Technol. B4, 1423 (1986)

43. J. C. Bean, Science 230, 127-131 (1985)

44. J. H. van der Merwe, J. Appl. Phys. 34, 123 (1962)

45. J. H. van der Merwe and C.A. Ball, in "Epitaxial Growth," J.W. Matthews, editor (Academic, New York, 1985) chapter 6

46. H. Temkin, J.C. Bean, T.P. Pearsall, N.A. Olsson and D.V. Lang, Appl. Phys. Lett. 49, 155 (1986).

47. H. Jorke and H.J. Herzog, J. Electrochem. Soc. 133, 998 (1986)

48. W. I. Wang, Appl. Phys. Lett. 44, 1149 (1984)

49. R. Fisher, H. Morkoc, D.A. Neumann, H. Zabel, C. Choi, N. Otsuka, M. Longerbone, L.P. Erickson, J. Appl. Phys. 60, 1640 (1986)

50. H. Kroemer, J. Crystal Growth 81, 193 (1987)

51. H. Shichijo, J.W. Lee, W.V. McLevige and A.H. Taddiken, IEEE El. Dev. Lett EDL-8, 121-123 (1987)

52. T. C. Chong and C.G. Fonstad, Appl. Phys. Lett. 51, 221-223 (1987)

53. G. B. Stringfellow, Technologies based on Organometallic Vapor Phase Epitaxy, in "Crystal Growth in Science and Technology," H. Arend, editor, lecture notes of the 13th International School of Crystallography, Erice, 1987 (see also this book).

LIQUID PHASE EPITAXY OF GARNETS

W. Tolksdorf

Philips GmbH Forschungslaboratorium Hamburg
Vogt-Kölln-Str. 30
D-2000 Hamburg 54, FRG

INTRODUCTION

Rare earth (R) iron garnets ($R_3Fe_5O_{12}$) have excellent properties for magnetic, magneto-optic and microwave applications. For magnetic bubble memory devices requiring single crystal films of several micrometer thickness an epitaxial technique had to be developed. The know-how concerning congruently melting non-magnetic garnet crystal growth from the melt for laser applications was used to grow gallium gadolinium garnet as suitable substrate crystals. On wafers of these crystals single crystal layers can be grown by liquid phase epitaxy (LPE) from very diluted high temperature solutions known from the flux growth of yttrium iron garnet (YIG) for microwave applications. Substrate crystals with garnet structure having a lattice constant in the range from 1.22 nm to 1.26 nm are now available [1]. This allows the growth of magnetic garnet layers with a wide variety of compositions to tailor the properties for various applications. LPE is a well suited process to study crystal growth phenomena, too. Some advantageous features of garnet LPE with respect to crystal growth will be discussed. Occasionally comparison to LPE of semiconductors is made.

SUBSTRATE

For the spontaneous formation of three-dimensional nuclei in a solution a certain supersaturation, respectively supercooling, is necessary to gain the nucleation energy. However, to continue the growth only the lower formation energy of two-dimensional nuclei has to be raised. That means the growth can proceed at supersaturations not allowing the formation of spontaneous nuclei. Such supersaturated solutions are used in LPE. The presence of an appropriate substrate surface allows the growth of layers.

LPE of garnets is mostly used for the growth of magnetic (actually ferrimagnetic) rare earth iron garnets on non-magnetic (actually dia- or paramagnetic) rare earth gallium garnets. Thus the structure of the subtrate and that of the growing garnet layer are identical, whereas the cation compositions are different.

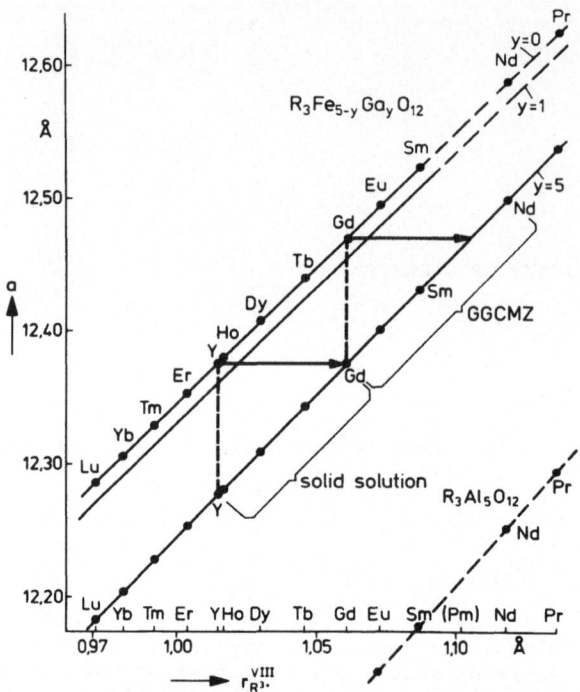

Fig. 1. Lattice constants of rare earth garnets.

A detailed review about the crystal chemistry of garnets has been published in [2], a review about the cubic structure in [3] and a monography about magnetic garnets has been published in [4]. The lattice constants of oxide garnets varies systematically with the substitution and can be calculated [5] from ionic radii. Figure 1 shows the relation of the lattice constants of rare earth iron, gallium and aluminium garnets as a function of the rare earth cation radii with coordination number VIII. From Figure 1 it is clear that gadolinium gallium garnet (GGG) is a good substrate for YIG. The misfit is defined as $\Delta a = a_s - a_f$ where a_s is the lattice constant of the substrate and a_f that of the film. Coherent growth of the epitaxial layer on the substrate is assumed [6]. Because of the high formation energy of dislocations no misfit dislocations are observed contrary to semiconductor epitaxy. Thus the layer remains strained as long as it is thin compared with the mostly 0.5 mm thick substrate. The effective misfit $\Delta a^{\perp} = a_s - a_f^{\perp} = 1.86\ \Delta a$ where $a_f^{\perp}$ is the strained lattice constant of the film in the growth direction [7]. These relations are compiled in Figure 2.

The thermal expansion coefficients of iron garnets are larger than those of gallium garnets. Thus, LPE is always performed with $\Delta a^{\perp} < 0$ that is under compressive stress [8]. The lattice constant of YIG and GGG at room temperature are 1.2376 nm [5]. YIG increases at 900°C to 1.2490 nm, GGG to 1.2472 nm, thus $\Delta a = -1.8\cdot10^{-3}$ nm. To grow 5 μm thick crack-free layers of substituted YIG on GGG the maximum relative strained misfit is: $\Delta a^{\perp}/a < +1.2\cdot10^{-3}$ at 20°C for tensile stress, changing to $-1.2\cdot10^{-3}$ at 900°C (compressive stress) and $\Delta a^{\perp}/a > -4.8\cdot10^{-3}$ at 20°C for compressive stress changing to $-7.2\cdot10^{-3}$ at 900°C.

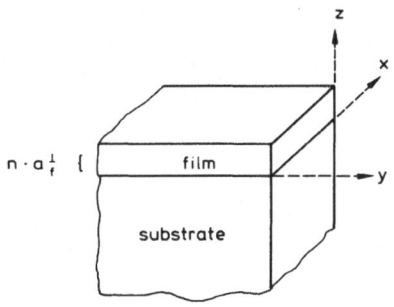

n = numbers of lattice planes

lattice parameters:

a_s : substrate (strainfree)

a_f : film (strainfree)

$a_f^\perp$: film in growth

direction z (strained)

$n \cdot a_f \ll$ thickness of substrate

$a_s = a_f^\bullet$ condition of coherence

effective misfit: $\Delta a^\perp = a_s - a_f^\perp$

misfit (strain free garnets): $\Delta a = a_s - a_f$

$$\Delta a^\perp = \frac{1+\mu}{1-\mu} \cdot \Delta a = 1.86 \, \Delta a$$

$\mu = 0.30$ Poisson number

Fig. 2. Epitaxy conditions and misfit for garnets.

MELT COMPOSITION

Nearly exclusively very diluted melts based on PbO and B_2O_3 are used with a Pb/B ratio ranging from 4 to 12, for GGG even melts free of boron oxide were used [9]. For crystallization of YIG a ratio of Fe/Y > 12 is necessary [10,11], GGG can be grown with a ratio Ga/Gd = 5/3 which corresponds to the chemical formula [9]. Table 1 gives examples for some melt compositions. Properties of the melts are discussed in [12–17]. A solubility model was proposed by van Erk [18] and studied for YIG [19], Bi substituted YIG [20,21] and for GGG [9,22]. For this "ion fluid model", that accounts for the different YIG components (Y^{3+}, Fe^{3+}, O^{z-}) being dissolved separately, the following equation can be formulated:

$$LnL = -\frac{\Delta H}{RT_s} + lnL_o$$

where $L = [Y]^3 \cdot [Fe]^5$, [Y] and [F] are cation fractions, the number of cations of Y and Fe divided by the numbers of all cations in the solution (Pb, B, Y, Fe). Not only the quantities of the garnet constituents in the melt determine the film composition but also the solvent components, e.g., the relation Pb/B, and the growth kinetics. The choice of the melt composition allows the realization of a large variety of layer

Table 1. Melt Compositions: Cation Fraction in %

Refs	Y [19]	Y	Y	YB [21]	YE [23]	YA [24]	GG [9]	GG [9]
Pb	73.85	76.21	70.30	70.52	73.66	80.23	77.80	75.74
Bi				7.32				
B	10.38	10.76	9.90	9.96	9.46	13.32	13.32	12.88
Fe	14.69	12.61	19.10	11.68	15.63	5.09	5.55	10.80
Y	1.08	0.42	0.70	0.52	1.25	1.36	3.32	1.08
T	930	802	922	1073	930	1090	977	957

T_s: saturation temperature (°C)
[1] Al; [2] Ga; [3] Gd; [4] Eu = 0.46, Yb = 0.79

compositions. Small amounts of substituents can be added readily from run to run. By an appropriate choice of the substrate the misfit Δa can be minimized. Thus series with increasing amount of substituents can be grown to study the physical properties of the layers [20,25-30].

GROWTH PROCEDURE

Since supersaturation of the melt can be maintained for a rather long time, LPE is preferably applied as an isothermal process using a large melt volume, contrary to semiconductor LPE [31]. For melt Y_o at a supercooling ΔT of 10 K the supersaturation C corresponds to 120 mg YIG for 100 g melt. The weight of a 5 μm thick YIG layer on both sides of a substrate with 3 cm in diameter is 36 mg, using ΔT of 15 K from kg of melt

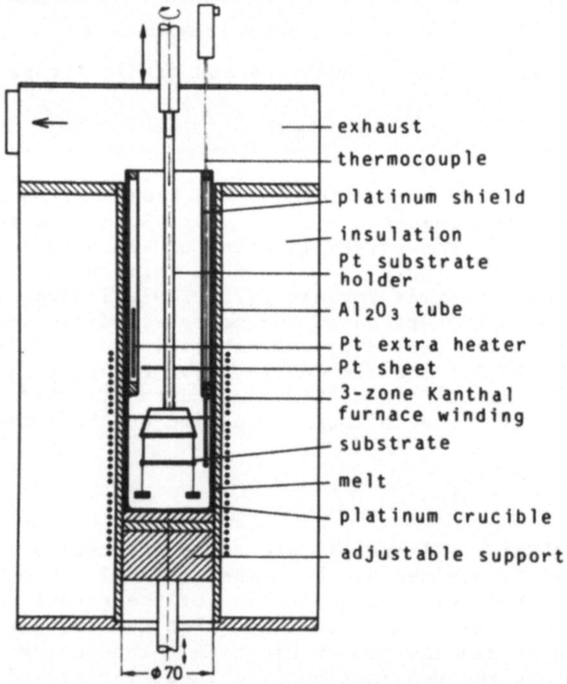
exhaust
thermocouple
platinum shield
insulation
Pt substrate holder
Al_2O_3 tube
Pt extra heater
Pt sheet
3-zone Kanthal furnace winding
substrate
melt
platinum crucible
adjustable support
Φ70

Fig. 3. Scheme of the LPE arrangement.

Y_0 the change of supersaturation $d\Delta C$ is less than 1% of ΔC. The melt of 2 kg has a volume of about 250 cm³ which is quite a handy size for research investigations. For better reproducibility a larger batch size of, for example, 1250 cm³ (about 10 kg) is required [25]. Figure 3 shows a scheme of the LPE arrangement and Figure 4 a scheme of the platinum crucible with the substrate in growth position as it is used in the Philips Research Laboratory in Hamburg. The crucible is made of platinum, the material having the best resistance against the PbO melt. A photograph of a Pt-crucible for 2 kg melt together with a substrate holder is shown in Figure 5. The crucible has a diameter of 70 mm, a height of 100 mm, a wall thickness of 1 mm, and a capacity of 2 kg melt. It has on top a recess of 10 mm tightly welded with a Pt-pipe of 200 mm length, wall thickness 0.3mm, to protect the ceramic parts of the furnace from the aggressive vapor of the melt. With an additional heater on top of the recess together with the three heating zones of the furnace isothermal conditions ($|\Delta T| < 0.5$ K) are established within the melt during growth.

A horizontal mode of dipping is applied for LPE: after homogenization for some hours above saturation temperature, the melt is cooled to growth temperature and the substrate is brought into a horizontal position 10 mm above the melt surface to adapt ambient temperature. In this position the

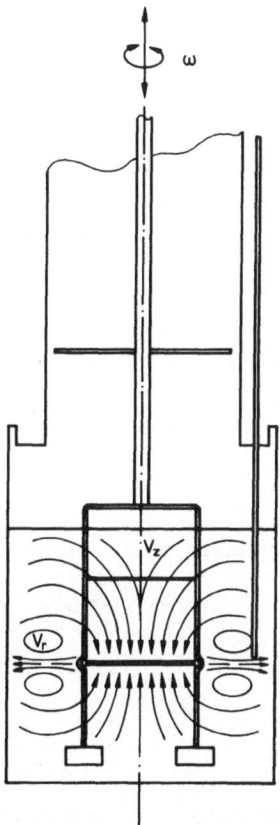

Fig. 4. Scheme of the crucible with the substrate holder in growth
         position. The arrows mark the convective flow due to oscillatory
         substrate rotation.

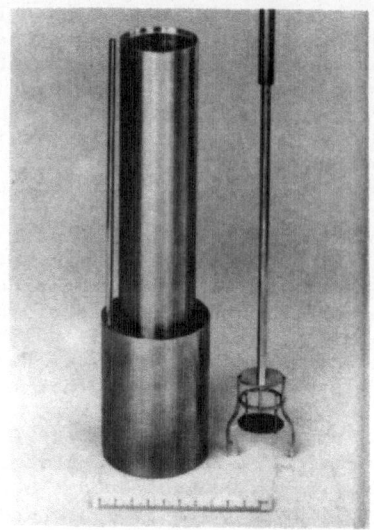

Fig. 5.   Photograph of the platin crucible and substrate holder.

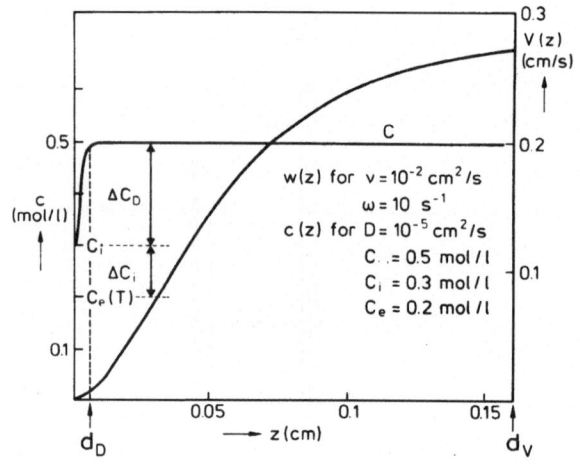

boundary layer:	diffusion	convection

$$\Delta C_i = C - C_e\,(T) - \Delta C_D\,(\omega)$$

$$\Delta C_D = C - C_i \quad \text{diffusion}$$

$$\Delta C_i = C_i - C_e \quad \text{integration}$$

$$\Delta T: \qquad \overline{\Delta C = C - C_e}$$

supercooling   supersaturation

Fig. 6.   Diffusion and convection boundary layer thickness and their relation to supersaturation.

stirring arms of the holder already are in the melt. Thus the heat flow within the melt is not changed when the substrate is lowered into growth position 30 to 40 mm below the melt surface. After growth the substrate is rapidly removed from the melt and rotated with 800 min^{-1} to spin off adhering residual amount of the melt. Especially with Bi containing melts the surface of the film has to be cleaned, moreover, by a hot dilute mixture of acetic and nitric acid after the sample has been taken out of the furnace [20,32].

GROWTH KINETICS

The rotating substrate induces controlled convection in the melt. A detailed description of this phenomenon is given in [33,34,35]. At the solid-liquid interface the melt is carried circumferentially by the substrate due to adhesion and friction and is thrown outwards due to centrifugal forces. This mass flow is compensated by a flow in axial direction towards the rotating substrate surface. Thus we have a fully three-dimensional mass flow, i.e., there exist velocity components circumferentially about the rotation axis (V_c), in the radial direction (V_r) and in the axial direction z (V_z). The flow direction of V_r and V_z is indicated in Figure 4. V_z is independent of the distance from the rotation axis and it decreases near the solid-liquid interface as shown in Figure 6. The distance from the interface at which V_z reaches 99% of the maximum value is defined as the convection boundary layer thickness (d_v) and depends on the square root ratio of the kinematic viscosity and the angular velocity as given in Eq. (1) of Table 2. The diffusion boundary layer thickness (d_D) given in Eq. (2) of Table 2 describes the distance from the interface at which the mass transport by convection (V_z) becomes small as compared to the mass transport by diffusion. The thickness of

Table 2. Diffusion d_D and Convection Boundary d_V Layer Thickness

$$d_V = 3.6 \ \nu^{1/2} \cdot \omega^{-1/2} \tag{1}$$

$$d_D = 1.61 \cdot D^{1/3} \cdot \nu^{1/6} \cdot \omega^{-1/2} \tag{2}$$

[16,23,33,34,36]

$Y_3Fe_5O_{12}$
LPE
900°C

$\left\{ \begin{array}{l} D(T) \approx 1 \cdot 10^{-5} [cm^2 \cdot s^{-1}] \text{ diffusion-coefficient} \\ \\ \nu(T) \approx 1 \cdot 10^{-2} [cm^2 \cdot s^{-1}] \text{ kinematic viscosity} \end{array} \right.$

	d_D [μm]	ω [min^{-1}]	ω [rad s^{-1}]
	79	40	4.2
$\dfrac{d_D}{d_V} = 0.04$	64	60	6.3
	50	100	10.5
	39	160	16.8

Relaxation time:

$d_V : \tau_1 \lessgtr \dfrac{\pi}{\omega} < 1$ rotation

$\quad \tau_1 \lesssim 0.3$ [s]

$d_D : \tau_2 \approx 3[s] \approx 10 \cdot \tau_1$

$\quad \tau_3 \approx 14$ [s] volume in 4 cm distance

Table 3

min^{-1}	10^{-6} $\frac{f}{cms^{-1}}$	Nu
40	2.17	1.8
90	2.81	1.2
160	3.15	0.9
ω-->	6.06	

both boundary layers is depending on ω. Assuming a diffusion coefficient and a kinematic viscosity [16] as given in Table 2, d_D can be calculated. The resulting d_D values are compiled in Table 2. When D(T) is one order of magnitude smaller, the values of d_D decreases by a factor of 2. The relaxation time (τ_1) for the hydrodynamic boundary layer (d_v) is short as compared to the relaxation time (τ_2) of the diffusion layer (d_D) [37]. Brice [38] estimated τ_1 for garnet melts to be less than one revolution, that is about 0.3 s for $\omega = 10$ rads^{-1} ($\omega \approx 100$ min^{-1}), τ_2 about a factor of 10 larger (about 3 s) and it will take 14 s (τ_3) to rotate the melt in 4 cm distance from the rotating interface. An experiment to test some of these assumptions is shown in Figure 7. From melt Y_O of Table 1 a layer is grown with $\Delta T = 50$ K and the lead incorporation (x) is measured as a depth profile by a secondary ion mass spectrometer (SIMS) [32]. For calibration of the SIMS signal electron probe analysis (EPMA) is used [39]. The first part of the layer (L_2) is grown with $\omega = 40$ min^{-1} with 2 s intersection with $\omega = 160$ min^{-1} and the second part (L_1) is grown with $\omega = 160$ min^{-1}, x follows the growth rate, the average growth rate for the applied ω is known from parallel experiments.

The supersaturation ΔC is the difference of the starting concentration C of the melt and the saturation concentration C_e at growth temperature. As a first approximation C and C_e can be considered as the yttrium cation concentration since a considerable surplus of iron cations is in the melt. Due to the growth process depletion of yttrium occurs at the interface and the growth rate decreases as indicated by the Pb concentration in this "transient" layer [33]. Steady state growth is

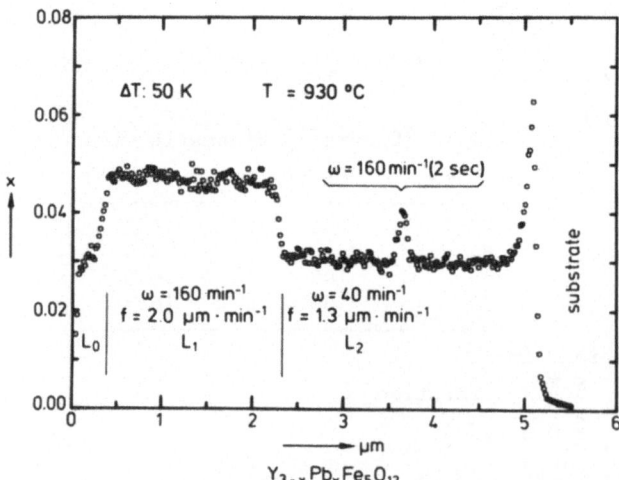

$$Y_{3-x}Pb_xFe_5O_{12}$$

Fig. 7. SIMS lead depth profile of a YIG layer grown with different rotation rates ω from melt Y_o.

404

established with the condition $\omega = 40$ min^{-1} as indicated in Figure 6. The driving force is now the interface supersaturation ΔC_i which is obtained from the difference between the interface concentration (C_i) and C_e. With increasing ω the diffusion boundary layer thickness (d_D) decreases and C_i increases and consequently the growth rate and x, as shown in Figure 7 at the transient zone from L_2 to L_1. Since in the first part of the layer the x value has reached already 0.41 within 2 s by changing from $\doteq 40$ $\omega = 40$ min^{-1} to $\omega = 160$ min^{-1}, it can be assumed that the time for changing steady state growth conditions from the lower to the higher rotation rate is not much longer than 3 s as estimated [38]. To prevent the disturbance from τ_3 the direction of rotation is reversed after some revolutions. Since τ_1 is smaller than one revolution, V_z and, therefore, also d_D is practically not effected. The decrease of x in the last part L_0 of the layer (top layer) is caused by undefined growth conditions, when the rotation is stopped and the sample is withdrawn from the melt. This top layer must be etched off for EPMA.

The easiest way to change the growth rate f is to increase the undercooling ΔT by lowering the growth temperature T_g thus decreasing C_e. When plotting $f(T_g)$ versus ΔT or versus T_g for a given rotation rate ω, the resulting relation is linear for smaller values of ΔT. However, at larger values of ΔT, $f(T_g)$ is sublinear, i.e., it reaches a maximum and then even decreases [16,19,21,40,41]. Excluding competitive, spontaneous nucleation this is explained rather speculatively by a decrease of the interface reaction rate, the change of the diffusion coefficient D and, with a weak influence only, the change of the viscosity. Independent of these influences the growth rate can be increased by increasing supersaturation. Since there is mostly a high surplus of iron ions in the melt composition, the rare earth ions concentration, e.g., from the melts of Table 1, can be used as a good approximation. Then C_e is the equilibrium concentration at T_s. By adding rare earth oxide stepwise to the melt, supersaturation $\Delta C = C - C_e$ is achieved and growth is performed at constant temperature (T_s of the starting composition). In Figure 8 the growth rates $f(C)$ for $\omega = 90$ min^{-1} plotted versus the relative supersaturation $\Delta C/C_e$ show a linear relation for melt Y_1 at 802°C and for melt Y_2 at 922°C whereas for melt Y_0: $f(T_g)$ versus the calculated values of $\Delta C/C_e$ is not linear. The undercooling ΔT, for melt Y_0 the growth temperature T_g, too, are given in a separate scale in Figure 8.

For (111) orientation a linear relation is found for f^{-1} at constant T_g and constant ΔC versus $\omega^{-1/2}$ [e.g., 14,19]. The slope and intercept of the linear regression straight lines $f^{-1} = f_0^{-1}(1 + Nu_1 \cdot \omega^{-1/2})$ might be used to separate two kinetic resistances: R_i being a sort of "interface resistance" to growth and R_D a "diffusional resistance", the relation R_D/R_i = Nu = $Nu_1 \cdot \omega^{-1/2}$, where Nu is the Nusselt number and the intercept f_0^{-1} is the growth rate with $\omega \longrightarrow \infty$ and $d_D \longrightarrow 0$ (see, for example, [16,19,42]).

For melt Y_0 and $\Delta T = 50$ K the growth rate $f(\omega)$ increases and Nu decreases with ω (see Table 3). So even for the high rotation rate $\omega = 160$ min^{-1} the growth rate is about 0.5 of the growth rate f_0 not limited by diffusion. Nu indicates that the two "resistances" are of the same order of magnitude.

The growth rate is dependent on the orientation (hkl). Very quick information about this dependence is achieved by using spheres as substrates [43]. Since no well defined forced convection can be applied $\Delta C_i/C_e$ is rather small even for high values of ΔT and thus this method is very sensitive to small morphological differences but it gives only qualitative results. Figure 9 shows the growth rate of (111) and (110) wafers in relation to $\Delta C/C_e$ for melt Y_1 [21]. The extrapolated (to $\omega = \infty$) growth rate f_0 is given as a solid line for (111) and (100) orientation

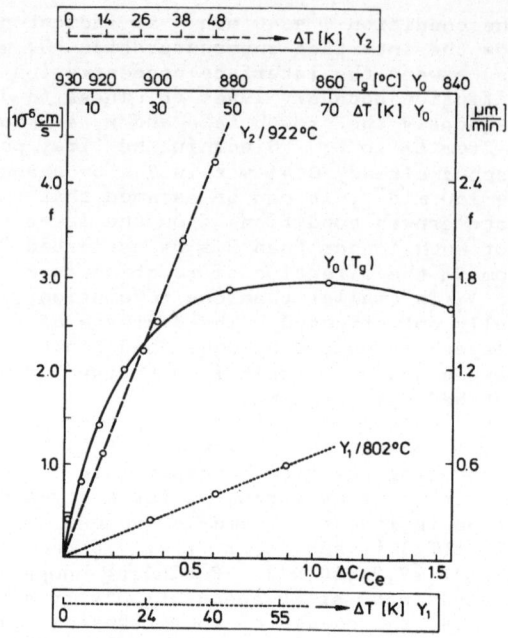

Fig. 8. Growth rate dependence of supersaturation with $\omega = 90$ min^{-1}.

and a linear relation is found. With increasing bismuth content even a superlinear relation is found. YIG on (111) substrates is growing atomically rough, thus continuous growth without a two-dimensional nucleation barrier is observed. YIG on (110) substrates is growing atomically smooth, since (110) is a growth form facet. At low ΔC two-dimensional nuclei have to be formed and growth occurs via a lateral layer mechanism. This difference is clearly indicated in Figure 9 where $f(\Delta C/C)$ is given for (110) orientation ($\pm 1^0$) with $\omega = 90$ min^{-1} and a superlinear relation is found. Growth on (110) is very sensitive to minor deviations

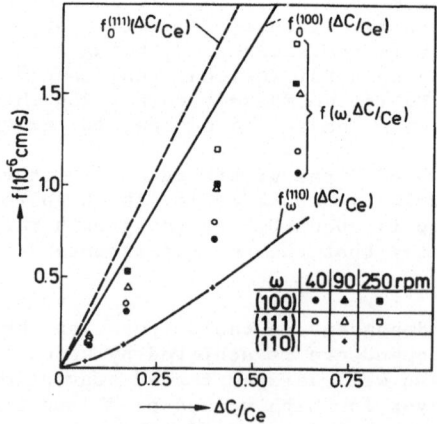

Fig. 9. Growth rate dependence of supersaturation, rotation rate ω and orientation (hkl) in melt YB_1.

from exact orientation and the study of the growth on (110) has to consider the misorientation. Detailed discussions about this are given in [16,41].

CONCLUSIONS

LPE of garnets is a typical example where the need of application has necessitated not only the improvement of the method but also has stimulated basic scientific investigations, which give general knowledge about the crystal growth process. The benefits of LPE of garnets are compiled in Table 4. For demonstration three examples of applications are given. Magneto-optical switching layers consisting of bismuth, gallium and aluminium substituted gadolinium iron garnet [19] were grown in large numbers on 3" (111) and (100) calcium, magnesium and zirconium substituted GGG substrates up to 4 pairs back-to-back in one run in 12 kg of a melt. The thickness was 5.5 µm varying only in the order of 1%. The magnetic compensation point of the three magnetic sublattices being very sensitive to compositional changes was within ± 1 K. This was made possible by using forced convection through oscillatory rotation in a large isothermal melt volume and by frequently refeeding the depleted material. The control of the interface supersaturation Δc_i by the rotation rate ω allows the growth of multi-layers, e.g., for planar magneto-optical waveguides [32,44].

The relatively thick diffusion boundary layer allows the growth on etched surfaces. Thus buried stripes for magneto-optical channel waveguides to be used, e.g., as optical isolators in communication systems, can be grown [45]. The choice of the growth temperature T_g by adjustment of the saturation concentration C allows to optimize lattice substitutions, such as by bismuth and lead both main components of the melt and allows to influence the growth induced magnetic and optical anisotropy [19,21,27,29,44]. Comparison of LPE layers with flux and melt grown crystals is possible and might lead even to an improvement of the latter.

Table 4. Liquid Phase Epitaxy of Garnets

Large Variety of Layer Compositions
   Homogeneous: Lateral and Vertical

No Temperature Gradient

Easy adjustable and constant:

growth temperature T_g
supersaturation $\Delta c = c-c_e$ $\qquad \frac{\delta \Delta c}{\Delta c} \approx 0.01$
and
supercooling $\Delta T = T_s-T_g$ $\qquad \frac{\delta \Delta T}{\Delta T} \approx 0.01$

Linear growth rate f
   free choice of orientation (hkl)

Easy change of f (Δc_i) by three parameters:

$$\Delta c_i = c - c_e(T) - \Delta c_D(\omega)$$

constant: T,ω c,ω c,T

Change of T_s T_g ω

Driving Force $\Delta T = T_s-T_g$

REFERENCES

1. D. Mateika, Substrates for epitaxial garnet layers, in: "Current Topics in Materials Science", E. Kaldis, ed., Vol. 11, North Holland, Amsterdam (1984).
2. S. Geller, Crystal chemistry of garnets, Z. Krist., 125:1 (1967).
3. W. Tolksdorf and W. Wolfmeier, in: Landolt-Börnstein, New Series Group III, Vol. 12, Part a, Springer Berlin (1978).
4. G. Winkler, "Magnetic Garnets", Vieweg, Wiesbaden (1981).
5. B. Strocka, P. Holst and W. Tolksdorf, An empirical formula for the calculation of lattice constants of oxide garnets based on substituted yttrium and gadolinium iron garnets, Philips J. Res., 33:186 (1978).
6. P. J. Besser, J. E. Mee, P. E. Elkins and D. M. Heinz, A stress model for heteroepitaxial magnetic oxide films grown by chemical vapor deposition, Mat. Res. Bull., 6:1111 (1971).
7. W. Tolksdorf, Preparation and imperfections of magnetic materials with garnet structure. Physics of Magnetic Garnets, LXX Corso Soc., Italiana di Fisica, Bologna, Italy (1978).
8. B. Knörr and W. Tolksdorf, Lattice parameter and misfits of gallium garnets and iron garnet epitaxial layers at temperatures between 294 and 1300 K, Mat. Res. Bull., 19:1507 (1984).
9. E. A. Giess, M. M. Faktor, R. Ghez and C. F. Guerci, Gadolinium gallium garnet liquid phase epitaxy and the physical chemistry of garnet molten solutions, J. Crystal Growth, 56:576 (1982).
10. S. L. Blank and J. W. Nielsen, The growth of magnetic garnets by liquid phase epitaxy, J. Crystal Growth, 17:302 (1972).
11. H. D. Jonker, Investigation of the phase diagram of the system $PbO-B_2O_3-Fe_2O_3-Y_2O_3$ for the growth of single crystals of $Y_3Fe_5O_{12}$, J. Crystal Growth, 28:231 (1975).
12. S. Knight, B. S. Hewitt, D. L. Rode and S. L. Blank, Measurement of the diffusion coefficient of rare earth species in $PbO-B_2O_3$ flux used for liquid phase epitaxial growth of magnetic garnet films, Mat. Res. Bull., 9:895 (1974).
13. J. E. Davies and E. A. Giess, The stability of the supersaturated state in isothermal fluxed melts used for magnetic garnet LPE, J. Crystal Growth, 30:295 (1975).
14. K. Fischer, D. Linzen, E. Sinn and S. Bornmann, Equilibrium reactions in oxidic high temperature solutions used for liquid phase epitaxy of garnets, J. Crystal Growth, 52:729 (1981).
15. K. Fischer and P. Görnert, On the solubility of garnets in PbO/B_2O_3 high temperature solvents, Crystal Res. & Technol., 17:775 (1982).
16. P. Görnert and F. Voigt, High temperature solution growth of garnets: theoretical models and experimental results, in: "Current Topics in Materials Science", E. Kaldis, ed., Vol. 11, North Holland, Amsterdam (1984).
17. E. Sinn, D. Linzen and K. Fischer, The importance of the oxygen ion concentration for oxidic high temperature solution, Cryst. Res. Technol., 20:965 (1985).
18. V. van Erk, A solubility model for rare earth iron garnets in a PbO/B_2O_3 solution, J. Crystal Growth, 46:539 (1979).
19. C.-P. Klages, W. Tolksdorf and G. Kumpat, The influence of excess iron oxide on the solubility of yttrium iron garnet and its growth kinetics on (111) substrates, J. Crystal Growth, 65:556 (1983).
20. C.-P. Klages and W. Tolksdorf, LPE growth of bismuth substituted gadolinium iron garnet layers: systematization of experimental results, J. Crystal Growth, 64:275 (1983).
21. C.-P. Klages and W. Tolksdorf, Segregation in garnet LPE, J. Crystal Growth, 79:110 (1986).

22. E. A. Giess, M. M. Faktor and F. C. Frank, The liquidus temperature and growth-dissolution kinetics of garnet liquid phase epitaxy, J. Crystal Growth, 46:620 (1979).

23. R. Ghez and E. A. Giess, Liquid phase epitaxial growth kinetics of magnetic garnet films grown by isothermal dipping with axial rotation, Mat. Res. Bull., 8:31 (1973).

24. J. M. Robertson, M. W. van Toll, J. P. H. Heynen, W. H. Smits and T. de Boer, Thin single crystalline phosphor layers grown by liquid phase epitaxy, Philips J. Res., 35:354 (1980).

25. P. Hansen, B. Hill and W. Tolksdorf, Optical switching with bismuth-substituted iron garnets, Phiips Techn. Rev., 41:34 (1983/84).

26. P. Hansen, K. Witter and W. Tolksdorf, Magnetic and magneto-optic properties of lead and bismuth substituted iron garnet films, Phys. Rev., B27:6608 (1983).

27. P. Hansen, K. Witter and W. Tolksdorf, Magnetic and magneto-optic properties of bismuth- and aluminium-substituted garnet films, J. Appl. Phys., 55:1052 (1984).

28. P. Röschmann and W. Tolksdorf, Epitaxial growth and annealing control of FMR properties of thick homogeneous Ga substituted yttrium iron garnet films, Mat. Res. Bull., 18:449 (1983).

29. C.-P. Klages, Site selectivity in praseodymium- and bismuth-substituted gadolinium gallium garnet epilayers, Mat. Res. Bull., 19:633 (1984).

30. T. Hibiya, Y. Morishige and J. Nakashima, Growth and characterization of liquid-phase epitaxial Bi-substituted iron garnet films for magneto-optic application, Jap. J. Appl. Phys., 24:1316 (1985).

31. H. J. Levinstein, S. Licht, R. W. Landorf and S. L. Blank, Growth of high-quality garnet thin films from supercooled melts, Appl. Phys. Letters, 19:486 (1971).

32. W. Tolksdorf, H. Dammann, E. Pross, B. Strocka, H. J. Tolle and P. Willich, Growth of yttrium iron garnet multi-layers by liquid phase epitaxy for single mode magneto-optic waveguides, J. Crystal Growth, 83:15 (1987).

33. V. G. Levich, "Physicochemical Hydrodynamics", Prentice Hall (1962).

34. H. Schlichting, "Boundary-Layer Theory", McGraw-Hill, New York (1979).

35. F. Rosenberger, "Fundamentals in Crystal Growth I", Springer, Heidelberg (1979).

36. J. A. Burton, R. C. Prim and W. P. Slichter, The distribution of solute in crystals grown from the melt: Part I, Theoretical, J. Chem. Phys., 21:1987 (1953).

37. S. Bruckenstein, M. I. Bellavance and B. Miller, The electrochemical response of a disk electrode to angular velocity steps, J. Electrochem. Soc., 120:1351 (1973).

38. J. C. Brice, Trends in liquid phase epitaxy, in: "1976 Crystal Growth and Materials", E. Kaldis and H. J. Scheel, eds., 571, North-Holland, Amsterdam (1977).

39. P. Willich, W. Tolksdorf and D. Obertop, Electronprobe microanalysis of epitaxial garnet films, J. Crystal Growth, 53:483 (1981).

40. R. Ghez and E. A. Giess, The temperature dependence of garnet liquid phase epitaxial growth kinetics, J. Crystal Growth, 27:221 (1974).

41. E. A. Giess and R. Ghez, Liquid phase epitaxy, in: "Epitaxial Growth, Part A", 183, Academic Press, New York (1975).

42. B. van der Hoek and W. van Erk, The interfacial and volume transport processes during LPE growth of garnets, J. Crystal Growth, 58:537 (1982).

43. W. Tolksdorf and I. Bartels, Facet formation of yttrium iron garnet layers grown epitaxially on spheres, J. Crystal Growth, 54:417 (1981).

44. H. Dammann, E. Pross, G. Rabe, W. Tolksdorf and M. Zinke, Phase matching in symmetrical single-mode magneto-optic waveguides by application of stress, <u>Appl. Phys. Lett.</u>, 49:1755 (1986).
45. W. Tolksdorf, I. Bartels, H. Dammann and E. Pross, Growth of buried garnet channel waveguides by liquid phase epitaxy, <u>J. Crystal Growth</u>, 84:323 (1987).

BIBLIOGRAPHY

A.   REVIEW ON THE HISTORY OF CRYSTAL GROWTH

1.   Nassau, K., Dr A. V. L. Verneuil: The Man and the Method, J. Crystal
     Growth, 13/14, 12-18, 1972.
2.   Brissot, J. J., A History of Crystals, Acta Electronica, 16, 285-290,
     1973.
3.   Bohm, J., Die Historiche Entwicklung der Kristallzüchtung – Eine
     Bibliographie, Crystal Research and Technology, 16, 275-292, 1981.
4.   Bohm, J., The History of Crystal Growth, Acta Physica Hungarica, 57,
     161-178, 1985.

B.   PHASE DIAGRAMS AND THERMODYNAMICS OF SOLIDS,
     SOLUBILITY DATA, SOLVENT CHEMISTRY

1.   Levin, E. M., Robbins, C. R. and McMurdie, H. F., Phase Diagrams for
     Ceramists, The American Ceramic Society, Vol. 1-5, 1964-1984.
2.   Lagowski, J. J., ed. of series, The Chemistry of Non-Aqueous
     Solvents, Academic Press, 1966.
3.   Gordon, P., Principles of Phase Diagrams in Materials Systems,
     McGraw-Hill, 1968.
4.   Haase, R. and Schönert, H., Solid-Liquid Equilibrium, Pergamon Press,
     1969.
5.   Reisman, A., Phase Equilibria, Basic Principles, Applications,
     Experimental Techniques, Academic Press, 1970.
6.   Swalin, R. A., Thermodynamics of Solids, Wiley, 1972.
7.   Alper, A. M., ed., Phase Diagrams, Materials Science and Technology,
     Vol. 1-5, Academic Press, 1970-1978.
8.   Kaufman, L. and Bernstein, H., Computer Calculation of Phase
     Diagrams, Academic Press, 1970.
9.   Franks, F., ed. of series, Water, a Comprehensive Treatise, Plenum
     Press, 1972.
10. Pelton, A. D. and Thompson, W. T., Phase Diagrams, Progress in Solid
     State Chemistry, Vol. 10, 119-155, Pergamon Press, 1976.
11. Freier, R. K., ed. of series, Aqueous Solutions, W. de Gruyter, 1976.
12. Linke, W. F., ed., Solubilities, Inorganic and Metal-Organic
     Compounds, Vol. 1-2, American Chemical Society, 1958-1965.
13. Stephen, H. and Stephen, T., Solubilities of Inorganic and Organic
     Compounds, Vol. 1, Pergamon Press, 1979.
14. Broul, M., Nyvlt, J. and Söhnel, O., Solubility in Inorganic Two-
     Component Systems, Elsevier Scientific Publishing Company, 1981.

C.   BASIC PRINCIPLES

1.   Volmer, M., Kinetik der Phasenbildung, Steinkopff, 1939.
2.   Honigmann, B., Gleichgewichts und Wachstumsformen von Kristallen,
     Steinkopff, 1958.

411

3. Van Hook, A., Crystallization, Theory and Practice, Reinhold, 1961.
4. Hirth, J. P. and Pound, G. M., Condensation and Evaporation, MacMillan, 1963.
5. Rutter, E., Goldfinger, P. and Hirth, J. P., eds., Condensation and Evaporation of Solids, Gordon and Breach, 1964.
6. Ubbelohde, A. R., Melting and Crystal Structure, OUP., 1965.
7. Powell, C. P., Oxley, J. H. and Blocher, J. M. jr., eds., Vapor Deposition, Wiley and Sons, 1966.
8. Jackson, K. A., Current Concepts of Crystal Growth from the Melt, in: Progress in Solid State Chemistry, Vol. 4, Pergamon Press, 1967.
9. Strickland-Constable, R. F., Kinetics and Mechanism of Crystallization, Academic Press, 1968.
10. Zettlemoyer, A. C., Nucleation, Dekker, 1969.
11. Parker, R. L., Crystal Growth Mechanisms – Energetics, Kinetics and Transport, in: Solid State Physics – Advances in Research and Applications, Vol. 25, Academic Press, 1970.
12. Ohara, M. and Reid, R. D., Modelling Crystal Growth Rates from Solutions, Prentice Hall, 1973.
13. Heimann, R. B., Auflösen von Kristallen, Springer, 1975.
14. Hannay, N. B., Changes of State, Plenum Press, 1975.
15. Rosenberger, F., Fundamentals of Crystal Growth (I), Springer, 1979, Vols. II and III to follow.
16. Brice, J. C., Crystal Growth Processes, Blackie Halsted Press, 1986.
17. Kurz, W. and Fisher, D. J., a) Fundamentals of Solidification; b) Solutions Manual, Trans Tech Publications, 1986.
18. Vere, A. W., Crystal Growth, Principles and Progress, Plenum, 1988.

D. GENERAL TREATISES

1. Buckley, H. E., Crystal Growth, Wiley and Sons, 1951.
2. Smakula, A., Einkristalle, Springer, 1962.
3. Gilman, J. J., The Art and Science of Growing Crystals, Wiley, 1963.
4. Chalmers, B., Principles of Solidification, Wiley, 1964.
5. Coll. Int. CNRS, Adsorbtion et Croissance Cristalline, Paris, 1965.
6. Knight, Ch. A., The Freezing of Supercooled Liquids, Van Nostrand, 1967.
7. Laudise, R. A., The Growth of Single Crystals, Prentice Hall, 1970.
8. Tarjan, I. and Matrai, M., Laboratory Manual on Crystal Growth, Akadecuiai Kiado Budapest, 1972.
9. Wilke, K. Th. and J. Bohm, Kristallzüchtung, Verlag Harri Deutch, Franfurt/Main, 1988.
10. Goodman, C. H. L., Crystal Growth – Theory and Techniques, Vol. 1-2, Plenum Press, 1976-1978.
11. Bond, W. L., Crystal Technology, Wiley and Sons, 1976.
12. Wilcox, W. R., Preparation and Properties of Solid State Materials, Vol. 1-4, in: Chemical Vapor Transport, Secondary Nucleation and Mass Transfer in Crystal Growth, Dekker, 1971-1979.
13. Bardsley, W., Hurle, D. T. J. and Mullin, J. B., Crystal Growth: A Tutorial Approach, North Holland Series in Crystal Growth, Vol. 2, Amsterdam, 1979.
14. Pamplin, B., Crystal Growth, Pergamon Press, 1980.
15. Holden, A. and Morrison, P., Crystals and Crystal Growing, MIT Press, 1982.
16. Chernov, A. A., Modern Crystallography III, Springer Series Solid-State Science 36, 1984.
17. Current Topics in Materials Science, North Holland, ca. 15 Volumes up to 1987.
18. Crystals, Springer, ca. 12 Volumes up to 1987.
19. Growth of Crystals, Plenum Press, ca. 12 Volumes up to 1987.
20. Preparation and Properties of Solid State Materials, Dekker, ca. 12 Volumes up to 1987.

21. Important Titles in Solid State Technology: a) Semiconductors and Metals, ca. 20 Volumes up to 1985; b) VLSI – Electronics, ca. 8 Volumes up to 1985; c) Physics of Thin Films, ca. 12 Volumes up to 1985, Academic Press.

E. SPECIAL METHODS OF CRYSTAL GROWTH

1. Pfann, W. G., Zone Melting, Wiley and Sons, 1958, 1966.
2. Park, N. L., Zone Refining and Allied Techniques, G. Newness, 1960.
3. Schäfer, H., Chemical Transport Reactions, Academic Press, 1964.
4. Schildknecht, H., Zonenschmelzen, Verlag Chemie, 1964.
5. Bockris, J. and O'M. Razummey, G. A., Fundamental Aspects of Electrocrystallization, Plenum Press, 1967.
6. Ovsienko, D. E., Growth and Imperfections of Metallic Crystals, Consultants Bureau, 1969.
7. Petrov, T. G., Treivus, E. C. and Kasatkin, A. P., Growing Crystals from Solutions, Consultants Bureau, 1969.
8. Henisch, H. K., Crystal Growth in Gels, Penn State University Press, 1970.
9. Lobachev, A. N., Hydrothermal Synthesis of Crystals, Consultants Bureau, 1971.
10. Brice, J. C., The Growth of Crystals from Liquids, North Holland, 1973.
11. Lobachev, A. N., Crystallization Processes under Hydrothermal Conditions, Consultants Bureau, 1974.
12. Faktor, M. M. and Garrett, I., Growth of Crystals from the Vapor, Chapman and Hall, 1974.
13. Elwell, D. J. and Scheel, H. J., Crystal Growth from High Temperature Solutions, Academic Press, 1975.
14. Matthews, J. W., Epitaxial Growth (Parts I and II), Academic Press, 1975.
15. Schneider, H. G., Ruth, V. and Kormany, T., Advances in Epitaxy and Endotaxy, Elsevier Scientific Publishing Company, 1976.
16. Eyer, A. and Zimmermann, H., Flüssig und Gaszonen Kristallization unter Schwerlösigkeit, Bundesministerium für Forschung und Technologie, Forschungsbericht W 77-12, München, 1977.
17. Lewis, B. and Anderson, J. C., Nucleation and Growth of Thin Films, Academic Press, 1978.
18. Rudolph, P., Profilzüchtung von Einkristallen, Akademie Verlag, 1982.
19. Ploog, K. and Graf, K., Molecular Beam Epitaxy of III-V Compounds, Springer, 1983.
20. Henisch, H. K., Crystals in Gels and Liesegang Rings, Cambridge University Press, 1988.

F. SPECIAL MATERIALS

1. Runyau, W. R., Silicon Semiconductor Technology, McGraw Hill, 1965.
2. Sittig, M., Semiconductor Crystal Manufacture, Noyes Development, 1969.
3. Kosalopova, T. Ya., Carbides: Properties, Production and Applications, Plenum Press, 1971.
4. Connolly, T. F., ed., Semiconductors: Preparation, Crystal Growth and Properties, Plenum Press, 1972.
5. Standley, K. J., Oxide Magnetic Materials, Clarendon Press, 1972.
6. Craik, D. J., Magnetic Oxides, Parts 1/2, Wiley and Sons, 1975.
7. Shay, J. L. and Wernick, J. H., Ternary Chalcopyrite Semiconductors – Growth, Electronic Properties and Applications, Pergamon Press, 1975.
8. Wunderlich, B., Macromolecular Physics – Macromolecular Crystals, Vol. 2, in: Crystal Nucleation, Growth, Annealing, Academic Press, 1976.

9. Lieth, R. M. A., Preparation and Crystal Growth of Materials with Layered Structures, Reidel Publ. Co., 1977.
10. Matkovich, V. I., ed., Boron and Refractory Borides, Springer Verlag, 1977.
11. Cullen, G. W. and Wang, C. C., eds., Heterostructure Semiconductors of Electron Devices, Springer Verlag, 1978.
12. Rooijmans, C. J. M., Crystals, Vol. 1, in: Crystals for Magnetic Applications, Springer Verlag, 1978.
13. Wunderlich, B., Selected Papers on Polymer Crystallization, North Holland, 1979.
14. Arizumi et al., Crystals, Vol. 4, in: Organic Crystals, Germanates, Semiconductors, Springer Verlag, 1980.
15. Belyaev, L. M., ed., Ruby and Sapphire, NBS, Washington DC., 1980.
16. Nassau, K., Gems Made by Man, Chilton Book Co., 1980.
17. Moss, T. S., ed., Handbook on Semiconductors, Vol. 3, S. P. Keller, ed., Materials and Preparation , North Holland, 1980-1981.
18. Nishizawa, J., ed., Semiconductor Technologies, Vol. I, North Holland Publ. Co., 1982.
19. Suematsu, Y., ed., Optical Devices and Fibers, Vol. 3, North Holland Publ. Co., 1982.
20. Kitagawa, T., ed., Computer Science and Technologies, North Holland Publ. Co., 1982.
21. McPherson, A., Preparation and Analysis of Protein Crystals, Wiley, 1982.
22. Nassau, K., Gemstone Enhancement, Butterworth, 1984.

G. INDUSTRIAL CRYSTALLIZATION

1. Bamworth, A. W., Industrial Crystallization, Leonard Hill, 1965.
2. Walton, A. G., The Formation and Growth of Precipitates, Interscience, 1967.
3. Matz, C., Die Kristallisation: Grundlagen und Technik, Springer, 1968.
4. Matz, C., Die Kristallisation in der Verfahrenstechnik, Springer, 1968.
5. Nyvlt, J., Industrial Crystallization from Solutions, Butterworth, 1971.
6. Randolpf, A. D. and Larson, M. A., Theory of Particular Processes, Academic Press, 1971.
7. Mullin, J. W., Crystallization, Butterworth, 1972.
8. De Jong, E. J. and Jancic, S. J., Industrial Crystallization, North Holland, 1979.
9. Nyvlt, J., Industrial Crystallization: The Present State of the Art, Verlag Chemie, 1982.

H. PROCEEDINGS OF INTERNATIONAL MEETINGS AND SUMMER SCHOOLS

1. Int. Meetings on Crystal Growth (ICCG), Published as separate Volumes of J. Crystal Growth, since 1968 every three years.
2. International Special Conference on Vapor Growth and Epitaxy, Published as separate Volumes of J. Crystal Growth, since 1970 every 2 to 3 years.
3. Reports of Summer schools organized by Int. Org. on Crystal Growth, since 1973 every 3 years, North Holland.
4. Reports on Multinational Meetings in Russia, since 1958 about 20 Volumes, Growth of Crystals, Consultants Bureau, in English.

I. PERIODICALS

1. Journal of Crystal Growth.
2. Materials Research Bulletin.

3. Crystal Research and Technology.
4. Progress in Crystal Growth and Characterization.
5. Soviet Physics: Crystallography.
6. Journal of Synthetic Crystals, Chinese edition with English abstracts
    and legends, quarterly.

J. NEWSLETTERS (NATIONAL)

1. British Association for Crystal Growth, Newsletter Secretary: J. G.
    Wilkes, Mullard Ltd., Southampton, Hampshire SO9 7BP, UK.
2. Group Français de Croissance Cristalline, GFCC, Secretary: J. J.
    Metois, CNRS, CRMC2, Campus de Luminy, Case 913, F-13288 Marseille
    Cedex 9, France.
3. Deutsche Gesellschaft für Kristallzüchtung und Kristallwachstum,
    DGKK, Ed.: G. Müller, Institut für Werkstoffwissenschaften VI,
    Universität Erlangen, Nürnberg, Martensstr. 7, D-8520 Erlangen,
    Germany.
4. Schweizerische Gesellschaft für Kristallographie, Sektion für
    Kristallwachstum und Materialforschung, Secretary: J. Hulliger,
    Institute of Quantum Electronics, Swiss Federal Institute of
    Technology, CH-8093, Zürich, Switzerland.

K. ABSTRACTS AND BIBLIOGRAPHY

1. Chemical Abstracts Selects: Crystal Growth, since 1980.
2. Bulletin Signalétique, Centre National de la Recherche Scientifique,
    France, 26 Rue Boyer, F-75971 Paris Cedex 20, France.
3. Keesee, A. M., Connolly, T. F. and Battle, G. C., jr., Crystal
    Growth Bibliography, Part A: Bibliography; Part B: Indexes,
    IPI/Plenum, 1979.

L. SURVEYS OF ACTIVITIES IN CRYSTAL GROWTH

1. International Directory of Solid State Materials - Production and
    Research, Research Materials Information Center (RMIC) by T. F.
    Connolly, Oak Ridge National Laboratory, PO Box X, Oak Ridge,
    Tennessee 37830, USA.
2. Sources of Single Crystals in the UK and Scandinavia, B. M. R.
    Wanklyn, Clarendon Laboratory, University of Oxford, UK.
3. Reports of Electronic Materials Unit, Royal Signals and Radar
    Establishment, St. Andrews Road, Malvern, Worcestershire WR14 3PS,
    UK.
4. Report by the Centre de Documentation sur les Syntheses Cristalline,
    Laboratoire de Physique Moleculaire et Cristalline, Faculté des
    Sciences, Place Eugene Bataillon, F-34000 Montpellier, France.
5. Information on Crystal Growth, Germany, Netherlands, Switzerland,
    Report of DKGG, A. Räuber and R. Nitsche, Universität und Fraunhofer
    Intitut, Freiburg i.Br., BRD.

Molecular beam epitaxy (continued)
  surface studies (continued)
    RHEED pattern formation,
      371-373
    RHEED pattern intensity
      oscillations, 374-375
    surface ordering, 373-374
  typical pressures during growth,
    362
Molten salts and oxides see Flux
    growth
Molybdates, flux systems, 134
Molybdenum, growth of sapphire, 279
Monte Carlo simulation, excess free
    energy of liquid clusters
    (illus.), 41
Mössbauer spectroscopy, 236
MSMPR crystallizer see Mixed
    suspension mixed product
    removal crystallizer
Muscovite,
  nucleation, 262
  polytypes, 199
Mutaftschiev's model, free energy of
    liquid/interface/solid system
    (illus.), 71

N-type dopants, 377-378
Neodymium YAG, grown by LHPG method,
    293
Newsletters (national), 414-415
Nickel, cubic isotropic crystals,
    160
meta-Nitroaniline, capillary growth,
    298
No-slip condition, vs `stagnant
    film' concept (illus.), 120
Nucleation,
  2-D `polymers', 56-58
  activation energy,
    and equilibrium shape, 34
    formula, 263
  chemical concept,
    capillarity approximation, 29,
      39-42
    equilibrium distribution of
      clusters, 34-39
  defined, 27
  `diffusion haloes', 42
  geological processes, 262
  growth, 2-D (illus.), 260-261
  homogeneous nucleation kinetics,
    steady state, 42-45
    time lag in nucleation, 45-47
  metastable, 263-264
  phase concept,
    equilibrium shape, 33-34
    saturated state, 27-28
    supersaturated state, 28-33
  scaled steady state nucleation
    rate (illus.), 47
  steady state, 42-45
  time lag, 45-47
  work, 31-33
Numbers see Grashof; Peclet; Schmidt

OMVPE see Organometallic vapor phase
    epitaxy
Onsager solution, 113
Open-tube techniques, II - VI
    compounds, 339-340
Organic crystals, see also Protein
Organic nonlinear optic crystals, in
    capillaries, 298
Organometallic vapor phase epitaxy,
  applications, 313
  chemical potential vs reaction
    coordinate, 306
  hydrodynamics and mass transport,
    303-309
  interface abruptness, 311
  purity of product, 309-310
  safety, 311-312
  summary, 313-314
  versatility, 310
Orientational domain states, 188-189
Orthoferrite, formation, 20
Ostwald ripening, in geological
    processes, 259
Ostwald's formula, Gibbs-Thomson
    effects, 16
Overgrowth, regular (illus.),
    143-144

p-n junctions, alloying technique
    (illus.), 318
P-type dopants, 376-377
Palladium, on tungsten, epitaxial
    growth mode II, 164
Paraffins, chainfolding, 272
Peclet number Pe, 92, 169
  equation, 170
  and Ivantsov model, 171
  low, alloys, 180-181
  and supercooling, 173
Pedestal growth method, fiber growth
    techniques, 290-291, 292, 294
Periodicals, bibliography, 414
Perovskite, metastable nucleation,
    264
PET (Polyethylene terephthalate),
    268
Phase, defined, 4
Phase boundary, defined, 4
Phase concept of nucleation, 27-34
Phase diagrams,
  bibliography, 411
  binary systems, applications,
    12-17
  cerium-hydrogen, 234
  lanthanum-hydrogen, 234
  yttrium iron garnet (YIG), 19
Phase dissociation, suppression, 233
Phase equilibria, a theoretical
    view, 1-23
Phase transitions,
  polymorphic, industrial
    crystallization, 222
  see also Structural phase
    transitions
Phosphine,
  pyrolysis in deuterium, 307-308
  safety considerations, 311-312